# THE SPERM CELL

# THE SPERM CELL

## FERTILIZING POWER, SURFACE PROPERTIES, MOTILITY, NUCLEUS AND ACROSOME, EVOLUTIONARY ASPECTS

Proceedings of the Fourth International Symposium on Spermatology, Seillac, France, 27 June-1 July 1982

*edited by*

JEAN ANDRÉ
Laboratoire de Biologie Cellulaire
Université Paris-Sud
91405 Orsay
France

1983

MARTINUS NIJHOFF PUBLISHERS
THE HAGUE / BOSTON / LONDON

*Distributors:*

*for the United States and Canada*

Kluwer Boston, Inc.
190 Old Derby Street
Hingham, MA 02043
USA

*for all other countries*

Kluwer Academic Publishers Group
Distribution Center
P.O. Box 322
3300 AH Dordrecht
The Netherlands

**Library of Congress Cataloging in Publication Data** CIP

International Symposium on Spermatology (4th : 1982 : Seillac, France)
The sperm cell.

Includes index.
1. Spermatozoa--Congresses. I. André, Jean, 1922- . II. Title. [DNLM: 1. Spermatozoa--Congresses. W3 IN922UU 4th 1982s / WJ 834 I608 1982s]
QP255.I567 1982 599'.016 82-22241
ISBN 90-247-2784-7

ISBN 90-247-2784-7



PRINTED IN THE NETHERLANDS

This volume contains the works presented at the *Fourth International Symposium on Spermatology*, held in the Domaine de SEILLAC (France) from June 27 to July 1st, 1982.

D. Schwartz (Villejuif) and Ch. Thibault (Paris) were honorary presidents.

The organizing Committees were :
the International Committee

B. Afzelius , Stockholm
J. André , Orsay
B. Baccetti , Siena
M. Bedford , New York
D.W. Fawcett , Nairobi
G. Meyer , Tübingen
H. Mohri , Tokyo

and the Local Committee

J. André , Orsay
M. Courot , Nouzilly
J.P. Dadoune , Paris
G. David , Kremlin-Bicêtre
M. Fellous , Paris
J.E. Fléchon , Jouy-en-Josas
R. Folliot , Rennes

The organization of this Symposium has been made possible owing to the help of :
the Ministère de la Santé
the Centre National de la Recherche Scientifique
the Institut National de la Recherche Agronomique
the International Union of Biological Sciences
and of the following firms : Grosse Verlag, Instruments de Médecine Vétérinaire, Jeol, Mérieux, Soro, Théramex.

# CONTENTS

# LIST OF PARTICIPANTS

ABOU-HAÏLA A., Laboratoire de Biologie Cellulaire, Université René Descartes, 45, rue des Saints-Pères, 75270 Paris, France.

AFZELIUS B.A., Wenner-Gren Institute, Norrtullsgatan 16, S-11345 Stockholm, Sweden.

ALNOT M.O., Laboratoire d'Histologie et Embryologie, C.H.U. Necker, 156, rue de Vaugirard, 75005 Paris, France.

ANDRÉ J., Laboratoire de Biologie Cellulaire 4, Université Paris-Sud, 91405 Orsay, France.

BACCETTI B., Institute of Zoology, Via Mattioli, 4, 53100 Siena, Italy.

BAWA S.R., Department of Biophysics, Panjab University, Chandigarh 160014, India.

BEDFORD J.M., Department of Obstetrics and Gynecology, Cornell University Medical College, New York, N.Y. 10021, U.S.A.

BELLVÉ A.R., Laboratory of Human Reproduction and Reproductive Biology, Harvard Medical School, Boston, Mass. 02115, U.S.A.

BERRUTI G., Institute of Zoology, Via Celoria 10, 20133 Milano, Italy.

BIELANSKA-OSUCHOWSKA Z., Instytut Fizjologii Zwierzat, S.G.G.W., Warszawa, ut. Nowoursynowska 166, Poland.

BILLARD R., Laboratoire de Physiologie des Poissons, Campus de Beaulieu, 35042 Rennes, France.

BROKAW C., Division of Biology, California Institute of Technology 156-29, Pasadena, Calif. 91125, U.S.A.

BROOKS D., Department of Animal Physiology, Waite Agricultural Research Institute, University of Adelaide, Glen Osmond, South Australia 5064, Australia.

BURRINI A.G., Institute of Zoology, Via Mattioli, 4, 53100 Siena, Italy.

BUSTOS-OBREGON E., Department of Cell Biology and Genetics, Faculty of Medicine, University of Chile, P.O. Box 6556, Santiago 7, Chile.

CALLAINI G., Institute of Zoology, Via Mattioli 4, 53100 Siena, Italy

CHARDON P., Laboratoire de Radiobiologie Appliquée, I.N.R.A., 78350 Jouy-en-Josas, France.

CHEVAILLIER P., Laboratoire de Biologie Cellulaire, Université Paris-Val-de-Marne, 94010 Créteil, France.

CHULAVATNATOL M., Department of Biochemistry, Faculty of Science, Mahidol University, Rama VI Road, Bangkok 4, Thailand.

CIBERT C., Laboratoire d'Histo-Embryologie et Cytogénétique, Hôpital de Bicêtre, 94270 Le Kremlin-Bicêtre, France.

CISNEROS P.L., Department of Biology, University of Houston, Houston, Texas 77004, U.S.A.

CLAVERT A., Institut d'Embryologie, Faculté de Médecine, 11, rue Humann, 67085 Strasbourg, France.

COHEN J., Department of Zoology and Comparative Physiology, Birmingham University, Edgbaston, Birmingham B1S 2TT, U.K.

COTELLI F., Institute of Zoology, Via Celoria 10, 20133 Milano, Italy.

COUROT M., Laboratoire de Fertilité Mâle, I.N.R.A., Station de Physiologie de la Reproduction, 37380 Nouzilly, France.

COURTENS J.L., I.N.R.A., Station de Physiologie de la Reproduction, 37380 Nouzilly, France.

CRAIG T., Department of Physics, University of Guelph, Guelph, Ontario N1G 2W1, Canada.

CROZET N., I.N.R.A., Laboratoire de Physiologie Animale, C.N.R.Z., 78350 Jouy-en-Josas, France.

CUMMINS J.M., Reproductive Biology Group, Department of Veterinary Anatomy, University of Queensland, St Lucia 4067, Australia.

DACHEUX J.L., I.N.R.A., Laboratoire de Physiologie de la Reproduction, 37380 Nouzilly, France

DADOUNE J.P., Laboratoire d'Histologie-Embryologie-Cytogénétique, Université René Descartes, 45, rue des Saints-Pères, 75270 Paris, France.

DA LAGE C., Laboratoire d'Histologie et Embryologie, C.H.U. Necker, 156 rue de Vaugirard, 75015 Paris, France.

DALLAI R., Institute of Zoology, Via Mattioli 4, 53100 Siena, Italy.

DAVID G., C.E.C.O.S., Laboratoire d'Histologie, Hôpital de Bicêtre, 94270 Le Kremlin-Bicêtre, France.

DE ALMEIDA M., Centre d'Immuno-Pathologie de l'INSERM, Unité 33, Hôpital Saint-Antoine, 75012 Paris, France.

DELPECH S., I.N.R.A., Station de Physiologie de la Reproduction, 37380 Nouzilly, France.

DEPEIGES A., Laboratoire de Génétique et Biologie Cellulaire, Université de Clermont-Ferrand, 63170 Aubière, France.

ESPONDA P., Instituto de Biologia Celular, Velazquez 144, Madrid-6, Spain.

FAIN-MAUREL M.A., Laboratoire de Biologie Cellulaire, Université René Descartes, 45, rue des Saints-Pères, 75270 Paris, France.

FELDMAN-MUHSAM B., Department of Medical Entomology, Hebrew University Medical School, P.O. Box 1172, Jerusalem 91-010, Israel.

FELLOUS M., Institut Pasteur, Departement d'Immunologie, 25 rue du Dr. Roux, 75011 Paris, France.

FENEUX D., Laboratoire d'Histologie-Embryologie, Hôpital de Bicêtre, 94270 Le Kremlin-Bicêtre, France.

FERRAGUTI M., Institute of Zoology, Via Celoria 10, 20133 Milano, Italy.

FLÉCHON B., I.N.R.A., Station de Physiologie animale 78350 Jouy-en-Josas, France.

FLÉCHON J.E., I.N.R.A., Station de Physiologie animale, 78350 Jouy-en-Josas, France.

FOLIGUET B., Laboratoire d'Histologie-Embryologie-Cytogénétique, Faculté de Médecine B, 54505 Vandoeuvre-les-Nancy, France.

FOLLIOT R., Laboratoire de Biologie Cellulaire, Université de Rennes I, 35031 Rennes Cedex, France.

FORD W.C., Department of Physiology and Biochemistry, The University, Whiteknights, Reading RG6 2AJ, U.K.

FRANKLIN L., Department of Biology, University of Houston, Houston, Texas 77004, U.S.A.

FRIEDLÄNDER M., Department of Biology, Ben Gurion University, Beer-Sheva, Israel.

GAGNON C., C.H.U.L., Molecular and Cellular Bioregulation Unit, 2705, bd. Laurier, Ste-Foy, Québec G1V 4G2, Canada.

GIBBONS I.R., Pacific Biomedical Research Center, University of Hawaii, Honolulu, Hawaii.

GIORGETTI C., CECOS Sud-Est, 165 rue Saint-Pierre, 13005 Marseill France.

GIOVANGRANDI Y., Service Gynécologie-Obstétrique, C.H.U. Bretonneau, 37000 Tours, France.

GIUSTI F., Institute of Zoology, Via Mattioli 4, 53100 Siena, Italy.

GONZALES J., Laboratoire d'Histologie-Embryologie, Hôpital Pitié-Salpetrière, 105 boulevard de l'Hôpital, 75651 Paris, France.

GUERIN J.F., Laboratoire d'Histologie, Faculté de Médecine, 69373 Lyon, France.

GUICHAOUA M., Laboratoire de Cytogénétique, Faculté de Médecine Nord, Boulevard P. Dramard, 13326 Marseille Cedex 15, France.

HAMAMAH S., I.N.R.A., Station de Physiologie de la Reproduction, 37380 Nouzilly, France.

HAMILTON D., Department of Anatomy, University of Minnesota, Minneapolis, Mn. 55455, U.S.A.

HARDING H.R., Department of General Studies, University of New South Wales, Kensington, 2033 N.S.W., Australia.

HARRISON R.A.P., A.R.C. Institute of Animal Physiology, Animal Research Station, 307 Huntingdon Road, Cambridge CB3 OJQ, U.K.

HAUSCHTECK-JUNGEN E., Zoologisches Institut, Winterthurerstr. 190, 8057 Zürich, Switzerland.

HECHT N.B., Department of Biology, Tufts University, Medford, Massachusetts, 02155, U.S.A.

HENDELBERG J., Department of Zoology, University of Göteborg, Box 250 59, S-400 31 Göteborg, Sweden.

HOLSTEIN A.F., Anatomisches Institut, Martinistr. 52, 2000 Hamburg 20, Germany

HOSHI M., Department of Biology, Nagoya University, Nagoya, Japan.

HUNEAU D., I.N.R.A., Station Centrale de Physiologie Animale, 78350 Jouy-en-Josas, France.

IBRAHIM M.A.R., University of Agriculture Science, Gödöllo, Hungary.

JESPERSEN Å., Institute of Comparative Anatomy, Universitetsparken 15, DK-2100 Copenhagen Ø, Denmark.

JEULIN C., Laboratoire d'Histologie-Embryologie-Cytogénétique, Hôpital de Bicêtre, 94270 Le Kremlin-Bicêtre, France.

JONES Roy, ARC Institute of Animal Physiology, Animal Research Station, 307 Huntingdon Rd., Cambridge CB3 0JQ, U.K.

JONES Russell C., Department of Biological Sciences, University of Newcastle, New South Wales, 2308, Australia.

JOUANNET P., Laboratoire d'Histologie-Embryologie, Hôpital de Bicêtre, 94270 Le Kremlin-Bicêtre, France.

JUSTINE J.L., Département de Biologie Animale, Faculté des Sciences, Dakar, Sénégal.

KANN M.L., Laboratoire de Physiologie de la Reproduction des Vertébrés, Université Pierre et Marie Curie, 4, Place Jussieu, 75230 Paris, France.

KATZ D., Department of Obstetrics and Gynecology, School of Medicine, University of California, Davis, Ca. 95616, U.S.A.

KLOTZ C., Centre de Cytologie Expérimentale, C.N.R.S., 67, rue Maurice Günsbourg, 94200 Ivry-sur-Seine, France.

KOEHLER J., Department of Biological Structure, University of Washington, Seattle, Washington 98195, U.S.A.

KVIST U., Department of Physiology, Karolinska Institutet, Box 60 400, 704 01 Stockholm, Sweden.

LEE W., Center for Bioengineering, Fl-20, University of Washington, Seattle, Wa. 98195, U.S.A.

LE LANNOU D., CECOS de l'Ouest, 1 bis rue de la Cochardière, Hôtel Dieu, 35000 Rennes, France.

LOIR M., Laboratoire de Physiologie de la Reproduction, INRA, 37380 Nouzilly, France.

LORNAGE J., Laboratoire d'Histologie-Embryologie-Cytogénétique, U.E.R. Médicale Grange Blanche, 8 avenue Rockefeller, 69373 Lyon, France.

McRORIE R.A., Department of Biochemistry, University of Georgia, Athens, Ga. 30602, U.S.A.

MAZZINI M., Institute of Zoology, Via Mattioli 4, 53100 Siena, Italy.

MOHRI H., Department of Biology, University of Tokyo, Komaba, Meguro-ku, Tokyo 153, Japan.

MORISAWA M., Ocean Research Institute, University of Tokyo, Minami-dai 1-15-1, Nakano, Tokyo 164, Japan.

MORISAWA S., St. Marianna University, School of Medicine, Japan.

MORTIMER D., Department of Obstetrics and Gynecology, University of Birmingham, Birmingham Maternity Hospital, Edgbaston, Birmingham B15 2TG, U.K.

MYLES D., Department of Physiology, University of Connecticut Health Center, Farmington, Connecticut 06032, U.S.A.

NAGANO T., Department of Anatomy, School of Medicine, Chiba University, Chiba 280, Japan.

NICANDER L., Department of Anatomy, Veterinary College of Norway, P.O. Box 8146, Dep. Oslo 1, Norway.

NICOTRA A., Istituto di Zoologia, Viala dell'Università 32, 00100 Roma, Italy.

OMOTO C.K., Division of Biology, California Institute of Technology 156-29, Pasadena, Calif. 91125, U.S.A.

ORTAVANT R., Laboratoire de Physiologie Animale, I.N.R.A., 37380 Nouzilly, France.

OVERSTREET J.W., Department of Human Anatomy, Room 3301, School of Medicine, University of California, Davis, Ca. 95616, U.S.A.

PAQUIGNON M., Laboratoire de Physiologie de la Reproduction, I.N.R.A., 37380 Nouzilly, France.

PAVLOK A., Institute of Physiology and Genetics of Animals Czechoslovak Academy of Sciences, CS - 277 21 Libechov, Czechoslovak.

PHILLIPS D.M., Population Council, 1230 York Avenue, New York, New York 10021, U.S.A.

PICHERAL B., Laboratoire de Biologie Cellulaire, Université de Rennes I, Campus de Beaulieu, 35042 Rennes Cedex, France.

PISSELET C., Laboratoire de Fertilité Mâle, I.N.R.A., Station de Physiologie de la Reproduction, 37380 Nouzilly, France.

PLÖEN L., Department of Anatomy and Histology, Faculty of Veterinary Medicine, Swedish University of Agricultural Sciences, S-750 07 Uppsala, Sweden.

PRIMAKOFF P., Department of Physiology, University of Connecticut Health Center, Farmington, Connecticut 06032, U.S.A.

RAO B., C.E.C.O.S., Centre Hospitalier de Bicêtre, 94270 Le Kremlin-Bicêtre, France.

REGER J.F., Department of Anatomy, University of Tennessee, 875 Monroe Ave., Memphis, Tennessee 38163, U.S.A.

RENIERI T., Institute of Zoology, Via Mattioli 4, 53100 Siena, Italy.

de REVIERS M., Station de Recherches Avicoles, 37380 Nouzilly, France.

ROUSSET V., Laboratoire d'Ichtyologie et Parasitologie Générale, Place Eugène Bataillon, 34060 Montpellier, France.

SCHWARTZ D., I.N.S.E.R.M., Unité de Recherches Statistiques, 16 bis, avenue Paul-Vaillant-Couturier, 94800 Villejuif, France.

SELLOS Laboratoire de Biologie Marine, 29110 Concarneau, France.

SELMI M.G., Institute of Zoology, Via Mattioli, 4, 53100 Siena, Italy.

SERRES C., Laboratoire d'Histologie-Embryologie, Centre Hospitalier de Bicêtre, 94270 Le Kremlin-Bicêtre, France.

SETCHELL B., A.R.C. Institute of Animal Physiology, Brabahein, Cambridge, U.K.

SHALGI R., Department of Embryology, Medical School, Tel Aviv University, Ramat Aviv, Israel.

SHAPIRO B.M., Department of Biochemistry, University of Washington, Seattle, Wa. 98195, U.S.A.

STOREY B.T., Department of Obstetrics and Gynecology, Division of Reproduction Biology, University of Pennsylvania, Med. Labs 339, Philadelphia, Pa. 19104, U.S.A.

SUBIRANA J.A., Unidad de Quimica Macromolecular del C.S.I.C., Escuela Ingenieros Industriales, Diagonal 999, Barcelona 28, Spain.

SUZUKI N., Department of Biochemistry, Teikyo University School of Medicine, 2-11-1 Kaga, Itabashi-ku, Tokyo 173, Japan.

TARKOWSKI A.K., Department of Embryology, Institute of Zoology, University of Warsaw, Krak. Przedm. 26/28, 00-927 Warsaw 64, Poland.

TERQUEM A., Laboratoire d'Histologie, Université René Descartes, 45 rue des Saints-Pères, 75270 Paris, France.

THIBAULT C., Laboratoire de Physiologie de la Reproduction des Vertébrés, Université Pierre et Marie Curie, 4 Place Jussieu, 75230 Paris, France.

VENDRELY E., Laboratoire d'Histologie et Embryologie, Université René Descartes, 45, rue des Saints-Pères, 75270 Paris, France.

WALT H., Institute of Pathology of the University of Zürich, University Hospital, CH-8091 Zürich, Switzerland.

WERNER G., Medizinische Biologie, Fachbereich Theoretische Medizin der Universität des Saarlandes, D-6650 Homburg/ Saar, West Germany.

WOOLLEY D., Department of Physiology, The Medical School, University of Bristol, Bristol BS8 1TD, U.K.

YASUZUMI G., Academy of Medical Technology and Nursing, Hanwa Memorial Hospital, 7 Karita Osaka city 558, Japan.

# PREFACE

Jean ANDRÉ
*Université de Paris XI, ORSAY, France.*

Sperm cells have long been considered as the most highly specialized of all living cells. They surely are, being very diverse, very complex, containing organelles which do not exist in any other cell -such as acrosome or crystallized mitochondria- and being endowed with a very unique behaviour, that is to meet and recognize the ovum, pierce its protective envelopes and inject into its cytoplasm a most precious deposit, the haploid genome of the species.

It is Baccio Baccetti's merit to have felt the need for a confrontation of the scientists working on sperm in order to clarify the apparent complexity of the enormous amount of knowledge accumulated on the subject. Thus, he successfully inaugurated the series of the *International Symposia on Spermatology*. The Seillac edition is the fourth in the series. After an initial stage during which morphology was predominant, our meetings have turned more and more towards function. It has been the will of the French Organizing Committee to devote this meeting mainly to Eutherians, and, among those, to man, in connection with the conflicting necessities to help the sterile couples and to control the population explosion at the surface of the world.

The papers have been presented under six main headings which constitute the six chapters of this book. Each one begins by one or two invited papers which review the actual state of knowledge ; then contributed papers follow, which present the original works of the participant laboratories.

The first chapter deals with *Evaluation and control of the fertilizing power of sperm*. The extensive use of artificial insemination and of storage of sperm would benefit greatly of such an evaluation. To this aim, new techniques have been developed , *in vitro* fertilization, rapid measure of sperm motility and mathematical analysis, among others. They have brought already a great deal of scientific and clinical data which are a good promise of improvements in diagnosis and therapeutics. For the time being however, due to the great complexity of gamete biology, the significance of the results obtained at the laboratory on a semen sample in terms of its fertilizing ability is not yet clear.

The second chapter deals with the *Surface properties of sperm cells*. The explosive development, during the last decade, of knowledge on molecular aspects of membranes has been of great benefit to the study of sperm cells. Membranes are now thought of as bidimensional spaces paved with a multitude of specific antennas. The use of advanced techniques, such as that of hybridoma antibodies, can disclose these receptors precisely and lead to an understanding at the molecular level of the unique behaviour of the male cell as it unites with the female cell.

The third chapter deals with the *Nucleus*, a topic which happened to be almost overlooked at the preceding three Symposia. The dramatic condensation of the haploid genome at the time of spermiogenesis, paralleled by its complete functional occultation, and its decondensation immediately after fertilization are followed in terms of the succession of basic proteins interacting with the DNA molecules. The amino acid sequence of several of these arginine-rich proteins is now known and shows an unexpected diversity, which contrasts with the uniformity of histones.

The fourth chapter concerns the most typical organelle of sperm cells, the *Acrosome*. In spite of the facts that it has been known for a long time and that the acrosomal reaction has been thoroughly scrutinized over the past decades for a number of species, the functions of the acrosomal enzymes, the locus and the timing of their deployment are questions still not clearly settled, so we felt that a general survey of the subject should be made.

*Sperm motility* is the subject of the fifth chapter. Two extreme, but complementary points of view on sperm motility are in rapid progress : at the molecular level and at the population level. At the molecular level, work has been initiated by Björn Afzelius and Ian Gibbons. The spatial arrangement and functioning of the major macromolecules that constitute a flagellum have now been extensively studied so that the dynamics and regulation of flagellar beat are beginning to be well understood at the molecular level. As far as the population level is concerned, a new, very ingenious tool, has radically changed the situation. This is the laser doppler velocimeter, which measures almost instantaneously a number of parameters of sperm motility in such a way that a semen sample can be characterized objectively. Also, the reaction of sperm cells to various media, to various metabolites or other active molecules can be measured rapidly with great accuracy. It can be augured that the data obtained will soon become precious to sperm banks.

The sixth and last chapter emerges from a long heritage : *Sperm structure in relation to function and phylogeny*. The basic approach of all the great problems in biology is comparison. Comparison of shape, dimensions, mass, and the series of the so-called physical parameters, comparison of composition, functioning, origin, and others. This is so at all levels of observation, from populations of individuals to molecules. Since sperm cells are so diverse, such comparisons are especially exciting and fruitful. They lead to an easier comprehension of general phenomena. A striking example is that of the quasi-universality of flagellar structure and function. Another benefit from extensive comparisons comes from the fact that evolution has sometimes masked in some species a structural or functional trait which has been kept conspicuous in others. An example of this situation is that of the acrosome reaction, a spectacular explosion in *Limulus*, yet very discrete in mammals, where it was recognized only long after that in sea urchins. So, the comparative and evolutionary aspects of spermatology are important. The analysis period, which leads to the synthesis period, is not yet completed and has still to be continued.

# CHAPTER 1 EVALUATION AND CONTROL OF THE FERTILIZING POWER OF SPERM

EVALUATION AND CONTROL OF THE FERTILIZING POWER OF SPERM

JAMES W. OVERSTREET
Department of Obstetrics and Gynecology, University of California,
Davis, Ca. 95616, U.S.A.

## 1. INTRODUCTION

In the great majority of mammals fertilization takes place in the ampulla of the oviduct. Fertilization is promoted by many biological mechanisms within the female tract which facilitate the transport of male and female gametes and their union. While a great deal of scientific information is now available on the biology of sperm transport in the female [reviewed by Overstreet (1)] and on fertilization in vivo [reviewed by Yanagimachi (2)], many of the experiments which examine the fertilizing power of sperm must necessarily be carried out in vitro.

The development of reliable laboratory techniques for in vitro fertilization has provided mammalian gamete biologists with useful tools for evaluating the fertilizing capacity of sperm cells (3). Since these techniques are also useful clinically in the diagnosis and treatment of human infertility, scientific attention has recently been focused on the human gametes. In this review, emphasis will be placed on the laboratory assessment of human sperm fertility and the biology of human gamete interaction in vitro. The scientific and clinical data now available suggest that this new technology does indeed offer great diagnostic and therapeutic promise (4, 5). However, the apparent complexity of human gamete biology cautions against simplistic interpretations and unrealistic expectations. The primary purpose of this chapter is to emphasize our current appreciation of this complexity.

## 2. ASSESSMENT OF HUMAN SPERM CAPACITATION AND THE ACROSOME REACTION

The sequence of events during gamete interaction in vivo and in vitro are probably quite similar, although the physiology of the processes may differ in certain details (2). The newly ovulated oocyte is surrounded by a thick acellular zona pellucida and several layers of granulosa cells.

Together these layers form the investments of the oocyte. To reach the surface of the vitellus, the spermatozoon must traverse each of the investing layers. This passage requires a series of precisely timed, coordinated functions of the sperm cell.

The term capacitation is used to describe physiological changes in the sperm cell which are required both for penetration of the oocyte vestments and for fusion with the vitellus. Capacitation may involve both alterations in the sperm surface and chemical changes within the sperm cell (2, 6). The capacitation phenomena have been studied in many species of mammals including humans, but the biology of these events remains poorly understood. Consequently, there is no direct method which is currently available to assess the progress of capacitation in a human sperm population. In the future, it may be possible to monitor surface changes in the sperm cells with lectins or other membrane probes (6). At the present time, however, the completion of capacitation can only be inferred from observation of the specific changes in the sperm cells which are known to be capacitation-dependent.

One distinct event which follows the completion of mammalian sperm capacitation is the acrosome reaction. The acrosome reaction involves a series of fusions between specific sperm membranes overlying the acrosome, an organelle within the sperm head (2). The resulting vesiculation of these membranes allows release of the acrosomal contents, including enzymes which may facilitate the passage of spermatozoa through the granulosa cell investments of the oocyte. The vesiculated acrosomal cap is ultimately lost from the sperm cell prior to its entry into the zona pellucida (2). The ultrastructural morphology of the acrosome reaction has been studied extensively, and these processes in human spermatozoa appear to be similar to those reported for other mammals (7).

In view of its critical importance in fertilization, occurrence of the acrosome reaction should be a sensitive indicator of the fertilizing potential of a spermatozoon. Transmission electron microscopy has been the only reliable method for detecting the acrosome reactions of human spermatozoa, but the complexity and laborious nature of the techniques have limited their application. Cytological techniques have been employed recently to assess the loss of the acrosomal cap from human sperm. One method employs fluorescein-conjugated lectin (8) and another utilizes a triple stain including trypan blue, Bismark brown and rose Bengal (9).

The latter technique offers the specific advantage of discriminating between live and dead sperm cells, thus improving its accuracy in detecting physiological loss of the acrosome (as opposed to post-mortem degeneration).

There also appears to be an alteration in the sperm behavior which requires capacitation but which is distinct from the acrosome reaction. Unique patterns of sperm movement, which are called activated motility may provide additional thrust for penetration of spermatozoa through the ovum vestments (10). The appearance of activated motility and its prevelance in a sperm suspension are potential indicators of the fertilizing capability of those spermatozoa (2). Unfortunately, a characteristic pattern of activated movement has never been identified in suspensions of capacitated human sperm. Using videomicrography, we have measured a consistent increase in the swimming speeds of human sperm following 4 hours of capacitation in vitro (11). Aitken and colleagues (12) have studied capacitated human sperm with time exposure photomicrography. These investigators identified specific patterns of human sperm movement, some of which were positively correlated with the penetration of zona free hamster eggs and some of which were negatively correlated with hamster egg penetration. It remains to be determined whether this type of motility assessment will have predictive value in relation to human sperm fertility.

## 3. ASSESSMENT OF HUMAN SPERM ZONA PELLUCIDA INTERACTION

Completion of sperm capacitation and the acrosome reaction is inferred from the penetration of oocytes in vitro. Direct observations of human fertilization in vitro would provide this information, but such experiments are ethically unacceptable to many individuals. There are also significant logistical problems associated with the recovery and culture of large numbers of viable human oocytes. The properties of the acellular, non-living zona pellucida, which are required for sperm-oocyte interaction, seem to be unaffected by the maturity or viability of the human vitellus (13). Furthermore, the acrosome reaction appears to be a consistent prerequisite for sperm penetration of this layer (13). Thus, we have been able to use non-viable and/or immature human oocytes for bioassay of the human sperm acrosome reaction and other sperm-zona pellucida interactions (13-16).

There is experimental evidence which suggests that specific receptors on the sperm surface may be required for sperm attachment and binding to the zona pellucida (17). The species specificity of sperm binding to the human zona pellucida (13) and the association of binding failure with certain cases of male infertility (15) lend support to this notion. The functions of the human sperm cell which are involved in initial interaction with the zona can be assessed directly by observation of spermatozoa on the zona surface (15, 16). In some mammalian species attachment to the zona pellucida must precede completion of the acrosome reaction, since spermatozoa without acrosomes are unable to bind to the zona (2). This may not be the case with human gametes, since both acrosome reacted and intact spermatozoa are found on the zona surface of penetrated oocytes (13, 16). However, very weak zona binding was observed under experimental conditions in which the oocyte was exposed to spermatozoa that had undergone acrosome reactions several hours earlier (16). This may indicate a requirement for completion of the acrosome reaction at or near the time of zona penetration.

After loss of the acrosomal cap, the inner acrosomal membrane of the sperm head is directly exposed to the zona substance. Lytic enzymes such as acrosin may be bound to this membrane and their digestive action could play a role in sperm penetration through the thickness of the zona (2). Measurements can be made of the acrosin content of human spermatozoa (18) and such assays could be useful in predicting the fertilizing potential of a sperm population.

A significant thrust of the sperm flagellum is also thought to be required for zona penetration (19). We have measured the movement characteristics of human spermatozoa on the surface of the human zona pellucida by high speed videomicrography (16). Under experimental conditions in which there was failure of zona penetration, a significantly lower flagellar beat frequency and a smaller flagellar amplitude were recorded. Interestingly, spermatozoa in these suspensions, while unable to penetrate the zona pellucida, fused readily with zona free hamster eggs in the same dishes (16).

Rigidity of the sperm head may also be required for shearing of the zona substance (20). Such rigidity may result from extensive disulfide bonding within the sperm nuclear chromatin (21). Unlike the situation in other mammals, the proportion of human spermatozoa with highly stabilized chromatin appears to vary among individual men (22). Assessment of this parameter may therefore prove useful in predicting the fertility of a population of sperm cells.

## 4. ASSESSMENT OF HUMAN SPERM FERTILITY WITH ZONA FREE HAMSTER EGGS

In 1976, Yanagimachi reported that the zona pellucida is the principal block to inter-species fertilization between human spermatozoa and hamster oocytes (23). Consequently, the removal of this vestment permits the fusion of human sperm with the hamster oolemma. We now appreciate that additional but less effective blocks to human sperm fusion are also present in the hamster vitelline membrane (2). Nevertheless, the zona free hamster egg system has been advocated as a practical assay for evaluating the fertilizing power of human sperm cells. The presence of decondensed sperm heads in the hamster ooplasm appears to be reliable evidence that capacitation and the acrosome reaction have occurred, since an acrosome reaction is thought to be required for sperm fusion with the hamster vitellus (23-25). Human sperm attachment to the surface of the hamster vitellus may also reflect the completion of the acrosome reaction, since only acrosome reacted sperm appear to be able to maintain contact with the vitelline surface (16, 25).

We do not know the true biological significance of human sperm fusion with the hamster vitellus or of sperm incorporation into the hamster ooplasm. During fertilization of homologous gametes, sperm incorporation into the ooplasm probably requires specific functions of the sperm cell which are distinct from those associated with either zona attachment or zona penetration, since a different region of the sperm head appears to be involved (26). It is reasonable to assume that observations of human sperm interaction with zona free hamster eggs could provide information on sperm fertility which is additional to other measures of capacitation or the acrosome reaction (15). However, there are no grounds for believing that the cellular mechanisms of interspecies fertilization are precisely the same as those involved in homologous human fertilization. In fact, the receptivity of the hamster vitellus to human sperm is significantly less than it is to sperm of its own species or even a closely related species such as the guinea pig (2).

Numerous clinical papers have demonstrated a strong association between male infertility and the failure of human sperm fusion with zona free hamster eggs (15, 24, 27-29). The utility of these assays for fertility diagnosis in individual cases is diminished by the lack of standardization in assay procedures and the significant risk of both false positive and false negative results (30).

A major technical problem with in vitro tests of sperm fertility is the identification of false negative results. That is, an apparent low fertility of a sperm population because of technical artifacts related to the culture medium or oocytes. We have recently developed a competitive in vitro assay of sperm fertilizing ability which utilizes contrasting fluorescent sperm markers (11). The dyes, fluorescein isothiocyanate and tetramethylrhodamine isothiocyanate (which fluoresce green and red, respectively) have no effect on human sperm motility or their ability to penetrate either the human zona pellucida or the zona free hamster egg (11). This technique allows the design of experiments in which an internal control population of spermatozoa is included with the oocytes and the sperm suspension being tested. The labels remain visible on the sperm mid-piece after sperm incorporation into the hamster ooplasm (11). Since the hamster vitellus exhibits no apparent block to polyspermic fertilization by human sperm (31), the number of penetrating sperm per vitellus as well as the percentage of penetrated eggs can be recorded.

## 5. BIOLOGICAL FACTORS WHICH MAY AFFECT THE FERTILIZING POWER OF HUMAN SPERM

### 5.1. Pathology of spermatogenesis

For several years we have been involved in research on human sperm pathology which may be associated with fertilization dysfunction. We have studied groups of infertile men using a mixed gamete assay in which the sperm suspension is incubated with human oocytes and zona free hamster eggs. Our initial series of 21 cases has been reported previously in detail (15, 30). We have subsequently studied an additional group of 10 selected cases using the same methodology (J.E. Gould, J.W. Overstreet, unpublished). All of these men had marriages with long-standing infertility and in every case the male contribution was proven by successful impregnation of the female using donor semen. In eight of the ten cases there was evidence of fertilization dysfunction. In three cases there was concomitant failure of zona penetration and of sperm incorporation into the vitellus. This pathology probably reflects an abnormality of the common sperm functions required for entry into the two gametes, i.e., capacitation and/or the acrosome reaction. In three cases with failure of zona penetration there was some evidence of sperm incorporation into the hamster ooplasm, although the percentage of penetrated vitelli was less than 10% (versus approximately 25% penetration for the two men with "normal" sperm function). This result suggests that capacitation and the acrosome reaction occurred, but that there were abnormalities in other functions of the sperm cell which were specifically required for zona penetration (e.g., sperm motility, enzymes). In the last two cases where there was zona penetration but no sperm fusion with the hamster eggs, an abnormality in the sperm functions related to membrane fusion must be suspected. In summary, these recent studies confirm our earlier observations that human male infertility may result from abnormalities in any one or several of the currently recognized steps in the fertilization process. Since other experiments (discussed below) suggest that some of these functions can be altered by in vitro manipulations, it may be possible to correct certain deficiencies in the sperm cell by careful control of in vitro conditions.

5.2. Biological variation in sperm function among fertile men

There is considerable evidence that spermatozoa from individual men can have significantly different responses to the laboratory conditions for in vitro fertilization. This means that sperm cells from a particular man may have unique requirements for complete expression of their innate fertilizing potential. Differences in the swimming behavior of seminal spermatozoa from fertile men have been appreciated for some time (32). However, it is also apparent that spermatozoa from fertile donors may differ significantly in the change of flagellar activity which follows capacitation (11, 12), in the time required for completion of capacitation (33) and in the "efficiency of capacitation" as reflected in the number of sperm necessary for penetration of the egg (12) or the overall percentage of eggs penetrated (11, 27, 28).

It is possible to examine the "relative fertility" of donors in a therapeutic artificial insemination program by recording the number of conceptions achieved in relation to the number of women inseminated with spermatozoa from each donor. We carried out experiments to determine whether there is any relationship between this relative fertility and the percentage of zona free hamster eggs penetrated either in the standard assay or in the competitive, double fluorescent label assay. We found no relationship between the donor's fertility in the AI program and either of the laboratory tests of sperm fertility (34). This result should not be surprising if we consider the many other sperm functions involved in producing a normal fetus (e.g., sperm transport, sperm longevity, embryonic survival, etc.), none of which are evaluated by these assays. The lack of close association between "fertility" in vitro and in vivo also seems reasonable in view of our growing awareness of the complexity of sperm behavior under in vitro conditions (see the next section).

## 6. TECHNICAL FACTORS WHICH MAY AFFECT THE FERTILIZING POWER OF HUMAN SPERMATOZOA

Our studies of infertility patients have clearly demonstrated the independence of the sperm functions associated with penetration of the human zona pellucida as opposed to those required for fusion with the hamster vitellus. The methods which are used for capacitating the sperm cells may preferentially enhance one or other of these functions. The traditional methods of preparing human spermatozoa for in vitro capacitation involve separation of the sperm cells from the seminal plasma and a series of washes, during which spermatozoa are sequentially concentrated by centrifugation and then diluted in fresh fertilization medium (15, 27). Capacitation can also be achieved without washing when whole semen is exposed to a column of ovulatory human cervical mucus in a capillary tube (14). As the spermatozoa swim out of the semen and through the mucus column much of the seminal plasma coating is presumably sheared from the sperm surface, thus obviating the need for additional washing procedures. When human oocytes were present in a compartment of fertilizing medium on the "downstream" side of the mucus column, the zona pellucida was penetrated within one hour of initial sperm-mucus contact (14).

The time required for sperm capacitation after cervical mucus penetration (i.e., the time elapsing between sperm departure from the semen and sperm penetration of the zona) appears to be significantly shorter in this system than in the traditional washing system (14). Furthermore, this system seems to mimic fairly closely the normal biological conditions of in vivo capacitation. We have recovered spermatozoa from mucus aspirated from the human cervix at various times after artificial insemination (35). The time required for zona penetration by these "in vivo capacitated" sperm and the sperm concentrations necessary to achieve such penetration were remarkably similar to those observed in the capillary tube system.

Surprisingly, when we compared the results of "capacitation" in the two systems (i.e., washing versus cervical mucus) by observing sperm fusion with zona free hamster ova, the relative efficiencies of the two methods were not the same. In eight paired experiments only 10% of 49 cryopreserved hamster vitelli were penetrated after exposure to mucus capacitated sperm, whereas 30% of 56 vitelli were penetrated after incubation with a similar number of washed capacitated sperm. Sperm numbers on the vitelline surface were more than 5 times higher in the latter group of gametes (36). More recent experiments utilizing mixtures of human oocytes and fresh hamster vitelli have confirmed the preliminary results. Thirty-seven percent of 78 hamster vitelli were penetrated after the sperm were capacitated by swimming through cervical mucus. In contrast, 65% of 82 vitelli were penetrated after incubation with a similar number of washed capacitated sperm. The results with human oocytes in the same sperm suspensions were 54% and 25% penetration, respectively (J.E. Gould, J.W. Overstreet and F.W. Hanson, unpublished). Although other interpretations are possible, the results of these experiments suggest that the biological events occurring during sperm capacitation in vivo and those occurring in vitro are not identical. They also serve to emphasize a recurrent theme, that different laboratory systems for preparation of the spermatozoa and/or for assessment of their function are likely to provide differing results.

The efficiency and kinetics of sperm capacitation in vitro can be affected by many variables including the composition of the media, the incubation temperature, the time of incubation, etc. A differential sensitivity of the sperm functions to these manipulations has been demonstrated by experiments in which aliquots of a semen sample were washed and capacitated in one of two preparations of the standard BWW medium. One preparation contained 0.3% bovine serum albumin and the second contained 3.5% human serum albumin (low and high albumin media, respectively). In previous studies by our laboratory and by others, the high albumin media have been used for relatively short-term capacitation intervals (approximately 4-6 hours), while the low albumin media have been used for longer, overnight incubation (generally 18-24 hours). We used a four combination experimental design in which sperm suspensions in high and low albumin media were incubated for both short and long intervals. When mixtures of human zonae and hamster vitelli were added to the sperm suspensions, we found that the penetration of hamster eggs increased as the incubation time increased, regardless of the albumin concentration. However, spermatozoa capacitated overnight in the high albumin media totally lost their ability to penetrate the human zona pellucida, even though 100% of the hamster vitelli in the same dishes were penetrated (16).

Study of the sperm in high albumin with electron microscopy indicated that the percentage of acrosome reactions increased from 8% after 4 hours incubation to 62% at the end of 24 hours incubation. It is likely that this change was the principal factor contributing to the higher rate of fusion with the hamster vitelli after extended incubation. At least two factors were identified which could have led to the failure of zona penetration. The sperm swimming speed and flagellar movements declined significantly during prolonged incubation, and these changes may have led to inadequate thrust for zona penetration. Careful ultrastructural study of the sperm after 24 hours of incubation revealed that more than 80% of the acrosome reacted sperm had changes in their acrosome characterized by partial or total loss of the equatorial segment (16). Similar alterations in the equatorial segment of the acrosome have been attributed to sperm senescence and have also been observed in hamster sperm suspensions after lengthy capacitation incubations (37). A similar failure of zona penetration was also observed in the animal studies. Thus, it appears that occurrence of the acrosome reaction at an early time in the incubation period, while deleterious to zona penetration, had no adverse effect on sperm fusion with the hamster vitellus.

A definition of the "optimum" conditions for in vitro fertilization is badly needed by investigators attempting to evaluate and/or control the fertilizing power of mammalian sperm. The assumption that such optimum conditions exist in the currently employed clinical tests is a serious misconception. There is an emerging view in the clinical literature that a normal range has been established for the percentage of zona-free hamster eggs penetrated by human sperm in vitro. The lower level of this range has been commonly set at 10-15% (27, 29), and it appears to be the practice of some laboratories to interpret any result below this range as an indication of male infertility. In addition to biological considerations which seriously challenge the validity of this logic (30), recent experiments have demonstrated that an alteration in experimental conditions can often change a negative test result to a positive one. By incubating the semen from infertility patients in an egg yolk buffer medium at 4°C for 24 hours or more, Binor and colleagues were able to improve the hamster egg penetration rates in 31% of 41 cases tested (38). Similar experiments carried out in our laboratory have led us to the same conclusion. In our series of 19 patients, 7 men had spermatozoa which penetrated less than 10% of hamster vitelli in the standard assay. After 48 hours of incubation in the egg yolk buffer medium at 4°C spermatozoa from 3 of the 7 men scored within the "normal range" including one case in which the penetration rate increased from 0% to 57% as a result of the experimental manipulation (J.R. Bolanos and J.W. Overstreet, unpublished).

## 7. CONCLUDING REMARKS

It is now possible to apply a number of sophisticated laboratory techniques for evaluation of the fertilizing power of mammalian spermatozoa. Although there is no method currently available for directly assessing the progress of capacitation in a sperm suspension, there is a strong possibility that methods which reliably monitor sperm surface changes may become widely available in the near future. The changes in sperm motility and acrosomal morphology which are required for fertilization can now be assessed with greater ease and precision than ever before. Direct studies of sperm-oocyte interaction in vitro can give information not only on the fertilizing power of the sperm suspension as a whole, but they can also be used to evaluate specific functions of the sperm cells which are involved in the fertilization process. However, in spite of our rapidly advancing knowledge of sperm biology in vitro, we still do not understand the precise relationship of in vitro tests to sperm function in vivo.

The application of our knowledge of sperm biology to human gametes requires appreciation of a number of special circumstances. In vitro tests of sperm fertility reveal significant differences in fertilizing power of spermatozoa among men of proven fertility. These differences do not appear to be related to the relative fertility of the individual men. The fertilizing power of spermatozoa from both fertile and infertile men can be altered in vitro by modifying the environment of the sperm cells. The complexity of this biology often leads to difficulty in making a confident diagnosis of male infertility on the basis of these tests alone. Nevertheless, as our knowledge of sperm biology accumulates, it may become possible to manipulate spermatozoa and achieve fertility in vitro even when we are unsuccessful in identifying or treating the responsible disorder of spermatogenesis.

## ACKNOWLEDGMENTS

The unpublished research described in this chapter was supported by research grants from the National Institutes of Health (HD 15149), the United States Environmental Protection Agency (R009089) and the National Institute for Occupational Safety and Health (OH 01148). The support of NIH Research Career Development Award HD 00224 is also acknowledged.

## REFERENCES

1. Overstreet JW. 1983. Transport of gametes in the reproductive tract of the female mammal. In: Mechanisms and Control of Fertilization. JF Hartman (Ed.). Academic Press, New York. In Press.
2. Yanagimachi R. 1981. Mechanisms of fertilization in mammals. In: Fertilization and Embryonic Development In Vitro. L Mastroianni, Jr. and JD Biggers (Eds.). Plenum Press, New York. pp. 81-182.
3. Bavister BD. 1981. Analysis of culture media for in vitro fertilization and criteria for success. In: Fertilization and Embryonic Development In Vitro. L Mastroianni, Jr. and JD Biggers (Eds.). Plenum Press, New York. pp. 41-60.
4. Johnston I, Lopata A, Speirs A, Hoult I, Kellow G, Plessis Y. 1981. In vitro fertilization: the challenge of the eighties. Fertil Steril 36:699-706.
5. Trounson A, Wood C. 1981. Extracorporeal fertilization and embryo transfer. Clin Obstet Gynaecol 8:681-713.
6. Koehler JK. 1981. Lectins as probes of the spermatozoon surface. Arch Androl 6:197-217.
7. Soupart P, Strong PA. 1974. Ultrastructural observations on human oocytes fertilized in vitro. Fertil Steril 25:11-44.
8. Talbot P, Chacon R. 1980. A new procedure for rapidly scoring acrosome reactions of human sperm. Gamete Res 3:211-216.
9. Talbot P, Chacon RS. 1981. A triple-stain technique for evaluating normal acrosome reactions of human sperm. J Exp Zool 215:201-208.
10. Katz DF, Overstreet JW. 1981. Mammalian sperm movement in the secretions of the male and female genital tracts. In: Testicular Development, Structure, and Function. A Steinberger and E Steinberger (Eds.). Raven Press, New York. pp. 481-489.
11. Blazak WF, Overstreet JW, Katz DF, Hanson FW. 1982. A competitive in vitro assay of human sperm fertilizing ability utilizing contrasting fluorescent sperm markers. J Androl 3:165-171.

12. Aitken RJ, Best FSM, Richardson DW, Djahanbakhch O, Lees MM. 1982. The correlates of fertilizing capacity in normal fertile men. Fertil Steril. In Press.
13. Overstreet JW, Hembree WC. 1976. Penetration of the zona pellucida of nonliving human oocytes by human spermatozoa in vitro. Fertil Steril 27:815-831.
14. Overstreet JW, Gould JE, Katz DF, Hanson FW. 1980. In vitro capacitation of human spermatozoa after passage through a column of cervical mucus. Fertil Steril 34:604-606.
15. Overstreet JW, Yanagimachi R, Katz DF, Hayashi K, Hanson FW. 1980. Penetration of human spermatozoa into the human zona pellucida and the zona-free hamster egg: a study of fertile donors and infertile patients. Fertil Steril 33:534-542.
16. Gould JW, Overstreet JW, Yanagimachi H, Yanagimachi R, Katz DF, Hanson FW. 1982. What functions of the sperm cell are measured by in vitro fertilization of zona-free hamster eggs? Fertil Steril. In Press.
17. Peterson RN, Russell L, Bundman D, Freund M. 1979. Sperm-egg interaction: evidence for boar sperm plasma membrane receptors for porcine zona pellucida. Science 207:73-74.
18. Mohsenian M, Syner FN, Moghissi KS. 1982. A study of sperm acrosin in patients with unexplained infertility. Fertil Steril 37:223-229.
19. Katz DF, Yanagimachi R. 1981. Movement characteristics of hamster and guinea pig spermatozoa upon attachment to the zona pellucida. Biol Reprod 25:785-791.
20. Bedford JM. 1974. Mechanisms involved in penetration of spermatozoa through the vestments of the mammalian egg. In: Physiology and Genetics of Reproduction, Part B. EM Coutinho and F Fuchs (Eds.). Plenum Publishing Corporation, New York. pp. 55-68.
21. Bedford JM, Bent MJ, Calvin H. 1973. Variations in the structural character and stability of the nuclear chromatin in morphologically normal human spermatozoa. J Reprod Fert 33:19-29.
22. Blazak WF, Overstreet JW. 1982. Instability of nuclear chromatin in the ejaculated spermatozoa of fertile men. J Reprod Fert. In Press.
23. Yanagimachi R, Yanagimachi H, Rogers BJ. 1976. The use of zona-free animal ova as a test-system for the assessment of the fertilizing capacity of human spermatozoa. Biol Reprod 15:471-476.
24. Barros C, Gonzalez J, Herrera E, Bustos-Obregon E. 1979. Human sperm penetration into zona-free hamster oocytes as a test to evaluate the sperm fertilizing ability. Andrologia 11:197-210.
25. Talbot P, Chacon RS. 1982. Ultrastructural observations on binding and membrane fusion between human sperm and zona pellucida-free hamster oocytes. Fertil Steril 37:240-248.
26. Bedford JM, Moore HDM, Franklin LE. 1979. Significance of the equatorial segment of the acrosome of the spermatozoon in eutherian mammals. Exp Cell Res 119:119-126.
27. Rogers BJ, Campen HV, Ueno M, Lambert H, Bronson R, Hale R. 1979. Analysis of human spermatozoal fertilizing ability using zona-free ova. Fertil Steril 32:664-670.
28. Hall JL. 1981. Relationship between semen quality and human sperm penetration of zona-free hamster ova. Fertil Steril 35:457-463.
29. Karp LE, Williamson RA, Moore DE, Shy KK, Plymate SR, Smith WD. 1981. Sperm penetration assay: useful test in evaluation of male fertility. Obstet Gynecol 57:620-623.

30. Overstreet JW. 1982. Evaluation of sperm function by tests of sperm ovum interaction in vitro. Proceedings of International Symposium on Human Fertility Factors, Cargese. In Press.
31. Binor Z, Sokoloski JW, Wolf DP. 1980. Penetration of the zona-free hamster egg by human sperm. Fertil Steril 33:321-327.
32. Katz DF, Overstreet JW, Hanson FW. 1981. Variations within and amongst normal men of movement characteristics of seminal spermatozoa. J Reprod Fert 62:221-228.
33. Perreault SD, Rogers BJ. 1982. Capacitation pattern of human spermatozoa. Fertil Steril. In Press.
34. Blazak WF, Overstreet JW. 1981. Competitive penetration of zona-free hamster eggs by spermatozoa from fertile donors. Fertil Steril 35:246-247. Abstract.
35. Gould JE, Overstreet JW, Hanson FW. 1981. Assessment of human sperm function after aging in vivo. Fertil Steril 35:240. Abstract.
36. Gould JE, Overstreet JW, Hanson FW. 1981. Penetration of zona free hamster ova as an assay of human sperm function. Clinical Research 29:27A. Abstract.
37. Barros C, Fujimoto M, Yanagimachi R. 1973. Failure of zona penetration of hamster spermatozoa after prolonged preincubation in a blood serum fraction. J Reprod Fert 35:89-95.
38. Binor Z, Ramaa R, Scommegna A. 1982. Further studies of dysfunctional human spermatozoa using the zona-free hamster oocyte penetration assay. Fertil Steril 37:300-301. Abstract.

# SPERMATID DIFFERENTIATION IN MAN DURING SENESCENCE

A.F.HOLSTEIN, Abt.f.Mikroskop.Anatomie, Universität Hamburg, Germany

Claims are to be found in the literature, that spermatogenesis continues into old age. The begetting of children by men who are over the age of 65 is authentic and verified in specific cases, but the question regarding the fertility of man during senescence seldom arises. Clinically relevant problems are rarely encountered, and therefore the investigation of spermatogenesis in man during senescence is oriented toward the aging process in the germinal epithelium and in the germ cells. We have examined the condition of the tissue, and the morphology of cells arising from a pool of constantly proliferating stem cells.

The testes of 140 men over the age of 65, who had undergone orchidectomy on account of prostate carcinoma, were at our disposal. Ninety percent of the men had mature spermatids in the seminiferous tubules. In relation to the number of cut seminiferous tubules containing mature spermatids, in each case a simple count of spermatogenesis was undertaken. On the basis of this count two groups of men were selected: One with the highest and one with the lowest count.

The first group comprised men who demonstrated intact spermatogenesis, and in the second were placed those who only had very few spermatids. The spermatids of the men in both groups were examined under the electron microscope. The senescent men with intact spermatogenesis had between 47% and 85% malformed spermatids. The defects were approximately equally distributed among all cell components. Acrosome deformities, and anomalies of the nucleus, the postacrosomal region, the neck, the middle piece, and the principal piece were observed in isolation and also in combination. There exists a certain pattern of the distribution of the malformations in the spermatids.

The group of senescent men with poorer spermatogenesis demonstrated

---

Supported by grants from Deutsche Forschungsgemeinschaft (Ho 388/5)

not only fewer mature spermatids in the seminiferous tubules but also between 80 and 95% of malformed spermatids as judged by electron microscope investigation. This percentage of malformed spermatids was higher than in the group of senescent men with intact spermatogenesis. Also in these cases all cell components were malformed with approximately equal frequency.The cellular patterns of deformity corresponded to that of group I.

For comparison with both these groups another group was examined which was composed of young adult men who had been orchidectomized of their own free will on account of sexual offences. The score count of these men was maximal, corresponding to intact spermatogenesis. Investigation with the electron microscope showed that between 37 and 55% of the spermatids were deformed. In these cases, however, deformity of the acrosome was dominant, while other cell components were more rarely deformed. The rate of deformity in the group of young men was, when considered on its own, surprisingly high, though lower than both the groups of older men. It is recognizable that men with intact spermatogenesis during senescence, in contrast to young adult men with intact spermatogenesis, demonstrated a considerably higher rate of deformity. The high deformity rate, and the cellular pattern of deformity of the spermatids were considered to be characteristic for the pattern of spermatogenesis of men during senescence.

These results might be linked to the fact that the used material was taken from senescent men suffering from prostate carcinoma. In response to this it must be said that 1. no other material on which to investigate spermatogenesis during senescence is available, and 2. that all patients investigated were in the early stages of the disease, and were orchidectomized before the onset of hormonal therapy. It was assumed that the disease had no significant influence upon the general condition, and especially not upon spermatogenesis. This assumption is supported by the claim of Harbitz (1973), Chowdhury et al. (1975) and Honoré (1978), who could not show any influence of a prostate carcinoma upon spermatogenesis. In order to substantiate this claim, we examined material from a fourth group of patients, who were between the ages of 41 and 60, and who were suffering from prostatic carcinoma. The spermatogenesis of these men was recognized by means of the score count as intact, and the rate of spermatid deformity lay in the same range as group III.

Senescent men with poorer spermatogenesis in almost all of the spermatids show defects on the acrosome, nucleus, the postacrosomal region, or also on the flagellum. The defects of the spermatids in old men are fundamentally similar to those in young men, with or without andrological diseases. In studies of testicular tissue from men over the age of 65, however, cytological details could be found which have not yet been observed in young men:

1. In many spermatids, contact with the Sertoli cells is disturbed. The microfilaments in the peripheral cytoplasm of the Sertoli cell, which normally surround the acrosome of the spermatid like a fibrous basket, have no relationship to the plasmalemma, instead they form a free loop in the cytoplasm. The cisterns of the endoplasmic reticulum in relation to the microfilaments are missing. The acrosome of the spermatid, without the fibrous basket of the Sertoli cell, is always deformed. The acrosome vesicle usually shows a one sided lateral thickening or invagination into the nucleus.

2. Where the acrosome faces the nucleus, normally the perinuclear cisterna is closed, and the nuclear envelope is thickened by a thin layer of electron dense substance. The spermatids of old men form thickenings of the nuclear envelope independent of acrosome contact, anywhere on other areas of the nuclear membrane. Such aberrant membrane thickenings apparently inhibit implantation of the flagellum on the nucleus, or may cause nuclear deformities.

3. Very often one finds an extreme elongation of the spermatid nucleus. In an approximately normal configuration of the acrosome, the postacrosomal region of the nucleus is thin and long. The redundant membrane exists of countless vesicles.

4. Unusual nuclear forms are numerous. They arise generally during the maturation of such spermatids which are without contact to the Sertoli cells.

5. Spermatids in senium often have the tendency to develop into giant cells. As the electron microscopic examination showed, spermatids of a clone loose their intercellular bridges and flow together to form a giant cell. In a series of sections up to 30 nuclei could be counted in a spermatid giant cell.

Also mature spermatids flow together with their perikarya. This process is different, however, from that in the immature spermatids. In the case of mature spermatids, neighboring cell membranes fuse with one another, while the intercellular bridges remain intact in the cytoplasm.

6. During the release of mature spermatids, often residual bodies are expelled from the germinal epithelium together with the spermatids. A contact with the Sertoli cells by means of junctional specializations does not exist.

From these findings it can be concluded that although in old age a differentiation of spermatids still takes place, the malformation quota is very high. Spermatids and spermatozoa with multiple deformations develop. In many cases the spermiation is also disturbed. Morphological details of the spermatids can be described that have not been observed in other stages of life, and not in patients with fertility disturbances. These manifestations in the spermatids of old men, among a series of histological changes of the germinal epithelium which have been reported elsewhere (Holstein, Hubmann 1981), belong to the phenomena which characterize the involution of spermatogenesis in senium.

REFERENCES

1. Chowdhury AR, Steinberger A, Steinberger E. 1975. A quantitative study of spermatogonial population in organ culture of human testis. Andrologia 7, 297-307
2. Harbitz TB. 1973. Testis weight and the histology of the prostate in elderly men. An analysis in an autopsy series. Acta path microbiol scand Sect A 81, 148-158
3. Holstein AF, Hubmann R. 1981. Spermatogonia in old age. In: Föhringer Symposium: Stem cells in spermatogenesis. AF Holstein, C Schirren eds. pp 77-88. Grosse,Berlin
4. Honoré LH. 1978. Aging changes in the human testis: a light-microscopic study. Gerontology 24, 58-65

# SPERM STRUCTURE AND FUNCTION IN 70 YEAR OLD HUMANS

B. BACCETTI, T. RENIERI, M.G. SELMI and P. SOLDANI
(Institute of Zoology, University of Siena, Italy)

## 1. INTRODUCTION

It is commonly accepted that aging drastically affects the male reproductive power, disturbing and preventing hormonal secretions, germinal cell proliferation and ejaculation (1,2, 3,4,5). Fine structural studies have been carried out on Leydig cells both in humans (6,7) and Stallions (8), all of them mainly dealing with functional parameters, and on human Sertoli cells (9). Leydig cells appear increased both in number and in volume, in ageing stallions, decreased in humans concomitantly with a declining androgen status. Sertoli cells resume a high capacity to divide, and frequently give rise to binucleate cells. No informations are available concerning spermatozoa and spermatogenesis.

## 2. PROCEDURE

### 2.1. Material and methods

We have selected a number of more than 70 year old men of proven fertility, studying their ejaculates and there, when present (20 individuals) spermatozoa with the routinary techniques of scanning and transmission electron microscopy. In some cases spermatogenesis on bioptical material by electron microscopy, sperm movement by microcinematography and egg penetration in zona free hamster eggs have been tested, in order to obtain a complete picture of human spermatozoa structure and function in aged individuals.

2.2. Spermatozoa in the ejaculate

Among the 20 selected aged men able to produce spermatozoa, some show in the ejaculate a high concentration of spermatozoa (up to 160 x $10^6$ cc), others lower concentration (ranging from 1 to 20 x $10^6$ cc), others are close to azoospermy, indipendently of age. In general, from 9 to 20% of the spermatozoa are motile; in a few cases all spermatozoa are immotile also indipendently of age and of sperm concentration.

2.3. Sperm morphology

At the scanning electron microscope level the heads of the ejaculated spermatozoa appear normal; the tails are thinner than those of young men, and sometimes the mid-piece has a reduced length (Fig. 1a). Immature forms appear very rare. Sections examined in transmission electron microscope reveal a normal structure in about 35% of the spermatozoa. The nucleus, the acrosomal complex, the mitochondrial helix, the axoneme and the accessory structures (accessory fibers, fibrous sheath) show normal morphology. About 65% of the spermatozoa show abnormal morphology at the level both of the head and of the tail. The acrosome is vesiculated, swollen and sometimes encircled by cytoplasmic droplets. The chromatin often appears granular but, examined with the test for lysine of Courtens and Loir (10) and with the test for S-S groups of Swift (11), shows a normal loss of lysine-rich nucleoproteins (Fig. 1b) and a high content of S-S bonds (Fig. 1c). The axoneme is surrounded by disordered mitochondria arranged in a short helix and by disordered accessory fibers. In many cases the classical structure of microtubules is altered and sometimes the central pair is missing (Fig. 2a).

2.4. Bioptic material

Sections of the biopsy samples reveal that the germinal

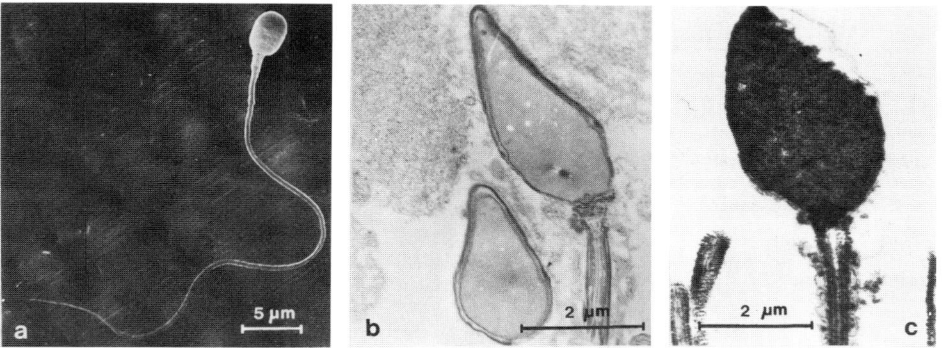


FIGURE 1. a) Scanning electron micrograph of a spermatozoon in a 72 year old man. b) Sperm head of an aged man with a reduced staining with PTA according to Courtens and Loir (10). c) Silver staining of S-S groups according to Swift(11) on the same material.

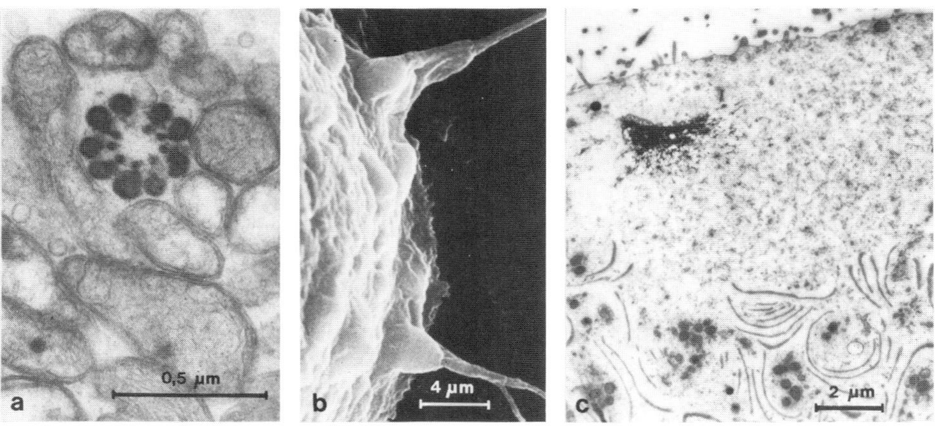


FIGURE 2. a) Cross section of the tail region in the same material. Mitochondrial helix is disordered and the two central axonemal tubules are lacking. b) Two aged human spermatozoa penetrating a zona free hamster egg, seen at scanning electron microscope. c) Thin section of a zona free hamster egg penetrated by an aged human spermatozoon with decondensing nucleus.

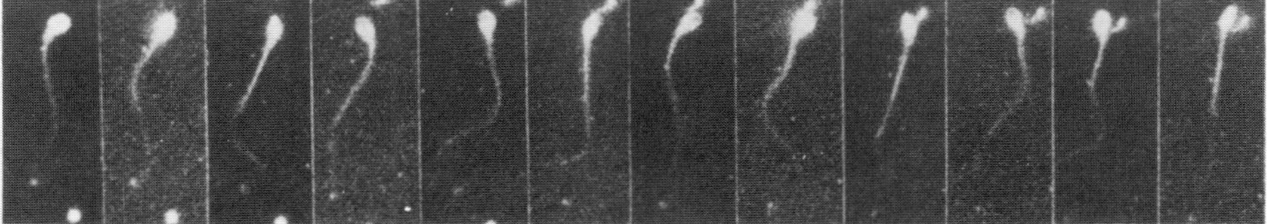

FIGURE 3. Consecutive frames of a film showing the sperm movement of a 72 year old man, taken at 18 frames per second.

cells are present in all stages of spermatogenesis even if in reduced number than in young men. Spermatid nuclei show a normal content of lysine, stained with the Courtens and Loir (10) technique. Some Sertoli cells are binucleate; the "freeze fracture" technique shows that the hemato-testicular barrier is almost normal with Sertoli-Sertoli cell junctions (tight and tight-gap) well developed. Leydig cells appear altered, for an evident vacuolization of the cytoplasm, an abnormally high presence of enormous Reinke's crystals and an impressive production of osmiophilic pigment granules.

2.5. Sperm movement

About 18% of the ejaculated spermatozoa of old men are motile. Examined in the dark field microscope (Fig. 3) by cinematographic recorder, they show a normal velocity. In some of them the amplitude and frequency of their waves are minor than those of the spermatozoa of young men.

2.6. Fertility test

An in vitro penetration assay utilizing spermatozoa of aged humans and zona free hamster eggs was employed in order to evaluate the fertilizing ability of aged individuals. Spermatozoa of the aged men having the best semen parameters were capable to complete some of the final stages of fertilization: fusion with the egg membrane (Fig. 2b) and decondensation of the sperm nucleus (Fig. 2c). Other aged men with poorer semen parameters failed to penetrate eggs. The system has been always used in presence of controls represented by spermatozoa of fertile young men, which consistently penetrated the eggs.

3.1. CONCLUSIONS

A few aged men (more than 70 year old) are able to produce spermatozoa. The general feature of these spermatozoa is quite conventional, the main characteristic is that immature forms are absent, the tail is meagre and thin, the mid-piece of reduced length. A fraction of the produced spermatozoa are motile (about 18%). The flagellum shows sometimes waves of reduced amplitude and frequency, but generally the velocity is normal.

The more conventional spermatozoa, belonging to the motile fraction, show normal nucleus, acrosome and axoneme. The number of mitochondrial gyres is frequently reduced. A major fraction of spermatozoa shows a number of malformations, and namely: swollen or vesiculated acrosomes, granular chromatin, perinuclear cytoplasmic droplets, disordered chondriome, disordered or deficient microtubular assembly in the axoneme. The nuclear maturation in spermatids and spermatozoa is normal; chromatin proteins are rich in S-S bonds and progressively devoid of histons. In these men spermatogenesis is obviously active, showing the normal stages. The hemato-testicular barrier is normal, with well developed Sertoli-Sertoli cell junctions. Leydig cells are altered by cytoplasmic vesiculations, pigment granules, abnormally frequent big Reinke's crystals.

In the fraction of motile spermatozoa the fertilization ability seems to be conserved: in the zona free hamster system penetration into the eggs has been obtained.

REFERENCES

1. Bishop MWH. 1970. J. Reprod. Fert. Suppl. 12, 65-87.
2. Stearns EL, McDonald JA, Kaufman BJ, Padua R, Lueman TS, Winter JSD, Faiman C. 1974. Am. J. Med. 57, 761-766.
3. Leathem JH. 1977. In: Johnson AD, Gomes WR (eds). The Testis, 4, Academic Press, N.Y.
4. Honoré LH. 1978. Gerontology 24, 58-65.
5. Harman SM. 1978. In: Schneider EL (ed). The aging reproductive system. Raven Press. N.Y.
6. Kaler LW, Neaves W.B. 1978. Anat. Rec. 192, 513-518.
7. Nankin HR, Lin T, Murono EP, Osterman J. 1981. J. Androl.

2, 181-189.
8. Johnson L, Neaves WB. 1981. Biol. Reprod. 24, 703-712.
9. Schulze W, Schulze C. 1981. Cell Tissue Res. 217, 259-266.
10. Courtens J, Loir M. 1975. J. Microsc. Biol. Cell. 24, 249-258.
11. Swift JA.1967. J.R. Microsc. Soc. 88, 449-460.

# IN VITRO STUDIES OF THE FERTILIZING ABILITY OF THE SPERM FROM ADOLESCENT, ADULT AND AGED MALE SYRIAN HAMSTERS

PAULINE L. CISNEROS
Department of Biology, University of Houston, Houston, Texas 77004, U.S.A.

## INTRODUCTION

There are indications that the male reproductive system is adversely affected as the animal ages. Both behavioral and physiological characteristics are involved, though variation among and within species is considerable. In bulls there is a general slowing down in mating behavior after 5-6 yr but some bulls do continue mating until 15 yr (Bishop,1970). Successful conceptions after artificial insemination decrease significantly with increasing age in bulls (Bishop,1964). In man a sexual peak is reached during the mid-teens; thereafter sexual activity gradually declines (Kinsey et al,1948) and the ease with which conception occurs decreases after age 25 (MacLeod & Gold, 1953). Success of crosses between male and female golden hamsters drops from 58% in young males to 26.9% in 15 mo males (Soderwall & Britenbaker,1955). When litter size from natural mating in rats was followed from male age 43d to 500d, the litter size peaked at 200d, then declined (Saksena et al,1979). The purpose of this study was to determine whether the fertilizing ability of the sperm per se is influenced by the age of the male. The results of in vitro fertilization by sperm from adolescent, adult and aged males plus results of natural mating (% successful conceptions and litter size) are described.

## MATERIALS AND METHODS

Animals were maintained on a reversed light cycle of 14h L:10h D/lights on 2100-1100. All animals were obtained from Engle Labs Inc.(Farmersburg, Indiana). Virgin females were used after showing 3 consecutive estrus cycles. The day of post-ovulatory vaginal discharge was designated Day 1 of the cycle (Orsini, 1961). Male hamsters were separated into four groups: adolescent, 2-4.5 mo; adult, 11-13.75 mo; aged I, 16-18 mo; aged II 22-25 mo. Male hamsters from each group were mated with a receptive female. Pairs were observed to insure that lordosis and mounting had occurred. Vaginal swabbings were made to detect the presence of sperm. The females were sacrificed by cervical dislocation 13-16d after mating to detect the presence or absence of pups. Male hamsters were sacrificed by cervical dislocation and the epididymides were removed. Sperm samples were obtained by laceration of the epididymides in 1 ml N/S. Sperm concentrations were determined spectrophotometrically and adjusted to $5x10^6$ and $10x10^6$. Incubation dishes were prepared containing 100ul Tyrode's buffer and 100ul human sera under paraffin oil. To each dish 50ul of sperm were added. All dishes were incubated at 37°C for 3h. Three females had been prepared for superovulation by injection of 20iu PMSG on Day 1 followed by 25iu HCG on Day 3 (Greenwald, 1962). Females were sacrificed 15.5h after the last injection, the oviducts were removed and ova recovered in the cumulus clots. Individual clots were transferred into hyaluronidase solution (1mg/ml) with BSA (3mg/ml) and rinsed in 3 washes of BSA. Ova from the 3 females were pooled together and evenly distributed among the dishes. After sperm incubation time of 6h, the ova were recovered from the dishes and mounted on slides under cover slips. Each egg was checked for penetration by sperm and was scored as penetrated when a sperm head or tail was in the perivitelline space or in the vitellus (Yanagimachi,1969). One way analysis of variance was used to determine differences with age for % penetration, litter size and organ weights. The G-test was used to compare % conceptions (Sokal & Rohlf,1969).

RESULTS

Results of in vitro fertilization indicate that the % of ova penetrated does not decrease significantly with age (Table 1). There is a slight decrease with age from adolescence(2-4.5mo) through aged I group (16-18mo), then a slight increase in the oldest age group (22-25mo). The trend is consistent in both sperm concentrations. Testing the % of conceptions in the different age groups yielded no significant decrease with age; however a definite pattern emerged (Fig.1). The % of conceptions is lower during adolescence, peaks during adulthood and decreases during senescence. The data for litter sizes follows a similar pattern. Litter sizes are smaller just after puberty, they are greatest when the males are approximately 1 yr and diminish as

Table 1. % Penetration(P) Determined by in vitro Fertilization at 4 Different Age Groups and 2 Sperm Concentrations

| age(mo) | n | #eggs | %P |
|---|---|---|---|
| sperm concentration: $5x10^6$/ml | | | |
| 2-4.5 | 9 | 153 | 41.5 |
| 11-13.75 | 6 | 136 | 28.3 |
| 16-18 | 9 | 166 | 19.1 |
| 22-25 | 6 | 162 | 30.9 |
| sperm concentration: $10x10^6$/ml | | | |
| 2-4.5 | 9 | 178 | 88.0 |
| 11-13.75 | 9 | 219 | 70.8 |
| 16-18 | 9 | 150 | 58.3 |
| 22-25 | 8 | 222 | 70.5 |

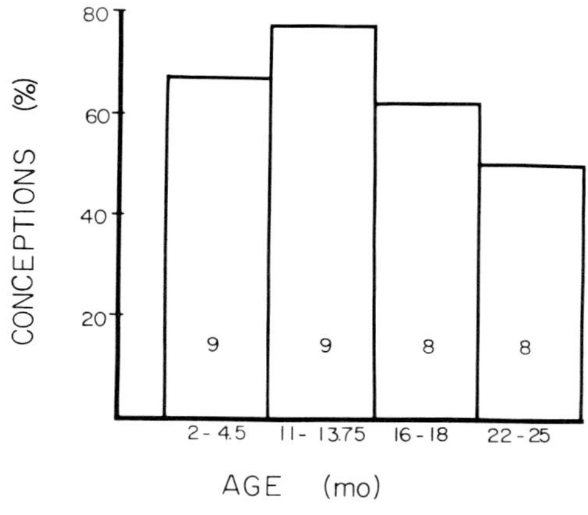


they age (Table 2). Body and organ weights-testis, epididymis-did not differ significantly with age.

Table 2. Litter size at four age groups

| age (mo) | n | litter size (mean±SEM) |
|---|---|---|
| 2-4.5 | 9 | 7.4 ± 2.2 |
| 11-13.75 | 9 | 9.6 ± 2.1 |
| 16-18 | 9 | 6.5 ± 3.0 |
| 22-25 | 9 | 5.4 ± 2.6 |

DISCUSSION

Despite numerous reports on the subject, the question of the influence of age on reproduction in the male is still unresolved. Studies cite age-related decreases in serum testosterone in man (Stearns et al,1974); decreases in Leydig cell response to HCG in the rat (Tsitouras et al,1979); decreases in successful matings in the hamster (Soderwall & Britenbaker,1955). Not all reports indicate adverse effects of aging on reproductive capacity. Engle(1952) reported that subjects of 70 yr had abundant sperm in more than half the tubules of the testes. Neither testicular weight nor plasma testosterone levels differed between mature and senescent mice (Nelson et al,1975). There is no deterioration of steroidogenic capabilities with age of isolated mouse Leydig cells; instead a suggestion is offered that the affinity of LH receptors increases rather than decreases in old age (Schafer et al,1982). Even the oldest male hamsters (24 mo) exhibited full copulatory behavior (Swanson et al,1982) when tested monthly from 12-24 mo. The findings of the present study indicate that the sperm per se loses little of its fertilizing ability with increasing age as determined by in vitro fertilization. Percentage of ova penetrated does not significantly differ in adolescent,adult and aged males. The slight decrease at 16-18 mo below that of the oldest group was unexpected but not significant. Though litter size showed no significant difference with age there was a definite pattern of larger size from adolescence to adulthood, then a gradual decline with increasing age. These findings agree with data from individual rats mated from 62d to 500d (Saksena et al,1979). There appears to be an optimal age for large litters from natural matings at one year which is not involved in in vitro fertilization. The % of fertilized ova for 12 mo males is equal to that of 22-25 mo males. No significant differences are found for % conceptions among the 4 age groups yet the pattern parallels that of litter size. The % conceptions in post-pubertal males is less than that in adult males (77.8%) and greater than that of the oldest males (50%). The number of animals used per group was small (n=8,9) compared to the study where difference in successful matings by young and senescent males was more apparent (Soderwall & Britenbaker, 1955). Analysis of weights showed no significant difference with age. This is in agreement with data for aging male rats (Lupo-diPrisco & Dessi-Falgheri,1980) and C57BL/6J male mice (Nelson et al,1975). Marked decreases in gonad weight was noted in cases of diseased mice (Nelson et al,1975) and hamsters dying of natural causes (Swanson et al,1982). When dealing with age's influence on reproduction, it is important to distinguish between age-related changes and effects due to age-related diseases. The male hamsters used in this study were chosen from the population as having good health and displaying full copulatory behavior. Use of cauda epididymal sperm eliminated possible age-related changes in accessory gland secretions which might affect the sperm's fertilizing ability. In conclusion, the present in vitro studies indicate that little of the sperm's fertilizing ability is lost with increasing age in the male hamster.

REFERENCES

1. Bishop,M.W.H.1964. J.Repro. Fertil. 7:383.
2. Bishop,M.W.H.1970. J.Repro. Fertil. Suppl. 12:65.
3. Engle,E.T.1952. In Cowdry's Problem of Aging, 3rd ed. Williams & Williams, Baltimore, p. 728.
4. Greenwald,G.1962. Endo. 71:378.
5. Kinsey, A.C.,Pomeroy,W.B. & Martin,C.E.1948. Sexual Behavior in the Human Male.Saunders, Philadelphia.
6. Lupo-diPrisco,C. & Dessi-Falgheri,F. Horm Res 12:149.
7. MacLeod,J. & Gold,R.1953. Fertil.Steril 4:194.
8. Nelson,J.F.,Latham, K.R. & Finch,C.E.1975. Acta Endocrin. 80:744.
9. Saksena,S.K.,Lau,I. & Chang,M.1979. Exp. Aging Res. 5:373.
10. Schafer,G.,Holstein,A.F. & Hilz,H.1982. Endo. 110:1362.
11. Soderwall,A.L. & Britenbaker,A.L.1955. J.Gerontol. 10:469.
12. Sokal,R.R. & Rohlf,F.J.1969. Biometry. Freeman & Co., San Francisco.
13. Stearns,E.J.,MacDonnell,J.A.,Kaufman,B.J.,Padna,R.,Lucman,T.S., Winter, J.S.D. & Faiman,C.1974. Am. J. Med. 57:761.
14. Swanson,L.J.,Dejardins,C. & Turek,F.W.1982. Bio.Reprod. 26:791.
15. Tsitouras,P.D.,Kowatch,M.A. & Harman,S.M.1979. Endo. 105:1400.
16. Yanagimachi,R.1969. J. Reprod. Fert. 18:275.

This work was supported in part by NIH grant HD 13236. Special thanks to Thomas Maurice Polette for statistical analysis data.

# APPARENT IMPROVEMENT IN HUMAN SEMEN AFTER LONG-TERM STORAGE

G. DAVID and F. CZYGLIK
CECOS BICETRE 78, rue du Général Leclerc 94270 LE KREMLIN BICETRE

The cryostorage of human semen has developed during recent years mainly for two reasons :

1) the convenience of cryostored semen in AID
2) the facility for autoconservation before vasectomy or either elective or therapeutic treatment leading to sterility.

Even if the primary indication does not require long-term storage, unlike cases of autoconservation in situations where sterilizing treatments are often administered to young men it would be desirable to be able to guarantee storage for at least 10 years. This poses the problem of the long-term maintenance of the fertilizing ability of spermatozoa, an aspect which has not previously been studied. In addition, the available results concerning post-thaw motility are contradictory (1, 2). We have therefore investigated the motility and fertilizing ability of donor semen after periods of 5-8 years' storage.

## MATERIAL AND METHODS

Semen from fertile men of less than 45 years of age was obtained after average periods of sexual abstinence of 3 days and frozen in 0.25 ml paillettes.

Sperm motility was studied in 100 ejaculates from 65 men, with the percentages of motile spermatozoa being assessed by the same observer. The initial post-thaw motility was determined 24 h after freezing. The results obtained from 9 donors (54 ejaculates) were used in the investigation of fertilizing ability. All ejaculates were produced within about 3 months for each donor. Some ejaculates (37) were used for AID less than 2 years after cryostorage, others (17) between 5 and 8 years after freezing. The characteristics of the first group of ejaculates were better : sperm concentration = 116.1 ( $\pm$ 50.7 s. d.) x $10^6$/ml, as compared to 91.1

($\pm$ 34.9 s. d.) ; percentage motile spermatozoa = 46.6 ($\pm$ 8.0 s. d.) vs 43.6 ($\pm$ 6.4 s. d.). These significant differences are explained by the fact that those ejaculates showing the better characteristics were used first. It should be noted that this bias favours the short-term storage samples.

For the present analysis we have considered all conceptions from these donors, irrespective of their outcome (pregnancy or spontaneous abortion). The mean success rate per cycle has been used as the indicator of success : this was calculated as the number of successes divided by the number of insemination cycles (of known issue). Furthermore, only those treatment cycles leading to a first pregnancy were considered as the success rate is higher for those leading to subsequent pregnancies (4).

## RESULTS

The mean post-thaw percentage of motile spermatozoa in the 100 ejaculates was found to remain constant : 41.3 ($\pm$ 12.9 s. d.) % at the outset, and 41.6 ($\pm$ 16.1 s. d.) % after 6-8 years'cryostorage.

Table I shows that there was not only a maintenance, but also an improvement, in the fertilizing ability of the spermatozoa.

TABLE I

Fertilizing ability after long-term cryostorage (9 donors)

| Duration of cryostorage | No. of ejaculates | No. of cycles | No. of conceptions | Mean success rate per cycle |
|---|---|---|---|---|
| $<$ 2 years | 37 | 448 | 37 | 8.3 |
| 5-8 years | 17 | 168 | 25 | 14.9 |

$p < 0.02$

## DISCUSSION

The long-term maintenance of the percentage of motile spermatozoa confirms the results of Sherman (1) who reported a stability within 25 ejaculates stored for 10 years.

The present study has, however, shown a maintenance, and perhaps even an improvement, in the fertilizing ability of ejaculates from the same group of donors used for AID over a period of several years. Indeed, the ejaculates were not in fact strictly comparable, and were not randomly selected ; the best ejaculates being used first. This would have put the long-term storage samples at a disadvantage.

Other than seminal factors may have been involved, such as :
1) a strong selection of the women, eliminating a large proportion of sterile and hypofertile cases
2) an improvement in the insemination technique on the part of the gynaecologists.

However, as shown in figure 1, the mean success rate per cycle remained stable for the first and subsequent treatments. This would seem to eliminate the above two objections.

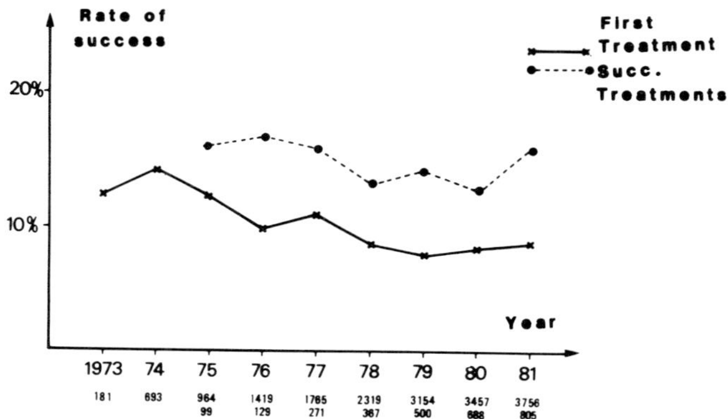


Figure 1

It may therefore be concluded that there is a maintenance of fertilizing ability with long-term cryostorage. The possibility of an improvement may also exist, but requires confirmation in a larger series.

REFERENCES

1. SHERMAN J.K.
Synopsis of the use of frozen human semen since 1964. State of the art of human semen banking. Fertil. Steril., 1973, 24, 397.
2. SMITH K.D. and STEINBERGER E.
Survival of spermatozoa in a human sperm bank. Effects of long term storage in liquid nitrogen. JAMA 1973, 223, 774.
3. DAVID G. et CZYGLIK F.
Tolérance à la congélation du sperme humain en fonction de la qualité initiale du sperme. J. Gyn. Obst. Biol. Repr. 1977, 6, 601.
4. DAVID G., CZYGLIK F., SCHWARTZ D. et MAYAUX M.J.
Results of A.I.D. for a first and succeeding pregnancies in "Human artificial insemination and semen preservation" DAVID and PRICE Ed., Plenum New-York, 1979, p. 221.

# SELECTION AMONG SPERMATOZOA

JACK COHEN

Department of Zoology and Comparative Physiology, University of Birmingham, P.O. Box 363, Birmingham B15 2TT, U.K.

## 1. WHAT IS SELECTION ?

There is no doubt that only a tiny proportion of mammalian spermatozoa are utilized for their primary biological function, fertilization. Whether their attrition is selective or random is controversial. It _is_ agreed, however, that any selection must imply discrimination: if all fail, or if _any_ one can succeed, selection cannot be occurring. Only if some fail and some pass a barrier, like cervix or phagocytes or zona, can selection have occurred. Even then it may not have; random access to a small "window" may have favoured the successful.

Some workers have supposed that differences in sperm survival reflect _intrinsic_ differences. Cohen (1), for example, supposed that failure in accuracy of crossing-over might result in spermatozoan susceptibility to antibody in female tract, while Mortimer (2) showed that diploidy largely excluded spermatozoa from passage to oviducts. On the other hand, Odeblad (3) believed human cervical mucus to select spermatozoa because the passages within it are tailored to fit only the "normal" spermatozoon. The utero-tubal junction of the rat permitted only living rat spermatozoa to pass - heterologous (or dead homologous) spermatozoa were generally excluded (4). Such _external_ constraint seems to contrast with the internally-determined 'suicide' of _Drosophila_ + spermatids segregating from SD. Nevertheless, during selection there must _both_ be difference between spermatozoa _and_ stringent environment to expose this crucially (5); of course, the crucial difference may not be the difference observed.

Many selective differences have been proposed or observed: IgG coating (6, 7), aneuploidy (8), abnormal morphology (9), motility (10), acrosome status (5), etc. Acrosome fragility or loss correlates with IgG-ve status (11) and it is unclear whether either is primary. This list has classically been treated either as a summative score ("good" and "bad" spermatozoa) or more logically as a series of (probably) independent lesions; unfortunately for logic, the $t^x/t^y$ mouse produces a whole suite of abnormalities - sperm-associated infertility in many dimensions (e.g. 12) - when there may be minimal genetic difference from a fertile $T/t^x$ or $T/t^y$ male.

Logic is not helpful, either, in distinguishing selection from change of the ascending minority; Johnson and Hunter (13) showed that the majority of oviducal spermatozoa of the rabbit resembled the minority in the ejaculate, antigenically, and proposed a progressive change in each spermatozoon, even though the oviducal majority was less than $\frac{1}{1000}$ of the ejaculate minority ! Siddiquey (unpubl.) has shown as few as 12-14 oviducal mouse spermatozoa/ fertilization, compared with about 250 epididymal spermatozoa in a similar system, and he cannot decide between selection and the excellence of 'natural' capacitation. A communication failure in Cohen's early explanations (e.g. 1, 14, 15), too, led to illogicalities. I proposed that the female tract prevented nearly all spermatozoa from attaining the fertilization site, but was widely quoted (e.g. 16, 17) as saying the contrary: that only a few spermatozoa could fertilize. Bavister's dramatic but puzzling demonstration (18) that hamster sperm/egg ratios close to unity can fertilize _in vitro_ was widely supposed to have disproved selective explanation when, of course, it is irrelevant: unless the 'unwanted' spermatozoa could fertilize as well, barriers in the female tract would be redundant !

## 2. EXPERIMENTAL AND OBSERVATIONAL EVIDENCE _FOR_ SELECTION

The major evidence is probably that accumulated from competition experiments, where oviducal spermatozoa are shown to be very fecund compared with capacitated ejaculate spermatozoa (7, 19, 20); some attempts (e.g. 21) have been misconceived, testing spermatozoa not thought to be selected. Heterogeneity has been supposed in the ejaculate (19) or the female has been supposed to segregate a small (but random) sperm population (7, 22).

The discovery that diploid (2) or supposedly aneuploid spermatozoa (8) are relatively depleted at fertilization also argues for discrimination in female tract. Apparently Mendelian segregation, followed by non-Mendelian transmission in the t-locus mouse has been the standard selectionist example (23), and MHC's are also supposed to show it (24).

Subtle evidence has been provided by Taylor (25): two buck rabbits of differing fecundity gave offspring proportions which were similar after a variety of order and interval between double matings. Even when the less fecund male mated one (or even four) hours after the better buck, he still contributed to the litter. At this time there was an enormous sperm-induced cervical leucocytosis (26), which the second ejaculate had to negotiate. Despite this, Taylor's ratios were not altered; the *effective* spermatozoa must be resistant to phagocytosis.

Heterogeneity among spermatozoa has, however, been best demonstrated by careful *in vitro* experiments (27). By using small insemination drops (10, 5 or 1µl) with the same sperm concentration, it was demonstrated that about 1000 epididymal spermatozoa (in 1 µl) could fertilize only 3 or 4 eggs, whether 5 or 15 were offered; in 5 µl, 5000 spermatozoa could fertilize 12 of 15. This showed that only 1/250 epididymal spermatozoa can fertilize in that system; in contrast, only 12-14 oviducal spermatozoa were needed/fertilization, even though a much larger drop (40 µl) was used.

## 3. EXPERIMENTAL AND OBSERVATIONAL EVIDENCE *AGAINST* SELECTION

The best experimental evidence, against the C-associated sperm heterogeneity which I suggested was responsible largely for sperm redundancy, is the Siddiquey and Cohen work just described. If only 1/250 mouse spermatozoa *can* fertilize, then the major part of the redundancy cannot be associated with female tract selection (although the high fecundity/low redundancy of oviducal spermatozoa looks suggestive).

The use of few spermatozoa for natural fertilization (5, 28), the ubiquity of Mendelian and sex ratios, and the lack of evident differences being discriminated (unlike gastropods and silkmoths), weigh against sperm selection in mammals. So do the apparent cytoplasmic continuity between spermatids (29) except for membrane-bound determinants (30), which might mediate t-locus and MHC discrimination. Finally, the IgG/phagocytosis/acrosome loss data could all be accommodated by a sperm 'senescence' model (31); unfortunately, when we tested whether the female tract had spermatozoa distributed according to age, the answer was 'no' (32).

## 4. CONCLUSIONS

While the immense spermatozoan attrition is agreed, global explanation is still lacking and summation of partial explanations (1, 33) does not persuade. It resembles the American comedian Burns' directions for weighing a hog: "balance a long plank very carefully, balance the hog with a carefully selected pile of rocks, then very carefully guess the weight of the rocks." !

Taylor has persuaded me of exposed contradictions in such sets of partial explanation: acrosomeless spermatozoa are both fertile and moribund; oviducal spermatozoa are both fecund when re-inseminated and 8 h later, and within-ejaculate age differences do not affect sperm success; female mice stop most spermatozoa reaching the oviduct, but only 1/250 of the ejaculate could (presumably) fertilize if it got there; some evidence shows correlation between sperm disabilities; other evidence shows independence.

Perhaps excess mammalian spermatozoa really do have a quite different secondary function. They may, for example, be immunological harbingers of the embryos to be immunologically tolerated later (34) - or is it only the non-gametic ones that are 'wasted' on these secondary functions, gametic spermatozoa existing as a special cryptic population up to the moment of fertilization ?

REFERENCES


1. Cohen J. 1975. Gametic diversity within an ejaculate. Functional morphology of the spermatozoon (Ed B. Afzelius. Wenner-Gren, Stockholm 329-339
2. Mortimer D. 1978. The distribution of diploid rabbit sperm in the female tract after AI. Animal Repro. Sci. 1: 245-250.
3. Odeblad E. 1969. Types of Human Cervical Secretion. Acta Europaea Fertilitas. 1: 99-116.
4. Marcus SL. 1955. The passage of rat and foreign spermatozoa through the utero tubal junction of the rat. Am. J. Obs. Gyn. 91: 985-989.
5. Taylor NJ. This Symposium.
6. Cohen J, Werrett DJ. 1975. Antibodies and sperm survival in the female tract of the mouse and rabbit. J. Reprod. Fert. 42: 301-310.
7. Cohen J, Tyler KR. 1980. Sperm populations in the female genital tract of the rabbit. J. Reprod. Fert. 60: 213-8.
8. Pearson PL, Pawlowitzki IH, Geraedts JPM, Van der Ploeg M. 1974. Chromosome studies in human male gametes. In "The Biology of the male Gamete". Biol. J. Linn. Soc. 6.
9. Krzanowska H. 1974. The passage of abnormal spermatozoa through the uterotubal junction of the mouse. J. Reprod. Fert. 38: 81-90.
10. Katz DF, Overstreet JW, Hanson FW. 1981. Variations within and amongst normal men of movement characteristics of seminal spermatozoa. J. Reprod Fert. 62: 221-228.
11. Taylor NJ. 1982. Equivalence of 'non-IgG binding' and 'acrosomeless' sperm population from the female genital tract of the rabbit. J. Reprod. Fert.

64: 181-184.

12. Tucker MJ. 1980. Explanation of sterility in $t^x t^y$ male mice. Nature 288: 367-8.
13. Johnson WL, Hunter AG. 1972. Seminal antigens: Their alteration in the genital tract of female rabbits and during partial in vitro capacitation with Beta-Amylase and Beta-glucoronidase. Biol. Reprod. 7: 332-340.
14. Cohen J. 1969. 'Why so many sperms? An essay on the arithmetic of reproduction;' Sci. Prog; Oxford 57: 23-41.
15. Cohen J. 1973. Crossovers, sperm redundancy and their close association. Heredity, 31: 408-413.

Wallace H. 1974. Chiasmata have no effect on fertility. Heredity, 33: 423-429.

Gwatkin RBL. 1977. Fertilization mechanisms in man and mammals. p. 41. Plenum Press. New York and London.

18. Bavister GD. 1979. Fertilization of hamster eggs in-vitro at sperm-egg ratios close to unity. J. Exp. Zool. 210(2): 259-264.
19. Cohen J, McNaughton DC. 1974. 'Spermatozoa: The probable selection of a small population by the genital tract of the female rabbit. J. Reprod. Fert. 39: 297-310
20. Overstreet JW, Katz DF. 1977. Sperm transport and selection in the female genital tract. In: Development in Mammals. 3: 31-66.
21. Fischer B, Adams CE. 1981. Fertilization following mixed insemination with 'cervix selected' and 'unselected' spermatozoa in the rabbit. J. Reprod. Fert. 62: 337-343.
22. Johnson MH. 1973. 'Physiological mechanisms for the immunological isolation of spermatozoa. Adv. Reprod. Physiol. 6: 274-325.
23. Bennett D. 1975. The T-locus of the mouse. Cell 6: 441-454.
24. Klein J, Juretic A, Baxevanis CN, Nagy ZA. 1981. The traditional and a new version of the mouse H-2 complex. Nature 291: 455-460.
25. Taylor NJ. 1981. Investigation of sperm-induced cervical leucocytosis in rabbits by a double mating study in rabbits. J. Reprod. Fert. (in press).
26. Tyler K. 1977. 'Histological changes in the cervix of the rabbit after coitus'. J. Reprod. Fert. 49: 341-45.
27. Siddiquey AKS, Cohen J. 1982. In vitro fertilization in the mouse and the relevance of different sperm/egg concentrations and volumes. J. Reprod. Fert. (in press).
28. Cheng P, Casida LE. 1948. Fertility in the rabbit as affected by the dilution of semen and the number of spermatozoa'. Proc. Soc. Exp. Biol. Med. 69: 36-39.
29. Fawcett, DW. 1972. Observations on cell differentiations and organelle continuity in spermatogenesis. In: The Genetics of the Spermatozoon. (eds Beatty KA and Gluecksohn-Waelsch S.) Copenhagen. 37-68.
30. Erickson RP, Lewis SE, Butley M. 1981. Is haploid gene expression possible for sperm antigens? J. Reprod.Immunol. 3: 195-217.
31. Hancock RJT. 1978. Sperm antigens and sperm immunogenicity. In Spermatozoa, Antibodies and Infertility (ed Cohen J. and Hendry WF). Blackwell. 1-9.
32. Tucker et al. This Symposium.
33. Cchen J. 1982. Chance ou choix: quel spermatozoide arrivera a destination? Contraception-fertilite-sexualite 10: 397-402.
34. Clarke A, Jecquier A. 1981. Unpublished research seminars.

# TRANSPORT OF SPERMATOZOA OF KNOWN AGE IN THE FEMALE GENITAL TRACTS OF THE RABBIT AND MOUSE, AND THEIR PERFORMANCE *IN VITRO*

M. J. TUCKER, N. J. TAYLOR, A. K. S. SIDDIQUEY and J. COHEN

Department of Zoology and Comparative Physiology, University of Birmingham, P.O. Box 363, Birmingham B15 2TT

## INTRODUCTION

It has frequently been suggested (e.g. 1) that sperm age may be important in sperm transport and fertilization. ($^3$H) arginine incorporation into late spermatids of rabbits and mice is specific to spermatid stages 11-15 (2, 3, 4). Therefore, following a single intratesticular injection of ($^3$H) arginine, any labelled spermatozoa in subsequent ejaculates would have been between spermatid stages 11-15 when the tracer was present; hence the spermatozoa would be age-labelled.

We studied the transport of spermatozoa of different ages in the female genital tracts of rabbits and mice following natural mating. Spermatozoa containing labelled populations of known age were also used to fertilize mouse eggs *in vitro*, to observe any possible discrimination on the basis of sperm age.

## MATERIAL AND METHOD

A stud buck rabbit (200 µCi/testis) and 20 C57BL/10 X CBA male mice (12.5 µCi/testis) were intratesticularly injected with ($^3$H) arginine monohydrochloride. Day of injection was designated day 0. Ejaculate samples from the buck were obtained on days 6-9 post-injection. On days 10-16 the buck was mated with parous Dutch does at 9 a.m.; on days 12-14 the buck was also mated with does at 9 p.m. Within 1 min *post coitum* (p.c.) a small sample of semen was obtained by atraumatically aspirating the posterior vagina. At 11h 30min p.c. mated does were killed and sperm samples were obtained from the vagina, uterus, isthmus and ampulla (5).

Male mice were mated with induced-oestrous females on days 7 and 10-18 post-injection. Mated females were killed at 4 or 8h p.c., and spermatozoa were obtained from the vaginal plug (= ejaculate), uterus, ampulla and the rest of the oviduct.

Other males yielded sperm suspensions on days 12-16 post-injection, and their ability to fertilize mouse eggs *in vitro* was tested by the procedure of Siddiquey and Cohen (6). Control fertilizations were obtained from uninjected and PBS-injected control males. 1μm serial sections of the *in vitro* fertilized eggs were made using ester-wax.

All air-dried sperm samples and egg sections were coated with autoradiographic gel emulsion and exposed for 8 weeks before staining and examination.

## RESULTS

The maximum proportions of ($^3$H) labelled spermatozoa were ejaculated on day 12 (rabbit: Fig. 1), and days 14 and 15 (mouse: Figs. 3 and 5). The proportion of labelled spermatozoa in the ejaculates when compared with the proportions found in the other regions of the rabbit and mouse tracts gave ratios that were randomly distributed positively and negatively, and the ratios only rarely attained statistical significance (Figs. 2 and 4).

Consistently high levels of fertilization *in vitro* were obtained with control spermatozoa (Fig. 5); however, as the proportion of labelled spermatozoa increased, so the level of fertilization *in vitro* decreased. Examination of the sections of fertilized eggs clearly showed the presence of the swollen fertilizing sperm heads, but no firm evidence for ($^3$H) labelled spermatozoa fertilizing eggs was found (75 eggs sectioned, 34 swollen sperm heads found).

## DISCUSSION

Previous work has shown peaks of labelled spermatozoa being ejaculated between days 12-14 (rabbit - 7), and days 14-15 (mouse - 2, 4) following arginine tracer injection, and our results have confirmed this. The appearance of labelled spermatozoa over a period from days 6-16 (rabbit) and 7-18 (mouse) indicates that the ejaculates contained spermatozoa of mixed ages (cf. 8).

The female rabbit tract, even at 11h 30min p.c. when any discrimination by the female tract would be expected to be most pronounced (9), is clearly unable to discriminate unlabelled from labelled spermatozoa of any age. There also appears to be no consistently significant discrimination on the part of the mated female mice at either 4 or 8h p.c.

It has been suggested that successful fertilization _in vitro_ is possible with ($^3$H) arginine labelled spermatozoa (4). Oddly though, a remarkable decrease in the proportions of ova fertilized occurred when using our sperm samples containing high proportions of ($^3$H) labelled spermatozoa. A difference exists here, however, in that our sperm populations were not 100% labelled. Nevertheless, the means by which the ($^3$H) labelled spermatozoa might disqualify themselves from fertilization, and also reduce the fecundity of their fellow unlabelled spermatozoa on days 14 and 15, is unclear to us at the moment. What is clear is that discrimination by the female genital tracts of the mouse and rabbit for _or_ against spermatozoa of a particular 'ripeness' (1) has been excluded by our results.

REFERENCES


1. Cohen J. 1975. Gametic diversity within an ejaculate. In: _The Functional Anatomy of the Spermatozoon_. 23rd Wener-Gren Symposium, pp. 329-339. Ed. Afzelius B. Pergamon Press, Oxford and New York.
2. Monesi V. 1964. Autoradiographic evidence of a nuclear histone synthesis during mouse spermatogenesis in the absence of detectable quantities of nuclear ribonucleic acid. _Exptl. Cell Res_. 36: 683-688.
3. Bellve AR, Anderson E. and Hanley-Bowdoin L. 1975. Synthesis and amino acid composition of basic proteins in mammalian sperm nuclei. _Dev. Biol_. 47: 349-365.
4. Kopecny V. and Pavlok A. 1975. Autoradiographic study of mouse spermatozoa arginine rich nuclear protein in fertilization. _J. Exp. Zool_. 191: 85-96.
5. Taylor NJ. 1982. Equivalence of 'non-IgG binding' and 'acrosomeless' spermatozoa from the female genital tract of the rabbit. _J. Reprod. Fert_. 64: 181-184.
6. Siddiquey AKS and Cohen J. (1982 - in press). _In vitro_ fertilization in the mouse and the relevance of different sperm/egg concentrations and volumes. _J. Reprod. Fert_.
7. Kopecny V. and Fulka J. 1975. Retention of basic nuclear protein in rabbit spermatozoa up to entry into the rabbit vitellus. _Ann. Biol. Anim. Bioch. Biophys_. 15: 119-122.
8. Mann T. 1964. _The Biochemistry of Semen and the Male Reproductive Tract_. Methuen, London.
9. Overstreet GW, Cooper GW and Katz DF. 1978. Sperm transport in the reproductive tract of the female rabbit. II. The sustained phase of transport. _Biol. Reprod_. 19: 115-132.

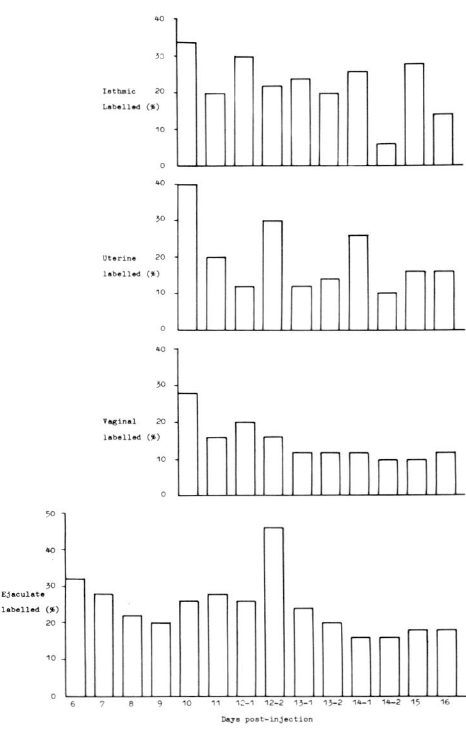


Figure 1. Proportion (%) of ($^3$H) arginine labelled rabbit spermatozoa recovered from various regions of the female tract on Days 10-16 following the tracer injection, and from the ejaculate on Days 6-16.

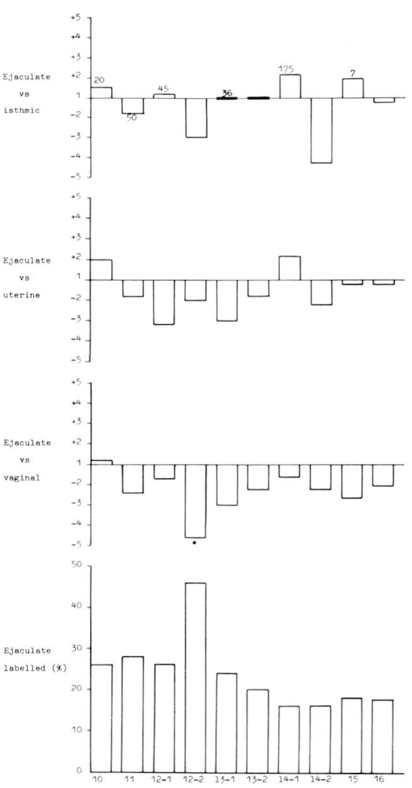


Figure 2. Proportion of ($^3$H) arginine labelled rabbit spermatozoa from the ejaculate on Days 10-16 post-injection, compared with the ejaculate vs tract ratio for each day. Ratio bars plotted above a midline of one indicate that the proportion of labelled spermatozoa recovered from the tract was greater than that found in the ejaculate for that mating, and vice versa. The numbers associated with the ejaculate isthmic ratios refer to the number of spermatozoa assessed. Unless otherwise stated all proportions were based on the assessment of 300 spermatozoa.
*This ratio is statistically significant from 1 (P=0.01-0.001).

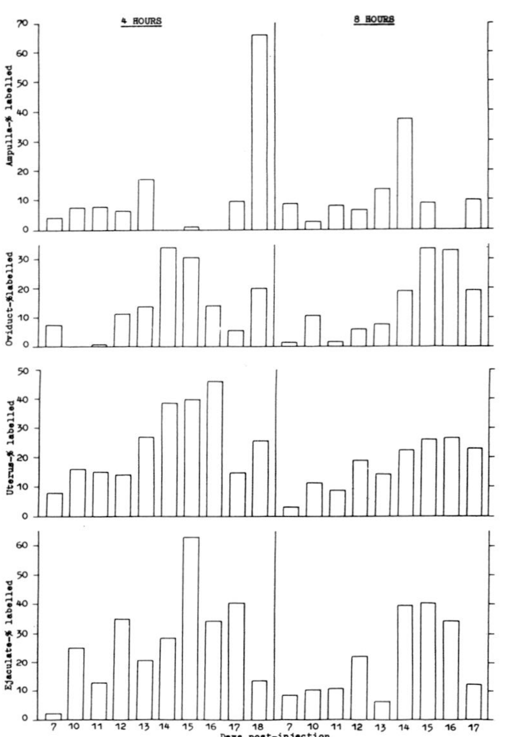


Figure 3. Proportion (%) of ($^3$H) arginine labelled mouse spermatozoa recovered from different regions of the female tract at 4 or 8h p.c. on days 7 and 10-18 post-injection.

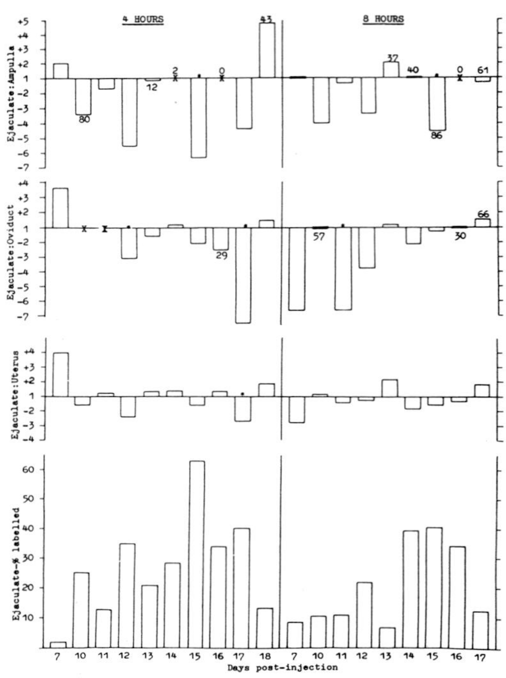


Figure 4. Proportion of ($^3$H) arginine labelled mouse spermatozoa from the ejaculate (cervical sample) on Days 7 and 10-18 post-injection, compared with the ejaculate vs tract ratio for each day. See caption for Fig.2 for an explanation of this format.
X Indicates where the ejaculate vs tract ratio tends to infinity.
* Ratio significantly different from 1 (P=0.01-0.001).

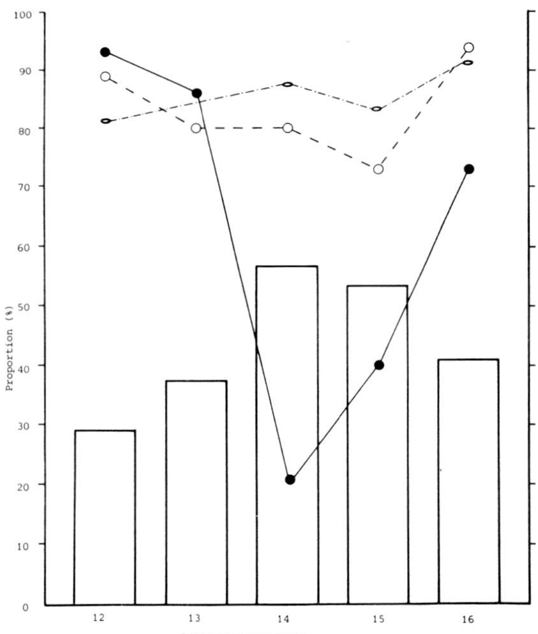


Figure 5. Results of the mouse in vitro fertilization experiments. Histogram bars represent the proportion of labelled spermatozoa in the suspensions used for in vitro fertilization. Open cicles and dotted lines represent the proportion of eggs fertilized on each day by spermatozoa from uninjected controls; ellipses and broken dotted lines represent the proportion fertilized by spermatozoa from PBS-injected controls. Closed circles and unbroken lines represent the proportion of eggs fertilized by the labelled spermatozoa.

# THE FECUNDITY OF SPERM POPULATIONS IN FEMALE RABBITS: DOES SPERM SELECTION EXIST ?

NICHOLAS J. TAYLOR

Department of Zoology and Comparative Physiology, University of Birmingham, P.O. Box 363, Birmingham B15 2TT

## INTRODUCTION

Cohen (1) produced an hypothesis which causally linked chiasma frequency and sperm redundancy as an explanation of gamete redundancy. An extension of this hypothesis is that the female genital tract normally only permits (selects) a small population of spermatozoa from within an ejaculate to fertilize. Cohen has been widely misinterpreted and thought to believe that most spermatozoa are incompetent gametes, unable to fertilize. However, if it was the case that most spermatozoa were non-functional, then clearly there would be no necessity for the female tract to select !

These ideas led to a search for experimental evidence of differences between ejaculated spermatozoa and spermatozoa recovered from various regions of the female tract. Cohen and McNaughton (2) produced dramatic evidence which showed that rabbit spermatozoa from regions high in the female tract were highly fecund, and Overstreet and Katz (3) found a similar result. Cohen and Werrett (4) found that mouse and rabbit oviductal spermatozoa were different from the vast majority of ejaculated spermatozoa in that they were incapable of binding serum IgG acrosomally, even after air-drying and acetone fixation. Cohen (5) suggested that this acrosomal binding of IgG was non-specific but selective, i.e. selection, not a change. Cohen and Tyler (6) made further investigations of the fecundity and IgG binding of early uterine and ejaculated sperm populations in rabbits and found that non-IgG binding sperm populations were highly fecund when transferred to the uteri of recipient does. So there seemed to be experimental evidence favouring the ideas of immunological sperm selection by the female genital tract. Unfortunately, these experiments are subject to criticism since a numerical bias in favour of 'selected' spermatozoa was introduced by the experimental design.

Taylor (7), using a new rapid acrosome stain for air-dried mammalian spermatozoa (8), showed that the failure of spermatozoa to bind IgG reflected acrosomal absence rather than acrosomal uncoatability. This was a paradoxical finding since these highly fecund spermatozoa had no acrosomes (at least when observed on the slides). Indeed, these spermatozoa are capable of fertilizing eggs in a recipient doe some 8h later. Further recent findings which contradict orthodoxy follow.

Recent experimental findings

It was considered possible that IgG binding to spermatozoa and the phenomenon of sperm-induced cervical leucocytosis (9) were part of a mechanism to remove excess spermatozoa from the female tract, so that those which remained were below an antigenic stimulus threshold (10). By utilizing a double-mating series of experiments with rabbits, Taylor (11) clearly demonstrated that this leucocytic response did not affect the relative fecundity of the same buck mating either first or second, even if it was the 'inferior' buck mating up to 4h after the first mating.

In a recent series of sperm transfer and ligation experiments (cf. 6), Taylor (12) has demonstrated that not only can some spermatozoa that attained the oviduct, or even the oviductal ampulla, within $1\frac{1}{2}$ h *post coitum*, fertilize *in vivo* some 9 h later, but also that early uterine spermatozoa are a highly fecund population (300 were required 'per fertilization') when compared with small numbers of ejaculated spermatozoa (7500 'per fertilization') *in vivo*. Interestingly, all these fertile populations had similar high proportions (>80%) of acrosome-intact spermatozoa (contra 4, 6).

DISCUSSION

Important contradictions have been made manifest by these findings. It is abundantly clear that spermatozoa with detached (or possibly very 'fragile') acrosomes fail to bind IgG, and that such spermatozoa seem to be selected by the female genital tract. It is also clear that populations containing high proportions of these spermatozoa are highly fertile. It is now almost totally accepted that spermatozoa must undergo an acrosome reaction in the vicinity of an egg in order to fertilize (13). So either the 10-20% (of 300 spermatozoa) with intact acrosomes are the fertilizing ones (*very* fecund), or the observed acrosome loss from spermatozoa removed from the oviducts does not occur *in vivo* to such an extent, i.e.

spermatozoa from high in the tract could have undergone capacitation changes which rendered their acrosomes 'fragile', so that they appeared to be 'non-IgG binding' in the studies of 4, 6, 7.

That sperm-induced cervical leucocytosis does not impair the transport of subsequent fertilizing spermatozoa is surprising in itself, but even more remarkable is that spermatozoa from the 'inferior' buck can fertilize when inseminated 4h after 'superior' spermatozoa. This contradicts explanations of 'sire-superiority' involving differences in capacitation time between ejaculates from different males and shows that spermatozoa _within_ an ejaculate must capacitate at different rates. The paradoxical results of Taylor (7) could also be explained if there were a constant generation of 'fertilizing' spermatozoa (with fragile acrosomes) from sperm populations in the female tract, especially in the isthmus; a _conversion_ of spermatozoa, that is to say not necessarily a selection. All these findings could have been explained on the basis of sperm age, i.e. only a few spermatozoa in any ejaculate could be 'ripe' at any one time. However, the findings of Tucker, Taylor, Siddiquey and Cohen (in this Symposium) have unfortunately removed this seductive possibility.

Some workers (e.g. 14) have asserted categorically that sperm selection by the female tract does not exist, and that observed differences between spermatozoa are a result of the spermatozoa 'selecting themselves'. This is like saying that soldiers select themselves on an assault course; but surely should we not be asking whether an army assault course selects soldiers ?

REFERENCES


1. Cohen J. 1975. Gamete redundancy - wastage or selection ? In: _Gamete Competition in Plants and Animals_, pp. 99-112. Ed: D. L. Mulcahy, North Holland, Amsterdam.
2. Cohen J. and McNaughton DC. 1974. Spermatozoa: the probable selection of a small population by the genital tract of the female rabbit. _J. Reprod. Fert._ 39: 297-310.
3. Overstreet JW and Katz DF. 1977. Sperm transport and selection in the female genital tract. _Development in Mammals_ 3: 31-66.
4. Cohen J. and Werrett DJ. 1975. Antibodies and sperm survival in the female tract of the mouse and rabbit. _J. Reprod. Fert._ 42: 301-310.
5. Cohen J. 1978. Non-specific (Fc) but selective attachment of antibody to sperms. _Proc. 3rd Int. Symp. Immunology of Reproduction_, pp. 234-238. Ed. K. Bratanov. Bulgarian Academy of Sciences, Sofia.
6. Cohen J. and Tyler KR. 1980. Sperm populations in the female genital tract of the rabbit. _J. Reprod. Fert._60: 213-218.

7. Taylor NJ. 1982. Equivalence of 'non-IgG binding' and 'acrosomeless' spermatozoa from the female genital tract of the rabbit. _J. Reprod. Fert._ 64: 181-184.
8. Taylor NJ. 1982. Eosin-Fast green as a rapid stain for air-dried mammalian spermatozoa. _Stain Technology_ (in press).
9. Tyler KR. 1977. Histological changes in the cervix of the rabbit after coitus. _J. Reprod. Fert._ 49: 341-345.
10. Johnson MH. 1973. Physiological mechanisms for the immunological isolation of spermatozoa. _Adv. Reprod._ 6: 279-323.
11. Taylor NJ. 1982. Investigation of sperm-induced cervical leucocytosis in rabbits by a double mating study in rabbits. _J. Reprod. Fert._ (in press).
12. Taylor NJ. 1982. _Populations of Spermatozoa from the Female Genital Tract of the Rabbit._ Ph.D. thesis, University of Birmingham.
13. Yanagimachi R. 1981. Mechanisms of fertilization in mammals. In: _Fertilization and Embryonic Development in vitro_, pp. 81-181. Ed. L. Mastroianni Jr. and J. D. Biggers. Plenum Publishing Corporation.
14. Mortimer D. 1978. Selectivity of sperm transport in the female genital tract. In: _Spermatozoa, Antibodies and Infertility_, pp. 37-54. Ed. J. Cohen and W. F. Hendry. Blackwell, London.

# THE CAPACITATION RATE OF RAT SPERM IN VITRO

RUTH SHALGI, RUTH KAPLAN AND LASLO NEBEL, DEPARTMENT OF EMBRYOLOGY AND TERATOLOGY, TEL-AVIV UNIVERSITY FACULTY OF MEDICINE, RAMAT-AVIV, ISRAEL.

## 1. INTRODUCTION

In the rat, as in many rodents where semen is deposited directly into the uterus, it is believed that the principal site for capacitation is the oviduct. In vivo, there is a temporal correlation between the migration of fertilizing sperm into the ampulla and the appearance of oocyte-cumulus complexes there. It has been suggested (Gwatkin 77) that in some rodents cumulus cells play an important part in capacitation while in others (Hoppe & Whitten 74, Niwa & Chang 74) their importance was not shown.

Even if cumulus cells support capacitation it is unknown whether these cells have any effect on the rate of capacitation.

We conducted a series of experiments to determine: 1) When rat sperm complete capacitation, acrosome reaction and are ready to penetrate the eggs. 2) Whether the presence of cumulus cells or the oocyte, play a role in determining the duration of the capacitation process. 3) Whether capacitation is necessary for dispersing the cumulus cells.

## 2. MATERIALS AND METHODS

### 2.1 Gametes

Semen was recovered from the uteri of mated adult rats. One drop of semen was placed in rat fertilization medium (RFM) as described previously (Shalgi et al 81). Spermatozoa were diluted in RFM to a final concentration of 2-6 x $10^5$ sperm/ml. Oocytes were obtained from superovulated immature rates which had been injected with 15 IU of PMSG followed 48-56 hr. later by 15 IU of hCG. Oocytes were isolated from the oviductal ampullae. For experiments involving cumulus intact eggs, the cumulus-oocyte complexes were transfered directly into the sperm suspension. Cumulus free oocytes were obtained by treating the complexes with hyaluronidase (Sigma type IV).

## 2.2 Culture Conditions

All studies were performed at $37^{o}$ in an atmosphere of 5% $CO_2$ in air. Sperm were capacitated in RFM under heavy paraffin oil. Oocytes were exposed to sperm in one of three ways: i) Intact oocyte-cumulus complexes were added to sperm suspension two hours after the sperm were suspended in culture medium. ii) Oocyte-cumulus complexes were introduced to sperm as soon as these were suspended in the culture medium. iii) Hyaluronidase treated oocytes denuded of cumulus cells were added to the sperm suspensions two hours after their suspension in RFM.

## 2.3 Determination of the duration of capacitation

Due to the difficulty in observing the acrosome reaction in rat sperm, the duration of capacitation was determined as the time period elapsing between sperm suspension in RFM and their penetration into the egg vitellus. The cultures were opened at half hour intervals commencing 3 hours after their suspension. The oocytes were isolated and examined by interference contrast microscopy. Ova were defined as penetrated when sperm were attached to the vitelline membrane or when the ovum contained an enlarged sperm head or pronuclei. The speed of penetration was calculated by regression analysis. The observed proportions $Pi$, were converted to arc sin $\sqrt{Pi}$.

## 3. RESULTS AND INTERPRETATION

The cumulus cells dispersed as soon as the cumulus-oocyte complexes were exposed to sperm. All oocytes were denuded from their surrounding cells within half an hour. This observation was constant regardless of whether the gametes were exposed to each other as soon as the sperm were suspended or two hours later. However, sperm attachment to the zona pellucida and subsequent penetration of the zona was not observed up to 3½ hr. after sperm suspension.

Table 1 shows that sperm attached to membrane could first be seen as early as 3½ hr. after sperm suspension in RMF (time 0). Fifty per cent of ova were penetrated after similar time periods in the three groups defined in the table. This similarity in the three groups was true for each separate experimental day.

Three sigmoid curves were calculated by pooling the data from 30 experimental days (Fig. 1). For this data 2505 ova were introduced to

Table 1. Incidence of Sperm Penetration

| | Group I | | | Group II | | | Group III | | |
|---|---|---|---|---|---|---|---|---|---|
| Time* hr. | Mean % | ± SE | No. of Oocytes | Mean % | ± SE | No. of Oocytes | Mean % | ± SE | No. of Oocytes |
| 3 | 0 | | 85 | 0 | | 65 | 1 | ± 1.0 | 100 |
| 3.5 | 9 | ± 9.0 | 97 | 11 | ± 48.0 | 91 | 9 | ± 6.0 | 86 |
| 4 | 48 | ± 16.0 | 124 | 25 | ± 11.9 | 104 | 49 | ± 14.0 | 145 |
| 4.5 | 51 | ± 14.0 | 129 | 63 | ± 12.7 | 100 | 63 | ± 12.6 | 137 |
| 5 | 75 | ± 7.9 | 60 | 72 | ± 9.0 | 74 | 65 | ± 12.6 | 65 |
| 5.5 | 65 | ± 14.0 | 67 | 90 | ± 8.5 | 67 | 71 | ± 17.5 | 73 |
| 50% penetration (hr.) | 4.61 ± 0.39 | | | 4.48 ± 0.14 | | | 4.55 ± 0.38 | | |
| "slope" @ 50% % per hr. | 39.84 | | | 49.22 | | | 37.30 | | |

Group I oocyte incubation with sperm 2 hours after time 0

Group II oocyte incubation time 0

Group III oocyte incubation 2 hr. after time 0 (Hyaluronidase treated)

Penetration rate in control cultures 96.3% ± 2.1 (324 oocytes)

*Time - No. of hours after sperm suspension

sperm suspension 2 hours after time 0, 1070 were introduced at time 0 and 506 were hyaluronidase treated eggs. In all three groups capacitation was asynchronous probably due to individual spermatozoa of the same ejaculate being in a more advanced stage of maturation than others.

By comparing the three experimental groups it appears that: a) Rat sperm can penetrate eggs as soon as 3 hr. after being dispersed in RFM medium. However, some sperm of the same ejaculate require at least 6 hr. This capacitation time was shorter than the 4 hr. that was found in vivo (Shalgi & Kraicer 78). This capacitation time was also shorter than the 5 hr. reported for in vitro capacitation by Toyoda & Chang in 74. b) Neither cumulus cells nor oocytes seem to have any effect on the time or rate of capacitation in vitro. In all three groups 50% of eggs were penetrated at a similar time and rate. c) Capacitation is not required for sperm to disperse cumulus cells. In this we concur with the work of Levin et al (81).

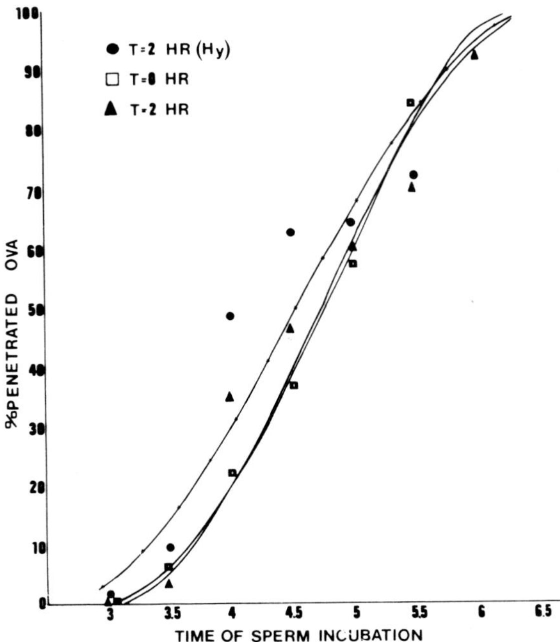


FIGURE 1. Cumulative incidence of eggs penetrated by sperm. *-*-* indicate the calculated sigmoid for the hyaluronidase treated group (Group III)

REFERENCES


1. Shalgi R, Kaplan R, Nebel L, Kraicer PF. 1981. The male factor in fertilization of rat eggs in vitro. J. Exp. Zool. 217: 399-402.
2. Lewin LM, Nevo Z, Gabsu A, Weissenberg R. 1982. The role of sperm-bound hyaluronidase in the dispersal of the cumulus oophorus surrounding rat ova. Int. J. Androl. 5: 37-44.
3. Shalgi R, Kraicer PF. 1978. Timing of sperm transport, sperm penetration and cleavage in the rat. J. Exp. Zool. 204: 353-360.
4. Hoppe PC, Whitten WK. 1974. Maturation of mouse sperm in vitro. J. Exp. Zool. 188: 133-136.
5. Toyoda Y, Chang MC. 1974. Fertilization of rat eggs in vitro by epididymal spermatozoa and the development of eggs following transfer. J. Reprod. Fert. 36: 9-22.
6. Niwa K, Chang MC. 1974. Various conditions for the fertilization of rat eggs in vitro. Biol. Reprod. 11: 463-469.
7. Gwatkin RBL. 1977. Fertilization in Man and Mammals, Plenum Press, New York.


* This work was supported in part by grant B-79-IX from the Population Council.

# INTERACTION OF ZONA-FREE HAMSTER EGGS WITH FRESH AND FROZEN RAM SPERMATOZOA IN VITRO

A. PAVLOK[1], J. PETELÍKOVÁ[2], J.-E. FLÉCHON[3]
Institute of Animal Physiology and Genetics, Czechoslovak Academy of Sciences, Liběchov, Czechoslovakia (1), Central Breeding Institute, Hradištko, Czechoslovakia (2), Station de Physiologie Animale, INRA, Jouy-en-Josas, France (3)

## SUMMARY

Zona-free hamster eggs were penetrated in vitro by fresh and frozen ram spermatozoa. Washing and in vitro preincubation were indispensable for induction of the ability of the spermatozoa to interact with the eggs. Eighteen frozen sperm samples from four fertile rams were tested for their ability to interact with zona-free hamster eggs. Major differences in penetration ability were observed not only between the ejaculates of the same animal (from 1.1 to 81.0 %), but also between the sperm of individual rams (from 25.0 to 63.3 %). These results are complemented with rather limited data of the fertilizing capacity of frozen sperm from the same animals in vivo.

## INTRODUCTION

As has already been shown in several papers, zona-free eggs of most mammalian species are penetrable in vitro by alien spermatozoa. Moreover some authors have already tried to utilize zona-free hamster eggs to evaluate the fertilization capacity of human spermatozoa and spermatozoa of some farm animals (for review see Yanagimachi, 1981).

The present paper reports the ability of fresh and frozen ram spermatozoa washed and preincubated in vitro to penetrate zona-free hamster eggs. Further experiments were carried out to discower whether there are differences in the penetrating ability of the frozen sperm among individual fertile rams, and if so their relationship to fertilizing capacity in vivo.

## MATERIAL AND METHODS

Mature golden hamster females were superovulated with 100 i.u. of PMSG Foligon (Intervet Internat. V.V. Boxmeer, Holland) or serum gonadotrophin (Bioveta, Ivanovice) and 100 i.u. HCG (Praedyn - Spofa, Praha) given approximatively 72 hrs later. 16 to 18 hrs after the injection of HCG, the female hamsters were killed, the oocytes collected and the zona pellucida

removed (Pavlok, 1981).
Freezing of spermatozoa was carried out at the Breeding Station Jevíčko (Petelíková et al., 1980). Thawing of sperm samples was performed by immersing the frozen metal sperm container directly in a water bath at 50°C for 20 sec. After 5 to 10 min. the sperm sample was diluted with two parts of culture medium (Pavlok, 1981). The diluted spermatozoa were centrifuged for 10 min. at approx. 1,000 g, the supernatant was discarded, and the spermatozoa washed twice with culture medium. Before fertilization, the spermatozoa (0.9 to 1.7 x $10^9$/ml) were preincubated in small glass tubes (100 to 200 µl) at 37.5°C. The sperm concentration for in vitro fertilization was 0.3 to 1.8 x $10^6$/ml. The fresh sperm samples were diluted and treated by the same method as the thawed samples. The details of insemination have been described (Pavlok, 1981).

RESULTS

The highest penetration rates were achieved with fresh spermatozoa preincubated for 5 to 6 hrs, and with frozen spermatozoa preincubated for 6 hrs independently on the level of penetrating ability of single sperm samples. This preincubation time was therefore used in the experiments in which the penetrating ability of frozen sperm from four rams was tested and compared with the conception rate of ewes in spontaneous heat also inseminated with the frozen sperm (Table 1). The rather limited data on overall conception rate are more less related to the penetrating ability of sperm in vitro ; however, Table 1 also show stricking differences in penetrating ability between individual ejaculates.
In order to evaluate the repeatibility of the present method, the results appear in Table 2. The penetrating ability of different sperm samples derived from the same ejaculate were tested on two different days. In three of four samples tested, the differences in penetration rate did not exceede 5 %.

Table 1. The penetrating activity of frozen sperm in vitro and its fertilizing capacity after insemination.

| Ram | No. of ejacultates | Total no. of eggs | Penetrated (%) | Min.-max.+ | Ewes inseminated (% pregnant) |
|---|---|---|---|---|---|
| SEK 2 | 5 | 322 | 204 (63.3) | 42.9-91.2 | 29 (65.8) |
| VEF 7 | 4 | 267 | 161 (60.3) | 32.0-91.9 | 13 (84.6) |
| SEK 3 | 6 | 403 | 147 (36.5) | 19.3-63.9 | 47 (34.0) |
| VEF 6 | 3 | 192 | 49 (25.0) | 1.1-81.0 | 51 ( 1.9) |

+ Minimal and maximal penetration rate of single ejaculates.

Table 2. The penetrating activity of sperm samples from identical ejaculates evaluated in two different experiments.

| Origin of sperm | Data of experiment | Total no. of eggs | Penetrated (%) |
|---|---|---|---|
| VEF 7 | 16. 6. 81 | 63 | 55 (87.3) |
| 31.10. 80 | 17. 9. 81 | 62 | 57 (91.9) |
| VEF 7 | 7. 1. 82 | 78 | 25 (32.0) |
| 11.12. 80 | 12. 1. 82 | 64 | 24 (37.0) |
| SEK 3 | 16. 6. 81 | 61 | 18 (29.5) |
| 31.10. 80 | 17. 9. 81 | 61 | 39 (63.9) |
| SEK 3 | 7. 1. 82 | 67 | 24 (35.8) |
| 3.11. 80 | 12. 1. 82 | 83 | 32 (38.5) |

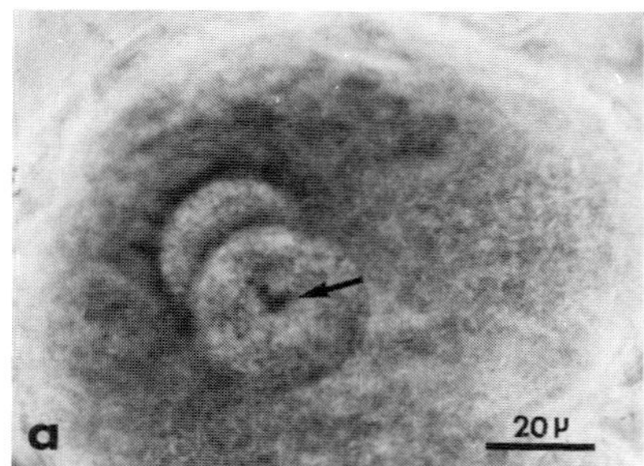


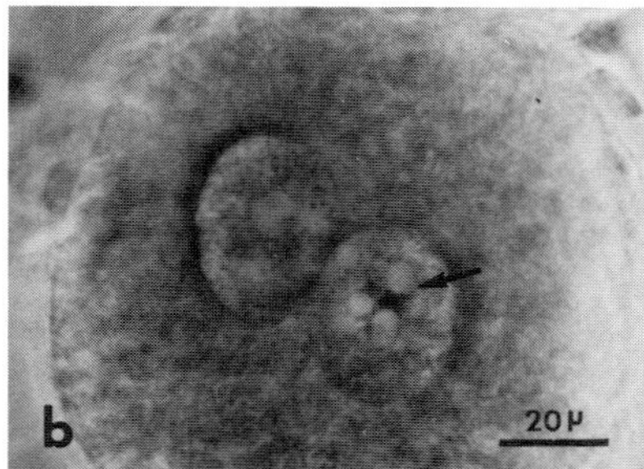


FIGURE 1. a,b) Two hamster zona-free eggs with female and male (ram) pronuclei (arrows).

The frequence of polyspermy was related to the penetrating rate with maximum of 2.1 sperm per egg in experiments with frozen spermatozoa and 1.7 sperm per egg after fertilization with fresh spermatozoa. Most of the penetrating sperm heads developed into pronuclei after 14 hrs but nearly all of them contained characteristic orcein positive spots (arrows) that probably represents not completely decondensed chromatin (Fig. 1a, b).

DISCUSSION

The present results show that ram sperm whashed and preincubated in vitro can fuse with and be successfully incorporated by zona-free hamster eggs without capacitation in the female genital tract. Hunter et al., (1980) obtained fertilized eggs from the transected oviduct of ewes not earlier than 8 hrs after insemination. This delay corresponds approximately to our results. Shams-Borhan and Harrison (1981) described ram sperm penetration into zona-free ovine oocytes after treatment with divalent cation ionophore A 23187 that artificially induced an acrosome reaction in less than one hour.

The relationship between sperm competence in vitro and fertilizing capacity in vivo are at present questionable because of high variability between the penetrating ability of frozen sperm samples originating from different ejaculates. As seen in Table 2, this variability is probably related more to the procedure of freezing than to the in vitro fertilization step. Since freezing procedures have detrimental effects mainly on the sperm plasma membrane, it seems reasonable to look more precisly at the extent of the damages of the sperm plasma membrane (Harrison and Fléchon, 1980) and to try to correlate this with the loss of sperm penetrating ability into zona-free eggs and ultimately with fertilizing capacity in vivo.

REFERENCES.

1. Harrison RAP, Fléchon JE. 1980. Reprod. Nutr. Develop. 20, 1801-1810.
2. Hunter RHF, Nichol R, Crabtree SM. 1980. Reprod. Nutr. Develop. 20, 1869-1875.
3. Pavlok A. 1981. Int. J. Fertil. 26, 101-106.
4. Petelíková J, Holas J, Kolářová H. 1980. Reprot of Central Breeding Institute, Hradištko pod Medníkem, no. 6. 1. 5.
5. Shams-Borhan G, Harrison RAP. 1981. Gamete Research 4, 407-432.
6. Yanagimachi R. 1981. In : Fertilization and Embryonic Development In Vitro (I. Mastrioani Jr, JD. Biggers eds.) Plenum Publishing Corp. pp. 81-182.

# GLYCOLYTIC ACTIVITY OF HUMAN SPERMATOZOA FROM NORMAL AND INFERTILE SEMEN

J.LORNAGE, J.F. GUERIN, J.C. CZYBA
Laboratoire d'Histologie, Faculté de Médecine,
69373 Lyon-Cedex 2, France.

## 1. INTRODUCTION

Glycolysis is the major energy producing pathway in human spermatozoa while oxidative metabolism looks characteristically low (1). It is generally assumed that most of ATP generated is needed to drive sperm motility. However the relationship between glycolytic activity and velocity of spermatozoa is not perfectly clear, in normal as well as infertile semen.For the latter metabolic alterations have been already reported (2,3).

In this work we studied the correlations between several biological und biochemical parameters of semen from men with normal spermiograms. We undertook a comparative study of glycolysis in spermatozoa from normal, oligozoospermic, asthenozoospermic and oligoasthenozoospermic subjects.

## 2. MATERIAL AND METHODS

### 2.1. Classification of subjects

Subjects were classified in different groups according with spermiogram results : group I was constituted of 37 healthy donors having no problems of fertility ; only subjects with sperm concentration above 50 millions/ml and initial motility above 50% were retained. Group 2 was constituted of patients consulting for couple sterility but having the same correct parameters of spermio gram. All the other subjects presented spermiogram abnormalities and were classified into 15 oligozoospermic (Group 3 : concentration below 40 millions/ml), 21 oligoasthenozoospermic (Group 4: concentration below 35 millions/ml, motility below 40%), and 31 asthenozoospermic. In this last case we made a distinction between severe asthenozoospermic with initial motility below 20% (Group 5) and moderate one where initial motility was correct but falled rapidly with time (group 6 : motility below 25% after 5 hours).

2.2. Procedure

One hour after ejaculation each semen was mixed with B2 medium (4) containing inorganic ions and organic compounds as glucose, lactate, acetate pyruvate and albumin. Spermatozoa were separated by centrifugation (400g for 10 minutes) washed again in B2 medium resuspended in this solution and then counted with a hemocytometer. This was time o of experiment. For groups 1 and 2, oxygen consumption was measured during 15 minutes by polarographic method;at time 0,30 and 60 minutes, 150 µl were prelevated and mixed with 300 µl of perchloric acid 0,6 N. The determinations of L lactate in these aliquots were made with a U.V. spectrophotometer. At the end of the experiment, both % of motility and velocity were objectively measured using Laser Doppler velocimetry technique (5). Statistical comparisons were made using student's test.

3. RESULTS

3.1. Correlations between biological and biochemical parameters

The only significant positive correlation was found between lactate and % of motility in group 1. (Table 1).

Table 1. Correlations in 2 groups of men having normal spermiogram results.

| Correlations | Normospermic Donors (Group 1) | Normospermic Patients (Group 2) |
|---|---|---|
| glycolysis - respiration | 0,12 | -0,05 |
| glycolysis-% of motility | 0,48 [b] | 0,24 |
| glycolysis [a]- velocity | 0,09 | -0,37 |

a : lactate production considering only motile spermatozoa.

b : $p<0,01$

In group 1 glycolysis and velocity appeared uncorrelated after considering only the metabolic activity of motile cells, while an unexpected negative but not significant correlation was observed in group 2. In both groups no relationship was found between glycolytic and oxidative metabolism.

3.2. Metabolic activity in the different groups

Mean results are summarized in Table 2.

Table 2. Comparison of glycolysis in the different groups.

| | Number | glycolysis (a) | glycolysis (b) |
|---|---|---|---|
| Group 1 | 37 | 0,73 ± 0,06 | 1,22 ± 0,12 |
| Group 2 | 21 | 0,57 ± 0,04* | 0,97 ± 0,09 |
| Group 3 | 15 | 0,77 ± 0,09 | 1,33 ± 0,17 |
| Group 4 | 21 | 0,48 ± 0,37 | — |
| Group 5 | 16 | 0,29 ± 0,07** | — |
| Group 6 | 15 | 0,47 ± 0,05** | 1,33 ± 0,17 |

a : lactate production : $\mu$moles/$10^8$ sperm/hr

b : lactate production : $\mu$moles/$10^8$ motile sperm/hr

Values are mean ± S.E.M. (*)$p<0,05$ ; (**)$p<0,01$v.s.normal group.

Lactate production was slightly weaker in group 2 than in group 1, while oxygen consumption was identical in both groups. The little elevation of glycolysis in oligozoospermic patients (group 4) was not significant;on the contrary lactate production was severely depressed in both groups of asthenozoospermic men. The glycolytic activity of spermatozoa in group 5 was not linear during one hour : most of the lactate production occured during the first 30 minutes of incubation and then pratically stopped. Data concerning oligoasthenozoospermic patients (group 4) showed a great heterogeneity in metabolic behavior : 14/21 exibited a lactate production very important in some cases ; but for 7/21 there was a paradoxal and highly variable consumption of lactate during the hour.

## 4. DISCUSSION

The absence of correlation found between oxygen consumption and glycolysis confirms the lack of regulation between these metabolic pathways in human sperm. The non-significant correlation between lactate production of motile cells and their velocity is unexpected. We can suppose that the transformation of ATP as potentiel energy to forward motility has an efficiency that varies according to each subject. Groups 1 and 2 could not be differentiated either by biological parameters (both had normal spermiograms results) or by metabolic ones. Nevertheless the metabolic behavior was not identical in both groups with regards of the correlation values. Moreover this group was not homogeneous since in some cases couple infertilities were probably due to wife infertility.

Alterations of spermiogram results were found in relation with modifications of glycolysis excepted in oligozoospermic group, that represents the rarest pathological situation. Our data concerning asthenozoospermic groups disagree with previous works (2,3) since we presently found a significant decrease of glycolysis even when initial motility was satisfactory. But our study confirms the observation that metabolic behavior is very different if we consider either asthenozoospermic or oligoasthenozoospermic samples ; moreover we could distinguish great metabolic variations inside that latter group, probably reflecting different pathogenic mechanisms.

## REFERENCES

1. Peterson RN, Freund M. 1976. Metabolism of human spermatozoa in "Human semen and fertility regulation in men", HAFEZ ESE,176-186
2. Pedron N, Giner J, Hicks JJ, Rosado A. 1975.Comparative glycolytic metabolism of sperm from normal, asthenospermic, and oligoasthenospermic men. Fertil Steril, 26,309-313.
3. Pedron N, Giner J, Hicks JJ. 1978. Lactate and pyruvate utilization by the spermatozoa of infertile human males. Int.J. Fertil. 23, 65-68.
4. Menezo Y. 1976. Milieu synthétique pour la survie et la maturation des gamètes et pour la culture de l'oeuf fécondé. C.R.Acad. Sci., 282, 1967-1970.
5. Jouannet P, Volochine B, Deguent P, David G. 1976. Study of human spermatozoa motility parameters by scattered light. Prog. Reprod. Biol. 1, 28-35.

# RESPIRATORY STIMULATION OF SEA URCHIN SPERMATOZOA IN A SYNTHETIC JELLY PEPTIDE AND THE FRAGMENTS

N. SUZUKI and S. ISAKA
Department of Biochemistry, Teikyo University School of Medicine, 2-11-1 Kaga, Itabashi-ku, Tokyo 173, Japan.

## 1. INTRODUCTION

The intensive studies on soluble factors of egg jelly of sea urchins stimulate the respiration and motility of sea urchin spermatozoa have been performed independently by two research groups. As the results, the factors responsible for those were elucidated to be oligopeptides ( 1-3 ). One of the peptides isolated from the egg jelly of the sea urchin *Hemicentrotus pulcherrimus* had the following sequence : Gly-Phe-Asp-Leu-Asn-Gly-Gly-Gly-Val-Gly ( 3 ). Recently, the same peptide was found in the egg jelly of the sea urchin *Strongylocentrotus purpuratus* ( 4 ).

On the other hand, it is reported that the majority of respiratory stimulation activity of factors was destroyed by treatment with subtilisin, thermolysin, pronase and papain while trypsin, pepsin, leucine aminopeptidase and α-chymotrypsin did not inactivate the factors ( 1-2 ). Taking account of structure of the peptide and enzyme specificities, it is assumed that some portions of the peptide is essential for the peptide to stimulate respiration of sea urchin spermatozoa, since thermolysin should breake the peptide bond at between aspartic acid and leucine and α-chymotrypsin is expected to attack the Phe-Asp bond in the peptide.

The present study was conducted to confirm the peptide structure established previously by chemical synthesis and also to confirm that which fragments produced by proteases' treatment have respiratory stimulation ability.

## 2. MATERIALS AND METHODS

*Hemicentrotus pulcherrimus* sea urchins were used in the present study. A peptide having the following sequence : Gly-Phe-Asp-Leu-Asn-Gly-Gly-Gly-Val-Gly was custom synthesized for us by Peptide Institute Inc. Osaka. Cellulose thin layer plates were products of Eastman Kodak. Thermolysin was purchased from Seikagąku Kogyo Co. Other chemicals were of analytical grade.

For the hydrolysis of the peptide with thermolysin and α-chymotrypsin, peptide was dissolved in 5 mM Tris-HCl ( pH 7.5 ) containing 5 mM $CaCl_2$ ( Buffer A ) in a final concentration of 5 mg/ml. Incubations were carried out at 24°C in sealed tubes with proteases for desired time and aliquots of reaction mixtures were tested for ability to stimulate sperm respiration after incubation. Incubation with thermolysin was performed in Buffer A for 2 h ( E/S = 1/500, w/w ) and incubation with α-chymotrypsin was carried out for 24 h and 53 h, respectively ( E/S = 1/20, w/w ). The reactions were stopped by placing the tubes in a boiling water bath for 20 min.

Thin layer chromatography of peptide and the fragments was performed as described previously ( 3 ). Thin layer electrophoresis was carried out on a cellulose thin layer plate ( 20 x 10 cm ) with pyridine/acetic acid/$H_2O$ ( 100/4/900, by volume, pH 6.5 ) or pyridine/acetic acid/$H_2O$ ( 1/1/48, by volume, pH 4.7 ; this solution was diluted to 1:10 with distilled water just before use ) at 900V for 10 - 20 min.

Amino acid analysis was performed on a Hitachi Model 835 amino acid analyzer with 10 to 15 nmol of samples. Respiration rates were determined using an Oxygen Consumption Recorder ( Yanagimoto Co. ) with a Clark type electrode at 20°C as described previously ( 5 ).

## 3. RESULTS AND DISCUSSION

After 24 and 53 h incubations with α-chymotrypsin, peptide retained the majority of respiratory stimulation ability toward spermatozoa. The hydrolysate was analyzed by thin layer chromatography. Two ninhydrin positive bands were obtained ( Fig. 1A ). One of the bands designated α-1 migrated at the same Rf position as intact peptide. The other band ( α-2 ) migrated slower than intact peptide. Peptides isolated from both bands α-1 and α-2 were further purified on thin layer electrophoresis. The peptide from α-1 was separated to two peptides ( α-1-1 and α-1-2 ) ( Fig. 1B, lower ) and the peptide from α-2 gave a single spot. Amino acid analysis data indicated that the peptide from α-1-1 was consisted of one residue each of Gly and Phe and the peptide from α-2 was lacked in one residue each of Gly and Phe ( Table 1 ). The peptide from α-1-2 had the same amino acid composition as that of intact peptide. Therefore, it is assumed that the sequence of peptide from α-1-1 is Gly-Phe and that of peptide from α-2 is Asp-Leu-Asn-Gly-Gly-Gly-Val-Gly.

After thermolysin digestion, peptide lost the most of respiratory stimulation activity. The hydrolysate gave one major band migrated at between

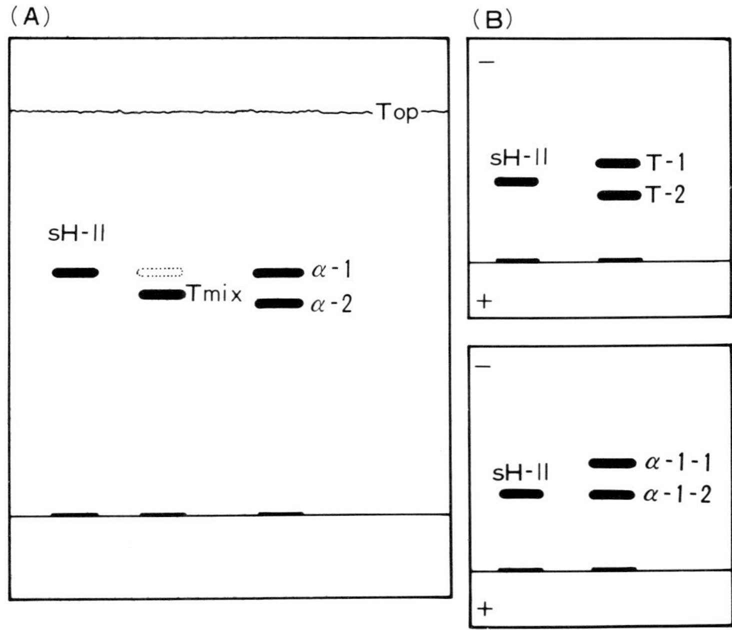


FIGURE 1. ( A ) Thin layer chromatography of a synthetic peptide ( sH-II ) and the thermolysin and α-chymotrypsin hydrolysates. ( B ) Thin layer electrophoresis of the fractions Tmix ( upper ) and α-1 ( lower ).

Table 1. Amino acid composition of peptide's fragments

| Amino acid | α-1-1 | α-2 | T-1 | T-2 |
|---|---|---|---|---|
| Asp | 0.00 | 1.92 ( 2 ) | 1.17 ( 1 ) | 1.00 ( 1 ) |
| Gly | 1.00 ( 1 ) | 3.84 ( 4 ) | 4.20 ( 4 ) | 1.23 ( 1 ) |
| Val | 0.06 | 0.93 ( 1 ) | 0.97 ( 1 ) | 0.09 |
| Leu | 0.00 | 1.00 ( 1 ) | 1.00 ( 1 ) | 0.09 |
| Phe | 1.19 ( 1 ) | 0.18 | 0.15 | 0.84 ( 1 ) |

intact peptide and α-2 on cellulose thin layer chromatography ( Fig. 1A ). Peptide fraction isolated from the band designated Tmix was then separated by thin layer electrophoresis. Two bands were obtained ( Fig. 1B, upper ). As shown in Table 1, amino acid composition of peptides isolated from bands designated T-1 and T-2 indicated that the sequence of peptide from T-2 was presumed to be Gly-Phe-Asp and that of peptide from T-1 was to be Leu-Asn-Gly-Gly-Gly-Val-Gly.

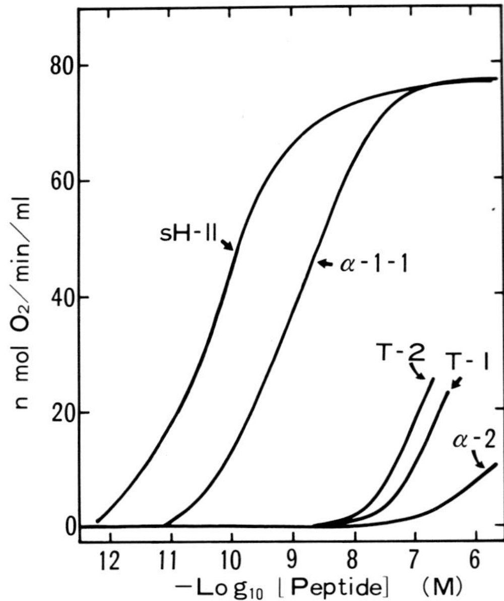


FIGURE 2. The relative potency of a synthetic peptide and the fragments.

Among four peptide' fragments, a fragment Asp-Leu-Asn-Gly-Gly-Gly-Val-Gly showed respiratory stimulation activity toward *H. pulcherrimus* spermatozoa, although the half-maximal respiratory responses of spermatozoa to the fragment was about one-hundredth of that of intact peptide ( Fig. 2 ).

A recent report by Garbers et al ( 4 ) indicates that three synthetic analogues of intact peptide consisting of 9 - 10 amino acid residues have respiratory stimulation ability toward *S. purpuratus* spermatozoa. Together this and our results, it is assumed that a certain length of peptide which may be consisted of more than 8 amino acid residues is required for respiratory stimulation action of peptide.

This work was supported in part by a Research Grant from the Ministry of Education, Science and Culture of Japan.

REFERENCES

1. Ohtake H. 1976. J. Exp. Zool. 198, 313 - 322.
2. Hansbrough JR, Garbers DL. 1981. J. Biol. Chem. 256, 1447 - 1457.
3. Suzuki N, Nomura K, Ohtake H, Isaka S. 1981. Biochem. Biophys. Res. Commun. 99, 1238 - 1244.
4. Garbers DL, Watkins HD, Hansbrough JR, Smith A, Misono KS. 1982. J. Biol. Chem. 257, 2734 - 2737.
5. Ohtake H. 1976. J. Exp. Zool. 198, 303 - 312.

# The Effect of Different Substrate Combinations on the Survival of Ram Cauda Epididymal Spermatozoa exposed to 1 mM-RS α-Chlorohydrin.

W.C.L. FORD and ANNE HARRISON

Department of Physiology & Biochemistry, The University, Whiteknights, Reading RG6 2AJ, UK.

## SUMMARY

The motility of ram cauda epididymal spermatozoa exposed to 1 mM-RS α-chlorohydrin and their ATP concentration declined dramatically after 15 min incubation with 5 mM-D glucose or 5 mM-D glucose + 10 mM-L lactate + 1 mM-pyruvate. By contrast the effects of α-chlorohydrin were relatively slight in the absence of glucose. The concerted action of α-chlorohydrin and glucose may dissipate energy in a 'futile cycling' reaction and this could explain the effective contraceptive action of the drug.

## INTRODUCTION

The reversible contraceptive effect of α-chlorohydrin and of 6-chloro-6-deoxysugars in the male has been ascribed to the inhibition of glyceraldehyde-3 phosphate dehydrogenase [E.C. 1.2.1.12] in spermatozoa (1,4). The spermatozoa are therefore unable to obtain energy from sugar metabolism but this lesion does not explain why they cannot support fertility by using substrates eg. lactate which 'by pass' the glycolytic pathway (see 2). We have examined the effect of different substrate combinations on ram epididymal spermatozoa exposed to 1 mM-RS α-chlorohydrin in order to resolve this question and to determine if the action of α-chlorohydrin supports the concept of a special role for glucose in sperm capacitation (see eg.5).

## METHODS

Spermatozoa were collected from the cauda epididymidis of ram testis, washed once and suspended in PBS buffer (4)(about $1.1 \times 10^8$ sperm/ml). A 1.0 ml sample was removed and mixed with 0.5 ml 1.0M-perchloric acid for metabolite assays and a 0.25 ml sample was taken to measure motility by a turbidimetric method (7). Two 24 ml portions were removed, 1.0 ml

25 m$\underline{\underline{M}}$-$\underline{RS}$ α-chlorohydrin was added to flask 1 and 1.0 ml PBS buffer to flask 2. Both flasks were incubated with shaking at 34°C for 10 minutes. Four 4.8 ml sub-portions were removed from each flask and the following substrate combinations were added:- (1) No added substrate (2) 5 m$\underline{\underline{M}}$-[2-$^3$H][U-$^{14}$C] $\underline{D}$ glucose (3) 10 m$\underline{\underline{M}}$-$\underline{L}$ lactate plus 1 m$\underline{\underline{M}}$-pyruvate (4) 5 m$\underline{\underline{M}}$-[2-$^3$H][U-$^{14}$C] $\underline{D}$ glucose plus 10 m$\underline{\underline{M}}$-$\underline{L}$ lactate plus 1 m$\underline{\underline{M}}$-pyruvate. The [2-$^3$H][U-$^{14}$C] $\underline{D}$ glucose contained [$^3$H] 0.5 Ci/mol and [$^{14}$C] 0.05 Ci/mol. Samples for metabolite assays were taken from all flasks 10, 30, 60 and 240 min after the addition of substrate. Samples of control spermatozoa were taken for motility assessment after 30, 60 and 240 min and of α-chlorohydrin treated spermatozoa after 15, 45 and 225 min. The perchloric acid extracts were neutralised and ATP, ADP, AMP and lactate were assayed (3,4). Tritiated water was measured after sublimation in Thumberg tubes.

## RESULTS

A high percentage (> 60) of control spermatozoa remained motile for 1h even in the absence of added substrate. After α-chlorohydrin treatment, with no added substrate the percentage of motile spermatozoa remained the same as controls for 45 min but declined markedly after 225 min. By contrast after 15 min only 20% were motile in the presence of 5 m$\underline{\underline{M}}$-$\underline{D}$ glucose and only 40% with 5 m$\underline{\underline{M}}$-$\underline{D}$ glucose plus 10 m$\underline{\underline{M}}$-$\underline{L}$ lactate plus 1 m$\underline{\underline{M}}$-pyruvate. (Table 1).

TABLE 1. The effect of 1 m$\underline{\underline{M}}$-$\underline{RS}$ α-chlorohydrin on the motility of ram cauda epididymal spermatozoa incubated with: (A) No added substrate, (B) 5 m$\underline{\underline{M}}$-$\underline{D}$ glucose, (C) 5 m$\underline{\underline{M}}$-$\underline{D}$ glucose plus 10 m$\underline{\underline{M}}$-$\underline{L}$ lactate plus 1 m$\underline{\underline{M}}$-pyruvate at different times after the addition of substrate. Mean ± SEM N=4. Significantly less than value for control spermatozoa with the same substrate measured 15 min later P=5%,*: 1%,†; 0.1%,‡ (Paired t-test).

| CONTROLS | | | | 1 m$\underline{\underline{M}}$-$\underline{RS}$ α-CHLOROHYDRIN | | | |
|---|---|---|---|---|---|---|---|
| Time (min) | Substrate Combination A | B | C | Time (min) | Substrate Combination A | B | C |
| 0 | 63 ± 4.8 | | | 0 | 81.2 ± 10.0 | | |
| 30 | 69 ± 10.6 | 82 ± 9.9 | 80 ± 7.8 | 15 | 63 ± 6.8 | 20 ± 2.8† | 40 ± 4.2† |
| 60 | 67 ± 9.3 | 70 ± 7.2 | 79 ± 8.4 | 45 | 60 ± 9.8 | 12 ± 1.7† | 33 ± 5.4† |
| 240 | 27 ± 2.9 | 71 ± 2.6 | 72 ± 4.3 | 225 | 8.8 ± 3.8* | 1 ± 0.3‡ | 11 ± 3.7‡ |

The ATP concentration in control spermatozoa changed very little throughout the incubations. α-Chlorohydrin treatment produced no significant decrease in ATP concentration during the first hour in the absence of added substrate, however a dramatic decrease occurred after 4h. In the two incubations where glucose was present a dramatic decline in ATP concentration occurred after 15 min. (Table 2). The ATP was converted to AMP and there was little change in total adenine nucleotide concentration, therefore there was a very marked fall in the energy charge. α-Chlorohydrin had little effect when 10 mM-L lactate + 1 mM-pyruvate were present in the absence of glucose.

Lactate production from 5 mM-[2-$^3$H] D glucose was completely inhibited by 1 mM-RS α-chlorohydrin. The release of tritiated water was markedly decreased but not completely prevented.

TABLE 2. The effect of 1 mM-RS α-chlorohydrin on the ATP concentration (n mol/$10^8$ spermatozoa) in ram cauda epididymal spermatozoa incubated with different substrates (see Table 1). Mean ± SEM N=3 . Significantly different from corresponding control P < 5%,*; < 1%,†; < 0.1%,‡. (Paired t-test after $\log_e$ transformation).

| Time (min) | CONTROLS Substrate Combination A | B | C | 1 mM-RS α-CHLOROHYDRIN Substrate Combination A | B | C |
|---|---|---|---|---|---|---|
| 0 | 29 ± 4.9 | | | 28 ± 6.7 | | |
| 10 | 30 ± 6.2 | 32 ± 8.6 | 35 ± 7.8 | 29 ± 6.8 | 12 ± 2.7* | 17(2)* |
| 30 | 32 ± 6.5 | 37 ± 11.2 | 34 ± 7.4 | 28 ± 6.5 | 10 ± 2.4† | 10 ± 2.4‡ |
| 60 | 34 ± 6.7 | 35 ± 7.2 | 35 ± 8.1 | 26 ± 5.3 | 8 ± 1.6† | 10 ± 1.7† |
| 240 | 29 ± 7.6 | 35 ± 7.1 | 38 ± 7.5 | 4 ± 2.5* | 4 ± 1.4† | 11 ± 2.0‡ |

## DISCUSSION

The rapid decline in the ATP concentration and motility of ram spermatozoa exposed to 1 mM-RS α-chlorohydrin when they were incubated with 5 mM-D glucose alone or in combination with 10 mM-L lactate plus 1 mM-pyruvate explains the contraceptive action of α-chlorohydrin. Spermatozoa are exposed to high concentrations of fructose in semen and glucose is present in semen and in the female reproductive tract so a similar effect is likely to occur *in vivo*. The mechanism is unknown, as yet, but it may involve 'futile cycling' reactions of glycolytic intermediates (see 6). This speculation is encouraged because whereas lactate production was totally inhibited the phosphorylation of glucose (measured by $^{3}$HOH release from [2-$^{3}$H] glucose) continued, albeit at a slower rate.

## ACKNOWLEDGEMENTS

Financial support was provided by the MRC (Grant No.9979/955/5B)

## REFERENCES

1. Brown-Woodman PDC, Mohri H, Mohri T, Suter D, White IG (1978) Mode of action of α-chlorohydrin as an antifertility agent. Biochem. J. 170, 23-37
2. Ford WCL (1982) The Mode of Action of 6-chloro-6-deoxysugars as antifertility agents in the male. In. Progress towards a male contraceptive. Ed. by SL Jeffcoate & M Sandler. In press. Chichester, John Wiley
3. Ford WCL, Harrison A (1981) Effects of 6-chloro-6-deoxysugars on adenine nucleotide concentrations in and motility of rat spermatozoa. J. Reprod. Fert. 63, 75-79
4. Ford WCL, Harrison A, Waites GMH (1981) Effects of 6-chloro-6-deoxysugars on glucose oxidation in rat spermatozoa. J. Reprod. Fert, 75-79
5. Fraser LR, Quinn PJ (1981) A glycolytic product is obligatory for initiation of the sperm acrosome reaction and whiplash motility required for fertility in the mouse. J. Reprod. Fert. 61, 25-35
6. Katz J, Rognstad R (1976) Futile cycles in the metabolism of glucose. Curr. Topics Cell Reg. 10, 238-289
7. Sokoloski JE, Blasco L, Storey BT, Wolf DP (1977) Turbidimetric analysis of human sperm motility. Fert. Steril. 28, 1337-1341

# CHAPTER 2 SURFACE PROPERTIES OF SPERM CELLS

## THE BEHAVIOR OF SPERM BEFORE FERTILIZATION

B.M. SHAPIRO, R.W. SCHACKMANN and R. CHRISTEN°

Department of Biochemistry, University of Washington, Seattle, WA 98195, U.S.A. and (°) Station Zoologique, 06230 Villefranche-sur-Mer, France.

### 1. INTRODUCTION

We know from a host of morphological studies that the sperm is linearly differentiated, with its acrosome, head, mitochondria and flagella spaced along the length of the cell. This allows not only for the cell's interaction with its environment but also for fusion with an egg of the appropriate species to initiate development of a new organism. The sperm has a limited set of cellular functions; it swims to, or is transported to an egg and fuses with the egg plasma membrane by exposing through exocytosis the specialized contents of the acrosomal granule. In mammals these events also include capacitation. Sperm motility in all species results from flagellar movement in which the chemical energy of ATP is transformed into motion by the dynein ATPase. In echinoderms, the acrosome reaction consists of exocytosis of the acrosomal granule and polymerization of actin into the acrosomal rod.

We have been studying the mechanisms by which sea urchin sperm motility is activated, the acrosome reaction is triggered and sperm become incorporated into the egg after membrane fusion. Each of these processes is a complex function involving specific components of the sperm. In the following brief discussion, we present some results and hypotheses to explain two aspects of sperm behavior prior to fertilization. We have focused specifically on the role of ionic changes in the regulation of motility and the acrosome reaction.

### 2. COORDINATED REGULATION OF SPERM MOTILITY, RESPIRATION AND VIABILITY ARE RELATED TO THE SPERM INTRACELLULAR PH

Since the early work of Gray (1), many different groups

in the ensuing half century have added to the evidence that both the respiration and motility of sea urchin sperm can be affected by the ionic composition of the seawater as well as by components in the egg jelly (2-10). In particular, Nishioka and Cross found that sperm diluted in the absence of $Na^+$ were immotile and that the addition of $Na^+$ initiated motility and caused acid release from the sperm. This suggested that intracellular pH ($pH_i$) might regulate motility (6). It was subsequently demonstrated (7,8) that increased intracellular pH was a common denominator in the initiation of sperm motility and respiration. Sperm diluted into seawater at a low external pH (5.5), without $Na^+$ or with elevated $K^+$ (200 mM) were found to have an intracellular pH at or below 7 as determined by uptake of radioactive alkylamines, methylamine or diethylamine. In each case there was no motility and little respiration. Conversely, when the $pH_i$ was increased by addition of $NH_4^+$, or by adding $Na^+$ to sperm in $Na^+$-free seawater, or by elevation of the extracellular pH, both motility and respiration were coordinately activated. An example of the respiratory activation by $NH_4^+$ is shown in Figure 1. Sperm in seawater containing 360 mM choline in place of $Na^+$ do not respire until $NH_4^+$ is added. Parallel measurements using uptake of diethylamine to estimate $pH_i$ reveal a shift from pH 6.9 to 7.6 occurs when $Na^+$ or $NH_4^+$ is added to sperm in $Na^+$-free seawater.

The coupling between respiration and motility is strong; any conditions which stimulate motility in quiescent sperm also stimulate respiration. The two processes have the capacity to be coordinately regulated since production of ADP (11) by the dynein ATPase of the flagellar apparatus could be a stimulus for respiration in tightly coupled mitochondria. If either process were inhibited, the other should also stop. If, for example, the dynein ATPase were sensitive to pH, then inhibition of this activity should lead to a decreased production of ADP and a decrease in respiration. In fact, dynein ATPase activity *in situ* has been shown to be significantly activated by an increase from pH 7.1 to 7.5 in sperm

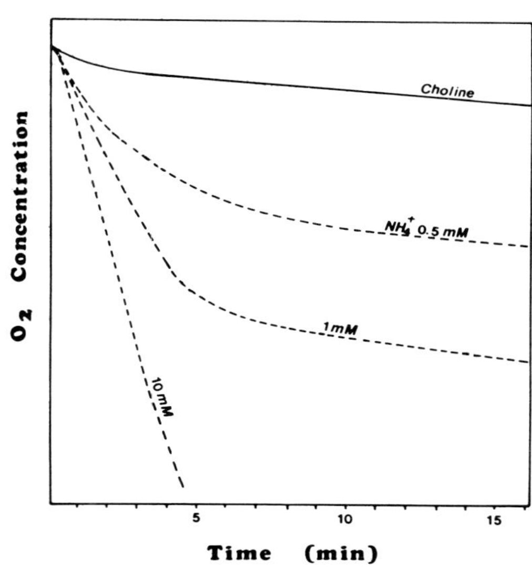


FIGURE 1. Sperm were diluted 200 fold into $Na^+$-free seawater (choline substituted for $Na^+$) at pH 8.0 containing varying concentrations (indicated in the figure) of $NH_4^+$ to stimulate respiration.

from the sea urchin *S. purpuratus*, and over a similar range in *T. gratilla* (12). It seems that the primary effect of pH is on motility rather than on respiration because when the $pH_i$ is rapidly lowered, the ATP levels in sperm tend to rise to a maximum. For example, as shown in Table 1, when sperm are diluted into seawater at pH 8, their ATP levels drop to a new steady state caused by a balance between consumption by the flagella and production by respiration, the latter not being sufficient at pH 8 to maintain the initially high level. When the extracellular pH is lowered to 5.5, motility stops, yet the concentration of ATP rises to the level found in dry, inactive sperm. These data suggest that the primary effect of decreasing $pH_i$ is to suppress ATP consumption and that the mitochondria are able to rephosphorylate any accumulated ADP.

Table 1. Sperm ATP levels and Extracellular pH.

| CONDITIONS | ATP (mM) | |
|---|---|---|
| | pH 8 | pH 5.5 |
| - Oligomycin | 3.5 | 5.5 |
| + Oligomycin (.5μM) | 0.6 | 0.7 |

Dry sperm were diluted 1000 fold into ASW (artificial seawater, 22) at pH 8 with or without oligomycin. The initial ATP concentration (1 min after dilution) was 5.6 mM. Values in the tables represent steady-state levels established within 10 min. after alterations. The pH of the sample was lowered from 8.0 to 5.5 with HCl. ATP concentrations were measured with the luciferase technique after extraction with boiling detergent (Christen, in preparation).

In collaboration with Dr. F. W. Dahlquist of the University of Oregon, we have found by $^{31}$P-NMR spectroscopy that our measurements of the $pH_i$ with radioactive diethylamine agree well with measurements by NMR. We have also found that sperm contain a creatine phosphate store that can be differentially depleted as a function of $pH_i$. Increased $pH_i$ decreases creatine phosphate levels with a corresponding increase in inorganic phosphate, suggesting that creatine phosphate serves as a phosphagen as it does in muscle. Moreover, sperm viability is also affected by conditions which alter the $pH_i$ and ATP levels. All conditions in which sperm viability is enhanced over that in seawater are characterized by high ATP levels and low $pH_i$. (Elevated ATP presumably results from a decreased usage of ATP by dynein, which is inactive at low pH, and continued function of the mitochondrion.) Table 1 shows that addition of oligomycin when sperm are under conditions of motility rapidly depletes ATP levels to approximately 10% of the initial levels.

The results summarized above indicate that sperm viability, respiration and motility are coordinately linked and that one of the agents that connects them is the intracellular pH. Thus, one mechanism by which the sperm regulates its behavior prior to fertilization is ionic and involves changes in, or regulation of, the intracellular pH by the ionic composition of the seawater. Work of others suggests that additional factors such as cyclic nucleotides may also be involved in the activation of motility (13).

Table 2. Intracellular pH and the Acrosome Reaction

| CONDITIONS | $pH_i$ |
|---|---|
| Experiment 1 | |
| ASW -jelly | 7.4 |
| ASW +jelly, 2' | 7.6 |
| ASW +jelly, 30' | 7.3 |
| ASW +jelly +16mM EGTA, 30' | 7.8 |
| Experiment 2 | |
| $Ca^{2+}$-free ASW +1mM EGTA -jelly | 7.5 |
| $Ca^{2+}$-free ASW +jelly, 5' | 7.6 |

Sperm were diluted 100 fold into ASW with or without $Ca^{2+}$ and incubated with tracer (2-10 μM) [$^{14}C$]$Et_2NH$. After an equilibrium was established, egg jelly or an equivalent amount of seawater was added and $pH_i$ values were determined at the designated times ( 8 ).

## 3. IONIC CHANGES INVOLVED IN THE ACROSOME REACTION

The acrosome reaction of invertebrate sperm involves exocytosis from an apical vesicle and assembly of actin into a filament. Jean Dan, some thirty years ago, initiated studies on the mechanism of this reaction with the observation that $Ca^{2+}$ is necessary (14). Since that time, many different laboratories have aided in defining the ionic requirements for the reaction (15-18). Both $Na^+$ and $Ca^{2+}$ are required for induction of the reaction by a component of the egg jelly coat, and the reaction results in release of $K^+$ and $H^+$ and accumulation of $Ca^{2+}$ and $Na^+$ by the sperm. Certain ion channel inhibitors such as tetraethylammonium (TEA), which blocks some potassium channels, and verapamil, which blocks some $Ca^{2+}$ and slow $Na^+$ channels, are inhibitory to the acrosome reaction (17). Several ionophores, such as nigericin (17) and A23187 (15,16,20) can spontaneously elicit this reaction, and these results suggest the acrosome reaction is under ionic regulation similar to stimulation of sperm motility and respiration.

When the acrosome reaction is triggered by addition of egg jelly to sperm, the associated acid release (17-19) suggests that an increase in $pH_i$ is one component of the reaction. Tilney and colleagues (19) and, more recently, Schroeder and Christen (21) demonstrated that actin polymerization occurs when the $pH_i$ is increased by addition of either $NH_4^+$ or ionophores. The intracellular pH has been measured by amine accumulation as described above and was found to elevate on addition of jelly, but this elevation is transient and the $pH_i$ soon returns to its previous level when the reaction is elicited under normal conditions (Table 2). However, if $Ca^{2+}$ in seawater is chelated immediately after induction of the reaction, the $pH_i$ elevates to a higher value and remains there. Additionally, if $Ca^{2+}$ is removed or chelated prior to jelly addition, egg jelly still causes a partial alkalinization of the sperm $pH_i$ (Table 2). Even removal of both $Na^+$ and $Ca^{2+}$ does not entirely prevent the alkaline shift of the intracellular pH upon addition of egg jelly. Thus, the pH changes that attend the induction of the acrosome reaction are complex. Part of the initial alkalinization appears to be independent of $Na^+$ and $Ca^{2+}$ and the subsequent reacidification seems to be related to the large uptake of $Ca^{2+}$ by the mitochondria that occurs after induction of the acrosome reaction.

The fact that the acrosome reaction results in multiple ionic changes within the sperm led us to speculate that alterations in sperm membrane potentials might also be associated. Using uptake of the lipophilic cation tetraphenylphosphonium (TPP) and lipophilic anion thiocyanate (SCN), we were able to show that sea urchin sperm have a potassium dependent, internally negative membrane potential (22). Sperm accumulate less radioactive TPP as the seawater potassium is increased and correspondingly exclude less radioactive thiocyanate. Upon addition of jelly, the sperm membrane potential is depolarized as measured by enhanced TPP efflux or thiocyanate uptake. A comparison of the two probes suggests the depolarization is largely limited to the plasma membrane potential. Thus, the acrosome reaction is associated with both increased intracellular pH and a depolarized membrane potential resulting from the multiple ion fluxes. The increased $pH_i$ may regulate actin polymerization and the $Ca^{2+}$ uptake is associated with exocytosis (17,19). $Ca^{2+}$ uptake and the complete acrosome reaction appear to require both increased $pH_i$ and membrane potential depolarization. For example, sperm placed in seawater with 180 mM $K^+$ will undergo the acrosome reaction and take up $^{45}Ca^{2+}$ if $NH_4^+$ is added to increase $pH_i$ as shown in Table 3. Sperm in artificial seawater or in $Na^+$-free seawater show no such comparable changes.

Table 3. $^{45}Ca^{2+}$ Uptake Stimulated by $NH_4^+$

| CONDITIONS | $^{45}Ca^{2+}$ Influx (nmole/$10^8$ sperm) |
|---|---|
| $Na^+$-free seawater, pH 7.8 | 3.3 |
| $Na^+$-free seawater + 10 mM $NH_4^+$ | 3.2 |
| 180 mM $K^+$ seawater, pH 7.8 | 2.4 |
| 180 mM $K^+$ seawater + 10 mM $NH_4^+$ | 17. |
| ASW, pH 7.8 | 2.7 |
| ASW + 10 mM $NH_4^+$ | 3.4 |

Sperm were diluted 500 fold into seawater of the specified ionic composition and incubated for 30 min with $^{45}Ca^{2+}$. Uptake was measured by centrifugation of sperm through silicone oil. Total seawater $Ca^{2+}$ was 10 mM. $Na^+$-free seawater substituted choline for $Na^+$. 180 mM $K^+$ seawater substituted $K^+$ for an equivalent amount of $Na^+$. ASW is our standard artificial seawater (22).

In summary, the acrosome reaction and sperm motility of the sea urchin S. purpuratus are regulated by the ionic components of the seawater. Alteration of the ionic composition of the external medium, which either depresses or stimulates motility, appears to do so by respectively decreasing or increasing the intracellular pH. The acrosome reaction is also associated with a change in pH and additionally allows for depolarization of sperm membrane potentials.

## REFERENCES

1. Gray, J. (1928) J. Exp. Biol. 5, 337-361.
2. Mohri, H. and Horiuchi, K. (1961) J. Exp. Biol. 38, 249-259.
3. Mohri, H. and Yasumasu, I. (1963) J. Exp. Biol. 40, 573-586.
4. Timourian, H. and Watchmaker, G. (1970) Dev. Biol. 21, 62-72.
5. Ohtake, H. (1976) J. Exp. Zool. 198, 303-312.

6. Nishioka, D. and Cross, N. (1978) In: Cell Reproduction (eds. E.F. Dirksen, D. Prescott and C.F. Fox) Acad. Press New York. pp. 403-413.
7. Lee, H.C., Schuldiner, S., Johnson, C. and Epel, D. (1980) J. Cell Biol. 87, No. 2 Pt. 2, 39a.
8. Christen, R., Schackmann, R.W. and Shapiro, B.M. (1983) J. Biol. Chem. In press.
9. Vacquier, V.A. (1979) Dev. Growth Diff. 21, 61-69.
10. Kinsey, W.H., SeGall, G.K. and Lennarz, W.J. (1979) Dev. Biol. 71, 49-59.
11. Chance, B. and Williams, G.R. (1955) J. Biol. Chem. 217, 409-428.
12. Gibbons, B.H. and Gibbons, I.R. (1972) J. Cell Biol. 54, 75-97.
13. Kopf, G.S., Tubb, J.D. and Garbers, D.L. (1979) J. Biol. Chem. 254, 8554-8560.
14. Dan, J.C. (1954) Biol. Bull. 107, 335-349.
15. Decker, G.L., Joseph, D.B. and Lennarz, W.J. (1976) Dev. Biol. 53, 115-125.
16. Collins, F. and Epel, D. (1977) Exp. Cell Res. 106, 211-222.
17. Schackmann, R.W., Eddy, E.M. and Shapiro, B.M. (1978) Dev. Biol. 65, 483-495.
18. Schackmann, R.W. and Shapiro, B.M. (1981) Dev. Biol. 81, 145-154.
19. Tilney, L.G., Kiehart, D.P., Sardet, C. and Tilney, M. (1978) J. Cell Biol. 77, 536-550.
20. Talbot, P., Summers, R.G., Hylander, B.L., Keough, E.M. and Franklin, L.E. (1976) J. Exp. Zool. 198, 383-392.
21. Schroeder, T. and Christen, R. (1982) Exp. Cell Res. In press.
22. Schackmann, R.W., Christen, R. and Shapiro, B.M. (1981) Proc. Nat. Acad. Sci. U.S.A. 78, 6066-6070.

# OOCYTE STRUCTURE AND THE DESIGN AND FUNCTION OF THE SPERM HEAD IN EUTHERIAN MAMMALS

J. Michael BEDFORD

Department of Obtetrics and Gynecology, Cornell University Medical College, New York, N.Y. 10021, U.S.A.

## 1. SUMMARY

Spermatozoa of eutherian mammals display novel features of form and function whose biological significance is not understood. They express a need for capacitation in the female before they can fertilize, the sperm head has become stabilized to an unusual degree, and is incorporated by the oocyte in a unique manner. These developments may be responses to functional and structural change in the vestments of the oocyte during evolution of the Eutheria.

Capacitation is expressed in development of an ability of the acrosome to react spontaneously in the presence of free $Ca^{2+}$, and in a hyperactivated form of motility. The first aspect may provide a means by which the spermatozoon can regulate the acrosome reaction in the apparent absence of any specific stimulation by the egg or other products of ovulation. The second aspect, hyperactivated motility, may augment the directed thrust of the spermatozoon to the level required to penetrate the unusually formidable vestments that characterize the eutherian oocyte. Integration of capacitation appropriate to an optimal fertilization rate in vivo may be effected in part by an innate heterogeneity among individual spermatozoa, compounded by the pattern of their transport through successive compartments of the female tract.

Imposition of the cumulus oophorus and a remarkable enhancement of the physical character of the zona pellucida in Eutheria may have changed the way that acrosomal enzymes are deployed during fertilization. The enzymes may enable the spermatozoon to pass through the cumulus oophorus. However, the striking stability and the design of the sperm head raise the likelihood that physical forces play a dominant role at the level of the zona pellucida. The novel way the sperm head is incorporated by the oocyte may result from this, specifically from a distinctive stability expressed by the inner acrosomal membrane. Its resilience may enable the membrane to resist frictional forces generated in oscillating interaction with the dense zona substance, but render it unable to fuse with the oolemma. That role has been assumed by an alternative site, the plasma membrane overlying the posterior equatorial segment of the acrosome which acts to preserve that membrane from involvement in the acrosome reaction. These and other considerations justify a re-examination of the part played by acrosomal enzymes in Eutheria.

## 2. INTRODUCTION

Ideas as to how the spermatozoa of eutherian mammals function during fertilization have tended to follow the concepts established for organisms that practice external fertilization, particularly the echinoderms and other invertebrates studied so fruitfully since the early years of this century and before (see 30). On the other hand, eutherian spermatozoa display several puzzling characteristics of function and structure, and broad extrapolation to Eutheria from invertebrate and even other vertebrate gamete systems may not be entirely justified. Eutherian spermatozoa must undergo a further period of capacitation in the female tract before they can fertilize, in vivo or in vitro. Moreover, the head of the eutherian spermatozoon displays several novel facets in its organization not seen even in the nearest evolutionary neighbors - the marsupials. Many of these new features are structural in nature; they enhance the stability of certain sperm head organelles, and they seem likely to reflect some change in the way the sperm head functions during fertilization. Indeed, the spermatozoon is incorporated by the eutherian oocyte in a puzzling complex sequence quite different from the simple mode seen in invertebrates and the few vertebrates studied in that respect (cf. 16, 20 and 6, 37).

There are no explanations as to why capacitation, a changed structural character of the sperm head, and an apparently unique mode of sperm incorporation by the oocyte, all have appeared as concomitants of gamete function in Eutheria. It seems important to consider the possible basis for these evolutionary changes and so their significance, not least for future efforts to interpret better the experiments directed to the 'what' and 'how' questions of mammalian gamete biology. The present evidence reviewed here seems largely consistent with the possibility that the special attributes of sperm form and function in

Eutheria are dictates of evolutionary change in the oocyte, in particular the character of its vestments.

3. WHY THE NEED FOR SPERM CAPACITATION?

There has been something of a 'catch 22' in efforts to understand sperm capacitation. A rather poor definition of the nature of capacitation has made it difficult to think critically about its biological significance. In turn, the absence of any real clue as to its role has made somewhat uncertain the interpretation of various experimental findings about sub-cellular changes induced in the spermatozoon. Now, it is reasonably certain that capacitation encompasses two functional elements - development of an ability to undergo the acrosome reaction and, in many species at least, change in the pattern of flagellar beat to one designated by Yanagimachi (50) as hyperactivated motility. If the special character of the eutherian oocyte and results obtained in a variety of experimental investigations are examined in parallel, it appears plausible that these changes of capacitation are functional responses to evolutionary change in the eutherian oocyte. Such a proposal (5) requires some consideration of hyperactivated motility and especially the acrosome reaction in eutherian spermatozoa.

a) Hyperactivated motility: A subtle change in the character of the sperm tail beat has been reported in at least eight eutherian species (see 17) to accompany onset of the ability to penetrate eggs. This state, hyperactivated motility, is expressed as a more rapid beat of the tail that, in a fluid medium, brings a greater undulation or arc traversed by the sperm head. It cannot be assumed that this will prove to be a major facet of capacitation in all Eutheria. Nor has it yet been calculated how different is the directed thrust the hyperactivated spermatozoon can mount against an opposing substrate. Nonetheless, the possibility that the hyperactivated state acts to maximize the forward thrust of the spermatozoon within the confines of the egg vestments, is a credible function for it. For, in Eutheria alone the spermatozoon is faced by an unusually formidable vestment it must penetrate in order to fertilize. The egg coat is trivial in reptiles and birds, and even monotreme and marsupial spermatozoa are faced only by a thin physically insignificant zona pellucida having a width of about 0.5 μ and 2.0 μ, respectively (28). Moreover, there is no cumulus oophorus around the ovulated egg. By contrast, in Eutheria the zona presents as a relatively rigid densely packed coat probably composed of at least three structural glycoproteins crosslinked to some extent by -S-S- bonds. In particular, it is much thicker (especially in relation to egg size) than in any other vertebrate group studied.

Experimental proof for the speculation that hyperactivated motility enhances the thrust needed to penetrate the eutherian oocyte, may be difficult to obtain. However, further comparative observation of the behavior of single epididymal and tubal spermatozoa of Didelphis virginana towards eggs, suggests that these marsupial spermatozoa may also undergo some functional change akin to capacitation (J.M. BEDFORD and J.C. RODGER, unpublished). In the context of the present suggestion, it would be of great interest to know whether hyperactivated motility appears as a facet of this functional change in marsupials whose zona pellucida is flimsy at best.

b) The acrosome reaction: An ability to undergo the acrosome reaction develops in eutherian spermatozoa as a concomitant of capacitation. The evidence from a variety of experiments is consistent with the possibility that the sub-cellular change(s) the sperm must undergo during capacitation actually represents a means by which the onset of the acrosome reaction can be regulated.

It is a matter of fundamental principle that in many other groups the reaction is induced and so is regulated by the cooperative action of egg coat substances and environmental $Ca^{2+}$ (19). In the case of echinoids, for example, the action on the sperm surface of a fucose sulphate-rich polysaccharide component of the egg coat ( 21 ) permits a calcium influx required for the sequence of biochemical events culminating in membrane fusion and then exocytosis of the acrosomal content. The thought/assumption has often been expressed that comparable factors from the oocyte may stimulate the acrosome reaction in eutherian spermatozoa. However, in mammals such as the guinea pig and hamsterwhere the acrosome can be visualized in the light microscope, it is apparent that spermatozoa in a physiological environment containing sufficient free $Ca^{2+}$ need only undergo capacitation to react there (51, 45). As a corollary, the experimental evidence of recent years suggests that neither the vestments of the egg, the egg itself, other products of ovulation, nor the site of fertilization in the Fallopian tube, critical for the onset of the acrosome reaction in vivo. Fertilization (and so the reaction), can be accomplished in vitro by spermatozoa capacitated in vitro (see 38), and in the isolated rabbit uterus when spermatozoa are given time to capacitate there (4). The granulosa cell vestment and matrix comprising the cumulus oophorus make no obvious contribution either. Cell-free ova are fertilized readily in vitro (31, 12, 38) and in the oviduct in the absence of follicular fluid (26, 32). Although activation of follicular fluid complement has been suggested as a mediator of the fusion reaction (14), eggs are penetrated readily in the oviduct of female rabbits and hamsters depleted of complement (10).

Studies with cell-free mouse oocytes have been interpreted as indicating that the sperm/zona surface interaction leads to the acrosome reaction in the fertilizing spermatozoon (40). It is not surprising that some mouse spermatozoa adhering to the naked zona undergo the acrosome reaction there in vitro. Nonetheless, the zona pellucida is not a necessary adjunct for induction of a normal acrosome reaction, and observation of spermatozoa where fertilization is occurring (33, 18) denies the idea that interaction between sperm plasma membrane and the zona surface is the inducing stimulus. Motile spermatozoa colonizing the intact cumulus oophorus have reacted before they contact the zona pellucida (1, 18) and, seen in the electron microscope, reacted rabbit spermatozoa are by far the most common type in the interstices of the cumulus and at the surface of the zona (3). After vaginal insemination, freely swimming reacted spermatozoa can be recovered from the ampulla (18) whether or not eggs are present (33). The point that the zona has no specific inductive role is made also in the ready occurrence of the reaction in a medium containing washed zona-free eggs that then incorporate the reacted spermatozoa normally (48, 53).

One cannot yet rule out the possibility that the eutherian oocyte has some residual ability to stimulate the acrosome reaction, but such a mechanism is unlikely. Since many spermatozoa probably react at the periphery of an intact cumulus mass, if not before, under physiological circumstances, a specific induction component would need to diffuse throughout zona and cumulus in significant concentration. Furthermore, indications that the eutherian vitellus does not provide such a stimulus are given by the ready penetration of the zona around long-dead oocytes (34, 52). That a product released by cortical granules may play such a role is equally unlikely since the reaction must occur in the fertilizing spermatozoon before their exocytosis. Moreover, spermatozoa readily penetrate primary oocytes in which exocytosis fails to occur (11, 32).

In summary, the mechanisms reflected in the change of capacitation may constitute an autonomous means of regulating the time of onset of the acrosome reaction in the absence of the control of this by the egg. The changes involved in capacitation apparently permit an influx of free calcium (42) required to initiate the sequence of biochemical events culminating in membrane fusion and dispersal of acrosomal content. It is clearly possible, therefore, that capacitation functions for eutherian spermatozoa as do the egg coat substances that stimulate the acrosome reaction in the well known invertebrate models. The key role of calcium is indicated in the fact that, where calcium is present, the ionophore A 23187 evokes a reaction even in non-capacitated spermatozoa (46). That capacitation becomes unnecessary when so driven in, reinforces the view that it functions primarily to regulate the moment at which the calcium influx can occur. The fact that the ionophore can substitute for egg jelly in invertebrate systems (21) again provides good reason to equate the action of egg coat components with that of capacitation (5).

c) Integration of capacitation for fertilization in vivo: The triggering of the acrosome reaction by the egg coat in invertebrates ensures the occurrence of that reaction in the close vicinity of the egg surface immediately before the anticipated moment of fertilization. Such coordination seems important, not least because invertebrate spermatozoa may cease to swim within a few seconds after the reaction has taken place. Eutherian spermatozoa may not fade for as long as two hours after the acrosome reaction (22). Even so, the autonomous control capacitation offers would seem imperfectly coordinated for consistently optimal fertilization in vivo. The time for completion of capacitation is largely a function of the specific character of the spermatozoon rather than of the capacitation environment (41). Nonetheless, the time between insemination and ovulation may vary by several hours even in induced ovulators, and by much longer in some spontaneous ovulators.

Two factors may act to ensure the availability of reacting spermatozoa in the oviduct over the hours that span the period of ovulation. First, whereas the reaction induced by egg substances may be synchronous within most of the sperm population (21), synchrony of reaction is not a characteristic of motile mammalian spermatozoa (e.g. 51, 45). In capacitating conditions, only perhaps 10% of the spermatozoa react after the minimal time required for capacitation, other fractions of the population reacting at successively later periods. Thus, there seems to exist an inherent heterogeneity within one population with respect to the time needed for capacitation under given conditions. Second, that innate heterogeneity may be compounded by the pattern of transport of individual spermatozoa through the different compartments of the female tract. Spermatozoa passing through the uterus on into the Fallopian tube display an ability to fertilize sooner than those spermatozoa experiencing only the uterus (or oviduct) alone (2, 4, 29). The potential for synergism between uterus and oviduct means that the timing of transport to the fertilization site also can operate to influence the moment individual spermatozoa complete the process of capacitation.

## 4. GAMETE STRUCTURE AND GAMETE INTERACTION IN MAMMALS

The spermatozoa of all invertebrate and non-mammalian vertebrate species studied appear to fuse with the oolemma by way of the inner acrosomal membrane, the sperm head then being drawn by its apex into the oocyte devoid of membrane other than nuclear envelope (16, 32a). The expectation that gamete interaction in eutherian mammals would mimic this logical sequence (16), has not been fulfilled. Not surprisingly, eutherian spermatozoa in the perivitelline space have been observed from time to time with the apex and thus inner acrosomal membrane indenting the oocyte surface. However, the inner acrosomal membrane does not take part in any fusion event. In marked contrast to other groups, the spermatozoa of Eutheria fuse first by way of a restricted area of persistent plasma membrane overlying the stable equatorial segment of the acrosome in the mid-region of the sperm head. Fusion induces an engulfment response by the eutherian egg cortex that then sequesters both inner membrane and equatorial segment of the acrosome for disposal within the ooplasm (see 6).

No explanation has emerged in the last 10 years to account for the fundamentally different way that spermatozoa enter the oocyte in Eutheria. One problem has been the fragmentary evidence for other vertebrate groups, particularly the monotreme and marsupial mammals. However, chicken spermatozoa fertilize in what might be regarded as the standard mode comparable to that in invertebrates (32a); and the basic similarity of the monotreme sperm head (7) makes it reasonable to anticipate a similar mode of fertilization in monotremes. Although its form has changed somewhat from the elongate ancestral therapsid mold displayed by chicken and monotreme spermatozoa, no truly novel features appear in the marsupial sperm head either (24, 47). In keeping with this, the fertilization pattern exemplified by that in the opossum, Didelphis virginiana, indicates that marsupials follow the common mode (37). In short, present evidence suggests that the way spermatozoa enter the eutherian oocyte is unique.

Comparison of the anatomy of both the male and female gamete among the three groups of the class Mammalia, brings to light some striking contrasts that bear on the abrupt departure of eutherian fertilization from what otherwise seems a basic format in biology. In essence, this suggests that the unique sequence in Eutheria may immediately result from an exaggerated stability of certain sperm head organelles determined in turn by the resilient nature of the oocyte vestment it must penetrate.

a) The oocyte: There exist obvious differences in the vestments of the unfertilized egg among the three mammalian sub-classes. In both monotreme and marsupial, the coat to be penetrated is trivial in a physical sense, protective mucus and shell coats being added after fertilization (28). The monotreme oocyte is covered only by a thin (0.5 μ) zona pellucida. That of the smaller marsupial oocyte, again the sole barrier between spermatozoon and oolemma, is at 2.0 μ somewhat thicker. Nonetheless, it presents as a flimsy structure having no elastic rigidity (37). The eutherian oocyte provides a striking contrast. A relatively massive zona pellucida composed of at least 3 structural glycoproteins maintains its intrinsic form as a spherical shell even in the absence of the oocyte. It is very much thicker with some variation according to eutherian species, although the egg is smaller. As seen in the electron microscope the zona material is more densely packed than that of the marsupial (5, 37).

The eutherian zona alone also is invested at fertilization by an additional barrier, the cumulus oophorus. This is a critical structure for consideration of the mechanisms of zona penetration. For, in the normal physiological circumstances that require early fertilization of the ovulated egg for optimal development, its added presence probably necessitates the dispersion of most of the (visible) content of the acrosome before sperm contact with the zona is established. The fact that the cumulus oophorus disperses relatively soon after ovulation apparently in species such as cow and pig, should not obscure the basic mechanisms involved in penetration of the egg. It is highly unlikely that the key elements of penetration differ from one eutherian to another. Since the fertilizing spermatozoon must negotiate the intact cumulus oophorus in several species studied, and be capacitated to do so, their situation seems the most appropriate in which to analyze the fundamental events involved.

b) The sperm head: Novel features of the sperm head in Eutheria have been revealed primarily through ultrastructural and chemical dissection. These features hint strongly at the existence of selection pressure for a design able to withstand the forces incurred in mounting a strongly directed thrust. The particular attributes of the sperm head appearing in Eutheria alone among the Mammalia are: (a) The presence of a perforatorium composed of -S-S- stabilized structural protein. Its form ensures the presentation of a minimal surface area at the sperm's leading edge; (b) A foreshortened sperm head flattened in one plane. Its profile seems appropriate to a cleaving action in the flat plane generated by oscillation through an arc delineated by the lateral limits of the perforatorium; (c) A highly keratinoid nucleus stabilized by -S-S crosslinks within the protamine matrix. Experiments in which spermatozoa were pre-treated with DTT before high-speed centrifugation through hyperosmolar (2.0 M) sucrose, provide some direct evidence that the -S-S crosslinked status of the sperm head lends a rigid quality to it (9); (d) Peri-nuclear material stabilized by disulphide bonds extending throughout the sperm head (15). This may act to cement the association between superficially placed organelles and the sperm nucleus underlying them; (e) An inner acrosomal membrane that, unable to resist strong ionic detergents (15), nevertheless appears significantly more stable than other sperm head membranes in the face of disruptive treatments (43, 49); (f) An inert posterior 'equatorial' segment of the acrosome whose distinctive stability probably rests in bridges of structural protein extending between its inner and outer membranes (38a).

c) Functional implications of gamete character in the Eutheria: The evolution of a highly stable sperm head in concert with the appearance of formidable vestments around the unfertilized egg suggests again (see the earlier section on hyperactivated motility) that physical interaction between them may be a primary element in penetration of the spermatozoon to the oolemma. The immediate determinant of the unusual mode of sperm fusion and incorporation may have been the evident stability of the inner membrane of the acrosome. This characteristic has not evoked previous comment from students of the anatomy of spermatozoa. However, its relatively stable nature may enable it to withstand frictional or shear forces created by sliding oscillation movement within the substance of the resilient zona pellucida. On the other hand, this may render it unable to perform the subsequent fusion role it plays in the other groups studied. Fusogenic membranes are necessarily organized such that their intrinsic proteins are free to move laterally in the plane of the membrane, and in no case do they manifest an unusual stability.

The alternative fusion site developed in Eutheria - the segment of plasmalemma apparently preserved by the stability of the equatorial segment beneath it (6a) - might also seem at similar risk during zona penetration, even in its more posterior location. However, additional features appear in the spermatozoa of at least some species, that would seem to minimize this (5). One, seen in hare, rabbit, bushbaby and hamster, for example, constitutes a shoulder of -S-S stabilized perinuclear material located immediately in front of the equatorial segment in a position that would divert the zona away from that surface. A second, represented to a variable degree in the dog, boar, musk shrew and in some human spermatozoa, is manifest in the sculpting of a waist in the nucleus, its recess being occupied by the equatorial segment. Both arrangements must minimize frictional interaction between the equatorial surface and the overlying zona.

The significance of the final element in the incorporation sequence, engulfment of the rostral region of the sperm head, is perhaps suggested in studies of sperm interaction with the rabbit primary oocyte (11). This immature cell often appears unable to mount such a cortical response, leaving much of the inner acrosomal membrane as part of the oocyte surface. In the normal absence of such a phagocytic step, the presence of a patch of stable sperm membrane in the oolemma may be disruptive for its function. A surface location of a patch of stable sperm membrane would also pose a more difficult situation for its ultimate disassembly by the egg.

The inference made in outline above that spermatozoa need to exert an unusual degree of directed thrust in the penetration of the eutherian zona pellucida bears finally on current concepts of the means by which spermatozoa are enabled to pass through it. As discussed below, the general view that sperm head lysins are the key to zona penetration may need to be re-evaluated.

5. THE MECHANISMS OF ZONA PENETRATION

It has long been accepted that acrosomal enzymes are used to digest a route through the egg vestments. This general concept is strengthened by the fact that where spermatozoa pass through a micropyle as in teleost fish, they lack an acrosome.

In examining the situation in mammals, it seems likely that penetration of the marsupial zona will indeed prove to depend on acrosomal protease(s), acting perhaps in concert with other enzymes. The level of several of possible relevance (acrosin, hyaluronidase, aryl sulphatase, N-acetylhexosaminidase) is as high and in some cases much higher in the acrosome of _Didelphis_, than in a typical eutherian such as the rabbit (37a). More important, because there is no cumulus, the marsupial can deliver the total content of the acrosome directly to the zona surface, and in the case of _Didelphis virginiana_, the surface area of the acrosome opposes the zona pellucida (37). The hole created by the penetrating spermatozoon is large, moreover, and the marsupial zona is highly susceptible to hydrolases of the type present in the acrosome. It dissolves in 2 - 3 seconds at 37°C in a 0.1% trypsin solution at pH 7.0, and in less than 20 minutes in an acrosomal extract that hydrolyses ≃ 0.5 nm of acrosin substrate per $10^6$ spermatozoa (37).

In the case of Eutheria, the point at which acrosomal enzymes facilitate sperm penetration may have changed. There seems no good reason on the basis of present evidence to doubt that hyaluronidase, possibly acting cooperatively with other enzymes (44, 37a) facilitates sperm passage through the cumulus oophorus. However, the role of acrosomal enzymes at the level of the eutherian zona seems questionable. Evidence in favour of their having a significant role in zona penetration rests primarily on suppression of fertilization in the presence of enzyme inhibitors (23, 35, 36). It seems important for such interpretation to establish, (a) that the inhibitors do not affect the motility of the spermatozoa in view of the possible importance of the role of hyperactivated motility for zona penetration, and, (b) that they do not influence the onset of the acrosome reaction or zymogen activation and so dispersion of acrosomal content.

Several independent observations now seem to call into question the view that acrosin is a 'zona pellucida-penetrating enzyme' in Eutheria. The suggestion implicit in the unusually stable nature of the sperm head that physical thrust has come to be a major factor in zona penetration, seems somewhat at odds with the concept of lytic digestion of a penetration slit. Moreover, since the presence of a cumulus oophorus means the visible content of the acrosome often is lost from the fertilizing spermatozoon before or as it adheres to the zona, and the eutherian zona pellucida is relatively far larger, it is curious that the acrosin content of the rabbit acrosome is if anything smaller than that of the marsupial *Didelphis* (37a). The belief that acrosin functions bound to the inner acrosomal membrane also is brought into question by direct observation, suggesting that it resides only in the matrix (27). Previous impression that acrosin remains bound to the inner acrosomal membrane after loss of the reacted acrosomal complex may rest in its non-specific adsorption to that and other areas of the sperm surface in the protocols used (25). Even were acrosin bound to the inner acrosomal membrane, in contrast to *Didelphis*, the form of the eutherian sperm head allows only a minor area overlying the apex of the perforatorium to oppose directly the zona substance in creating the penetration slit (8). There is no concentration of enzyme at the apex of the perforatorium, and so it is difficult to envision how lysins located on the dorso-ventral surface might operate at the anterior margin of the sperm head. Further concern arises in the demonstrated inability of ram acrosin to affect the homologous zona pellucida (13). Finally, it is apparent that rabbit spermatozoa penetrate a more trypsin/acrosin-resistant zona pellucida, whether created by fertilization (33a) or by a lectin reagent (8), equally as easily as the control zona. Neither do protease inhibitors interfere with fertilization in the mouse when zona binding is first allowed to occur (39).

In view of the great interest in acrosomal enzymes shown by biochemists and physiologists alike, the evidence cited would seem to justify a careful re-examination of their function in eutherian mammals.

REFERENCES


1. Austin CR, Bishop MWH. 1958. Proc. Roy. Soc. B. 149, 241-248.
2. Adams CE, Chang MC. 1962. J. Exp. Zool. 188, 203-210.
3. Bedford JM. 1968. Am. J. Anat. 123, 329-358.
4. Bedford JM. 1969. J. Reprod. Fert. Suppl. 8, 19-26.
5. Bedford JM. 1982. In Reproductive in Mammals, Vol. 1. Gametes and Fertilization, 2nd Eds Austin, C.R. Short R.V., Camb. Univ. Press. In press.
6. Bedford JM, Cooper GW. 1978. In Membrane Fusion - Cell Surface Reviews, Vol. 5. Eds, Poste A. Nicolson GL. Elsevier, Amsterdam, pp. 65-125.

6a. Bedford JM, Moore HDM, Franklin LE. 1979. Exp. Cell Res. 119, 119-126.

7. Bedford JM, Rifkin JM. 1979. Am. J. Anat. 156, 207-230.
8. Bedford JM, Cross NL. 1977. J. Reprod. Fert. 54, 385-392.
9. Bedford JM, RIfkin JM. Unpublished results.
10. Bedford JM, Witkin SS. Unpublished results.
11. Berrios M., Bedford JM. 1979. J. Cell Sci. 39, 1-12.
12. Brackett BG, Killen DE, Peace MD. 1971. Fertil. Steril. 22, 816-828.
13. Brown CR. 1982. J. Reprod. Fert. 64, 457-462.
14. Cabot CL, Oliphant G. 1978. Biol. Reprod. 19, 666-672.
15. Calvin HI, Bedford JM. 1971. J. Reprod. Fert. Suppl. 13, 65-75.
16. Colwin HI, Colwin al . 1967. In Fertilization, Vol. 1 eds Metz CB, Monroy A. Academic Press. pp. 296-367.
17. Cummins JM. 1982. Gam. Res. 6, 53-63.
18. Cummins JM, Yanagimachi R. 1982. Gam. Res. 5, 239-256.

19. Dan JC. 1954. Biol. Bull. 107, 203-218.
20. Dan JC. 1967. In Fertilization, Vol. I, eds Metz CB, Monroy A. Academic Press, pp. 237-293.
21. Decker GL, Joseph BD, Lennarz WJ. 1976. Dev. Biol. 53, 115-125.
22. Fleming AD, Yanagimachi R. 1982. J. Exp. Zool. 220, 109-116.
23. Fraser LR. 1982. J. Reprod. Fert. 65, 185-194.
24a. Harding HR, Carrick FN, Shorey CD. 1979. In The Spermatozoon. Eds Fawcett DW, Bedford JM. Urban and Schwarzenberg, Baltimore, pp. 289-304.
25. Harrison RAP, Fléchon JE, Brown CR. 1982. J. Reprod. Fert. 66, 349-358.
26. Harper MJK. 1970. J. Exp. Zool. 173, 47-62.
27. Huneau D, Fléchon JE, Harrison RAP. 1983. See this volume.
28. Hughes RL. 1977. In Reproduction and Evolution, eds Calaby JH, Tyndale-Biscoe CH. Aust. Acad. Sci. Canberra, pp. 281-291.
29. Hunter RHF, Hall JF. 19 . J. Exp. Zool. 188, 203-210.
30. Lillie FR. 1919. Fertilization, Chicago Univ. Press.
31. Miyamoto H, Chang MC. 1972. J. Reprod. Fert. 30, 309-312.
32. Moore HDM, Bedford JM. 1978. Biol. Reprod. 19, 879-885.
32a. Okamura F., Nishiyama H. 1978. Cell Tiss. Res. 190, 89-98.
33. Overstreet JW, Cooper GW. 1979. J. Exp. Zool. 209, 97-104.
33a. Overstreet JW, Bedford JM. 1974. Dev. Biol. 41, 195-192.
34. Overstreet JW, Hembree WC. 1976. Fertil. Steril. 27, 815-831.
35. Perreault S., Zaneveld LJD, Rogers BJ. 1980. J. Reprod. Fert. 60, 461-467.
36. Reddy JM, Joyce C., Zaneveld LJD. 1980. J. Androl. 1, 28-32.
37. Rodger JC, Bedford JM. 1982. J. Reprod. Fert. 64, 171-179.
37a. Rodger JC, Young RJ. 1981. Gam. Res. 4, 507-514.
38. Rogers BJ. 1978. Gam. Res. 1, 165-223.
38a. Russell L., Peterson RN, Freund M. 1980. Anat. Rec. 198, 449-459.
39. Saling PM. 1981. Proc. Nat. Acad. Sci. USA 78, 6231-6235.
40. Saling PM., Storey BT. 1979. J. Cell Biol. 83, 544-555.
41. Saling PM, Bedford JM. 1981. J. Reprod. Fert. 63, 119-123.
42. Singh JP, Babcock DF, Lardy HA. 1978. Biochem. J. 172, 549-556.
43. Srivastava PN, Munnel JF, Yang CH, Foley CW. 1974. J. Reprod. Fert. 36, 363-372.

44. Srivastava PN, Farooqui AA. 1979. Biochem. J. 181, 331-337.

45. Talbot P, Franklin LE. 1976. J. Exp. Zool. 198, 163-176.

46. Talbot P, Summers RG, Hylander BL, Keough EM, Franklin LE. 1976. J. Exp. Zool. 198, 383-392.

47. Temple-Smith PD, Bedford JM. 1976. Am. J. Anat. 147, 471-500.

48. Toyoda Y., Chang MC. 1968. Nature 220, 589-591.

49. Wooding FBP. 1973. J. Ultrastr. Res. 42, 502-516.

50. Yanagimachi R. 1981. In Fertilization and Embryonic Development _in vitro_. Eds Mastroianni 1., Biggers JD. Plenum Press, NY, pp. 81-182.

51. Yanagimachi R., Usui N. 1974. Exp. Cell Res. 89, 161-171.

52. Yanagimachi R, Lopata A, Odom CB, Bronson RA, Malin CA, Nicolson GL. 1979. Fertil. Steril. 31, 562-574.

# ROLE OF THE CUMULUS IN ZONA PENETRATION

RUTH SHALGI[1] and DAVID M. PHILLIPS[2]. [1]Department of Embryology, Tel Aviv University Medical School, Ramat Aviv, Israel; and [2]The Population Council, 1230 York Avenue, New York, New York USA

We have recently shown that there are differences in the mechanics of sperm entry between *in vivo* and *in vitro* fertilization in hamster and rat. In these studies we chose to examine the problem with the SEM where we can examine many ova. The SEM allowed us to see the entire sperm head and get a more accurate 3-dimensional perspective of the spatial relationship between the male and female gametes. These observations have revealed that there are striking differences between the ultrastructural events which occur during fertilization *in vivo* as opposed to those which occur *in vitro* in both the hamster and the rat. In cycling animals or hCG-PMSG superovulated artificially inseminated animals, the spermatozoon which had undergone acrosome loss appeared to contact egg microvilli initially by the equatorial segment (Fig. 1, Shalgi and Phillips, 1980a; Phillips and Shalgi, 1982). Soon after fusion of the oolemma and sperm

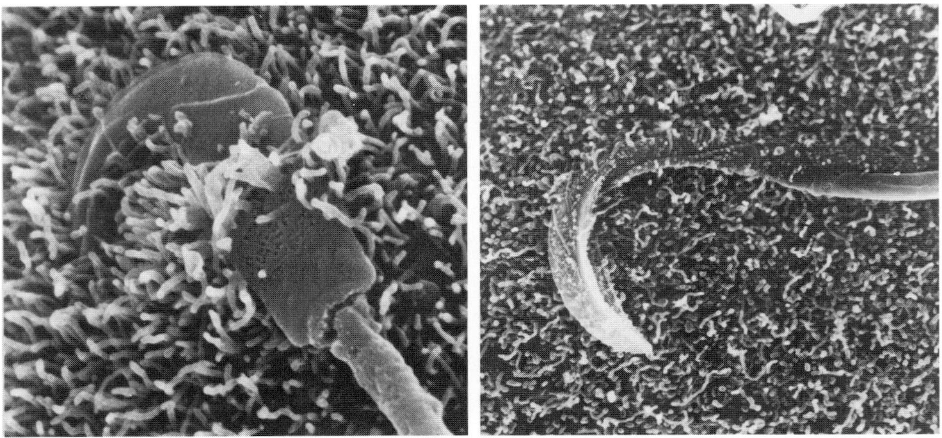

FIGURE 1. Early stage of sperm incorporation *in vivo* in hamster (left) and rat (right).

plasma membrane, microvilli charasteristic of the oocyte are observed over the broad side of the sperm head. The sperm head undergoes a pronounced flexure as it is incorporated into the ooplasm. On the other hand, when eggs with intact zonae were fertilized in vitro and prepared in exactly the same way for the SEM, the spermatozoon was observed to contact the ovum by the most anterior tip of the sperm head. Subsequently, microvilli encircled the sperm head as it was incorporated into the ooplasm tip first (Fig. 2, Shalgi and Phillips, 1980b; Shalgi and Phillips, 1982). Therefore, the morphology of sperm penetration in hamster or rat ova fertilized

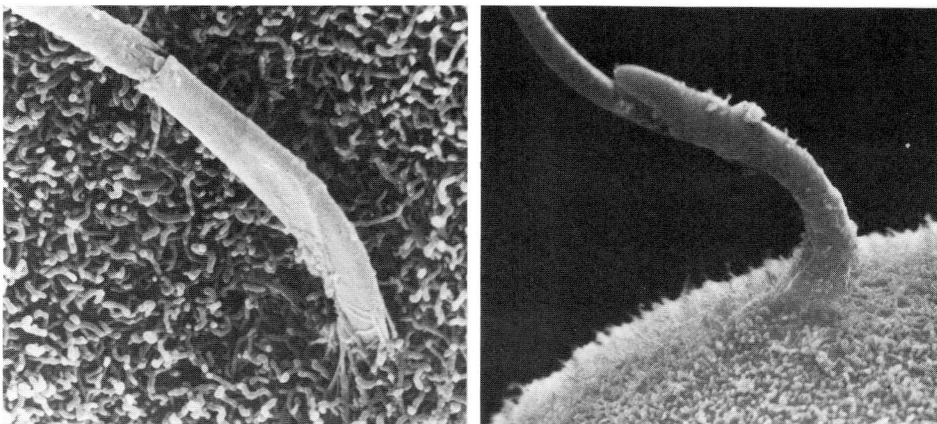

FIGURE 2. Early stage of sperm incorporation into zona intact eggs in vitro in hamster (left) and rat (right).

in vivo is very different from the morphology of penetration in zona intact hamster or rat ova fertilized in vitro. In living rat ova fertilized in vivo or in vitro, we can observe the same types of differences. It is clear in these images that in vivo the rat sperm traverse the zona at an oblique angle whereas in vitro spermatozoa penetrate the zona at a right angle to the zona surface (Figures 3 and 4). This observation and other unpublished data have allowed us to speculate that the cumulus may play an important role in zona penetration.

It is important to consider that the cumulus oopherous is absent in ova fertilized in vitro but present in ova fertilized in vivo. Presumably, acrosomal hydrolases dispersed from the high numbers of spermatozoa present in the in vitro fertilization system disperse the cumulus. High numbers of spermatozoa are not present at the site of fertilization in vivo.

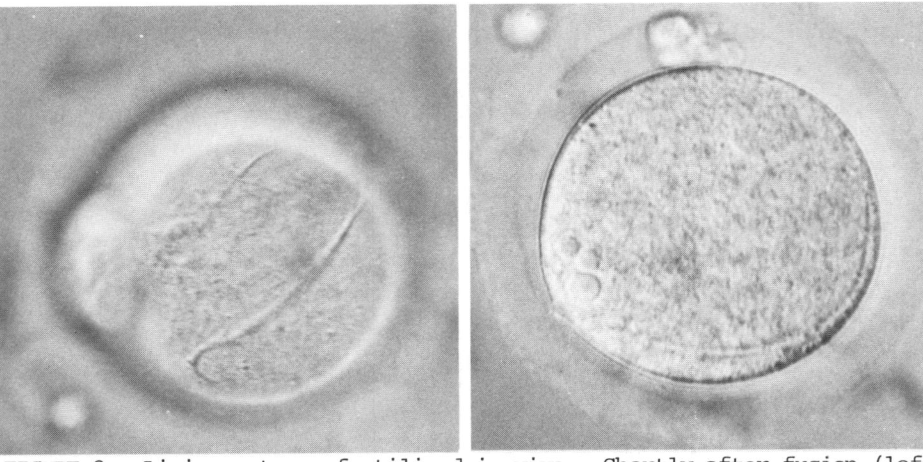

FIGURE 3. Living rat ova fertilized *in vivo*. Shortly after fusion (left); pronuclear stage (right).

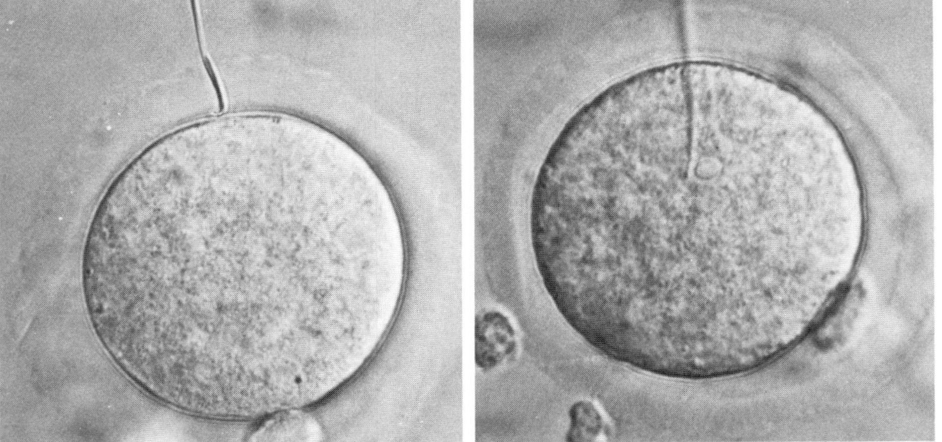

FIGURE 4. Living rat ova fertilized *in vitro*. Shortly after fusion (left); pronuclear stage (right).

Jim Cummins (unpublished) has recently found that hamster spermatozoa penetrate the zona more rapidly when the cumulus is present than when the cumulus is absent. In recent studies with R. Yanagamachi we have found that acrosome reacted hamster sperm penetrate the zona-free ovum *in vitro* in a similar manner to the way that they penetrate ova *in vivo* (Phillips and Yanagamachi, in press).

Taken together, the above evidence suggests that the cumulus may cause sperm to penetrate the zona at an acute angle. When the cumulus is absent spermatozoa penetrate the zona at a right angle, perhaps because they do not have cumulus cells to push against during penetration. This causes the sperm to associate with the ova tip first. When both cumulus and zona are removed, the sperm is not directed at right angles to the ovum and can associate with the ovum with the equatorial segment of the acrosome as is the case *in vivo*. We suggest that zona-intact *in vitro* systems using very few spermatozoa should be developed as they may facilitate penetration similar to the natural situation.

REFERENCES

Shalgi R, Phillips DM. 1980a. Mechanics of sperm entry in cyclic hamster. J Ultrastr Res 71:154-161.

Shalgi R, Phillips DM. 1980b. Mechanics of *in vitro* fertilization in the hamster. Biol Reprod 23:433-444.

Shalgi R, Phillips DM. 1982. Sperm penetration into rat ova fertilized *in vitro*. J Androl (in press).

Phillips DM, Shalgi R. 1982. Sperm penetration into rat ova fertilized *in vivo*. J Exptl Zool 221:373-378.

Phillips DM, Yanagamachi R. 1982. Differences in the manner of association of acrosome-intact and acrosome-reacted hamster spermatozoa with egg microvilli as revealed by scanning electron microscopy. Biol Reprod (submitted).

# FREEZE-FRACTURE OBSERVATIONS ON THE INTERACTION OF HUMAN SPERMATOZOA WITH ZONA-FREE HAMSTER OOCYTES

JAMES K. KOEHLER, IVAN de CURTIS, MORTON A. STENCHEVER AND DIANNE SMITH
Department of Biological Structure, University of Washington, School of Medicine, Seattle, Washington 98195, U.S.A.

## 1. INTRODUCTION

The use of zona-free oocytes, particularly those of the golden hamster, in fertilization studies has grown steadily since the discovery that zona removal greatly diminishes the block to interspecies fertilization (2). Such studies have added to our knowledge of the basic structural and functional relationships involved in sperm-egg interactions as well as providing a useful clinical tool in the assessment of male fertility (3-5). We report here on ultrastructural studies of the interaction between human sperm and zona-free hamster oocytes. Emphasis is placed on the results of freeze-fracture studies which provide views of internal membrane structure of the interacting sperm and egg. Oocyte microvilli play a role in this interaction and will be a focus for the discussion.

## 2. PROCEDURE

### 2.1. Culture Methods

Sperm obtained from men of known fertility were mixed with 4 parts BWW (6) containing 0.4% bovine serum albumin and Hepes buffered to pH 7.4. Thrice washed sperm (300 g, 6 min each) were incubated at a concentration of $1x10^7$/ml at 36°C in a 1% $CO_2$ atmosphere. Superovulated hamsters (*Mesocricetus auratus*) were sacrificed, oviducts removed by dissection and ova treated with hyaluronidase and trypsin to remove granulosa cells and the zona pellucida, respectively. Twenty-five to 30 oocytes were cultured with a 0.1 ml aliquot of preincubated human sperm under mineral oil. Samples were removed periodically for electron microscopy.

### 2.2. Electron Microscopy

Sperm-egg cultures obtained after ½ to 3 hour co-incubation were washed in fresh medium and fixed in 1.5% glutaraldehyde in 0.15 M sodium cacodylate buffered to pH 7.2 for 1 hour. Cells were then rinsed in buffer and post-osmicated in 1% $OsO_4$ in 0.15 M cacodylate, dehydrated in graded ethanol and embedded in Epon 812. Sections were cut with a Dupont diamond knife on a PBII ultramicrotome and observed with a Philips 201 electron microscope.

Freeze-fracture specimens were rinsed after glutaraldehyde fixation and glycerolated (20%) for several hours at 4°C before rapid freezing in liquid-solid nitrogen slush. Some sperm samples were frozen and freeze-fractured without prior fixation or cryoprotection.

## 3. RESULTS AND DISCUSSION

The initial interaction between human sperm and zona-free hamster oocytes involves the tips of egg microvilli and various portions of the sperm head surface. On acrosome reacted sperm, microvilli can be seen attached to the inner acrosomal membrane, delimiting the anterior portion of the sperm head, the equatorial segment, postacrosomal region and even the neck piece of the flagellum. The usual orientation of interacting sperm at the egg's surface is a lateral one (Fig. 1). Nonacrosome reacted sperm can also be found attached to the egg surface by microvillar interactions, however, in contrast to the report by Soupart and Strong (7), such sperm have not been seen to penetrate the hamster oocytes. A freeze-fractured image of a reacted sperm interacting with microvilli at the tip of the sperm head can be seen in Fig. 2. In some specimens the microvilli appear to encompass or surround the sperm head (Fig. 3). Such impressions are reinforced by scanning electron microscope observations from our laboratory; the studies of Talbot and Chacon (8) and the work of Shalgi and Philips (9) using the homologous hamster sperm-egg system. Very often, the most extensive microvillar attachments are seen clustered in the medial region of the sperm head (equatorial segment). In many cases the intramembranous particles of the sperm membrane have become clustered into patches. Such clustering is particularly prominent on the neck piece and flagellar membrane and less so over the sperm head. Whenever microvilli have been observed attached to such "patched" membranes, they are invariably localized to the particle-rich zones of membrane (Fig. 4). Although the patching or clustering of particles is probably artifactual, it suggests that there is a specific affinity between oocyte microvillar membranes and transmembrane protein or glycoprotein components of the sperm surface. Despite the observation that microvilli appear to attach to a variety of different sites over the sperm head and flagellum, we have observed specific fusion of the gametes only in the medial (equatorial or postacrosomal) regions of the sperm head.

Several ultrastructural studies of the human sperm-zona-free hamster system have appeared in the literature since the original description by Yanagimachi *et al.* (3). In that study it was shown that microvilli are involved in the early interactions of the gametes and that the fusing sperm had undergone the acrosome reaction. The exact fusion site was not determined; however, it was indicated that the locus was in the posterior portion of the sperm head. Barros *et al.* (10) confirmed that the penetrating spermatozoa were acrosome reacted and suggested that the postacrosomal region was involved in the membrane fusion process. Talbot and Chacon (8) have recently reported that initial binding of sperm by oocyte microvilli may be toward the anterior portion of the head and that subsequent interactions proceed more posteriorly and involve the equatorial segment in the fusion process. They further support the notion that only acrosome reacted sperm are capable of fusing with ova. Considering this general concurrence of data and opinion, the observations of Soupart and Strong remain enigmatic since they claimed that nonacrosome reacted sperm were seen penetrating a human ovum. In the homologous hamster system, Moore and Bedford (11) concluded that the equatorial segment was the initial site of sperm-egg fusion, a result generally in line with the best data available for the human sperm-hamster egg system. From the information gathered thus far it seems that this heterologous system mimics quite accurately the interactions of homologous gametes at least in the *in vitro* situation.

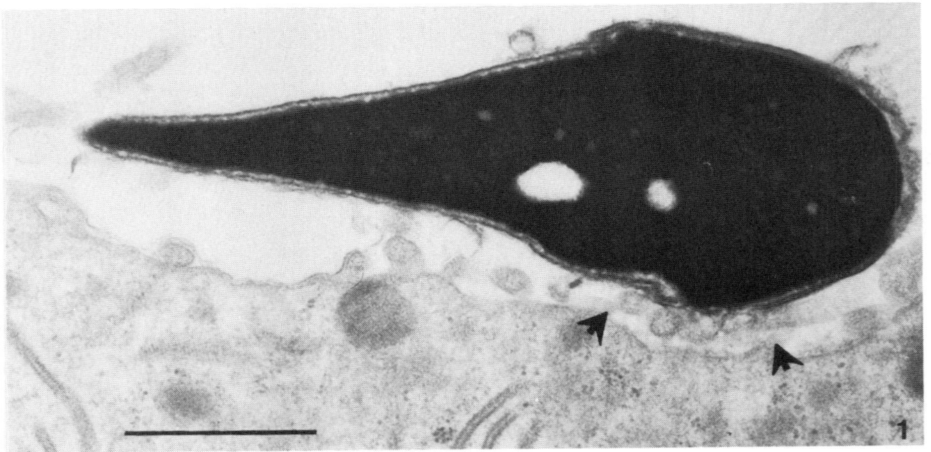


FIGURE 1. Human spermatozoon making contact with a zona-free hamster oocyte. Microvilli are clustered adjacent to the equatorial segment region of the sperm head (arrows). Note that the acrosome has reacted and the anterior portion is covered by inner acrosomal membrane.

FIGURE 2. Freeze-fracture preparation of human sperm (P-face) in initial attachment to hamster oocyte. Microvilli are attached at several sites at sperm anterior tip (arrows). Sperm has undergone the acrosome reaction, leaving a ridge (arrows) at the equatorial segment.

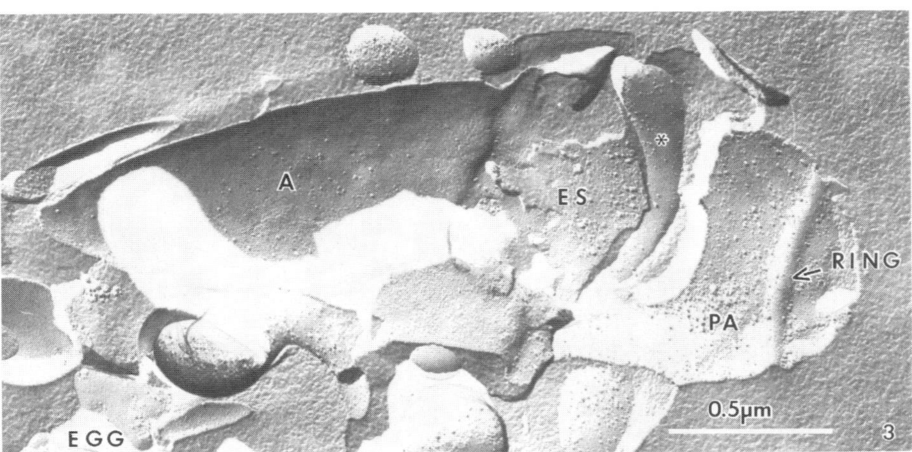


FIGURE 3. Sperm in close proximity to oocyte surface. Egg microvilli encircle the sperm especially in the equatorial region (*). A: acrosomal region (E-face), ES: equatorial segment, PA: postacrosomal region.

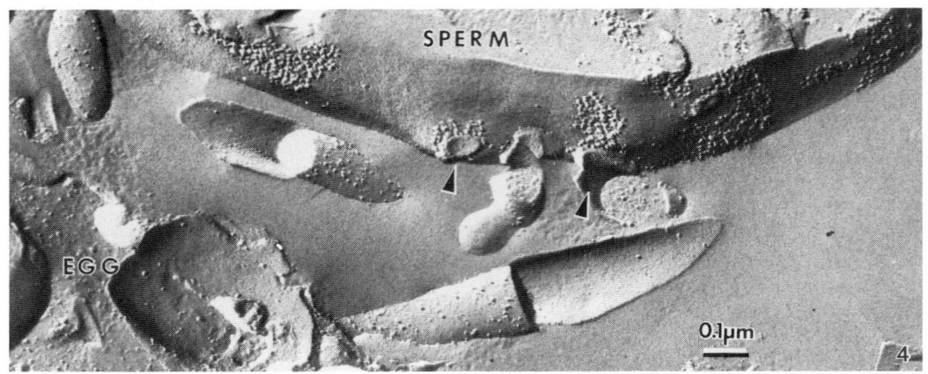


FIGURE 4. Sperm and egg in close proximity. Note that fragments of attached microvilli are attached to the sperm surface exclusively at sites of intramembranous particle clusters (P-face).

REFERENCES

1. Chang MC, Hancock JL. 1967. Experimental hybridization. In Comparative Aspects of Reproductive Failure. Benirschke K, Ed. Springer Verlag, New York, pp. 206-217.
2. Yanagimachi R. 1972. Penetration of guinea pig spermatozoa into hamster eggs in vitro. J. Reprod. Fertil. 28:477-480.
3. Yanagimachi R, Yanagimachi H, Rogers BJ. 1976. The use of zona-free animal ova as a test system for the assessment of the fertilizing capacity of human spermatozoa. Biol. Reprod. 15:471-476.
4. Binor Z, Sokoloski JE, Wolf DP. 1980. Penetration of the zona free hamster egg by human sperm. Fertil. Steril. 33:321-327.
5. Karp LE, Williamson RA, Moore DE, Shy KK, Plymate SR, Smith D. 1981. Sperm penetration assay: useful test in evaluation of male fertility. Obstet. Gynecol. 57:620-623.
6. Biggers JD, Whitten WK, Whittingham DF. 1971. The culture of mouse embryos in vitro. In Methods of Mammalian Embryology. Daniel JC, Ed. Raven Press, New York, p. 101.
7. Soupart P, Strong PA. 1975. Ultrastructural observations on polyspermic penetration of zona pellucida-free human oocytes inseminated in vitro. Fertil. Steril. 26:523-537.
8. Talbot P, Chacon RS. 1982. Ultrastructural observations on binding and membrane fusion between human sperm and zona pellucida-free hamster oocytes. Fertil. Steril. 37:240-247.
9. Shalgi R, Phillips D. 1980. Mechanics of sperm entry in cycling hamsters. J. Ultrastruct. Res. 71:154-161.
10. Barros C, Gonzalez J, Herrera E, Bustos-Obregon E. 1974. Human sperm penetration into zona free hamster oocytes as a test to evaluate the sperm fertilizing ability. Andrologia 11:197-210.
11. Moore HDM, Bedford JM. 1978. Ultrastructure of the equatorial segment of hamster spermatozoa during penetration of oocytes. J. Ultrastruct. Res. 62:110-117.

This research was sponsored by a grant, PCM-77-01138, from the NSF, and a grant to Dr. DeCurtis from the Italian National Research Center.

# INFLUENCE OF INITIATION OF FORWARD MOTILITY ON THE FERTILIZING ABILITY OF IMMATURE BOAR SPERMATOZOA IN *in vivo* HOMOLOGOUS and *in vitro* HETEROLOGOUS SYSTEMS OF INSEMINATION

DACHEUX J.L.* & PAQUIGNON M.
I.N.R.A. Station de Physiologie de la Reproduction Nouzilly 37380 Monnaie
* Lab. Physio. Comparée, Fac. Sciences, 37200 TOURS, FRANCE

## 1. INTRODUCTION

The fertilizing ability of spermatozoa is acquired in the epididymis simultaneously with the acquisition of forward motility, and is maximal in fully motile spermatozoa taken from the cauda of the epididymis of the boar (1). The induction of forward motility by rabbit testicular sperm with caffeine has been shown to improve ovum activation *in vitro* (2). Similarly, after their motility has been increased by treatment with caffeine, the spermatozoa from guinea-pig caput epididymis become as fertile *in vivo* as the distal cauda sperm (3). In view of these observations, it was proposed to evaluate the fertilizing ability of boar testicular and caput epididymal spermatozoa following the initiation of forward motility with caffeine, either by an homologous oviduct insemination *in vivo* or by an heterologous fertilization of zona-free hamster eggs *in vitro*.

## 2. MATERIALS and METHODS

### 2.1. Collection of spermatozoa

Testicular spermatozoa were obtained from rete testis fluid collected from unanesthetized boars (4) (5). Epididymal spermatozoa were collected from several regions of the caput, corpus and cauda of the epididymis after slaughter or castration. Ejaculated spermatozoa were collected manually.

### 2.2. Induction of forward motility

Spermatozoa (testicular, epididymal and ejaculated) were suspended in citrate egg-olk buffer containing 6.2 mM sodium citrate, 228 mM glucose, 15.2 mM KCl, 16 mM potassium bicarbonate, 15 mM $MgCl_2$ and 5% (V/V) egg-yolk. The buffer had been previously centrifuged at 10,000 g for 30 min. The concentration of spermatozoa was adjusted to $1.10^8$ spz./ml. Forward motility was induced with 2.5 mM, caffeine (6). After incubation for 30-60 min, the spermatozoa were used for insemination either *in vivo* or *in vitro*.

2.3. Capacitation of the spermatozoa

The medium used for manipulation of gametes was the same as above described, but the forward motility was induced with caffeine, after capacitation *in vitro* (7).

2.4. Incubation with epididymal protein.

Boar testicular and caput epididymal sperm were incubated at 37°C for 15 min in fluid extracted from corpus epididymus, and then for 15 min in caudal epididymal fluid. Motility was then induced with caffeine in egg-yolk medium.

2.5. Insemination

- Homologous insemination *in vivo*: The fertilizing ability of spermatozoa was tested by surgical insemination (0.2 ml - $2x10^7$ spermatozoa) into the oviducts of oestrous adult Large White sows. The eggs were recovered 48 hours after insemination by flushing the oviduct with 10 ml of dulbecco medium, and were examined under phase contrast for evidence of spermatozoa on the pellucida and segmentation.

- Heterologous insemination *in vitro*: A previously described technique for binding boar spermatozoa to zona-free hamster eggs *in vitro* was used (7-8). When sperm (testicular and epididymal) were tested, the zona-free eggs were introduced into a sperm suspension diluted with citrate egg-yolk buffer ($1.5x10^6$ spz./ml) with or without caffeine. A gradient in the binding ability of sperm from the epididymis was established by incubating the sperm with eggs in Krebs ringer bicarbonate (KRB) medium. After 4 hours of incubation at 37°C the eggs were picked up, washed twice in KRB medium and fixed with glutaraldehyde (2.5%). The number of linked spermatozoa to ova was counted by using phase-contrast microscopy.

3. RESULTS

3.1. Stimulation of motility

Forward motility was induced in 30 to 40% of the testicular spermatozoa and 60% of the caput epididymal spermatozoa observed after 30 min of incubation with caffeine. Without caffeine, 30 min incubation at 37°C, testicular sperm were not motile, but 10-15% of caput epididymal spermatozoa showed progressive motility.

3.2. Fertilizing ability

There were no fertilized eggs after insemination *in vivo* (Table 1), with either testicular or caput epididymal spermatozoa. Neither stimulation with caffeine nor capacitation *in vitro* influenced the results. When the ova

Table 1. Effects of progressive motility induction on the fertilizing ability of boar sperm.

| Sperm origin | Stimulated motility | Capacitation | Number of boars | Number of sows | Number of recovered eggs | % fertilized eggs |
|---|---|---|---|---|---|---|
| Testis | + | - | 3 | 6 | 57 | 0 |
| | + | + | 4 | 4 | 20 | 0 |
| Anterior Caput Ep. | + | - | 4 | 8 | 35 | 0 |
| | - | - | | | 30 | 0 |
| | + | + | 2 | 2 | 10 | 0 |
| Ejaculate | + | - | 1 | 2 | 14 | 100 |
| | + | + | 1 | 2 | 10 | 100 |

Table 2. Effect of epididymal medium on the fertility of motile induced sperm from testis and epididymal anterior caput.

| Sperm origin | Epididymal medium | Number of boars | Number of sows | Number of recovered eggs | % fertilized eggs |
|---|---|---|---|---|---|
| Testis | + | 2 | 2 | 9 | 0 |
| Epididymal caput | + | 5 | 8 | 27 | 11 |
| | - | | | 50 | 0 |

Table 3. Boar sperm binding to hamster zona-free egg according to their origin and the induction of progressive motility.

| Sperm origin | | Stimulated motility | Number of boars | Number of eggs | 0 spz. | 1-10 spz. | 10-100 spz. | >100 spz. |
|---|---|---|---|---|---|---|---|---|
| Testis | | - | 2 | 23 | 100 | | | |
| | | + | 2 | 21 | 100 | | | |
| Epidid. | | | | | | | | |
| Zona | 1 | - | 3 | 26 | 100 | | | |
| Caput | 2 | - | 3 | 27 | 100 | | | |
| | | + | 2 | 32 | 100 | | | |
| | 3 | - | 3 | 30 | 93 | 7 | | |
| | 4 | - | 3 | 34 | 60 | 34 | 6 | |
| Corpus | 5 | - | 3 | 35 | 18 | 76 | 6 | |
| | 6 | - | 3 | 36 | | 17 | 76 | 7 |
| | 7 | - | 3 | 38 | | | 21 | 79 |
| | 8 | - | 3 | 36 | | | | 100 |
| Cauda | 9 | - | 3 | 15 | | | | 100 |
| | 10 | - | 3 | 20 | | | | 100 |

were observed after 48 h, no spermatozoa were fixed on the pellucida. When ejaculated sperm were used as a control, all the ova were fertilized and spermatozoa were found on the pellucida.

The presence of epididymal medium (Table 2) increased the in vivo fertilizing ability of caput epididymal sperm by a small amount but did not affect testicular sperm.

With heterologous insemination in vitro (Table 3), the ability to bind to the zona-free hamster eggs appeared in spermatozoa collected from the corpus of the epididymis. No binding was observed when testicular or caput epididymal spermatozoa were used and the induction of the forward motility, in these gametes did not improved the binding to ova. The sperm binding was maximal when caudal epididymal sperm was used.

## 4. DISCUSSION

The induction of a forward motility in vitro did not improve the fertilizing ability of testicular and caput epididymal boar spermatozoa. These results confirm those obtained when sperm motility was initiated with several days of either epididymal ligature (9) (10) or storage at 4°C (11). Capacitation of caffeine treated sperm had no influence on fertilizing ability. Activation by immature motile spermatozoa of the ovum in vitro (2) could not be reproduced in vivo. Motility of testicular and epididymal sperm did not enhance their ability to bind to zona-free hamster eggs. The intensity of heterologous binding increased in the epididymal corpus region where the fertilizing ability appears (1). The presence of epididymal proteins is not sufficient to induce fertility of motile testicular sperm and the slight augmentation obtained with caput epididymal sperm seemed to be related more to survival of the sperm than to increases in fertility.

## REFERENCES


1. Holtz W, Smidt D. 1976. J. Reprod. Fert. 46: 227-229.
2. Brackett BG, Hall JL, OH YK. 1978. Fert. Ster. 29: 571-582.
3. Shilon M, Paz G, Homonnai ZT, Schoenbaum. 1978. Internat. J. Androl. 1978. 1: 416-423.
4. Voglmayr JK, Scott TW, Setchell BP, Waites GMH. 1967. J. Reprod. Fert. 14: 87-99.
5. Dacheux JL, O'Shea T, Paquignon M. 1979. J. Reprod. Fert. 55: 287-296.
6. Dacheux JL, Paquignon M. 1980. Testicular development, structure and function ed. A. Steinberger and E. Steinberger. Raven Press. NY. 513-521
7. Imai H, Niwa K, Iritani A. 1979. J. Reprod. Fert. 56: 489-492.
8. Imai H, Niwa K, Iritani A. 1977. J. Reprod. Fert. 51: 495-497.
9. Bedford JM. 1967. J. Exp. Zool. 166: 271-282.
10. Orgebin-Crist MC. 1967. Ann. Biol. anim. Bioch. Biophys. 7: 373-389.
11. Voglmayr JK, White IG, Parks RP. 1978 Theriogenology. 10: 313-321.

# ACQUISITION OF ZONA BINDING STRUCTURES BY RAM SPERMATOZOA DURING EPIDIDYMAL PASSAGE

S. Fournier-Delpech, S. Hamamah, G Colas, M. Courot
I.N.R.A. Station de Physiologie de la Reproduction. Nouzilly, 37380 MONNAIE

The non-capacitated spermatozoa require passage through the epididymis before they are capable of binding to the zona pellucida of homologous (mouse: 8; rat: 7) or heterologous (ram: 4) eggs.

It seemed pertinent to assess the capacity of testicular epididymal or ejaculated spermatozoa of rams to bind to the zona of the ewe ova. Since epididymal maturation of the spermatozoa occurs under the control of the androgens (6), the zona binding ability was also studied on spermatozoa recovered from the tail of the epididymis in rams castrated with or without endocrine supplementation by androgens from the contralateral testis; because of the difficulties to obtain sheep oocytes in large quantities, this study was completed using rat ova to provide sufficient data, rodent ova permitting analysis of initial contact between spermatozoa and eggs investments as it is now well ascertained (1, 9). To eliminate parameters due to motility in binding between ova and collising spermatozoa experiments were made on immobilized spermatozoa at 0°C.

## MATERIALS AND METHODS

Medium was Toyoda and Chang (10) mixture added v/v with milk diluent (2)

A. Preparation of oocytes

Ewe's oocytes: 50 hours after the PMSG injection in ewes at induced ovulation (3), oocytes were recovered from the large follicles and granulosa cells removed mechanically.

Rat oocytes: Denuded eggs were prepared using hyaluronidase (5). Eggs were stored for 1 to 2 hours at 0°C.

B. Preparation of spermatozoa

a) Normal testicular, epididymal or ejaculated spermatozoa: 6 rams were castrated and spermatozoa recovered by micropuncture of the rete testis or epididymal tubule at the different levels of the epididymis (head, body, tail); ejaculated spermatozoa were obtained using artificial vagina.

b) Spermatozoa from castrated epididymis: 3 rams were orchidectomized, the epididymides remaining in place for 9 days.

c) Spermatozoa from castrated epididymis with endocrine testicular supplementation: 3 rams were submitted to unilateral orchidectomy; the castrated epididymis received for a further 9 days the androgenic secretion from the contralateral testis.

C. Binding assay:

All dilutions, washing and incubations were performed at 0°C.

. Homologous inseminations: diluted spermatozoa ($5 \times 10^6$/ml) were distributed in aliquots of 100 ùl under paraffin oil added with one or two sheep oocytes.

. Heterologous insemination: epididymal distal spermatozoa from castrated epididymis supplemented or not by the contralateral testis were initially diluted to $140 \times 10^6$/ml and serially diluted to provide a range of concentration of spermatozoa (0.01 to $140 \times 10^6$/ml); suspensions of sperm cells were then distributed in aliquots of 100 µl and added with 10 to 15 rat oocytes. At the end of incubation (60 minutes) eggs were washed and mounted to be examined under the phase contrast microscope.

## RESULTS AND DISCUSSION

### A. Development of sperm binding ability to the homologous ova in the normal rams.

Ram testicular spermatozoa did not bind to zona pellucida except a few sperm cell which adhered by the head and flagella to the zona pellucida, suggesting non specific attachment, with 2 or 3 spermatozoa per egg on 3/10 eggs. In contrast oocytes mixed with epididymal spermatozoa had numerous bound spermatozoa with the acrosome directed against the zona pellucida. In the head of the epididymis, half of the oocytes had more than 10 bound spermatozoa per egg. In the body of the epididymis the attachment of spermatozoa to the zona reached a maximum with values similar to those of the tail of the organ or of ejaculated sperm (table 1). This observation

Table 1. Acquisition of zona recognition sites by ram spermatozoa

| Origin of spermatozoa | N° eggs oocytes | N° oocytes with bound spermatozoa 0 spz | 1-3 spz | >10 spz |
|---|---|---|---|---|
| Rete testis | 10 | 7 | 3 | 0 |
| Epididymis | | | | |
| head | 10 | 5 | 0 | 5 |
| body (middle) | 6 | 0 | 0 | 6 |
| tail | 19 | 0 | 0 | 19 |

suggest that the capacity of the spermatozoa to recognize and attach to the zona is achieved before they have acquired the capacity to produce viable lambs which is achieved only the tail of the organ (4).

B. Endocrine regulation of this function

The sperm zona binding structures that develop on the surface of spermatozoa in the epididymis are regulated by endocrine activity of the testis: 3/9 ewe's ova had bound spermatozoa when sperm were recovered from the tail of the castrated epididymis without testicular supplementation versus 8/10 when sperm were taken from the tail of castrated epididymides supplemented by the contralateral testis. These experiments were carried out at a concentration of $5 \times 10^5$ spermatozoa by incubation and were at the limit of the significance ($p = 0.05$).

Using rat ova in sufficient quantities to provide a valid comparative study, the endocrine regulation of the zona binding structures acquired by the spermatozoa into epididymis appear well established (Fig. 1).
The proportion of eggs having bound spermatozoa increased with the number of sperm incubated and reached a saturated level of 100% for spermatozoa from castrated supplemented epididymides. The saturated level was less than 50% when spermatozoa from non supplemented epididymides were used. Furthermore, the number of spermatozoa bound per ova was poor in these incubations: 2 or 3 per egg versus more than 10 per egg when spermatozoa from supplemented epididymides were incubated with rat ova.

FIGURE 1. Endocrine regulation of the zona binding structures acquired by the spermatozoa into the epididymis. (Heterologous test).

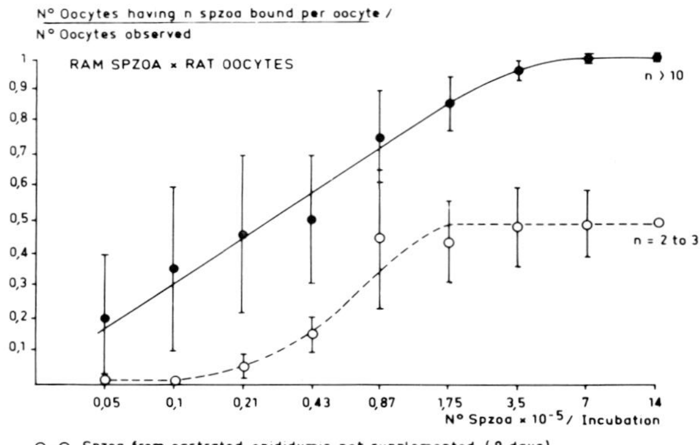


IN CONCLUSION, spermatozoa acquire in the epididymis zona binding structures which are independent of progressive motility. This process is influenced by endocrine activity of the testis and is a further example of the dependance of sperm epididymal maturation on androgens as decribed by ORGEBIN-CRIST (6).

REFERENCES

1. Bedford J.M. 1977. Anat. Rec. 188: 477-488.
2. Colas G, Dauzier L, Courot M, Ortavant R, Signoret J.P. 1968. Ann. Zootech. 17: 47-57.
3. Colas G. 1975. Ann. Biol. anim. Bioch. Biophys. 15: 317-327.
4. Fournier-Delpech S, Colas G, Courot M, Ortavant R, Brice G. 1979. Ann. Biol. anim. Bioch. Biophys. 19: 591-605.
5. Fournier-Delpech S, Courtens J.L., Pisselet Cl, Delaleu B, Courot M. 1982. Gamete Res. 5: 403-408.
6. Orgebin-Crist M.C., Danzo B.J., Davies J. 1975. Handbook of Physiology, sect 7, (D.W. Hamilton and R.O. Greep eds.) American Physiological society Bethesda. M.D. pp 319-338.
7. Orgebin-Crist M.C. and Fournier-Delpech S. 1982. Int. J. Androl. In press.
8. Saling M. 1982. Biol. Reprod. 26: 429-436.
9. Schmell E.D., Gulyas B.J. 1980. Biol. Reprod. 23: 1075-1085.
10. Toyoda Y, Chang M.C. 1974. Reprod. Fert. 36: 9-22.

ACKNOWLEDGEMENTS

We thank Dr MONTGOMERY for help in final reading of this manuscript.

# IS SPERM α-L-FUCOSIDASE RESPONSIBLE FOR SPERM-EGG BINDING IN *CIONA INTESTINALIS* ?

M.HOSHI[1,2], R. DE SANTIS[1], M.R. PINTO[1], F. COTELLI[1,3] and F. ROSATI[1,4]

[1]Stazione Zoologica di Napoli, Italy; [2]Department of Biology, Nagoya University, Japan; [3]Dipartimento di Biologia, Università di Milano, Italy, and [4]Istituto di Zoologia, Università di Siena, Italy.

## 1. INTRODUCTION

Species-specific and self-non self binding of spermatozoa to the vitelline coat is an obligatory step for fertilization in *Ciona intestinalis* (1,2). It is known that terminal fucose residues of the glycoproteins of the outer surface of the vitelline coat play the major role in the binding (3,4). However, it is yet to be proved that the counterpart on the sperm surface is a fucose-binding protein. This paper presents evidence that α-L-fucosidase on the sperm surface is responsible for the binding.

## 2. EXPERIMENTAL

2.1. *Animals and gametes*. Gametes of *Ciona intestinalis* from the Gulf of Naples, were collected by a previously described method (1). Spermatozoa of *Phallusia mammillata* were collected in a similar manner and assayed for glycosidase activities. Glycerol-treated eggs (1) whose vitelline coat maintains the specificity of binding properties, were used to assay the binding.

2.2. *Assay of sperm glycosidases in situ*. The activities of various glycosidases in the spermatozoon were fluorometrically assayed by using 4-methylumbelliferyl glycosides (4-MU-glycosides) as substrates (5). The substrates were placed in buffered sea water (total volume 0.5 ml), and incubated at 25°C. The reaction was stopped by adding 2.0 ml of 0.2M glycine-NaOH buffer, pH 10.4. After spinning down the cells the quantity of free 4-methylumbelliferone, was spectrofluorometrically determined (Ex 365 nm; Em 395 nm).

Table I summarizes the profile of glycosidase activities in the intact spermatozoon. In Ciona the activity of α-L-fucosidase was much higher than

Table I. Activities of glycosidases in ascidian spermatozoa.

| | Ciona | | Phallusia |
|---|---|---|---|
| | pH 3.9 | pH 8.2 | pH 3.9 |
| α-L-Fucosidase | 100⁺ | 2.0 | 9 |
| β-L-Fucosidase | N.D. | N.D. | N.D. |
| β-D-Fucosidase | 0.3 | 0.3 | N.D. |
| α-L-Arabinosidase | N.D. | 0.2 | 2 |
| α-D-Glucosidase | 2.0 | 0.1 | 2 |
| β-D-Glucosidase | 0.3 | 0.1 | 1 |
| α-D-Galactosidase | 0.2 | 0.1 | 1 |
| β-D-Galactosidase | 0.5 | 0.1 | 6 |
| β-D-N-Acetylglucosaminidase | 4.4 | 0.5 | 100 |
| β-D-N-Acetylgalactosaminidase | 8.2 | 0.4 | 32 |

Activities for each species are expressed as the percentage of the highest one at pH 3.9. - N.D. Not detectable.

any other glycosidases assayed.

In Phallusia β-D-N-acetylglucosaminidase exhibited the highest activity. The optimum pH of α-L-fucosidase in Ciona spermatozoa was 3.9 and the activity at pH 8.2 was 2% of the maximum activity.

Sperm glycosidases were solubilized by the addition of 0.1% Triton X-100 or by sonication.

2.3. Purification and characterization of α-L-fucosidase from Ciona spermatozoa. α-L-fucosidase of Ciona spermatozoa was solubilized by sonication. The enzyme was 5400-fold purified from the supernatant by centrifuging at $10^5$ x g using ammonium sulphate precipitation (25-50 % saturation) gel filtration on a BioGel A5M column and affinity chromatography on a column of Sepharose-ε-aminocaproylfucosylamine.

The optimum pH and Km values of the purified enzyme were the same as those in the intact spermatozoon. The purified enzyme was also competitively inhibited by 4-MU-β-L-fucoside.

2.4. Binding assay. Sperm binding to the glycerol-treated eggs was inhibited by millimolar concentration of p-nitrophenyl-α-L-fucoside, p--nitrophenyl-β-L-fucoside and 4-MU-α-L-fucoside (Table III).

Table II. Inhibition of sperm-binding to the vitelline coat by fucose and fucosides.

| Addition | % sperm bound |
|---|---|
| None | 40 |
| L-Fucose, 50 mM | 17 |
| pNP-α-L-Fuc, 2.5 mM | 12 |
| pNP-β-L-Fuc, 2.5 mM | 14 |
| pNP-β-D-Fuc, 2.5 mM | 44 |

Free L-fucose also inhibited binding (3) but only at much higher concentration.

Spermatozoa bound to the vitelline coat in normal sea water (pH 8.2) at 20°C, detached within 7 minutes of decreasing the pH to 5 or 4. At 0°C the binding was maintained, even if the pH was decreased.

## 3. CONCLUSIONS

The binding of spermatozoa to the vitelline coat is inhibited by synthetic substrates as well as by competitive inhibitors of sperm α-L-fucosidase. Detachment of spermatozoa from the vitelline coat at low pH is temperature-dependent. All these data suggest that sperm α-L-fucosidase works as a fucose-binding protein which is responsible for the sperm binding to the vitelline coat in normal sea water. It is suggested that α-L-fucosidase on the sperm surface binds to the sperm receptor on the vitelline coat upon formation of enzyme-substrate complex. Binding will be maintained for a rather long period, since the physicochemical conditions (particularly pH) do not favour hydrolysis. In <u>Phallusia</u>, β-D-N-acetyl-glucosaminidase is the most active sperm glycosidase, and N-acetylglucosamine has been indicated as the terminal sugar of the vitelline coat essential for sperm-egg binding (6).

Thus, glycosidase involvement in sperm-egg binding may be a general mechanism at least in the ascidians.

## REFERENCES

1. Rosati F & De Santis R. Exptl. Cell Res. 112, 111-119, 1978.
2. De Santis R, Jamunno G & Rosati F. Dev.Biol. 74, 490-499, 1980.

3. Rosati F & De Santis R. Nature 283, 762-764.
4. Pinto MR, De Santis R., D'Alessio G & Rosati F. Exptl.Cell Res. 132, 289-295.
5. Robinson D. Biochem.J. 63, 39-44, 1956.
6. Honneger TG. Exptl.Cell Res. 138, 446-451, 1982.

# SPERM/ZONA PELLUCIDA REACTIONS LEADING TO FERTILIZATION OF MOUSE EGGS IN VITRO.

H.M. FLORMAN[†], P.M. SALING[*], and B.T. STOREY[†]

[†]Depts. of Obstetrics & Gynecology and Physiology, School of Medicine, University of Pennsylvania, Philadelphia, PA U.S.A. and
[*]Dept. of Anatomy, School of Medicine, University of North Carolina, Chapel Hill, NC U.S.A.

## 1. INTRODUCTION

In the dance of sperm and egg which leads to fertilization in mammals, the two partners first meet head-to-zona. The subsequent sequence of steps, which appears applicable to hamster (7,8,13) and mouse (16), if not to guinea pig (11), is the following: binding of capacitated sperm to the zona; sperm acrosome reaction at the zona surface; sperm penetration of the zona; and gamete membrane fusion. It is the first two of these reactions in mouse gametes which we have been examining in our studies; this paper presents a brief account of our findings and interpretations.

## 2. EXPERIMENTAL SYSTEM

Mouse gametes were handled in modified Krebs-Ringer bicarbonate medium containing bovine serum albumin (BSA) at 20 mg/ml. This is a culture medium (CM) capable of capacitating mouse sperm and sustaining in vitro fertilization of mouse eggs (12,20). Inseminations were carried out in medium CM at 37° with sperm capacitated by preincubation in CM for 30-60 min. Full details of methods are given in refs. 3-5, 14-18.

## 3. THE REACTIONS

### 3.1 Sperm Binding to Zona

The binding of sperm to the zonae of cumulus-free, zona-intact egg was quantified by centrifugation through a step gradient of Dextran-containing CM: the "Stop-Fix" method (18). The reaction is rapid, having no lag phase and a half-time of 6 min post-insemination (p.i.) (4,18). It requires $Ca^{2+}$; $La^{3+}$ can substitute for $Ca^{2+}$, but not $Sr^{2+}$ (10,18). The binding occurs between a site on the intact plasma membrane of the sperm and the zona pellucida; sperm which have undergone the acrosome reaction do not bind to the zona (16). This site on the sperm plasma membrane becomes functional as the sperm mature in the epididymis (15). The zona pellucida site has been shown by Bleil and Wassarman (2) to be the zona glycoprotein ZP3. The reaction between the zona and sperm sites requires only simple collisional contact: sperm reversibly immobilized in the presence of $La^{3+}$ but absence of $Ca^{2+}$ bind normally, if the sperm/egg suspension is gently agitated to bring the gametes into contact (15).

Specific inhibitors of the enzyme trypsin inhibit the binding reaction if added at the time of insemination (14). Pre-treatment of eggs is without effect: a sperm plasma membrane site must interact with the inhibitors. If sperm binding is allowed to proceed for 20 min, when binding is nearly maximal, these compounds no longer affect binding. The $Ca^{2+}$-specific chelator EGTA reverses the binding reaction if added within the first 10 min of sperm/egg contact (18). But if EGTA addition is delayed 20 min after sperm/egg contact, the binding reaction becomes insensitive to EGTA (6). Taken together, these observations suggest: the first step of the binding reaction is largely electrostatic in nature, involving both $Ca^{2+}$ and positively charged residues on ZP3, analogous to the initial binding of substrates to trypsin; the second step involves formation of a covalent bond, analogous to the acylation step which follows the binding of substrates (1). Once the second step is complete after 20 min, trypsin inhibitors and EGTA are excluded from the binding site.

### 3.2 Acrosome Reaction at the Zona

3-Quinuclidinyl benzilate (QNB) at 50 μM inhibits _in vitro_ fertilization of cumulus-intact zona-intact and cumulus-free zona-intact eggs, but not of zona-free eggs (3). QNB is a specific inhibitor of zona penetration: its mode of inhibition is blocking the acrosome reaction in sperm bound to the zona. That the zona is the inducer of this reaction is shown by occurrence of the acrosome reaction in sperm bound to isolated zonae or treated with solubilized zona protein; in both cases QNB inhibits the acrosome reaction completely (4). Wassarman and Bleil (19) have implicated the sperm receptor zona protein ZP3 as also being the specific acrosome reaction inducer. QNB does not affect the zona; it binds to specific sites on the sperm plasma membrane (5). The acrosome reaction induced by solubilized zona protein and that induced by isolated intact zonae or zona-intact eggs differ in one respect: the former proceeds on addition of zona protein with no lag; the latter proceeds only after a lag period of 30 min p.i.

QNB loses its potency as an inhibitor of mouse _in vitro_ fertilization if added at increasing times p.i. (6). Inhibition is complete if QNB is added up to 20 min p.i. At longer times, QNB becomes less effective; at 60 min p.i., it is without effect on fertilization. The onset of insensitivity to QNB appears to be due to occlusion of the sperm binding site for QNB, as shown by the following experiment (6). If insemination is allowed to proceed for 30 min in CM free of QNB, then the transfer of eggs with bound sperm to sperm-free CM containing 50 μM QNB has no effect on the percentage fertilization, compared to QNB-free controls. But if insemination proceeds for 30 min in CM-QNB, fertilization is strongly inhibited, even after transfer of eggs with bound sperm to QNB-free medium. The half-time of dissociation of QNB from unbound sperm in CM is 22 min (3), yet inhibition by QNB persists in QNB-free CM over the subsequent 330 min required for maximal fertilization in this system (4). This suggests that a reaction occurs between the sperm bound to the zona and zona protein(s) which occludes the sperm QNB binding site, such that bound QNB cannot dissociate from it and free QNB cannot associate with it. As this reaction proceeds, _in vitro_ fertilization becomes increasingly less sensitive to inhibition by QNB added at longer times p.i. The time course of the onset of insensitivity to QNB inhibition of fertilization is well resolved from the time course of the acrosome reaction in sperm bound to the zona pellucida. Insensitivity to QNB has become complete at 60 min, while the acrosome reaction requires 240 min p.i. to reach its maximal extent (4).

### 3.3 Time Course of the Reactions

The four reactions identified above between sperm and zona, which lead to zona penetration and hence to fertilization of mouse eggs _in vitro_, form a reaction sequence plot (Fig. 1A). Shown for comparison on a longer time scale (Fig. 1B) is the reaction sequence plot of the acrosome reaction and fertilization. The first and second stages of the binding reaction are essentially complete before the third reaction at the sperm QNB site begins. The time course of the acrosome reaction, both with regard to time lag before onset and rate after onset, is that expected for a reaction stemming directly from the third reaction as precursor. This reaction between zona pellucida and sperm plasma membrane, which occludes the sperm QNB site, is therefore postulated to be the physiological "trigger" for the fourth reaction, the acrosome reaction occurring at the zona.

## 4. ENVOI

Time course and inhibitor studies have identified four reactions, each one a precursor to the next, which comprise the early reactions in fertilization of mouse eggs. The mechanisms of these reactions can now be approached within the constraints of these kinetics. As the outline of the steps in the dance of sperm and egg begin to be perceived, so also do their intricacies become better appreciated.

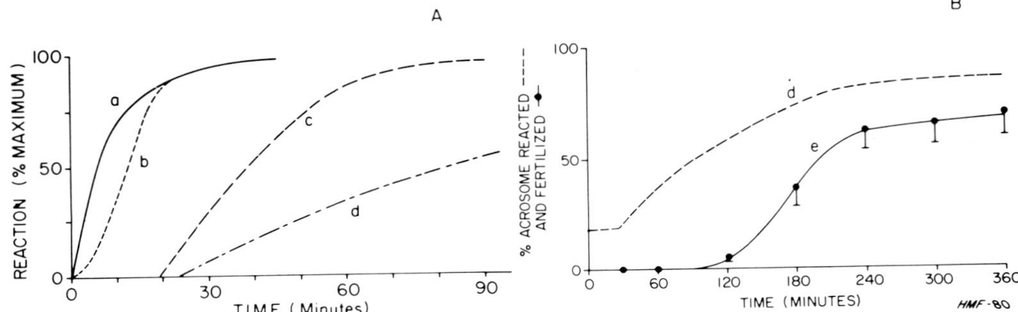


FIGURE 1. Reaction sequence plots for the four sperm/zona reactions leading to zona penetration (A) and for the acrosome reaction and fertilization (B). The reactions in (A) are: a: EGTA-reversible, $Ca^{2+}$-mediated binding of sperm to zonae; b: onset of insensitivity to EGTA reversal of binding; c: onset of insensitivity to QNB inhibition of fertilization; d: onset of acrosome reaction in sperm bound to zonae. Reaction d has been corrected for sperm which score as partially acrosome-reacted but bind to the zona from earliest time p.i. All reactions in (A) are normalized to maximal extent of reaction. In (B), reaction d has not been normalized to maximal extent nor corrected for partially acrosome-reacted sperm, for comparison with (A). Note the different time scales in (A) and (B): The acrosome reaction d is only 60% complete on the time scale in (A). Reaction e is the fertilization of cumulus-free intact eggs under the same conditions as for the other reactions. The criterion for fertilization was the presence of a sperm nucleus, either decondensed or in pronuclear phase, and tail within the ooplasm.

REFERENCES


1. Bender, ML, Kezdy, FJ. 1965. Annu. Rev. Biochem. 34: 49-76.
2. Bleil, JD, Wassarman, PM. 1980. Develop. Biol. 78: 185-202.
3. Florman, HM, Storey, BT. 1981. J. Exp. Zool. 216: 159-167.
4. Florman, HM, Storey, BT. 1982. Develop. Biol. 91: 121-130.
5. Florman, HM, Storey, BT. 1982. J. Andrology 3: 157-164.
6. Florman, HM, Saling, PM, Storey, BT. 1982. J. Andrology (in press).
7. Gwatkin, RBL. 1976. Fertilization Mechanisms in Man and Mammals. New York: Plenum Press.
8. Hall, MV, Franklin, LE. 1981. J. Androl. 2: 14.
9. Heffner, LJ, Storey, BT. 1982. J. Exp. Zool. 219: 155-161.
10. Heffner, LJ, Saling, PM, Storey, BT. 1980. J. Exp. Zool. 212: 53-59.
11. Huang, TTF, Fleming, AD, Yanagimachi, R. 1981. J. Exp. Zool. 217: 287-290.
12. Inoue, M, Wolf, DP. 1975. Biol. Reprod. 13: 340-346.
13. Phillips, DM, Shalgi, RM. 1980. J. Exp. Zool. 213: 1-8.
14. Saling, PM. 1981. Proc. Natl. Acad. Sci. U.S.A. 78: 6231-6235.
15. Saling, PM. 1982. Biol. Reprod. 26: 429-436.
16. Saling, PM, Storey, BT. 1979. J. Cell Biol. 83: 544-555.
17. Saling, PM, Sowinski, J, Storey, BT. 1979. J. Exp. Zool. 209: 229-238.
18. Saling, PM, Storey, BT, Wolf, DP. 1978. Develop. Biol. 65: 515-525.
19. Wassarman, PM, Bleil, JD. 1982. In: Cellular Recognition (Glaser, L, Gottlieb, D, Frazier, W, eds.) New York: Alan R. Liss, in press.
20. Wolf, DP, Inoue, M. 1976. J. Exp. Zool. 196: 27-37.

# A FREEZE-FRACTURE STUDY ON EPIDIDYMAL AND EJACULATED SPERMATOZOA OF *MACACA FASCICULARIS*

J.F. REGER, M.A. FAIN-MAUREL and J.P. DADOUNE[+].
Laboratoire de Biologie Cellulaire et Laboratoire d'Histologie, Université René Descartes, 75270 Paris, France.

## 1. INTRODUCTION

Observations on the distribution and organization of particles in spermatozoa from the epididymis have been made on the opposum, *Didelphis virginiana* (4), the rat (6) the boar (7) and the brush-tailed opposum (8). To complement these observations the following study was made on a primate, *Macaca fascicularis*.

## 2. MATERIAL AND METHODS

The epididymis and ejaculates taken from several animals were studied. Tissue and ejaculates were fixed, washed and equilibrated in 30% glycerol. Replicas were obtained in a cryofract 190 (Reichert-Jung).

## 3. OBSERVATIONS *(for details see figure explanations)*

Spermatozoa from four epididymal regions and from the ejaculate always present the same particle organization for a given sample. They contain homogeneously distributed, 7-8 nm size, plasmalemmal, PF-face particles throughout the head, mid-piece and principal-piece. The smallest number of particles are present in the post-acrosomal region. A unique, particle arrangement occurs in the plasmalemmal, PF-face in spermatozoa from proximal levels of the body of the epididymis only. In this region, domains of 4-6 nm size, PF-face, plasmalemmal particles arranged in square arrays are visible. Ejaculate spermatozoa exhibit similar particle organization to that of spermatozoa found in the distal portions of

+ *The authors wish to express their appreciation to M. Jacques Escaig for the use of the cryofract-190 (Reichert-Jung) in his laboratory.*

the epididymis. In terms of particle organization a second type of spermatozoon was found in a seperate animal. This specialization consists of aligned groups of 7-9 nm size particles originally termed "cords" (Koehler, 1970). The "cords" are situated just anterior to the striated ring of the post-acrosomal region.

## 4. DISCUSSION

In terms of the transitory particle domains in the peripheral head level only, and only in spermatozoa situated in the proximal body of the epididymis, the results here mimic those of the rat (6). In the boar (7), however, similar particle domains were seen in spermatozoa at much lower epididymal levels. In all three species the particle domains are transitory. The presence of such arrays so selectively distributed raises interesting functional questions. The subsequent interaction of the plasmalemma overlying the acrosome, as well as the attainement of fertilization capacity in the epididymis is well known. It may be that the timing and the localization of these domains reflect presumptive plasmalemmal-acrosomal interaction, or perhaps the particle arrays are precursor particles of the plasmalemmal PF-face particles ultimately seen in mature sperm. Our observations on the ejaculates of several animals demonstrate two, different particle patterns in the posterior segment of the post-acrosomal region. It is difficult to assess the significance of this difference. It may reflect either an age, or a species difference, or another unknown parameter. "Cords", originally described in rabbit (2), have also been seen in other mammalian spermatozoa (1, 3, 5). Pedersen (5) has commented on certain other post-acrosomal differences observed in thin sections of human and monkey (*Macaca arctoides*) spermatozoa. He interpreted the differences as a reflection of species variation. The differences observed here, in particulate composition of the post-acrosomal region in the monkey may reflect intra species variation.

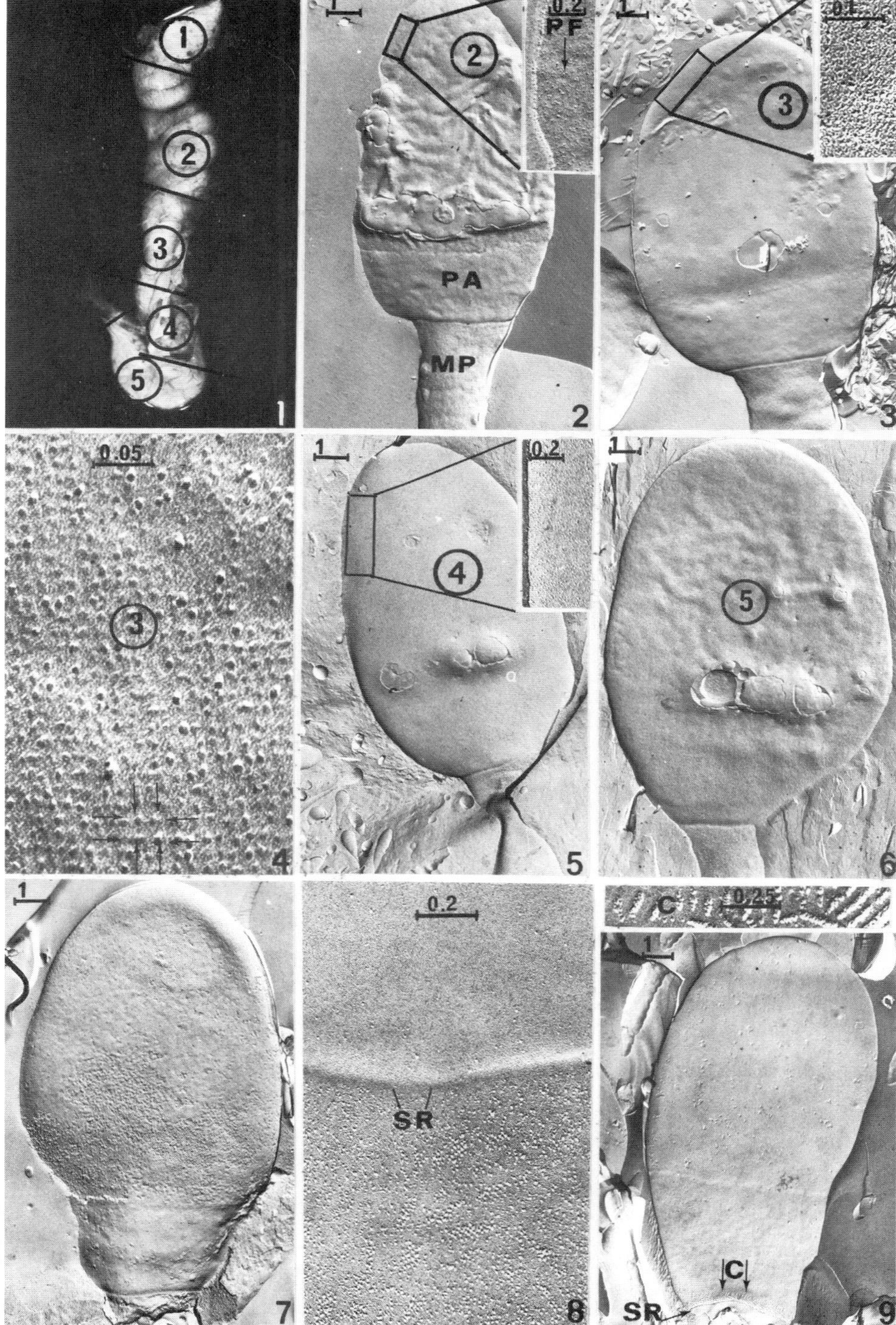
1
2
3
4
5
1
PF
0.2
PA
MP
0.1
0.05
0.2
0.2
SR
C
0.25
SR
C
6
7
8
9

## REFERENCES


1. Friend D.S. and Fawcett D.W. 1974. Membrane differentiations in freeze-fractured mammalian sperm. J. Cell Biol., 63, 641-664.
2. Koehler J.K. 1970. A freeze-etching study of rabbit spermatozoa with particular reference to head structures. J. Ultrastruct. Res., 33, 598-614.
3. Koehler J.K. 1972. Human head ultrastructure; a freeze-etching study. J. Ultrastruct. Res., 39, 520-539.
4. Olson G.E. 1980. Changes in intramembranous particle distribution in the plasma membrane of *Didelphis virginiana* spermatozoa during maturation in the epididymis. Anat. Rec., 197, 471-488.
5. Pedersen H. 1972. The post acrosomal region of the spermatozoa of man and *Macaca arctoides*. J. Ultrastruct. Res., 40, 366-377.
6. Suzuki F. and Nagano T. 1980. Epididymal maturation of rat spermatozoa studied by thin sectioning and freeze-fracture. Biol. Repr., 22, 1219-1231.
7. Suzuki F. 1981. Changes in intramembranous particle distribution in epididymal spermatozoa of the boar. Anat. Rec., 199, 361-376.
8. Temple-Smith P.D. 1981. Redistribution of intramembranous particles in the mid-piece plasma membrane of the brush-tailed possum spermatozoon during epididymal transit. J. Anat. (Brit.), 55, p. 301.


## FIGURE EXPLANATIONS (scale is in microns)

Fig. 1. Epididymis. Numbered regions denote areas chosen for spermatozoon observation.

Fig. 2. Spermatozoon (head) (2, diagram). PF-face particles (PF). Post acrosomal area (PA). Mid-piece (MP).

Fig. 3. Spermatozoon (proximal body) (3, diagram).

Fig. 4. Spermatozoon (proximal body (3, diagram). Arrows outline square order domains.

Fig. 5. Spermatozoon (distal body) (4, diagram). Note lack of square array particles (Inset).

Fig. 6. Spermatozoon (proximal tail) (5, diagram).

Fig. 7. Spermatozoon (ejaculate, experimental animal).

Fig. 8. Spermatozoon (ejaculate; experimental animal). Enlarged area of post-acrosomal area. Note lack of cords just anterior to striated ring (SR). Compare with Fig. 9 (Inset).

Fig. 9. Spermatozoon (ejaculate, non experimental animal). Note cords (C) distal to striated ring (SR).

# SURFACE OBSERVATIONS ON SOME MAMMALIAN SPERMATOZOA AFTER RAPID-FREEZING AND DEEP-ETCHING.

Y. TOYAMA, T. NAGANO and F. SUZUKI

Department of Anatomy, School of Medicine,
Chiba University, Chiba 280, Japan

Specialized structures of the sperm membranes have been studied by surface replica and freeze-fracture (1-5). Domains of the sperm plasma membrane are also reported (2, 5). We attempted to reveal surface membrane structures of the boar and rabbit spermatozoa by deep-etching and rotary shadowing, and obtained new findings that gathered particles of 15 nm in diameter with subunits appear on the ES face of the postacrosomal region of the boar sperm.

PROCEDURE: Fresh semen and epididymal contents of the boar and rabbit were frozen either in a metal contact device with liquid nitrogen or by dipping into liquid propane directly. After being fractured and etched, replicas were made in the FD-2 type machine (Eiko Engineering Co., Japan). Observations were focused on the postacrosomal region in this study.

RESULTS AND DISCUSSION: Near the posterior end of the postacrosomal region, the "cords" consisting of aggregations of particles are recognized. They are identical with those in the fixed materials (4). It has been reported that the postacrosomal sheath shows regional differentiation and substructures underneath the plasma membrane (6). On the PF face of the membrane of this region, striations are found to run along the axis of the spermatozoon (Fig. 1). The periodicity of the stripes is about 15 nm. This feature may correspond to the striations in the same region of the fixed preparations (4). On the EF face in this region, faint stripes with 15 nm periodicity are also observed. In deep-etched preparations, part of the PF face of the plasma membrane had been torn off forming "windows". Through the window, a periodic regular structure can be found

(Fig. 2). It consists of parallel rows of particles of 6 nm in diameter. The orientation of the rows coincides with that of the striations on the PF face in this region. This structure appears to represent the en face view of the postacrosomal sheath lining the plasma membrane reported by Pedersen (6).

Some areas of the ES face of the region of the boar spermatozoon are occupied by closely packed particles. Each particle is 15 nm in diameter and consists of about six subunits. The particles show a patch distribution in this area and are arranged in hexagonal pattern (Figs. 2,3). It seems that the particle-free regions between the patches are much wider than in similar regions of the guinea pig spermatozoon, whose acrosomal membrane shows prominent domain structures (2,5). Although the functional significance of the 15 nm particles in the boar spermatozoon is not known, the particles resemble to those in the urinary bladder epithelium (7,8). The particles consisting of subunits have not been found in the rabbit spermatozoon so far studied.

REFERENCES:

1. Phillips, D.M. 1977. Surface of the equatorial segment of the mammalian acrosome. J. Ultrastr. Res. 16:128-137.
2. Friend, D.S. and Fawcett, D.W. 1974. Membrane differentiation in freeze-fractured mammalian sperm. J. Cell Biol. 63:641-664.
3. Koehler, J.K. 1978. The mammalian sperm surface: Studies with specific labeling techniques. Int. Rev. Cytol. 54:73-108.
4. Suzuki, F. 1981. Changes in intramembranous particle distribution in epididymal spermatozoa of the boar. Anat. Rec. 199:361-376.
5. Friend, D.S. 1982. Plasma-membrane diversity in a highly polarized cell. J. Cell Biol. 93:243-249.
6. Pedersen, H. 1972. The postacrosomal region of the spermatozoa of man and Macaca arctoides. J. Ultrastr. Res. 40:366-377.
7. Staehelin, L.A., Chlapowski, F.J. and Bonneville, M.A. 1972. Luminal plasma membrane of the urinary bladder. I. J. Cell Biol. 53:73-91.
8. Chlapowski, F.J., Bonneville, M.A. and Staehelin, L.A. 1972. Luminal plasma membrane of the urinary bladder. II. J. Cell Biol. 53:92-104.

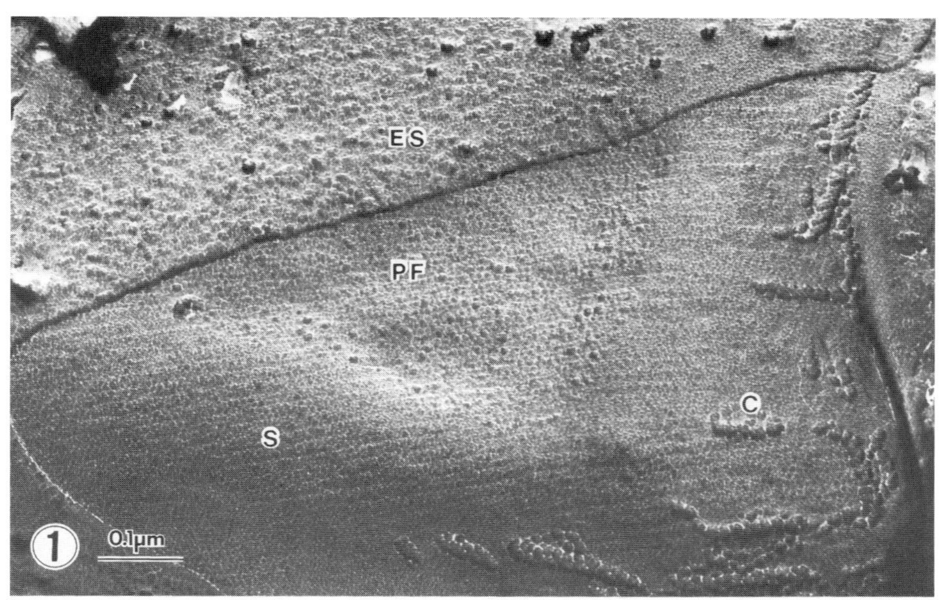
ES
PF
C
S
1
0.1µm

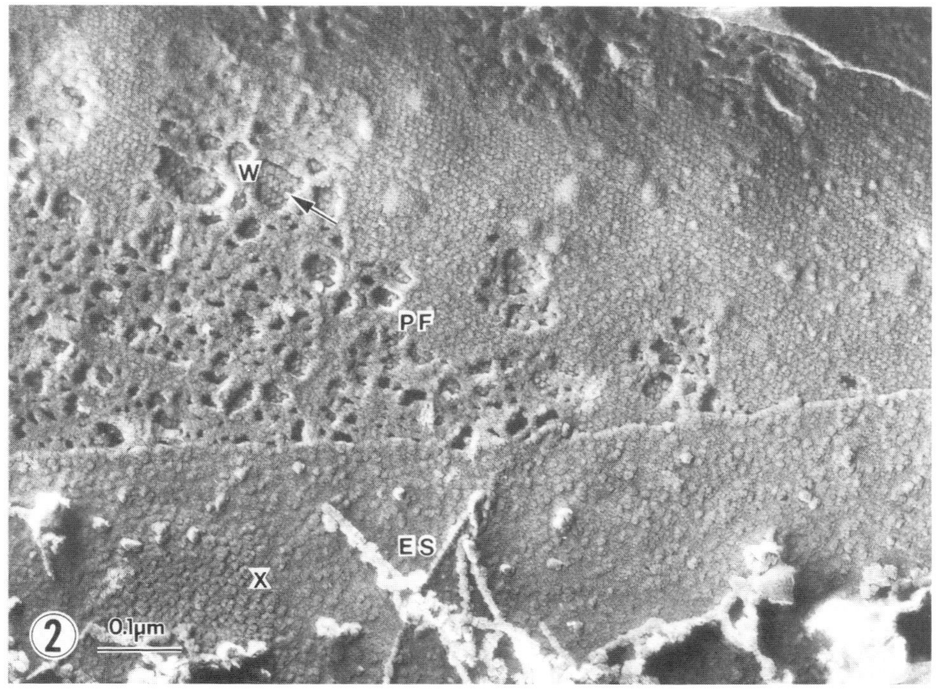
W
PF
ES
X
2
0.1µm

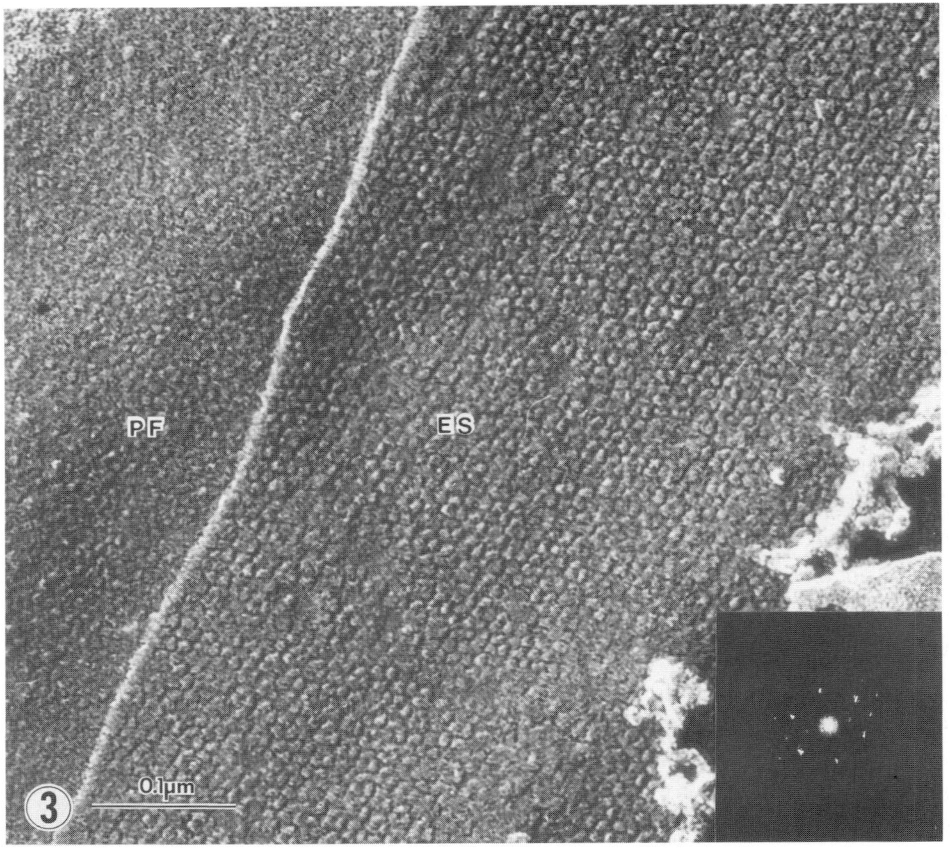


EXPLANATIONS OF FIGURES:

Fig. 1. Showing the PF (PF) and ES (ES) faces of the postacrosomal region of the boar spermatozoon. The cords (C) are seen as linear aggregations of the particles. The striations (S) are found in this region.

Fig. 2. Showing the PF (PF) and ES (ES) faces of the boar postacrosomal region. On the PF face, the striations are evident. The windows(W), produced by tearing off the membrane during the preparation processes, can be seen. Through the window, the striations appear to consist of rows of particles (arrow). On the ES face, a patch of the gathered particles of 15 nm in diameter is found (X).

Fig. 3. High magnification of the particles on the both faces (ES and PF) of the postacrosomal region of the boar. Each particle on the ES is 15 nm in diameter with a central pit and consists of subunits. On the PF face, the particles are smaller than those on the ES and may indicate that the particles are dumbell in shape. The optical diffraction pattern indicates a hexagonal arrangement of the particles on the ES (inset).

# A LOCALIZED SURFACE ANTIGEN OF GUINEA PIG SPERM IS PRECIPITATED BY MONOCLONAL ANTIBODIES TO AN APPARENTLY "INTERNAL", HEAT-STABLE COMPONENT

Paul PRIMAKOFF and Diana G. MYLES, Department of Physiology, University of Connecticut, Farmington, CT. USA

## INTRODUCTION

Using monoclonal antibodies (Mabs), we have constructed a map of the surface of guinea pig sperm (Primakoff and Myles, in preparation). This map establishes that (1) there are at least five distinct regions on the sperm surface in which particular antigens are localized and (2) several antigens can be localized in a single region. One region of antigen localization is the anterior head (AH) surface which lies over the acrosome and equatorial segment. Using immunoprecipitation, we have identified five distinct antigens that are localized on the AH surface and are recognized by Mabs.

One of the antigens found on the anterior head is comprised of three polypeptides of molecular weight 62,000 (62K), 52,000 (52K), and 38,000 (38K), (see below). Fifteen of the 27 anterior head-specific antibodies that we generated recognize this antigen (Primakoff and Myles, in preparation). In the present paper, we show that some of the 15 antibodies that precipitate this localized surface antigen apparently bind to an antigenic determinant found inside the cell.

## MATERIALS AND METHODS

The hybridoma production, radioactive solid phase assay, immunofluorescence, surface labeling and gel electrophoresis procedures have been described (1). Names are assigned to the antibodies according to the surface region to which they bind followed by a number. Antibodies numbered 1-100 are from cloned lines and antibodies numbered 1001-1100 are from uncloned lines. Thus, for example, AH-20 is an antibody from a cloned line that binds exclusively to the anterior head region. Antibodies from cloned lines are also given affectionate names for everyday use in the lab. AH-20 is Henry, AH-21 is Harvey, AH-30 is Eraserhead, and so on.

## RESULTS

The 15 antibodies that recognize the 62,52,38 K antigen can be divided into two groups. Ten of the antibodies recognizing this antigen are in Group 3

(AH-20,-21,-1022, -1023,-1024,-1025,-1026,-1027,-1028,-1029) and five in Group 3a (AH-30,-31,-1032, -1033,-1034)(Primakoff and Myles, in preparation). Several important differences have thus far been found between Group 3 and Group 3a. The most critical difference is that Group 3 antibodies bind to live sperm whereas Group 3a antibodies do not. Group 3a antibodies can bind to sperm only if they are first treated with saponin, frozen and thawed once, or fixed for 20 minutes in 1% (or higher) formaldehyde. The determinants to which Group 3a antibodies bind are apparently inaccessible in the intact, live cell. This inaccessibility is not a trivial consequence of a large size of the 3a antibody molecules, since all five Group 3a (as well as most of the Group 3 antibodies) are of the IgG class (Table 2). The other differences between Group 3 and Group 3a antibodies are compiled in Table 1.

Table 1

| | Group 3 | Group 3a |
|---|---|---|
| (1) | bind to live sperm | do not bind to live sperm; bind to treated sperm |
| (2) | smooth fluorescent pattern on the AH (Fig. 1A, B) | uneven fluorescent pattern on the AH (Fig. 1C, D) |
| (3) | precipitate less of the 3 surface polypeptides 62,52,38K (Fig. 2A) | precipitate more of the 3 surface polypeptides 62,52,38K (Fig. 2B) |
| (4) | bind to heat-labile antigenic determinant (cf. Table 2) | bind to heat-stable antigenic determinant (cf. Table 2). |

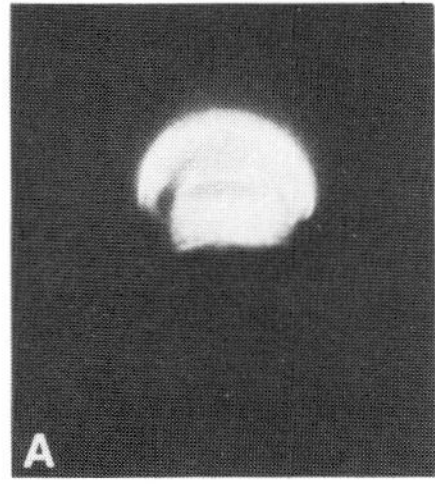

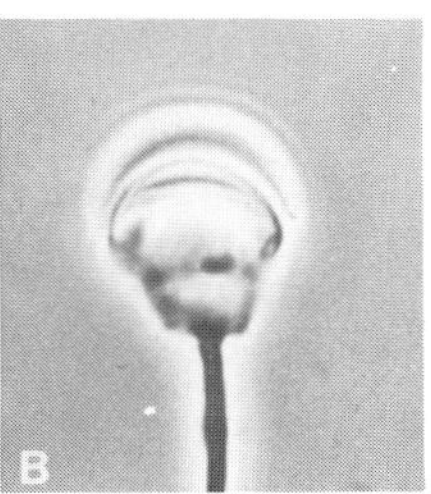

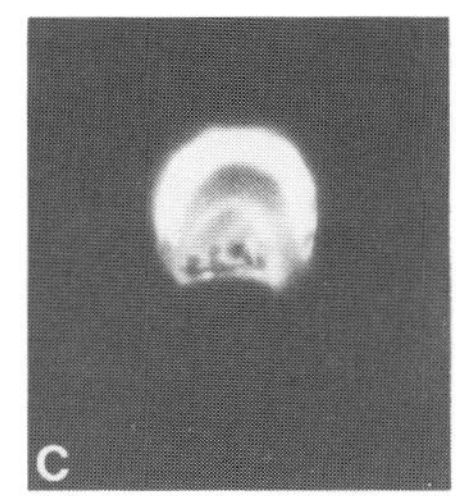

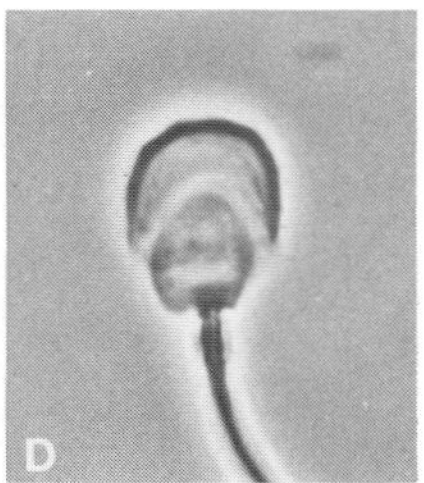


FIGURE 1. Indirect immunofluorescence of AH-21 and AH-30. (A) Fluorescence of AH-21 on live cells, post-fixed in 1% formaldehyde; (B) same cell, phase contrast; (C) Fluorescence of AH-30 on cells, pre-fixed in 1.5% formaldehyde; (D) same cell, phase contrast.

Table 2

Class and Subclass of Antibody and Heat Lability of Antigenic Determinants

| Group 3 | antibody class and subclass † | cpm bound boiled/untreated |
|---|---|---|
| AH-20 | IgG2b | 0/1127 |
| AH-21 | IgG2b | 0/1376 |
| AH-1022 | IgG1 | 233/1468 |
| AH-1023 | IgG1,IgM * | 102/1710 |
| AH-1024 | n.d. | 31/1348 |
| AH-1025 | IgG2a | 96/1562 |
| AH-1026 | n.d. | 0/1492 |
| AH-1027 | IgM | 3/902 |
| AH-1028 | IgG1 | 64/950 |
| AH-1029 | IgG1,IgM * | 6/350 |
| Group 3a | | |
| AH-30 | IgG1 | 2132/2434 |
| AH-31 | IgG1 | 2667/3536 |
| AH-1032 | IgG1 | 2462/2558 |
| AH-1033 | IgG1 | 1789/2052 |
| AH-1034 | IgG2b | 680/818 |

† Determined using class and subclass specific anti-mouse Ig antisera on Ouchterlony plates.

* These two uncloned lines are apparently producing two classes of antibodies.

n.d. Not determined.

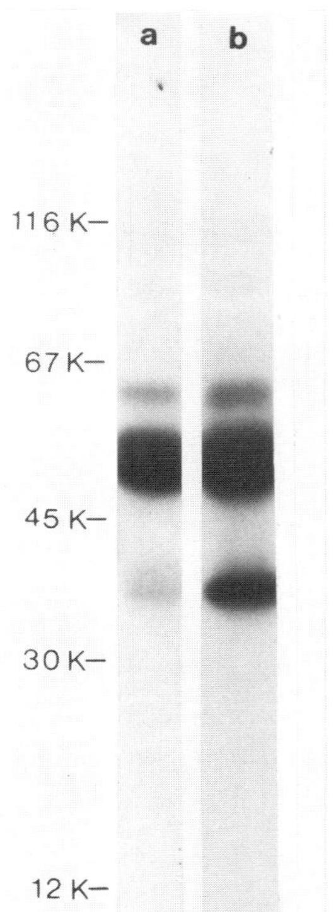


FIGURE 2. Autoradiograph of immunoprecipitate obtained with Group 3 and Group 3a antibodies. AH-1022, Group 3, 28 day exposure, lane a; AH-31, Group 3a, 8 day exposure, lane b.

## DISCUSSION

A key observation in this study is that 3a antibodies do not bind to live sperm. After saponin or freeze-thaw treatment, or formaldehyde fixation of the cells, Group 3a antibodies can bind. Such treatments may result in the cells being permeabilized, allowing antibodies to bind to an internal component. Formaldehyde fixation of other cell types results in access of antibodies to internal components (2). Thus, an important conclusion concerning widely used fixation techniques for sperm (1% or higher formaldehyde) is that they appear to permeabilize cauda epididymal sperm and allow antibodies to reach components inside the cell.

There are several possible explanations for the properties of the two antibody Groups. One simple explanation is that in treated cells 3a antibodies are binding to the surface antigen identified by Group 3 antibodies. This suggests that the surface antigen may be a transmembrane protein and that 3a antibodies bind to a cytoplasmic portion of this protein. Alternatively, they could bind to a cytoplasmic structure (cytoskeletal?) to which this transmembrane protein is itself tightly bound. For some unknown reason, the cytoplasmic antigenic determinant is heat-stable.

One alternative explanation is that even in treated cells, the recognized antigenic determinants on the 62, 52, 38K surface antigen are not accessible to Group 3a antibodies. These determinants only become accessible during the immunoprecipitation procedure when the cells are completely lysed with 1% Triton, allowing Group 3a antibodies to precipitate the 62, 52, 38K polypeptides. In this alternative, Group 3a antibodies in treated cells show their fluorescent pattern by binding to an internal sperm molecule. This internal molecule would fortuitously share an antigenic determinant with the surface antigen and would also fortuitously be located in the anterior head. Such a molecule could be, for example, in the outer acrosomal membrane. This interpretation could explain the heat stability of the Group 3a antigenic determinants assuming they reside in a carbohydrate moiety existing in common on molecules in the two membranes.

The feasibility of isolating the recognized antigen(s) when a Mab is available should allow discrimination between the different interpretations.

This work was supported by NIH grant HD 16580. P.P. holds an American Heart Association Established Investigatorship Award.

REFS: 1. Myles, D.G., Primakoff, P. and Bellve, A.R. 1981. Cell 23, 433-439.
2. Satake, M., McMillan, P.N. and Luftig, R.B. 1981. PNAS 78, 6266-6270.

# ESTABLISHMENT OF SPERM SURFACE TOPOGRAPHY DURING DEVELOPMENT ANALYZED WITH MONOCLONAL ANTIBODIES

Diana G. MYLES and Paul PRIMAKOFF. Department of Physiology, University of Connecticut Health Center, Farmington, CT., USA

## INTRODUCTION

The establishment of sperm surface topography may occur at different times in sperm development. For instance, it might be superimposed on the shaping of spherical spermatid cells into asymmetric sperm during spermiogenesis or the topography may change after the sperm leaves the testis, during passage through the epididymis. Establishment of final topography could result from surface molecules being directed to a particular region as they appear on the surface. Alternatively, molecules may originally appear on the surface in a pattern that is altered during subsequent stages of maturation.

Using seven monoclonal antibodies that are directed against different antigens (Primakoff and Myles, in preparation), we have asked when the antigens first appear on the surface and if the localization pattern is altered during subsequent stages of sperm maturation.

## MATERIALS AND METHODS

Monoclonal antibodies were produced as previously described (1). Testicular sperm and round spermatids were obtained by chopping testes of Hartley guinea pigs in the medium of Han and Tung (2) with ganged razor blades. Live cells were used and kept at 4$^{o}$C during staining. Sperm were obtained from the cauda epididymis and stained as previously described (1) using a F(ab')$_2$- or Fab-FITC antimouse antibody (Cappel Laboratories). Micrographs were taken with a Zeiss Universal Microscope using a 63x Planapo lens, N.A. 1.4.

## RESULTS

### Localized antibody binding patterns on cauda epididymal sperm

All of the seven monoclonal antibodies in this study bind in one of four specific patterns on the cell surface of sperm from the cauda epididymis with 94-100% of the sperm showing the same binding pattern with each antibody. These four binding patterns are defined as follows: (1) AH-40, AH-21 bind to the anterior head - the region of the cell surface over the acrosome including the equatorial segment, (2) PH-10, PH-21 bind to the posterior head - the surface region posterior to the anterior head region up to the beginning of the tail, (3) WH-1, WH-30 bind to the whole head, and (4) PT-1 binds to the posterior tail, including all of the principal piece and end piece.

Timing of appearance of surface antigenic determinants

Three different stages of sperm development were tested to determine when the antigen was first detectable on the surface of the cells. The presence of antigen on the surface was determined by indirect immunofluorescence on live, intact cells. The stages were as follows: (1) Round spermatids, including all spermatids where a clearly defined acrosome could be observed in association with the nucleus and prior to stages when the spermatid has begun to change shape; (2) testicular sperm, obtained from the testes by chopping testicular tissue; and (3) cauda epididymal sperm, including sperm from approximately epididymal segments IV through VI (3).

Of the seven antibodies examined two, PH-20 and PT-1, bound to round spermatids in addition to the other two stages tested (Table 1). Three additional antibodies, WH-1, WH-30, and AH-21, bound first to testicular sperm. Two of the antibodies, AH-40 and PH-10, bound only to sperm taken from the cauda.

Table 1. Appearance of Surface Antigenic Determinants During Development

| Antibody | Polypeptides in Immunoprecipitate $M_R \times 10^{-3}$ | Round Spermatids | Testicular Sperm | Cauda Epididymal Sperm |
|---|---|---|---|---|
| AH-40 | 70,62,46,25,18 | - | - | AH |
| AH-21 | 62,52,38 | - | WH - AH | AH |
| PH-10 | 58,48 | - | - | PH |
| PH-20 | 66,48,41 | + | WH+ | PH |
| WH-1 | 42 | - | WH | WH |
| WH-30 | 89,45 | - | WH | WH |
| PT-1 | none | + | PT | PT |

Changes in surface antigen localization between testicular and cauda epididymal sperm

The antibody-binding patterns on cauda sperm were compared to patterns on testicular sperm to see what changes occurred in antigen surface localization between these two developmental stages (during epididymal passage). Figure 1 shows the observed changes in diagrammatic form. In the first case no change occurs between the binding patterns on testicular and cauda sperm; this is shown with WH-1 (Fig. 2), WH-30, and PT-1. In the second type the antibody binding first appears during epididymal passage as exhibited by AH-40 (Fig. 3) and PH-10. In the third case, there is an alteration in the binding pattern between testicular and cauda sperm. This is observed for PH-20 (Fig. 4) and AH-21 where the binding becomes restricted during the transition between these two developmental stages.

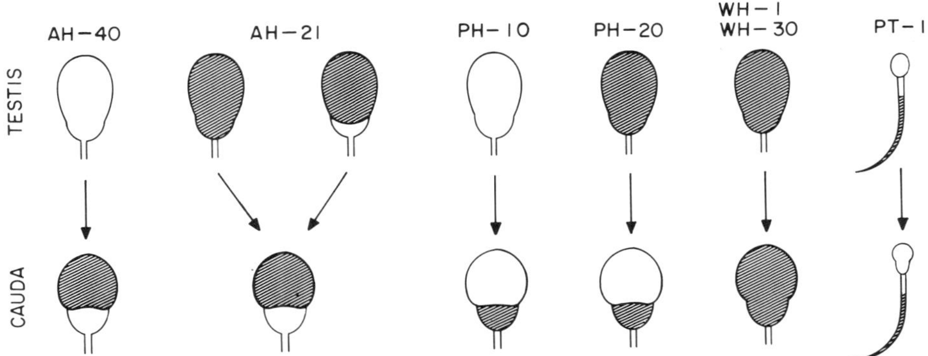


FIGURE 1. Changes in antibody binding patterns between testicular and cauda sperm.

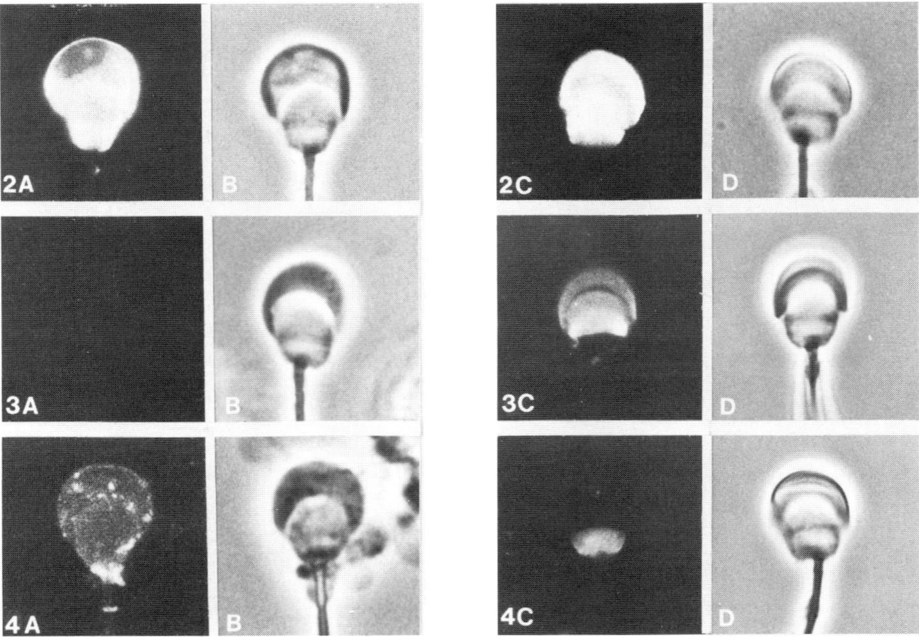


FIGURES 2-4. Antibody-binding patterns using indirect immunofluorescence. Testicular sperm: Fluorescence (A) and same sperm with phase contrast (B). Cauda epididymal sperm: Fluorescence (C) and phase contrast (D). FIG. 2, WH-1; FIG. 3, AH-40; FIG. 4. PH-20.

Antibody binding to washed cauda epididymal sperm

Cauda epididymal sperm (7 x $10^7$) were washed three times in 200 ml $Mg^{2+}$-Hepes medium (1) to determine if any "loosely associated" antigens would be removed. All seven antibodies bound to washed sperm and the binding patterns were not altered.

DISCUSSION

The establishment of surface domains must result from each localized molecule eventually being directed to a particular surface region. Our results indicate that the recruitment of antigens into domains is an ongoing process in development and that different surface molecules are separated into the distinct domains found on cauda epididymal sperm at different times. This can occur during the formation of testicular sperm (WH-1, WH-30, PT-1), or during epididymal passage (alteration of PH-20 and appearance of PH-10 and AH-40). Potentially a third timing of localization, during testicular sperm development in the testis, may be demonstrated by AH-21 that shows heterogeneous binding on testicular sperm.

Alteration in surface patterns during development could result from different phenomena. Those antigens that first appear in one pattern and then show an altered distribution (AH-21, PH-20) may be actually moving in the plane of the membrane, or they may be modified or covered up in one particular region. Whatever the process is, it must be selective since the binding of the whole head antibodies WH-1 and WH-30 is not altered in the anterior head region during epididymal passage, while binding of PH-20 changes from a whole head to a posterior head binding pattern.

The appearance of new antibody binding sites during epididymal passage may be due to uncovering, modification, or new insertion of antigens either from inside the cell or from epididymal fluid. Simple washing experiments demonstrate that these antigens are not easily removed, as can be shown in some cases of sperm surface molecules (4). Therefore, if the testicular to cauda alterations that we observed here result from the addition of molecules on the surface from epididymal fluid (see Ref. 5 for evidence in the mouse and for review), the added molecules must adhere tightly to integral membrane components or actually be inserted into the lipid bilayer.

This work was supported by NIH grant HD 16580 and an AHA Investigatorship to P.P.

REFERENCES

1. Myles, D.G., Primakoff, P. and Bellve, A.R. 1981. Cell 23, 433-439.
2. Han, L.-P.B. and Tung, K.S.K. 1979. Biol. Reprod. 21, 99-107.
3. Hoffer, A.P. and Greenberg, J. 1978. Anat. Rec. 190, 659-678.
4. Schmell, E.D., Yuan, L.C., Gulyas, B.J. and August, J.T. 1982. Fertil. Steril. 37, 249-257.
5. Vernon, R.B., Muller, C.H., Herr, J.C., Feuchter, F.A. and Eddy, E.M. 1982. Biol. Reprod. 26, 523-535.

# SPERM MATURATION ANTIGENS

R. JONES, S.J. GAUNT, C.R. BROWN & B.P. SETCHELL
ARC Institute of Animal Physiology, 307 Huntingdon Road, Cambridge CB3 0JQ, UK.

## INTRODUCTION

The acquisition of motility and fertilizing capacity by spermatozoa during their passage through the epididymis is accompanied by morphological and biochemical changes to the nucleus, acrosome and plasma membrane (1). Many of these changes are likely to be mediated by interaction with the secretions of the epididymal epithelium but the extent and nature of these processes and the mechanisms regulating them are poorly understood. This communication describes the interaction of several androgen-dependent epididymal secretory proteins (mol. wts. 18,500, 19,000 and 32,000) (2) with the plasma membrane of rat spermatozoa during their passage through the epididymis, and further documents maturation changes in surface antigens using monoclonal antibodies.

## MATERIAL AND METHODS

Spermatozoa were collected from the testis by puncturing the rete and from different sites (Fig.1) on the epididymis by micropuncture (3). Galactose and N-acetylgalactosamine residues on glycoproteins on washed spermatozoa and on secreted proteins were labelled with galactose oxidase/$NaB^3H_4$ (3). Proteins solubilized from spermatozoa with 0.2% deoxycholate (DOC) were separated on denaturing SDS polyacrylamide gels and those labelled detected by fluorography. In some instances, proteins were transferred to nitrocellulose paper by 'Western Blotting' (4). Proteins with mol. wts.

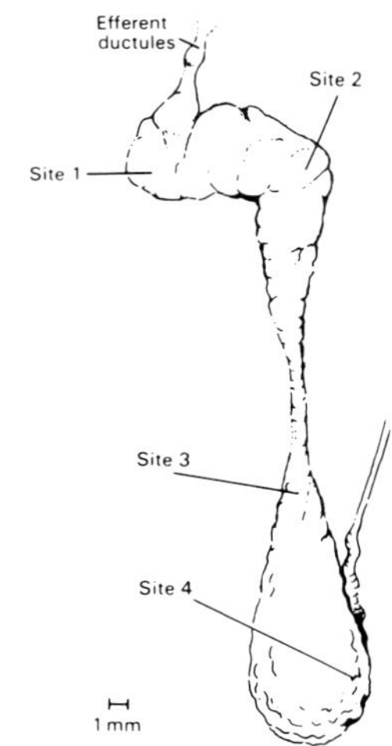


Fig. 1. *Sketch of rat epididymis (testis removed) illustrating the sites from which spermatozoa were removed by micropuncture for labelling experiments*

of 18,500, 19,000 and 32,000 were purified from cauda epididymidal plasma (CEP) (2) and antisera prepared in rabbits. Antisera against the 18,500 mol. wt. protein cross-reacted with the 19,000 mol. wt. protein and vice versa, suggesting a high degree of similarity between the 2 proteins. The distribution of antigens on spermatozoa was investigated by indirect immunofluorescence using a Zeiss epifluorescent microscope and proteins on 'Western Blotts' detected by an immunoperoxidase reaction (4). Immunoprecipitation of labelled proteins was carried out using Staphylococcus aureus cells (5). Monoclonal antibodies to surface antigens on cauda epididymidal spermatozoa were prepared (6) and the presence of the antigen on live spermatozoa detected by indirect immunofluorescence.

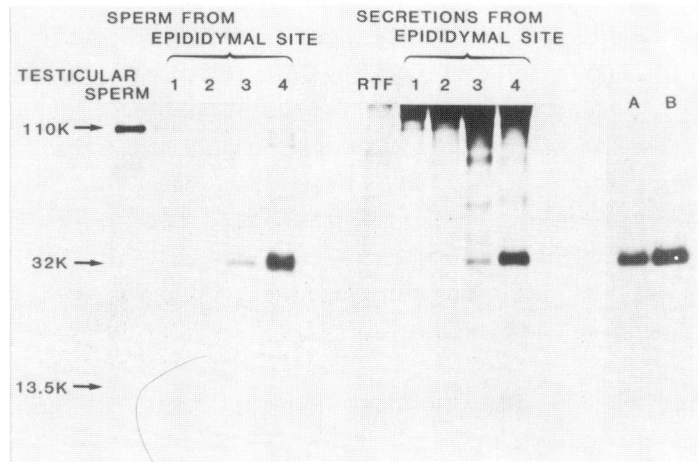


Fig.2. Fluorographs of galactose oxidase/$NaB^3H_4$ labelled proteins on spermatozoa and epididymal secretions.
A. DOC extract of labelled cauda sperm.
B. As for A + antiserum against secreted 32,000 mol. wt. proteins + protein A.

## RESULTS AND DISCUSSION

The major surface glycoprotein which labels on testicular spermatozoa with galactose oxidase/$NaB^3H_4$ migrates on SDS polyacrylamide gels with a mol. wt. of 110,000 (Fig.2). As spermatozoa pass through the initial segment of the epididymis labelling of this protein decreases sharply and is no longer detectable on spermatozoa removed from regions of the duct distal to the caput flexure. Instead, an increasing amount of incorporated $NaB^3H_4$ becomes associated with two proteins with an average mol. wt. of 32,000. Labelling of these proteins is first detectable on spermatozoa from site 2 and reaches a maximum on spermatozoa from site 4 (Fig.2). Transient labelling of a protein with a mol. wt. of 13,500 is also apparent as spermatozoa pass through the cauda epididymidis.

When glycoproteins in epididymal secretions are incubated with galactose oxidase/NaB$^{3}$H$_{4}$, the major labelled macromolecular component migrates with a mol. wt. of 32,000 (Fig.2). An antiserum raised in rabbits against the purified secreted 32,000 mol. wt. proteins immunoprecipitates the labelled 32,000 mol. wt. proteins solubilized from spermatozoa with deoxycholate (Fig.2B), suggesting that the secreted and membrane bound forms are homologous.

Homology was also demonstrated between the secreted 18,500 + 19,000 mol. wt. proteins and 2 proteins in deoxycholate extracts of cauda epididymidal spermatozoa (Fig.3). When secreted proteins in RTF, CEP, and deoxycholate extracts of testicular and cauda epididymidal spermatozoa are separated on SDS polyacrylamide gels, blotted onto nitrocellulose paper, and the presence of the 18,500 + 19,000 mol. wt. proteins tested by an immunoperoxidase technique (4), then a positive reaction is found only with CEP and extracts of cauda epididymidal spermatozoa. Immunofluorescence revealed that the 18,500 + 19,000 mol. wt. proteins are restricted to the midpiece of mature spermatozoa (Fig.3b).

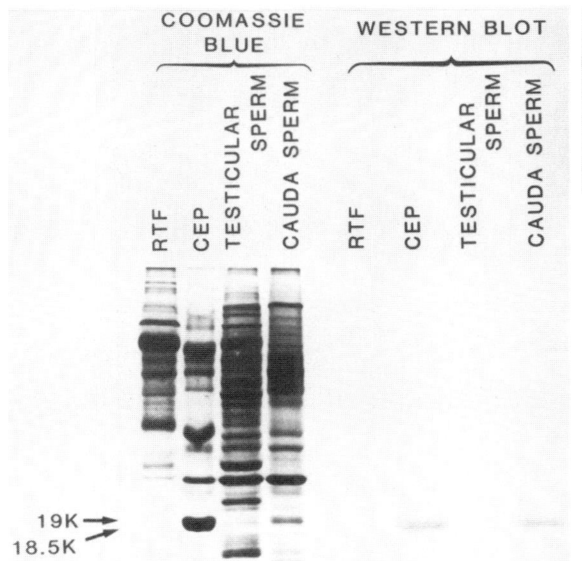


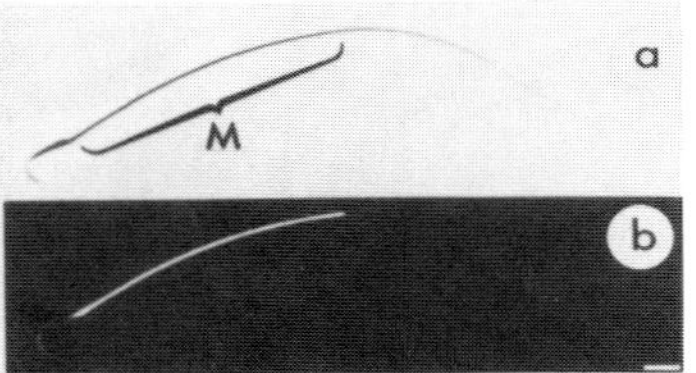


Fig.3. Detection of 18,500 + 19,000 mol. wt. proteins in DOC extracts of cauda sperm and on whole cells by immunofluorescence.
a. Phase contrast
b. uv illumination
Bar = 5μ

Four monoclonal antibodies, recognising antigens on different morphological domains of cauda epididymidal spermatozoa, are shown in Fig.4. The antigens restricted to the midpiece and principal piece are also found on testicular spermatozoa but the antigens recognised on the acrosome and postacrosomal regions develop during sperm maturation.

In conclusion, we have shown using whole cell labelling techniques and monoclonal antibodies that there are considerable differences in surface membrane molecules between rat testicular and cauda epididymidal spermatozoa, differences which must have developed at some stage during sperm maturation. The major androgen-dependent epididymal secretory proteins (mol. wts. 18,500, 19,000 and 32,000) are present on the plasma membrane of mature but not immature spermatozoa, suggesting that they become 'coupled' to receptors or 'partition' into the membrane. The antigens recognised by the monoclonal antibodies may arise by a similar mechanism, but it is also possible that they appear as a result of processing of the carbohydrate side chains on existing membrane glycoproteins.

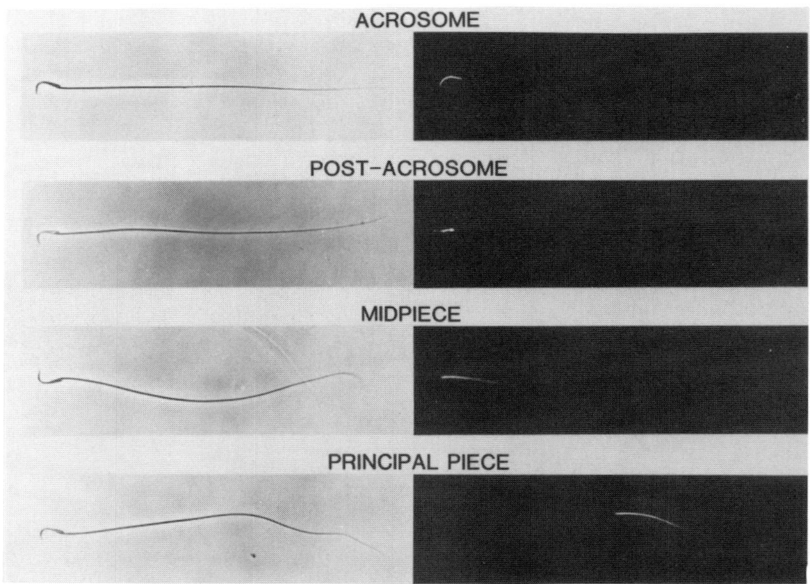


Fig.4. Detection of monoclonal antibodies on cauda sperm by immunofluorescence. LHS - phase contrast. RHS - uv illumination.

REFERENCES

1. Bedford JM. 1975. In Handbook of Physiology, vol.5, pp.301-307. Eds. DW Hamilton & RO Greep. Baltimore, Williams & Wilkins Co.
2. Jones R, von Glos KI, Brown CR & Parker MG. 1980. Biochem J, 188, 667-676.
3. Jones R, Pholpramool C, Setchell BP & Brown CR. 1981. Biochem J, 200, 457-460.
4. Towbin H, Staehelin T & Gordon J. 1979. Proc Natl Acad Sci USA, 76, 4350-4354.
5. Kessler SW. 1975. J Immunol, 115, 1617-1624.
6. Gaunt SJ. 1982. Dev Biol, 89, 92-100.

# DISTRIBUTION OF FIBRONECTIN ON CELLULAR SURFACE DURING THE RABBIT SPERMIOGENESIS

G. YASUZUMI, N. YABUMOTO, Y. TAKAHASHI, M. TADANO and M. KATO
Academy of Medical Technology and Nursing Hanwa Memorial Hospital, Osaka City, Japan.

## 1. INTRODUCTION

Spermiogenesis is in itself dramatic structural reorganization keeping with the intimate intercellular communication which occurs through the extracellular matrix. Recently, it has become apparent that a large molecular weight (450,000) glycoprotein, fibronection (7), also known as the large external transformation-sensitive (LETS) protein (3) or cell surface protein (9) is also a major component of such a matrix. The aim of the present work was to investigate in detail at the ultrastructural level the immunocytochemical distribution of the gkycoprotein, fibronectin (FN) during the spermiogenesis of adult rabbit using the peroxidase conjugated IgG fraction goat anti-rabbit FN. Previous reports (4, 5, 10) have suggested that there is a correlation between the presence of cell surface FN and the ability of spermiogenesis. In the present studies, we have examined the expression of intracellular and extracellular FN on rabbit spermatids differing in their metamorphic potentials. Although we have found for the first time differences in the intracellular and extracellular FN during spermiogenesis, our results in this experimental rabbit testis indicate that changes in FN expression are not related to any simlpe way to metamorphic potentials.

## 2. PROCEDURE

### 2.1. Materials and methods

Adult rabbit testis tissues were sliced into 2 mm in thickness and immersed in periodate-lysine-paraformaldehyde (PLP) fixative or 8% paraformaldehyde (PFA) for 24 hrs at 4° C. They were washed at 4° C with several changes of phosphate-buffered saline (PBS) containing sucrose as follows: 1) PBS containing 10% sucrose overnight; 2) PBS containing 15% sucrose overnight; 3) pBS containing 20% sucrose overnight; and 4) PBS containing 20% sucrose and 5% glycerin for onr hr. The slices were frozen by using OCT compound and dry-ice-acetate and stored at -70° C until used. Four- or six-micron frozen sections obtained using a cryostat were placed on egg white albumin coated slide, dried for 30 min at room temperature, and washed in cold PBS for 5 min. Each section was exposed for 40 min at room temperature to a 1:20 or 1:50 dilution of peroxidase (PO) conjugated IgG fraction goat anti-rabbit FN serum (Cappel Laboratories, USA). In control experiments, the rabbit anti-FN serum was replaced either with normal rabbit serum, with antiserum absorbed with purified plasma Fn (50 µg/ml), or with PBS. The slides were then washed with 3 chnges of PBS (5 min each) and immersed for 20 min in $5.0 \times 10^{-5}$ M diamino-benzidine (DAB) containing 0.01% $H_2O_2$. The slides were washed with 3 changes of cold PBS (5 min each), then fixed in 2.5% glutaraldehyde, postfixed in 1.0% osmium tetroxide, and embedded in epoxy Epon resin by routine methods. Each fixative was adjusted to pH 7.2 with 0.1 M cacodylate buffer, abd fixztion was carried out for 60 min at 4° C. Ultrathin sections were out on an LKB Ultrotome and examined without staining with any metals with an electron microscope, model LKM-2000.

## 2.2. Observations

The basal lamina and collagen fibril bundles surrounding the seminiferous tubule were found as FN-localized sites (Figs. 1 and 2). An intensive FN-PO immunoreactivity was present on the plasme membrane of the developing spermatid (Fig. 3). When the proacrosomal granule appears attached to the anterior pole of spermatid nucleus, the reaction product was detectable on its nucear surface and the Golgi vesicle remnants but not within their lumen. The reaction product of nuclear surface was localized on the nuclear envelope facing the acrosomal vesicle as the acrosomal vesicle gradually develops (Fig. 4). In a more advanced stage of acrosimal formation, the reaction product was on the outer membrane of acrosamal vesicle (Fig. 5). The long cone-shaped, premature spermatid nucleus showed the same endogeneous electron-density in its corresponding control section (Fig. 6). Although such a dense nucleus was enveloped with a less dense acrosomal cap free of the reaction product, the plasma membrane was still a localization site of FN (Fig. 7). It was clearly identified that the spermatozoon detached from the Sertoli's cell was completely devoid of the cell surface FN (Fig. 8). It is noteworthy that the proximal centriole as well as the Jensen's ring (Fig. 9) and the basal plate (Fig. 10) exhibited the similar electron-density which was not seen in their corresponding control sections (Fig. 6). In the principal flagellum, the fibrous sheath showed FN-PO reaction deposits in a small amount (Fig. 11). We were conveniently able to observe the tangential section through the nuclear envelope of the Sertoli's cell to which the developing spermatids attached. Faint staining of the nuclear pore complex was detectable as seen in Fig. 12.

## 3. DISCUSSION

In the present study the expression of FN-PO during spermiogenesis of the rabbit could be investigated by the PO-labelled antiboty method with the help of electron microscopy. This fact depends on that the enzyme-labeled antibodies were able to penettaye into tissues and cells, to form stable immune complexes with intracellular or extracellular antigens. FN glycoprotein, being one of the outer components of the cell surface, is a good candidate to play a role in cell-cell adhesion. There are, in fact, several observations that suggest an involvement of FN glycoprotein in cellular adhesion (6, 8, 9). The FN mediates adhesion of acrosomal vesicle to the nuclear envelope in the developing spermatid. The abundance of FN in the basal lamina and collagen fibril bundles is in agreement with such a role. Furthermore, the presence of FN on the surface of premature spermatozoa appearing implanted in the Setoli's cell makes the spermatozoa possible easily to adhere the Sertoli's cell.

The classical cytologists considered that the Jensen's ring or ring centriole arises in close relation to the distal centriole of the early spermatid. On the basis of the present immunocytochemical analysis, the Jensen's ring seems to have been originated from the distal centriole, because the ring and the centriole demonstrated the similar expression of FN.

The sperm head basal cavity, implantation fossa, is covered by the double layers of nuclear envelope. Its outer layer is covered by a thick layer, half moon-shaped basal plate. The biological role of the basal plate has been poorly understood. It is interesting that FN-PO was detected on this structure. It is well known that the basal body of epithelial cells is essential for motility and works for as the site of the motility of the beat. Although the basal body is lost in mammalian spermatozoa, it has been considered that the obliquely placed proximal centriole could probably serve as a kinetic center for their activities. Considering the profound effects of FN on cell motility (1, 2), the presence of FN-PO in the proximal centriole, Jensen's ring, basal plate and fibrous sheath would be of siginificance to the biological behavior of spermatozoa.

The presence of FN-PO in the nuclear pore complex of Sertoli's cell may have relebance to the regulatory function of the glycoprotein in the nuclear activity. Accordingly, the Sertoli's cell is not only a mechanically supporting cell for developing spermatids, but the Sertoli's cell probably participates in nutrition of the developing germ cells.

## 4. SUMMARY

The peroxidase-labeled antibody method was employed for immunocytochemical localization of intracellular and extracellular FN in developing spermatids of the rabbit. FN was dispersed all over the cell surface of early spermatids, but disappeared in the mature spermatozoon. However, FN was detectable in the Jensen's ring and the proximal centriole, justifying the concept that the Jensen's ring arises in close relationship to the distal centriole. The presence of the FN albeit at a small amount on the basal plate, proximal centriole, Jensen's ring and fibrous sheath would play the pivotal role in the activities of spermatozoa. The PN-PO mediated adhesion of acrosomal vesicle to the nuclear envelope in the developing spermatid. The presence of

PN on the surface of premature spermatozoa made them possible easily to adhere the Sertoli's cell. All the results mentioned above indicated that expression or release of FN *per se* is not a determinant of spermiogenesis, although it might be a facter in certain steps of the spermiogenesis.

REFERENCES


1. Ali IU, Hynes RQ. 1978. Effects of LETS glycoprotein on cell motility. Cell, 14, 439-446.
2. Ali IU, Mautner V, Lanza R, Hynes RO. 1977. Restoration of normal, orphology, adhesion and cytoskeleton in transformed cells by addition of a transformation-sensitive surface protein. Cell, 11, 115-126.
3. Hynes RO. 1976. Cell surface proteins and malignant tranformation. Biochem. Biophys. Acta 458, 73-107.
4. Kornblatt MJ, Knapp A, Levine M, Schachter H, Murray RK. 1974. Studies on the structure and formation during soermatogenesis of the sulfoglycerogalactolipid of rat testis. Can. J. Biochem. 52, 689-697.
5. Letts PJ, Meistrich ML, Bruce WR, Schachter H. 1974. Glycoprotein glycosyltransferase levels during spermatogenesis in mice. Biochem. Biophys. Acta 343, 192-207.
6. Pearlstein E. 1976. Plasma membrane glycoprotein which mediates adhesion of fibroblasts to collagen. Nature (London) 262, 497-500.
7. Vaheri A, Mosher DF. 1978. High molecular weight, cell surface-associated glycoprotein (fibronectin) lost in malignant transformation. Biochem. Biophys. Acta 516, 7-25.
8. Yamada KM, Kennedy DW, Kimata K, Pratt RM. 1980. Characterization of fibronectin interactions with glycosaminoglycans and indentification of active proteolytic fragments. J. Biol. Chem. 255, 6055-6063.
9. Yamada KM, Pastean I. 1976. Cell surface protein and neoplastic transformation. Trends Biochem. Sci. 1, 222-224.
10. Yasuzumi G. 1979. Spermatogenesis in animals as revealed by electron microscopy. 30. Some modifications of cell surface in developing spermatids of the grasshopper. Monitore zool.

ital. 13, 265-277.

EXPLANATION OF FIGURES

Electron micrographs of immunocytochemical localization of intracellular and extracellular FN in developing spermatids of the rabbit.

FIGURE 1. FN deposits on the basal lamina surrounding the seminiferous tubule.

FIGURE 2. Similar deposits on the collagen fibril bundles surrounding the seminiferous tubule.

FIGURE 3. The FN expression can be seen on the plasma membrane of a developing spermatid.

FIGURE 4. FN deposits are detectable on the nuclear envelope facing the acrosomal vesicle (arrow).

FIGURE 5. The reaction product is visible on the outer surface of acrosomal vesicle (arrows).

FIGURE 6. Premature spermatid nuclei of the control sample show endogeneous electron density, without showing proximal centriole and basal plate.

FIGURE 7. The plasma membrane surrounding a developing spermatid nucleus demonstrates the FN reaction product in a small amount.

FIGURE 8. A premature spermatid shows FN reaction product on the plasma membrane attached to the cell, but does not on the free one.

FIGURE 9. The proximal centriole (long arrow) and the Jensen's ring (short arrows) exhibit clearly the reaction products.

FIGURE 10. The similar reaction appears on the basal plate (arrow).

FIGURE 11. The fibrous sheath shows FN reaction deposits in a small amount.

FIGURE 12. Faint reaction products are detectable on the nuclear pore complex of the Sertoli's cell

(The scale indicates 1 μm.)

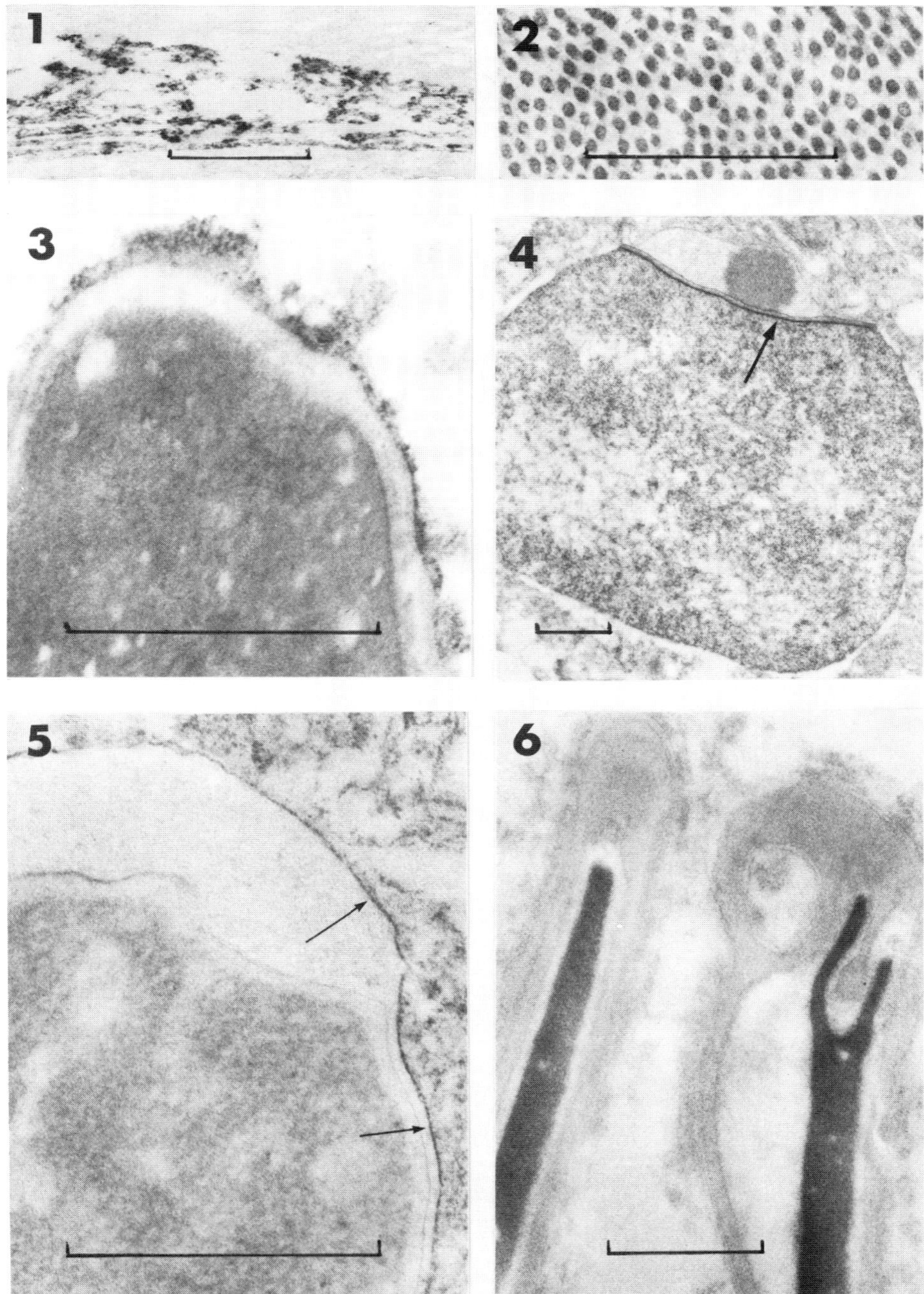
1
2
3
4
5
6

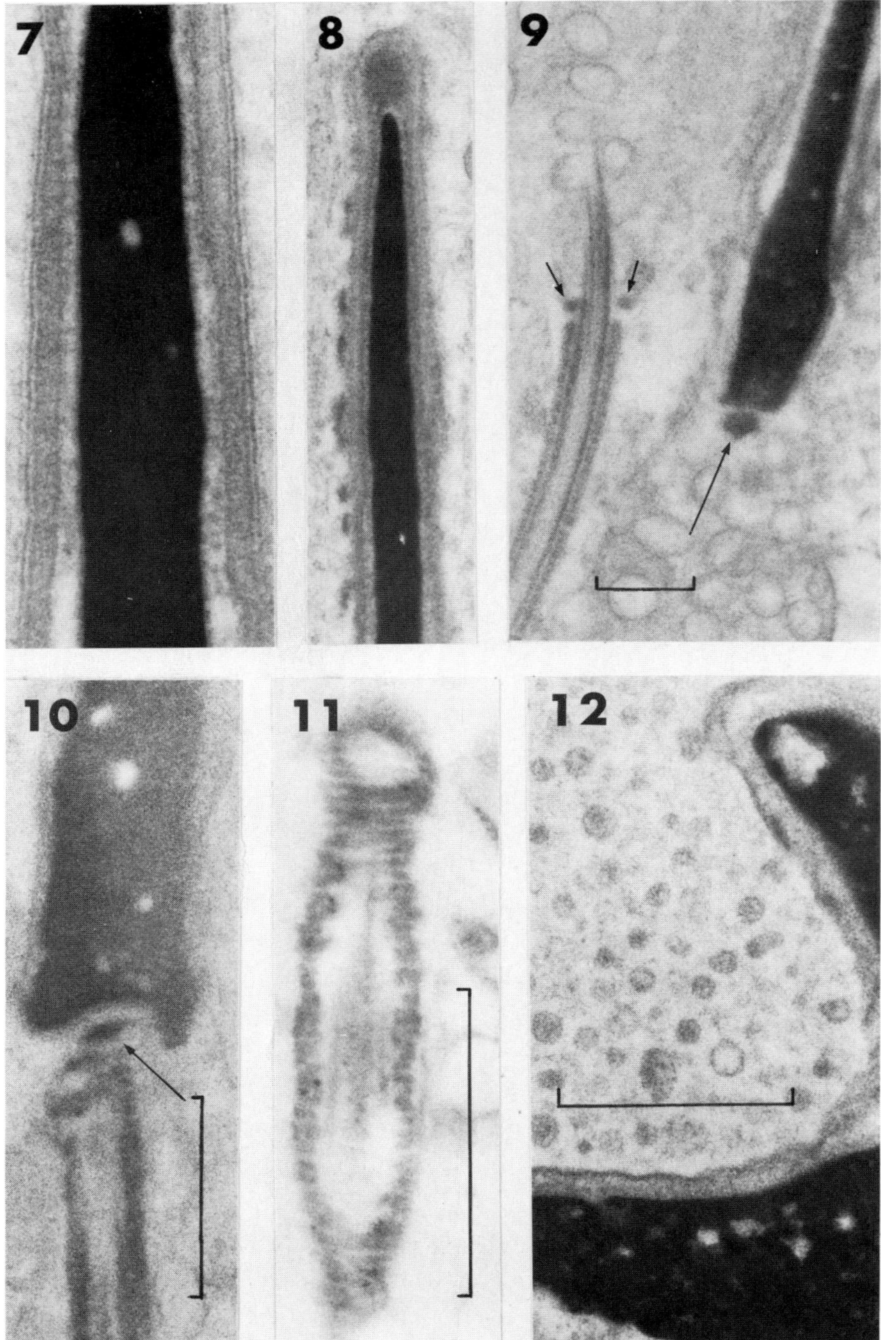
7
8
9
10
11
12

# INTERACTION OF GUINEA-PIG SPERMATOZOA WITH IMMUNOGLOBULINS : EVIDENCE OF Fcγ-SPECIFIC MEMBRANE RECEPTORS ON SPERMATOZOA.

Marta DE ALMEIDA, Thérèse NEVEU and Brigitte MARQUANT-LE GUIENNE

(Centre d'Immuno-Pathologie et d'Immunologie Expérimentale de l'INSERM (U 23), du CNRS (LA 289) et de l'Association Claude-Bernard (C 16), Hôpital Saint-Antoine, 75012 PARIS, France).

## 1. INTRODUCTION

A receptor specific for a site on the Fc portion of the immunoglobulin molecule (Fc R) has long been known to exist on the surface membrane of a wide variety of cells (1,2,3) as well as on the yolk sac and the placental membranes (4). Most of these receptors were defined by their ability to bind the IgG which has been complexed with the antigen or has been heat-aggregated. However, some Fc R have shown to be capable of binding monomeric IgG (5).

Considering the apparent non-specific binding of immunoglobulins to spermatozoa in vitro, we can postulate that this cell may possess Fc R. In this report we show the presence of these receptors on guinea pig spermatozoa using homologous monomeric and aggregated IgG. Furthermore we shall try to define the Fc R isotype specificity by means of immune complex binding.

## 2. MATERIALS AND METHODS

2.1. Spermatozoa. Guinea-pig spermatozoa were collected from excised epididymis by retroperfusion of the vas deferens and epididymis cauda.

2.2. Preparation of monomeric and aggregated IgG. The IgG fraction of guinea-pig hyperimmune anti-DNP-BGG(dinitrophenylated bovine gammaglobulin) sera was obtained by affinity chromatography on protein A-Sepharose CL4B column (6) and was centrifuged 60 minutes at 130,000 g in order to eliminate any aggregates. A sample of this IgG preparation was aggregated by being heated at 63° C for15 minutes.

2.3. Preparation of the immune complexes. Guinea-pig IgG1 and IgG2 were separated by chromatography on DEAE 23 from purified IgG anti-DNP antibodies. F(ab')2 fragments were isolated from IgG2 by pepsine digestion (7). Antigen-antibody complexes were prepared as described in Fig. 2.

2.4. Demonstration of IgG bound on spermatozoa. This demonstration was based on the ability of sheep red cells coated with protein A of Staphylococcus aureus (Sp.A) to form rosettes with IgG bearing cells (8). Twenty-five µl of washed guinea-pig spermatozoa suspension (3 x $10^6$ per ml) were incubated with 25 µl of several dilutions of the

IgG preparations at 4° C and at 37° C, for different periods of time, with or without sodium azide (0.02 M). After three washings 25 µl of a 1 % suspension of Sp.A-coated red cells was added and left over night. A minimum of 200 spermatozoa were counted and a rosette was taken as a spermatozoa with five or more red cells adhering to it.

## 3. RESULTS AND DISCUSSION

At 4° C, with or without azide, an incubation time of 4 hrs produced a maximal rosette number - 50 % for monomeric IgG and 31 % for the heat-aggregated form (Fig. 1). At 37° C, without azide, the rosettes were very unstable. In the presence of azide, the highest percentages of rosette were obtained within 1/2 hr of incubation - 44 % for monomeric IgG and 32 % for heat-aggregated IgG. There was no "spontaneous" rosetting of Sp.A-red cells with spermatozoa. This means that at the epididymal level spermatozoa are not covered with IgG. The rosettes observed with aggregated IgG were exclusively associated with the sperm head. After monomeric IgG binding the rosettes extended from the head to the middle piece.

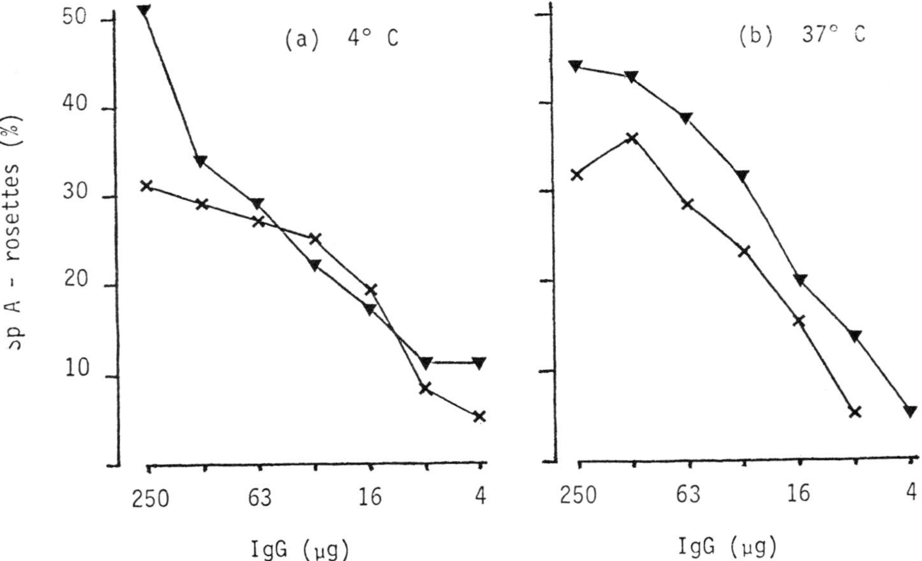


Fig. 1 - Binding of monomeric (▼) or aggregated (×) guinea-pig IgG to guinea-pig spermatozoa at 4° C for 4 hr (a) or at 37° C for 1/2 hr (b). The percentages are the mean of 3 experiments.

These results demonstrate the presence of Fc R for monomeric IgG on spermatozoa and confirm the existence of Fc R for aggregated IgG recently described on spermatozoa of different species (9, 10). In general single IgG molecules bind to Fc R with low affinity (5). Only Fc R on macrophages showed high affinity for monomeric IgG (11). Surprisingly the same situation seems to be true on guinea-pig epididymal spermatozoa, at least in our experimental conditions. This could explain why, in vivo, spermatozoa can carry IgG on their surface even in the absence of antigen(s) (12).

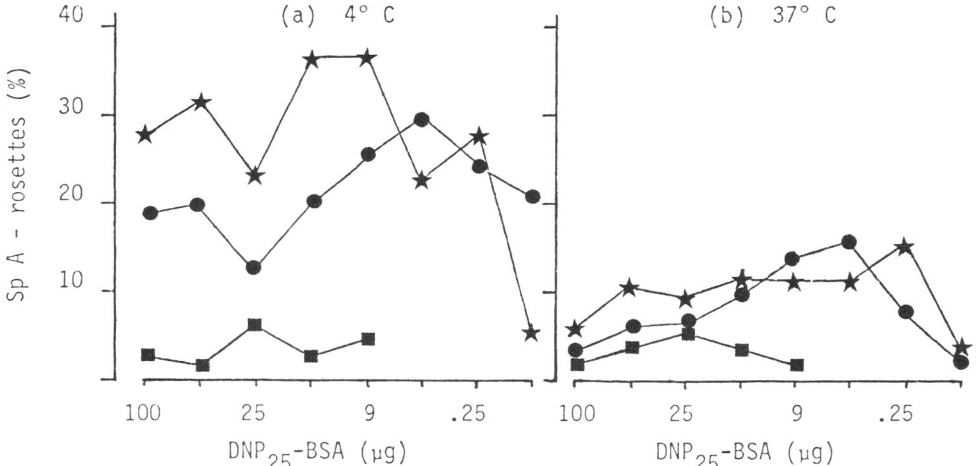


Fig. 2 - Binding of 25 μg of IgG1 (★), IgG2 (●) and F(ab')2 (■) anti-DNP complexed with variable amounts of $DNP_{25}$-BSA to guinea-pig spermatozoa at 4° C for 4 hr (a) or at 37° C for 1/2 hr (b).

The results in Fig. 2 show that both IgG1 and IgG2 immune complexes were bound to the spermatozoa surface. Higher rosette percentages were obtained at 4° C - 37% for IgG1 and 30 % for IgG2 - than at 37° C. F(ab')2 complexes gave less than 7 % rosettes. These data demonstrates that guinea-pig spermatozoa bear Fc R with specificities for homologous IgG1 and IgG2 isotypes.

What is the biological function, if any, of these IgG Fc R on spermatozoa ? We can speculate that 1) The association of IgG with spermatozoa results in the masking of antigenic sites on sperm-head surface ; 2) spermatozoa coated with IgG are more prone to be phagocytosed in utero (13) ; 3) ejaculated spermatozoa may differ in their affinity for IgG in such a way that a small non-coated population will be able to reach the oviduct and fertilize the ovocyte (12).

## REFERENCES

1. Kerbek RS, Davies AJS. 1974. Cell, 3, 105.
2. Myhre EB, Kronvall G. 1980. Infect. Immun. 27, 808.
3. Torpier G., Capron A., Quaissi MA. 1979. Nature, 278,447.
4. Schlamowitz M. 1977. Immunology of Receptors (ed. Cinader B.), Immunology Series, 6, 253.
5.. Stout RD. 1981. J. Immunol. Meth. 40, 7.
6. Hjelm H., Hjelm A., Sjöquist J. 1972. FEBS Lett. 28, 73.
7. Leslie RGQ, Melamed MD, Cohen S. 1971. Biochem. J. 121, 829.
8. Ghetie V, Medesan C, Sjöquist J. 1976. Scand. J. Immunol. 5, 1199.
9. Sethi KK, Brandis H. 1980. Eur. J. Immunol. 10, 964.
10. Witkin SS, Shahani SK, Gupta S, Good, RA, Day NK. 1980. Clin. exp. Immunol. 41, 441.
11. Dorrington KJ. 1977. Immunology of Receptors (ed. Cinader B.), Immunology Series, 6, 183.
12. Cohen J, Werret DJ. 1975. J. Reprod. Fert. 42, 301.
13. Symons DBA. 1967. J. Reprod. Fertil. 14, 163.

# IDENTIFICATION AND LOCALIZATION OF TWO BOAR SPERM-MEMBRANE PROTEINS

Marianne Klint, Anita Fridberger, Per A. Peterson and Leif Plöen*
Dept. of Cell Research, University of Uppsala and *Dept. of Anatomy and Histology, Faculty of Vet.Med., Swed. University of Agricultural Sciencies, Uppsala, Sweden

The plasmalemma of the mammalian spermatozoon contains a number of specific proteins, some of which have a clear regional distribution (see ref. 1-5). It is reasonable to assume that one or more sperm membrane proteins are involved in the initial recognition events of fertilization. We have isolated a number of boar sperm membrane glycoproteins and in this communication two of them will be described in some detail.

## Isolation of sperm membrane proteins

Ejaculated boar spermatozoa were washed three times in PBS (phosphate-buffered saline, pH 7.4) and solubilized in a Tris-NaCl buffer containing 20 mM Tris-Cl, pH 8.0, 50 mM NaCl, and 15 mM sodium deoxycholate. After centrifugation at 100,000 xg the supernatant was passed over a Lens Culinaris-Sepharose column (6) for enrichment of glycoproteins. The column was then washed and bound material was eluted using $\alpha$-methylmannoside. About twenty different proteins with molecular weights ranging from 15,000 to about 200,000 could be identified by SDS-polyacrylamide (PAGE) gradient gel electrophoresis (7). The proteins were isolated using preparative SDS-PAGE gel electrophoresis. Protein bands were stained with Coomassie brilliant blue and as soon as they became apparent they were cut out and eluted from the gel. The proteins were not boiled nor reduced prior to separation in an attempt to preserve as much as possible of their native three-dimensional structure. The purity of the isolated proteins was tested on analytical SDS-gels.

## Specificity of antisera

Antibodies were raised against the isolated protein fractions in female rabbits. Antisera against two antigens with apparent molecular weights of 43,000 and 36,000, respectively, were examined in some detail. To demonstrate the specificity of these antisera Fab-fragments were prepared (8) from the sera of the rabbits prior to and after immunization. The Fab-fragments were coupled to cyanogen-bromide activated Sepharose (9). Samples of detergent solubilized boar spermatozoa were passed over the immunosorbents. The columns were then washed and eluted with a 4M urea solution, pH 3.0, containing 0.01% Tween 80. The desorbed proteins, immediately dialysed to remove the urea and to raise the pH, were concentrated by ultra-filtration (10) and analyzed by SDS-PAGE (fig. 1). The Fab fragments against the 43,000 dalton component were highly specific and those against the 36,000 dalton protein recognized this component well over the background of non-specifically binding proteins.

To ascertain that the glycoproteins were sperm-specific and not present on cells from other porcine tissues iso-antisera against ejaculated and epididymal sperm were separately raised in gilts. Fab-fragment from the antisera were prepared and immobilized on Sepharose-beads and used as immunosorbents. Several solubilized sperm components bound to these columns. Among the iso-antigens identified two had apparent molecular weights of 43,000 and 36,000 respectively. The iso-antigens were labelled with $^{125}I$ (11) and separately immunoprecipitated with the rabbit antisera raised against the isolated 43,000 and 36,000 dalton proteins. The results imply that these two components are derived from boar sperm membranes, that they are sperm specific; and that they are present on both epididymal and ejaculated spermatozoa.

Localization of sperm membrane proteins

The localization of the two glycoproteins described above was studied by indirect immunofluorescence. Fab-fragments from the immunized rabbits were used as the primary antibodies and fluorescein-conjugated Fab-fragments (goat anti-rabbit IgG) as the secondary antibody.

Incubations were made on fresh epididymal and Percoll (12) washed ejaculated boar spermatozoa.

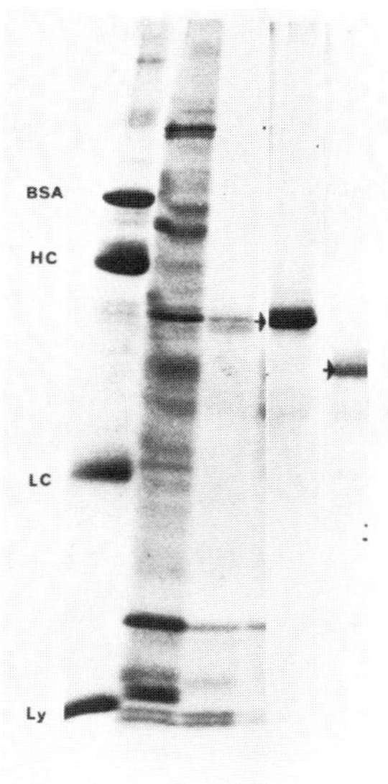


Fig. 1 Solubilized spermatozoa were passed over immunosorbents (see text). Desorbed proteins were analyzed by SDS-PAGE M=markers: BSA - bovine serum albumin (69,000) HC=IgG, heavy chain (53,000); LC=IgG, light chain (23,000) and L=lysozyme (14,000). LE=Lens eluate of solubilized boar sperm. NRS=protein bound to the column containing rabbit Fab-fragments. The molecular weights of the proteins obtained with the anti 43,000 and the anti 36,000 component Fab fragments, respectively, are indicated.

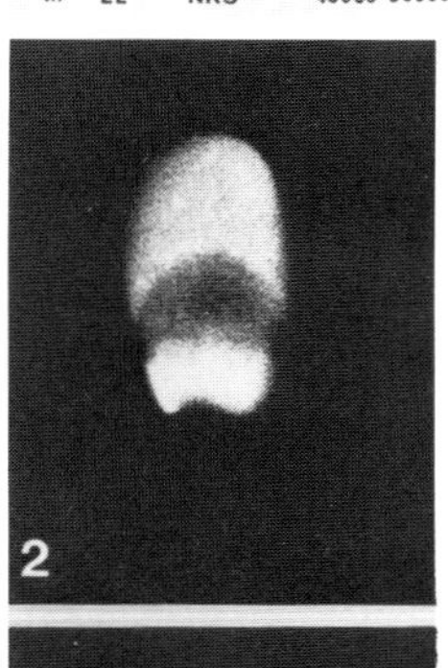


Fig. 2 Indirect immunofluorescence of ejaculated boar spermatozoon using Fab-fragments of the anti 43,000 dalton component antibodies as the primary reagent. The entire head is stained but the staining is weaker over the equatorial segment of the acrosome.

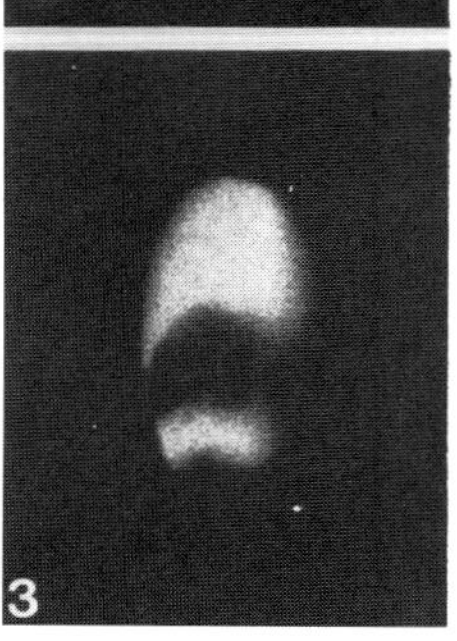


Fig. 3 Indirect immunofluorescence of ejaculated boar spermatozoon using Fab-fragments of the anti 36,000 dalton component antibodies as the primary reagent. The equatorial segment of the acrosome and the anterior part of the postacrosomal sheath are unstained. The remainder of the head is stained.

The 43,000 dalton component seemed to occur on the entire head regions but the equatorial segment of the acrosome was more weakly stained (Fig. 2). The 36,000 dalton component had a similar localization but on most spermatozoa no or very little staining was observed in the equatorial segment and the anterior part of the postacrosomal sheath (Fig. 3). Both components appeared to have the same surface distribution in epididymal and ejaculated spermatozoa.

In conclusion, the boar spermatozoon contains specific surface antigens some of which have a clear regional distribution.

Acknowledgements

This study was supported by the Swedish Medical Research Council and by grant A 5837/B4145 from the Swedish Council for Forestry and Agricultural Research.

REFERENCES

1. Millette, CF. 1977. Immunobiology of Gametes, Cambridge University Press, Cambridge. (Eds. M. Editin, MH Johnson) (51-71)
2. Fellous M, Gachelin G, Buc-Caron MH, Dubois P, Jacob F, 1974 Dev. Biol 41, 331-337
3. O´Rand MG, Romrell LJ. 1981, Dev. Biol. 84, 322-332
4. Gold Myles D, Primakoff P. Bellvé AR. 1981 Cell 23, 433-439
5. Feuchter FA, Vernon RB, Eddy EM. 1981, Biol. Reprod. 24, 1099-1110
6. Hayman MJ, Crumpton MJ. 1972. Biochem.Biophys.Res.Comm. 47, 923-930
7. Blobel G, Dobberstein B. 1975. J. Cell. Biol. 67, 852-862
8. Porter R.R. 1959 Biochem J 73, 119-126
9. Cuatrecasas P. 1970. J. Biol. Chem. 245, 3059-3065
10. Berggård I. 1962 Ark. Kemi 18, 291-326
11. Hunter WM, Greenwood FC. 1962. Nature 194, 495-496
12. Pertoft H, Rubin K, Kjellén L et al. 1977. Exp. Cell. Res. 110, 449-457

# ANTIGENIC DETERMINANTS ON THE SPERM CELL SURFACE.

P. CHARDON*, J.E. FLÉCHON**, H. LEVEZIEL ***
Laboratoire de Radiobiologie Appliquée *, Physiologie Animale**,
Génétique Biochimique***, INRA, 78350 Jouy-en-Josas, France.

## 1 - INTRODUCTION.

The presence of HLA antigens on human sperm has been demonstrated by cytotoxic assay, immunofluorescence and inhibition-of-motility assay (FELLOUS et al., 1970; HALIM et al., 1974; ARNAIZ-VILLENA et al., 1978). Similar results were found for SLA antigen on boar sperm (VAIMAN, 1974). The present study was undertaken to detect BoLA antigens on washed ejuculated bull sperm. Difficulties in obtaining reproductible and significant results led us to experiment with a large variety of reactives. In parralel and using the same material, we tried to procure positive results for other types of surface antigens using 1) different anti-bovine IgG antibodies and protein A 2) several antibovine blood group determinants from the B,J and S system 3) mouse sera anti bull whole sperm. Lastly, instead of anti-BoLA alloserum, we used a monoclonal antibody anti-human β2-microglobulin (β2m), cross-reacting with bovine β2-microglobulin (TEILLAUD et al., 1982), but with no better success.

## 2 - MATERIAL AND METHODS.

Animals. The bulls were typed for BoLA specificity and blood group system.

Sperm preparation. The suspension of sperm was washed according to HARRISSON (1976), then layered over ficoll Triosyl (7.5 %) and centrifuged (500g, 5 min; 1200g, 10 min).

Antibodies. The sera anti-blood group determinants were produced and analyzed by the "Service d'Analyse des Groupes Sanguins - INRA". Anti-BoLA allosera were raised by skingrafting and identified according to the last international BoLA workshop. Monoclonal anti-human β2m(M18) was kindly supplied by Dr. J. KALIL (U.95 INSERM, Hôpital St Louis, Paris, France).

Immunofluorescence. $2.10^6$ sperm cells were incubated for 30 min at room temperature in 1 ml of diluted specific antibodies. After two washings, the spermatozoa were incubated with labelled (Fab')2 anti-IgG

(FITC or TRITC) and washed twice.

Cytotoxicity. After the preincubation of $2.10^6$ of sperm cells in 50 µl of diluted antibodies for 1 hr, 100 µl of rabbit or horse complement was added. Cytotoxicity was evaluated 1 hr later using Acridine Orange and Ethidium Bromide as live/dead stain.

Absorption. $2.10^8$ sperm cells were incubated in 50 µl of different dilutions of antibodies (1 hr at room temperature). Absorbed reagents were then tested against bovine peripheral blood lymphocytes (PBL) by a conventional microcytotoxicity test.

3 - RESULTS.

Immunoglobulins. Sperm surface IgG were characterized immunofluorescently by FITC and TRITC labelled anti-bovine IgG and their (Fab')2 fragments. For five independent experiments with 4 different bulls, we noticed positive labelling on the acrosomes of fresh ejaculated sperm (Fig. 1). FITC protein A gave a similar although broader distribution of IgG (fig 2). In some experiments, we observed a decrease in the fluorescence when we increased the number of washings. Fluorescence intensity also decreased with the use of frozen sperm. Tests performed on the latter material were completely negative in the majority of cases.

Blood group determinants. B system. Antigenic determinants controlled by the B system in cattle were checked on an O1-Y-G'- positive bull. Control were performed with anti B1 and anti G1. For the positive blood group determinants, significant labelling was observed with FITC (Fab')2 anti-bovine IgG. The fluorescence, localized on the whole sperm, was usually more intense than with the use of (Fab')2 anti-IgG alone.

J system. Two kinds of tests were made on two positive bulls. In one with fresh sperm, positive labelling was observed on the acrosome and post-acrosomal region ; in the other with frozen sperm, no labelling was seen on these areas (fig 3).

S system. An S and H'-positive and UI-negative bull was tested. Significant fluorescence was observed with the first two reagents. The controls using using UI were negative (Fig 4).

Histocompatibility antigens. BoLA antigens. Allosera against BoLA antigens were negative in cytotoxicity, absorption and fluorescence tests for fresh sperm from 4 different bulls.

β2m. Monoclonal antibodies against human β2m were also negative for the fresh or frozen sperm of 5 different bulls. Fig 6 shows the typical absorption of anti-β2m cytotoxic activity against PBL. With $2.10^8$ lymphocytes, all cytotoxic activity was removed. Immunofluorescent tests performed as control with mouse anti-bull whole sperm were positive.

4 - DISCUSSION.

Conflicting results for HLA antigens have been obtained in human sperm by different research teams. LAW and BODMER's (1978) experiments showed few, if any, HLA antigens on the sperm, and assumed that β2m was absorbed on sperm from the seminal plasma. On the contrary, HALIM (1982) in a recent paper showed strong evidence for the presence of HLA antigen on sperm by cytotoxicity tests. In immunocytochemical studies, SÖDERSTROM (1982) describes the decrease of AgB determinants with rat sperm differentiation.

Negative results for the BoLA antigen detection were surprising as compared to our previous results on boar sperm. They could be interpreted as a species difference or as a divergence in the sensitibity and specificity of the allosera. The BoLA antigens in the bull could be masked in ejaculated sperm by the IgG we found on the surface of the heads. Our future results will deal with epididymal sperm. Positive fluorescence results obtained for antigens other than BoLA or β2m show that our immunocytochemical techniques are efficient. The detection of blood group antigens, particularly of the S system, confirms the work of PARASH (1974) obtained by absorption tests.

REFERENCES.

ARNAIZ-VILLENA A. & FESTENSTEIN H., (1976) - Lancet ii, 707.
FELLOUS M. & DAUSSET J. (1970) - Nature 255. 191.
HALIM A.; ABBASSI K. & FESTENSTEIN H (1974) - Tissue Antigens 4.1.
HARRISON R.P.A. (1976) - J. Reprod. Fert 48:367.
LAW H.Y; and BODMER W.F. (1978) - Tissue Antigens, 12:249.
SODERSTROM K.O., SEGE C., ANDERSON C.C. (1982) - J. of Imm. 128:1971
TEILLAUD J.L.CREVAT D., CHARDON P. KALIL J., GOUJET-ZALE C., MAHOUY G., VAIMAN M., FELLOUS M., PIOUS D. (1982) - Immunogenetics 15 : 377.
PRAKASH C.(1974) - Ann. Genet. 6:1.
HALIM K.,WONG O.M., MITTAL K.K. (1982) - Tissue antigens 19:90.
VAIMAN M., FELLOUS M., WIELS J., RENARD C., LECOINTRE J., DUMESNIL du BUISSON F., and DAUSSET J., (1978) - J. Immunogen. 5 : 135.

1
2
3
4

CYTOTOXICITY TEST
% BOVINE P.B.L. KILLED
100
50
ABSORBED WITH:
2x2.10^9 spz
2x5.10^8 pbl
2x10^8 pbl
1:2
1:4
1:8
1:16
1:32
1:64
1:128
1:256
ANTI BETA 2 M DILUTION

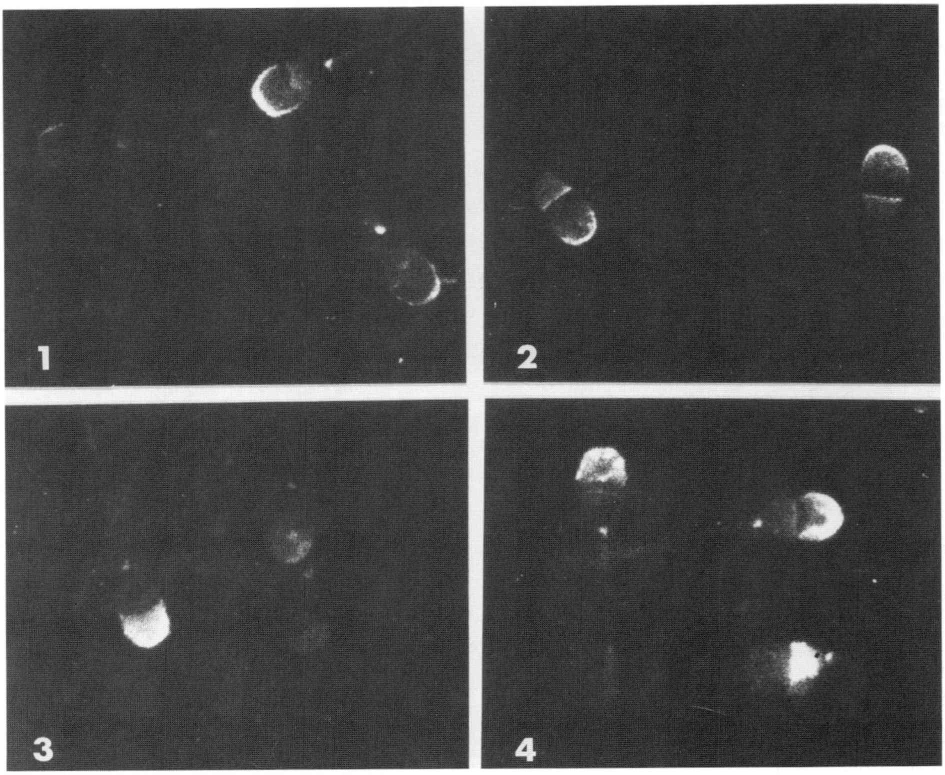

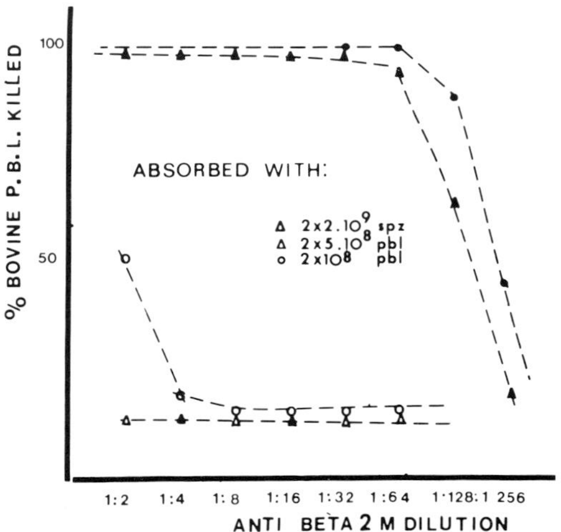

# RELATIONSHIP BETWEEN Y-CHROMOSOME AND H-Y ANTIGEN EXPRESSION IN HUMAN SPERMATOZOA

C. JEULIN[1], J. WIELS[2], M. CASANOVA[3], and M. FELLOUS[3]

## 1. INTRODUCTION

Antibodies against the H-Y antigen which is coded for or controlled by a gene on the Y-chromosome (1-4), have been produced since 1971 (5). In the presence of H-Y antigen, the gonad is induced to form a testis, while in its absence ovarian differentiation occurs. In a sperm population heterogenous for X or Y-chromosomes, H-Y antigen should be present on the surface of only those cells which carry Y-chromosome-if haploid expression exists. Specific antibodies directed against the H-Y antigen should selectively kill those cells carrying the corresponding antigen. The relatively minor shift in the sex ratio of progeny in mice inseminated with spermatozoa treated with H-Y antiserum (6), indicated only relatively small quantitative differences in H-Y distribution between the two types of spermatozoa. It may be that the sperm membrane has been conserved from the diploid state (6) or that there have been contributions by Sertoli cells (7).

We describe a method which permits the simultaneous detection of antibody and complement-mediated cytotoxicity (through the uptake of trypan blue dye or eosin yellowish), and of the Y chromosome in the surviving spermatozoa (those not stained with vital dye) after fluorescent staining ("F-bodies").

## 2. PROCEDURE

### 2.1. Material and methods

2.1.1. Semen samples. 14 human semen samples whose parameters all fell within the normal ranges, were obtained from volunteers (sperm concentration $>50\times10^6$/ml, $>60\%$ motile forms, $>70\%$ vital forms).

(1) Lab. Histologie-Embryologie, CHU KREMLIN-BICETRE, FRANCE
(2) Lab. Immuno-biologie des tumeurs, Institut G. ROUSSY, VILLEJUIF, FRANCE
(3) Lab. Immunologie, Institut PASTEUR, FRANCE.

2.1.2. Antisera. 1) H-Y antiserum was raised in inbred "Lou" female rats by 6-7 weekly intraperitoneal injections of male spleen cells from the same inbred strain. Heat inactivated sera (56°C, 30 min) was absorbed on female human erythrocytes. A monoclonal H-Y antibody was also tested in the experiments. 2) Anti-WI38 antiserum (WI38 - human embryonic cell line). This antiserum was cytotoxic for spermatozoa. 3) Control serum was raised in female rats and absorbed on erythrocytes. 4) Lyophilized rabbit complement (Merieux).

2.1.3. Media. RPMI 1640 medium (Gibco) supplemented with 10% fetal calf serum and phosphate buffer saline (PBS) pH=7.2 were used.

2.1.4. Microcytotoxicity assay. Freshly collected semen samples (<1h) were washed with PBS (600g, 20 min, 20°C). The supernatant was discarded and the spermatozoa resuspended in RPMI medium. The sperm pellets ($1.5 \times 10^6$ cells) were first incubated (30 min, 20°C) with either 50 μl of H-Y antiserum (tested sample), 50 μl of RPMI medium, Anti WI38 antiserum, or control serum (control samples). Pure rabbit complement (50 μl) was then added and incubation continued for another 45 min (20°C). The reaction was stopped by adding 1 ml of RPMI medium.

2.1.5. Vital staining. The percentage of vital spermatozoa in the various samples was determined by staining an aliquot from each, with acridine orange (AO) and ethidium bromide (EB : ref.8). 200 spermatozoa were counted per sample under fluorescence microscope.

2.1.6. Double staining - Presence of Y chromosomes in vital spermatozoa.
Expt n°1. Trypan blue (TB) and quinacrine dihydrochloride (QD) staining. After cytotoxicity testing, the spermatozoa were centrifuged (800g, 5 min) and the supernatant discarded. The pellet was resuspended in 0.3 ml of 1% TB in PBS for 10 min at 20°C (9). Spermatozoa were then washed twice with PBS (800g, 5 min). Following the 2nd wash, the cells were smeared on slides, air dried and fixed for 20 min in absolute ethanol.

The Y-bearing spermatozoa were identified by staining for 20 min with a 1% (w/v) solution of QD in a buffer of 100 mM sodium citrate and 200 mM disodium phosphate, pH=5.5 (10). The Y-counts (300 spermatozoa : 0, 1 or 2 F-bodies scored in the vital cells) were performed blind on randomised and coded slides.

Expt n°2. Eosin yellowish (EY) and QD staining. After the cytotoxicity testing, the pellet was resuspended in 0.2 ml of a 1% EY (Sigma), solution in 0.9% (w/v) NaCl for 1 min at 20°C. Smears were made and fixed for 10 min in a vital staining fixative see ref.11 (saturated picric acid, 40% formaldehyde, saturated ammonium molybdate, 3:1:1), after which the Y-bearing spermatozoa were counted as above.

2.1.7. Statistical analysis. The results of experiments Nos 1&2 were analysed with analyses of variance and Student's t-tests.

2.2. Results

Table. Presence of the Y-chromosome in vital spermatozoa. Values are mean±s.d

| | Experiment n°1 n=8 | | Experiment n°2 n=6 | |
|---|---|---|---|---|
| Sample | % vital cells | % Y-count | % vital cells | % Y-count |
| Washed spermatozoa(WS) | 71.1± 5.7 | 47.7± 5.2 | 64.1± 8.1 | 41.4± 4.8 |
| W.S + C | 62.6± 8.5 | 43.5± 6.3 | 59.5±11.1 | 42.4± 7.8 |
| W.S + Anti H.Y antiserum + C | 59.8± 8.6 | 44.3± 6.1 | 40.9± 9.8*** | 27.1± 9.8*** |
| W.S + Anti WI38 antiserum + C | 59.3±10.3 | 39.0*± 4.7 | — | — |
| W.S + control serum + C | — | — | 59.7± 7.7 | 32.7± 8.2* |
| Non treated sample | — | 48.6± 2.8 | — | 43.0± 3.5 |

*$p < 0.05$ ***$p < 0.001$

2.2.1. Expt n°1. Cytotoxicity of H-Y antiserum could not be demonstrated in this experiment as TB was found to be unsuitable as a vital stain. Anti WI38 antiserum was found to be slightly cytotoxic for Y-bearing spermatozoa.

2.2.2. Expt n°2. 1) % vital spermatozoa. The analysis of variance showed there to be a significant ($P \leq 0.001$) difference between treatments, although there was significant ($P < 0.01$) heterogeneity between samples. After treatment with anti H-Y antibody, % vital cells was reduced : 40.9% vs 64.1% in the washed spermatozoa. 2) % Y-bearing spermatozoa. Analysis of variance showed there to be a significant ($P < 0.001$) difference between treatments, although there was again a significant ($P < 0.01$) heterogeneity between samples, and also an interaction between semen samples and treatments. After treatment with H-Y antibody, % Y-bearing spermatozoa was reduced to 27.1% vs 43.0% in the non treated controls. Treatment with control serum also affected this percentage, but to a lesser extent reduction to 32.7%.

DISCUSSION

The present data have shown that the incidence of Y-bearing spermatozoa can be selectively reduced by anti H-Y antibody in a complement-dependent microcytotoxicity assay. These results corroborate those previously described in mice (6), where artificial insemination with H-Y antibody-treated spermatozoa caused a slightly, but statistically significant, higher prevalence of females among the progeny. A similar effect was found when does were inseminated with rabbit spermatozoa + anti cock spermatozoa antibodies ; the percentage of males produced by does so treated being significantly reduced (12). In addition, our results suggest that some Y-bearing spermatozoa are not H-Y positive. The distribution of H-Y antigen on human spermatozoa may be determined by factors other than the Y chromosome, and Ohno (7) has recently proposed that the H-Y antigen present on the plasma membrane could be contributed by Sertoli cells.

REFERENCES


1. Koo GC, Wachtel SS, Breg WR, Miller OJ. 1975. Mapping the locus of the H-Y antigen. Cyt. Cell Genet. 16, 1-5.
2. Ohno S, Wachtel SS, and Koo GC. 1976. Hormone-like role of H-Y antigen in bovine free - martin gonad. Nature 261, 597-599.
3. Wachtel SS, Koo GC, Breg WR, Elias S, Boyse EA, and Miller OJ. 1975. Expression of H-Y antigen in human males with two Y chromosomes. N. Engl. J. Med. 293, 1070-1072.
4. Bennett D, Boyse EA, Lyon MF, Mathieson BJ, Scheid M, and Yanagisawa K. 1975. Expression of H-Y (male) antigen in phenotypically female Tfm/Y mice. Nature 257, 236-238.
5. Goldberg EH, Boyse BA, Bennett D, Scheid M, and Carswell EA. 1971. Serological demonstration of H-Y (male) antigen on mouse sperm. Nature 232, 478-480.
6. Bennett D, Boyse EA. 1973. Sex ratio in progeny of mice inseminated with sperm treated with H-Y antiserum. Nature 246, 308-309.
7. Ohno S. 1982. Expression of X and Y linked genes during mammalian spermatogenesis, in "prospects for sexing mammalian sperm". Edited by R.P Amann and G.E Seidel JR, Colorado State University, Fort Collins Colorado U.S.A.
8. Festenstein H, Halim K, Arnaiz-Villena A. 1978. HLA antigens on human spermatozoa 11-16. In Cohen J and Hendry WE, Spermatozoa, antibodies and infertility, Blackwell Sci. Publ. London.
9. Shalev A, Berezi I, and Hamerton JL. 1977. Demonstration of fluorescent X and Y Bodies after antibody - and complement - mediated cytotoxicity. Immunogenetics 5, 405-414.
10. Barlow P and Vosa CG. 1970. The Y chromosome in human spermatozoa. Nature, Lond. 226, 961-962.
11. Dalion. 1963. Techniques histologiques et anatomo-pathologiques. P.425. Traité de Biologie Appliquée. HR Olivier. Tome III, Diagnostics microscopiques 2e partie. Maloine Paris.
12. Hancock RJT. 1978. Comparaison of effects of normal rabbit sera and anti-cock sperm sera on rabbit sperm, including comparaison of effects on the sex ratio. Biol. Reprod. 18, 510-515.

# SURFACE CHANGES IN MONKEY SPERMATOZOA DURING EPIDIDYMAL MATURATION AND AFTER EJACULATION.

M.A. FAIN-MAUREL, J.P. DADOUNE and J.F. REGER[+].
Laboratoire de Biologie Cellulaire et Laboratoire d'Histologie, Université René Descartes, 75270 Paris, France.

## 1. INTRODUCTION

Studies on surface changes in mammalian spermatozoa during epididymal transit, as revealed by cytochemical technics or lectin-binding, has mainly involved domestic and laboratory animals (1-8). As an extension of these investigations a systematic study was made on a primate, *Macaca fascicularis*.

## 2. MATERIAL AND METHODS

Samples were obtained during the spring as follows: epididymal spermatozoa after castration from five successive regions (initial segment, caput, proximal corpus, distal corpus, cauda) and ejaculated spermatozoa after electro-ejaculation.

After washings in PBS, the surface coat was studied using 1) iron hydroxyde labeling techniques at pH: 1.8 and pH: 9 after fixation in glutaraldehyde (Cf. 2) 2) incubation for the demonstration of concanavalin A (Con A) and wheat germ agglutinin (WGA) receptors with ferritin-lectin conjugates before fixation (4) 3) cationized ferritin (CF) after washing in an isotonic Ficoll solution (5).

Controls include: methylation before electro-positive colloidal iron; acetylation before electro-negative colloidal iron (2); neuraminidase before CF; Con A + α-methyl-D-mannoside; WGA + N-acetyl-D-glucosamine.

## 3. RESULTS AND DISCUSSION

Examination demonstrates marked variations in spermatozoon membrane surface characteristics during epididymal transit.

+ *We wish to thank Mr N. Marchi et Mrs M. Baurès for their skilful technical assistance.*

After colloidal iron at pH: 1.8 (Figs 1-3), sperm electro-negative charges strongly increase on the tail during epididymal transit as shown by other studies (2,3,5,8). Moreover, in *Macaca fascicularis*, the plasma membrane overlaying the post-acrosomal region is strongly labeled in spermatozoa from the distal corpus. After ejaculation, the plasma membrane of the connective piece remains intensively labeled. Methylation inhibits labeling.

Cationized ferritin was also utilized for labeling negatively charged groups in ejaculated spermatozoa (Figs. 4-6). The cell surface is uniformly labeled. Neuraminidase treatment has no apparent effect. The presence of sialic acid groups is, however, not excluded as has been demonstrated in quantitative study on ram and bull (5).

After colloidal iron at pH: 9 (Fig. 7), as in ram (2), the region of the plasma membrane covering the apex of the acrosome is reactive. But, in *Macaca fascicularis*, electro-positive charges are strongly labeled on the plasma membrane of the post-acrosomal cap and on the flagellum of spermatozoa from the initial segment. This labeling decreases and totally disappears at the level of the distal corpus. All the electro-positive charges are blocked by acetylation.

It is necessary to note that, in the proximal epididymis, electro-positive and electro-negative charges coexist on the plasma membrane of spermatozoa, which is in agreement with the concept of a mosaic composition.

Ferritin-WGA in the caput epididymidis more strongly labels the cell surface of the head and the principal piece, than the middle piece (Figs. 8,9). N-acetyl-D-glucosamine controls are negative. On the other hand, as described in the rabbit (6), the labeling disappears during spermatozoon passage through the lower regions of the epididymis.

Ferritin-Con A labeling demonstrates a high degree of variability in the surface distribution of receptors according to the species studied (2,4,6,7). In the *Macaca*, Con A binding occurs in the distal corpus of the epididymis, mainly in the head of spermatozoa. In the cauda, the density of receptors increases in the principal piece and, after ejaculation, in the middle piece. Controls with α-methyl-D-mannoside are negative (Figs. 10,11).

In summary, the changes in charge distribution and lectin-binding sites on the spermatozoon plasma membrane occur in the corpus epididymidis.

## REFERENCES


1. Bedford J.M. 1975. Maturation, transport and fate of spermatozoa in the epididymis. In: Handbook of Physiology. Sec. 7, vol. 5 (D.W. Hamilton and R.O. Greeps, eds). Amer. Physiol. Soc., Washington, D.C. 303-317.
2. Courtens J.L. and Fournier-Delpech S. 1979. Modifications in the plasma membranes of epididymal ram spermatozoa during maturation and incubation *in utero*. J. Ultrastruct. Res., 68, 136-148.
3. Flechon J.E. 1975. Ultrastructural and cytochemical modifications of rabbit spermatozoa during epididymal transport. In: The Biology of Spermatozoa. Transport, survival and Fertilizing ability (E.S.E. Hafez and C.G. Thibault, eds). Karger, Basel, 36-45.
4. Gordon M., Dandekar, P.V. and Bartoszewicz W. 1975. The surface coat of epididymal, ejaculated and capacitated sperm. J. Ultrastruct. Res., 50, 199-207.
5. Holt W.V. 1980. Surface-bound sialic acid on Ram and Bull spermatozoa: deposition during epididymal transit and stability during washing. Biol. Reprod., 23, 847-857.
6. Nicolson G.L., Usui N., Yanagimachi R., Yanagimachi H. and Smith J.R. 1977. Lectin-binding sites on the plasma membranes of rabbit spermatozoa. J. Cell Biol., 74, 950-962.
7. Olson G.E. and Danzo B.S.,1981. Surface changes in Rat spermatozoa during epididymal transit. Biol. Reprod., 24, 431-443.
8. Yanagimachi R., Noda Y.D., Fujimoto M. and Nicolson G.L. 1972. The distribution of negative surface charges on mammalian spermatozoa. Am. J. Anat., 135, 497-520.


## FIGURE LEGENDS

Figs. 1,2,3: Colloidal Iron at pH: 1.8 in epididymis caput (1), distal corpus (2) and ejaculate (3).

Figs. 4,5,6: Cationized ferritin after neuraminidase treatment in ejaculate.

Fig. 7: Colloidal Iron at pH: 9 in initial segment.

Fig. 8,9: Ferritin-WGA in epididymis caput.

Figs. 10,11: Ferritin-Con A in distal corpus (10) and ejaculate (11).

All figures are x 20,000 except Figs. 5 and 11 which are x 40,000.

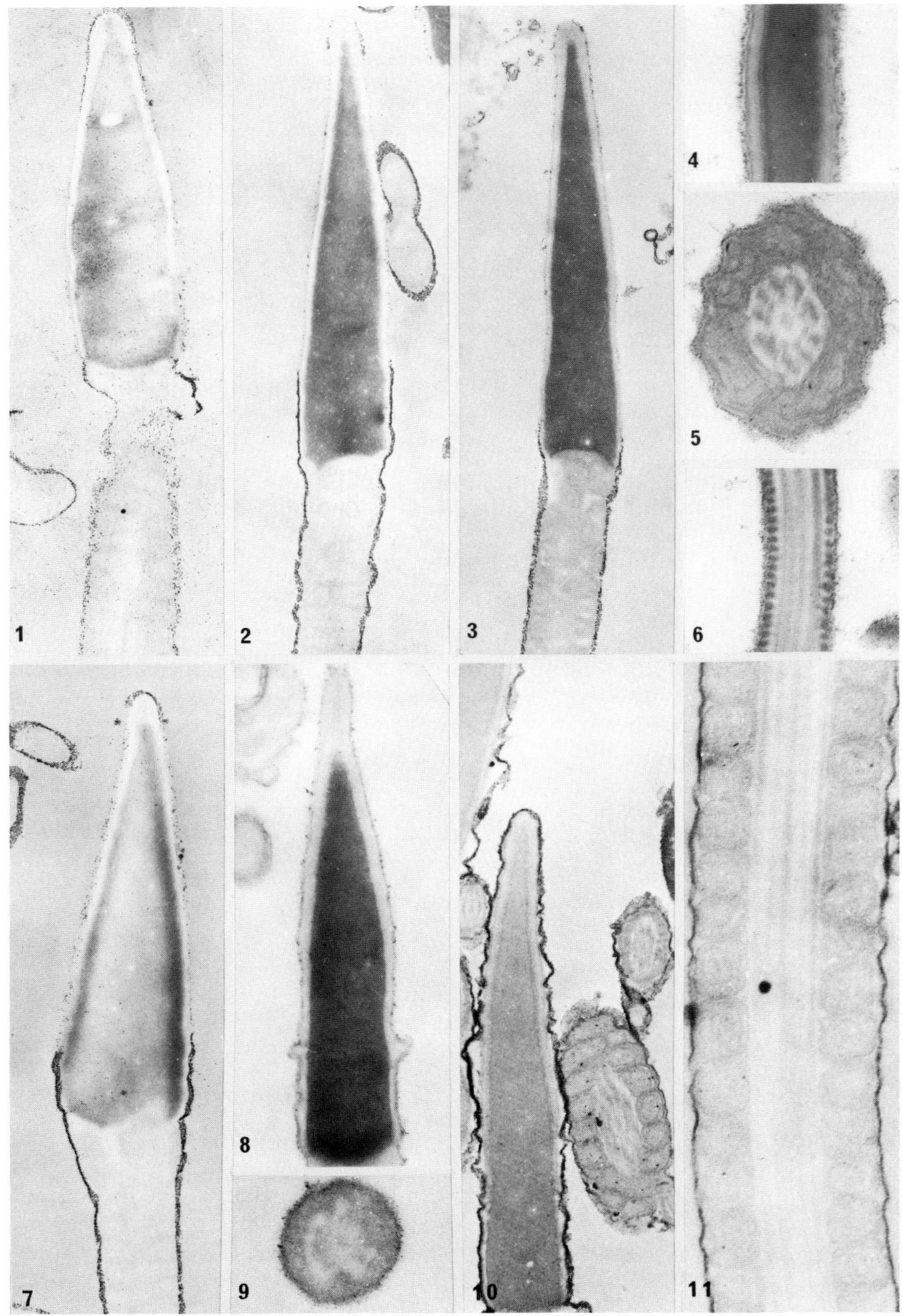
1
2
3
4
5
6
7
8
9
10
11

# E.M. LOCALIZATION OF SUGARS HAVING AFFINITY FOR LECTINS ON THE RAM SPERMATOZOA

S. Fournier-Delpech, S. Hamamah, B. Delaleu, J.L. Courtens, Cl. Pisselet

I.N.R.A. Station de Physiologie de la Reproduction. Laboratoire de Fertilité mâle. Nouzilly, 37380 MONNAIE.

In the epididymis, zona pellucida affinity develop on the membrane of the head of the spermatozoa (2) under the control of the endocrine activity of the testis which regules the epididymal sperm maturation (7). Referring to experiments in which lectins were used to analyse the structure of cell membrane (5,6), we have studied glycoconjugates on the surface in the head of the ram spermatozoa, the role of the epididymis in the regulation of these conjugates, and the possible role of lectin-like proteins in the induction of zona binding in testicular sperm.

## MATERIALS AND METHODS

### A. Electron Microscopic localization:

Testicular or distal epididymal sperm were diluted respectively as v/v or v/3v in saline, after extraction by micropuncture from the rete testis or the tail of the epididymis of normal rams, or orchidectomized epididymis for 9 days (castrated epididymis) or from the epididymis remained in place for 9 days after unilateral orchidectomy (castrated supplemented epididymis). Testicular spermatozoa were also treated with epididymal secretion as follows: 100 µl of testicular sperm recovered by cannulation of the rete testis were centrifuged in duplicate; the packed spermatozoa were then resuspended in 50 µl of testicular plasma, or in 50 µl of epididymal secretion recovered from the tail of the castrated epididymis supplemented for 3 months (ie: free of spermatozoa) with the contralateral testis were incubated for 15 minutes at room temperature and then 100 µl of saline was added.

Spermatozoa were absorbed on formwar carbon-coated films mounted on E.M. grid and were labelled with Concanavalin or Asparagus or Wheat germ (WGA) or Soyabean agglutinins linked with colloidal gold as previously described (1). Prior to this, the wheat germ agglutinin had been attached to Bovine serum albumin (3). To assess the specific binding of the lectins to their hapten sugars, we observed the effect on the labelling of the spermatozoa with 0.5 M solutions of αméthylmannose (CON A), or fucose (Asparagus A), or N.acetylglucosamine (wheat germ) or Nacetylgalactosamine (Soyabean A). The lectins were displaced and we obtained unlabelled spermatozoa with all except the soyabean agglutinin.

B. Zona binding assay of testicular treated with lectins:

$5 \times 10^5$ spermatozoa were suspended in Toyoda and Chang mixture (6) free of glucose and were incubated for 15 minutes with the four lectins in concentrations ranging from 0.01 to 2 mg/ml. They were washed and then used to inseminate 10 to 15 denuded rat oocytes under paraffin oil at 0°C for 45-60 minutes; eggs were recovered washed and mounted to be observed under a phase contrast microscope.

## RESULTS AND DISCUSSION

### A. E.M. Localization

The results of the E.M. study are shown in the figure 1. Con A receptors were located on the susacrosomal plasma membrane of the testicular spermatozoa, and on the postacrosomal area of the epididymal distal spermatozoa. This localization could be the results of epididymal activity induced by androgens since there was more Con A susacrosomal receptors on the spermatozoa from the castrated epididymis than on the spermatozoa from the castrated epididymis supplemented by the contralateral testis.

Wheat germ agglutin receptors exist on the entire surface of the head of the spermatozoa and do not change markedly in the epididymis.

Asparagus A receptors were localized on the susacrosomal surface of the testicular spermatozoa and further into the postacrosomal area of the spermatozoa in the distal epididymis. There were few WGA and Asparagus receptors on the surface spermatozoa of the castrated epididymis, but they were maintained in castrated epididymis which had been supplemented by the contralateral testis.

Soyabean A affinity was localized on the susacrosomal surface on the testicular and epididymal spermatozoa but the binding was not specific. The significance of these heavy deposits remains to be elucitated since they were not displaced by the competition with their hapten sugar.

Incubation of testicular spermatozoa with epididymal secretion resulted in a strong decrease in the labelling with Concanavalin, Asparagus, Wheat germ agglutinins. This suggests a masking of the saccharides by epididymal products during the in vitro incubation.

FIGURE 1. Localization on testicular or epididymal spermatozoa of binding sites for Concanavalin, Wheat germ and Asparagus agglutinins.

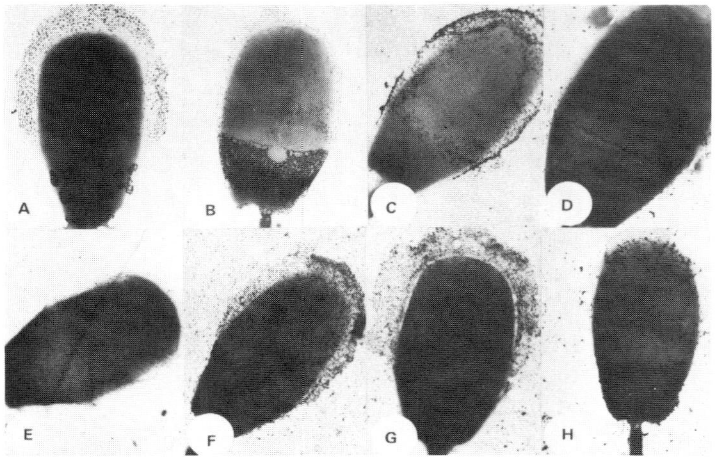


A: Con A Receptors of spermatozoa from testis
B: Con A Receptors of spermatozoa from tail of normal epididymis
C: Con A Receptors of spermatozoa from tail of castrated epididymis (9d)
D: Con A Receptors of spermatozoa from testis castrated supplemented Ep. (9d)
E: Con A Receptors of spermatozoa from testis labelled after their incubation with epididymal secretion.
F: WGA Receptors of spermatozoa from testis or epididymal tail
G and H: Asparagus A receptors of respectively testicular and epididymal spermatozoa.

B. Zona binding assay of testicular spermatozoa treated with lectins

The results are shown in the figure 2.

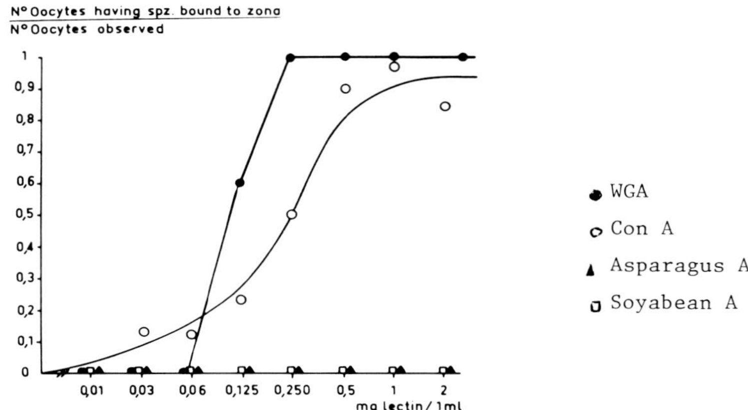


Testicular spermatozoa treated with Asparagus or Soyabean Agglutinins did not bind to rat zona. WGA induced binding to the eggs by the entire surface of the spermatozoa as well as in the testicular spermatozoa. Con A induced oriented binding of the zona pellucida identical to that which occurs between mature ram spermatozoa and rat oocyte (2). The nature of the attachement of the sperm to the zona remains to be elucided: due to saccharidic residues (N-acetyl glucosaminyl or Mannosyl) of the zona or to a remodeling by the lectins of the membrane spermatozoa. This indication suggests that the binding of an epididymal lectin (as Con A) to the sperm cells allows them to commence the initial stages of the attachment to the zona.

IN CONCLUSION, mannosyl an fucosyl residues of the surface of the head of the ram spermatozoa appear of particular interest since their localization is controlled by the epididymis; furthermore, Con A receptors could mediate testicular (rat) zona attachment similar to that of the mature sperm.

REFERENCES

1. Courtens J.L., Rozinek J, Fournier-Delpech S. 1982. Andrologia in press
2. Fournier-Delpech S, Courtens J.L., Pisselet Cl, Delaleu B, Courot M. 1982. Gametes Research. 5: 403-408.
3. Horisberger M. and Rosset J. 1976. Experienta. 32:8-998.
4. Koehler I.K. 1981. Archives of Androl. 6: 197-217.
5. Olson G.L. and Orgebin-Crist M.C. 1982. Ann. N. Y. Acad. Sc. In press
6. Toyoda Y. and Chang M.C. 1974. Reprod. Fert. 36: 9-22
7. Orgebin-Crist M.C., Danzo B.J., Davies J. 1975. Handbook of Physiology Sect. 7 (D.W. Hamilton and R.O. Greep eds). American Physiological society, Bethesda M.D. 319-338.

# THE JOINT SPERMATOZOA OF DYTISCIDAE

G. WERNER
Medizinische Biologie, Universität des Saarlandes, D-6650 Homburg/Saar, West-Germany.

The paired or grouped spermatozoa of Dytiscidae have been thoroughly described by Ballowitz (1895), who studied no fewer than 12 species. Already in the light microscope this author gave a precise account of the most peculiar forms of the heads, which may even vary to a large extent within different species of this family. The heads of the spermatozoa of Acilius, Dytiscus, and Hydaticus are flat, of triangular shape, the somewhat thicker side is indented by a deep pouch. Conjugation is achieved in the way that two spermatozoa come to lie antiparallely with these parts of the heads while the free thin sides are wound around the corresponding partner. In Colymbetes Ballowitz found elongated heads, the posterior parts of which beeing hood-shaped and enclosing the beginning of the tail. These spermatozoa are joined to groups of three or more. The results of Ballowitz have been confirmed and extended by electron microscopy by Bawa (1975) for Dytiscus and Mackie and Walker (1974) for Dytiscus and Colymbetes. The latter authors found the cement to be PAS-positive material. The origin of this glue and the process of the pairing however still remained unknown. Further information to these questions may be found by following the development during late spermiogenesis.

For Acilius, Dytiscus, and Hydaticus results have already been published elsewhere (Werner, 1976). For Rhantus whose spermatozoa closely resemble those of Colymbetes new results will be presented here. Besides morphological diversity the unexpected results were that the origin of the cement and the formation of the conjugation in the spermatozoa of Dytiscidae exhibit remarkable differences.

The results of the foregoing study may be briefly summarized for Dytiscus: The membrane of the spermatid becomes different from the remainder in certain places of the head and anterior part of the tail and gets a special coating. Later the heads of the spermatids are found in deep indentations of the cells of the follicular wall. Vesicles are pinched off from the Golgi-apparatus of these cells. They release their tubular contents into the invagination. This PAS-positive material is deposited to the membrane of the spermatid preferentially to that side to which later during pairing the second spermatozoon will be attached. Moreover, fibrillar material is found around the so-called barb which is also released by the cell of the follicular wall. After leaving the invagination of the wall cells the maturing spermatozoa first remain single. Not before they reach the lower part of the vas deferens pairing will appear. The tubular filaments between the spermatozoa are transformed into granules, around the borders of the heads a cloud of fine fibrils appears. The fibrillar material around the "barb" also spreads and fills the distance between the joint spermatozoa. While in Dytiscus only the membrane is changed, the cement itself is made by cells of the follicle wall.

In Rhantus the formation of groups of spermatozoa is found to be a process of still higher complexity. It is accomplished by at least three different components which must be added in a certain sequence. They are apparently made by the spermatid itself. There is no evidence for participation of other cells.

During development of the late spermatid the cell membrane fuses with the envelope of the nucleus beginning on two sides (fig. 1). This compound membrane bears a coat

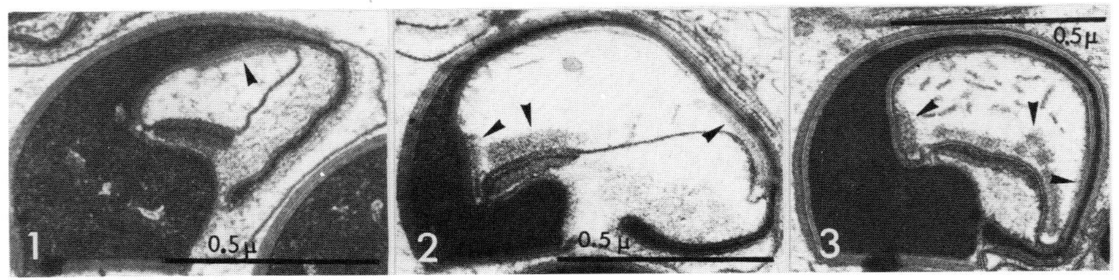


of rodlets. The slightly concave face of the head will later be covered by the cement. The fusion of the cell membrane with the nuclear envelope is continued around the convex face of the head (fig. 2, 3). This compound membrane is coated as the fusion proceeds, the later appearing coat, however, is not as thick and complete as the first. The posterior part of the head is formed as a hood. Thereby in one place of the inner side the cell membrane also fuses with the nuclear envelope. And here a first coat is seen as well though not so firmly attached as on the outer side (fig. 1). Little later a second coat is found on the membrane covering the beginning of the tail. A third coat distinctly separate from the second one is nearby (fig. 2, arrows). Besides these coats a few membrane-bound vesicles and membrane-like lamellae are found in the hood. The uniform coat covering the membrane at the beginning of the tail dissolves itself into small aggregations of fine filaments (fig. 3, 4). Opposite to this a crystalline material appears for the first time (fig. 4). As a third component a spongy material becomes visible, possibly derived from the membrane-like lamellae, which

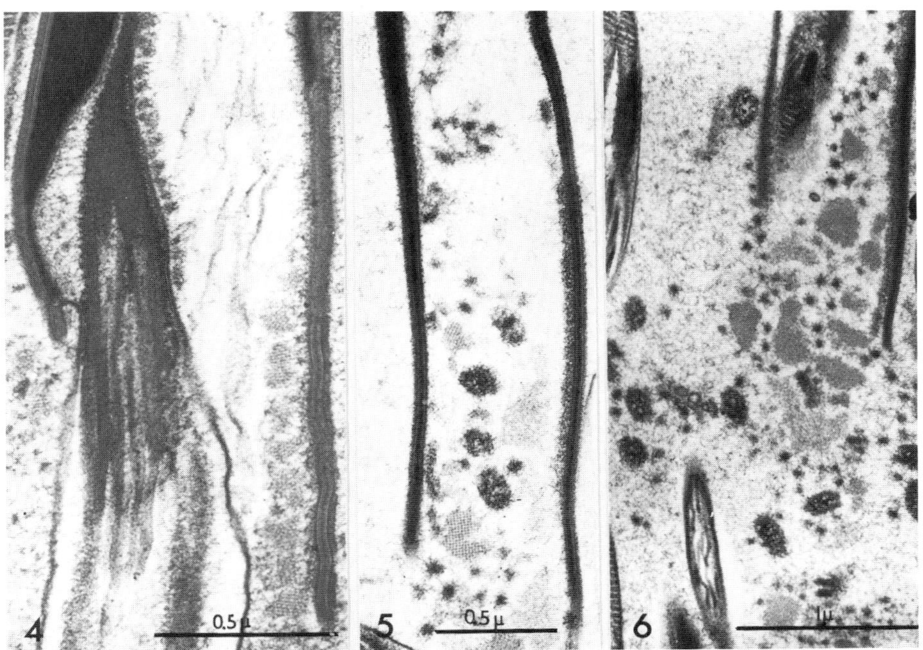


have disappeared then (fig. 5). These three components are simultanously delivered from the hood between the spermatozoa (fig. 6). First the spongy material is attached to the formerly concave now almost straight face of the head which ultimately becomes even convex (fig. 7). Then the fine filamentous substance is superimposed to the lateral half (fig. 8). Now a transformation takes place, after which the former distinctly discernible components have disappeared. Instead a crescent-like condensation is visible surrounded by fine fibrillar material (fig. 9). Not before this stage the conju-

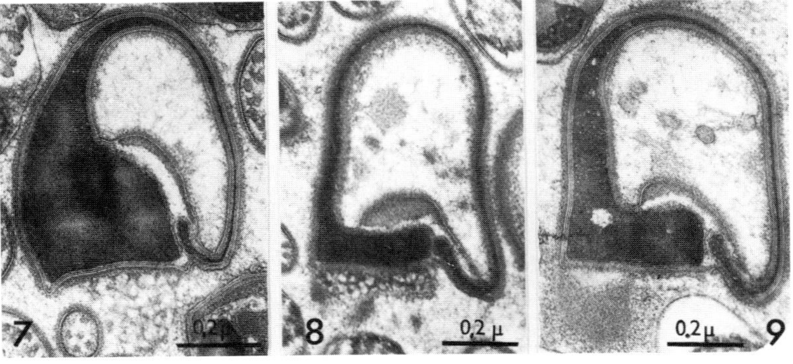


gation of the spermatozoa can take place. The conjugation is formed between a group

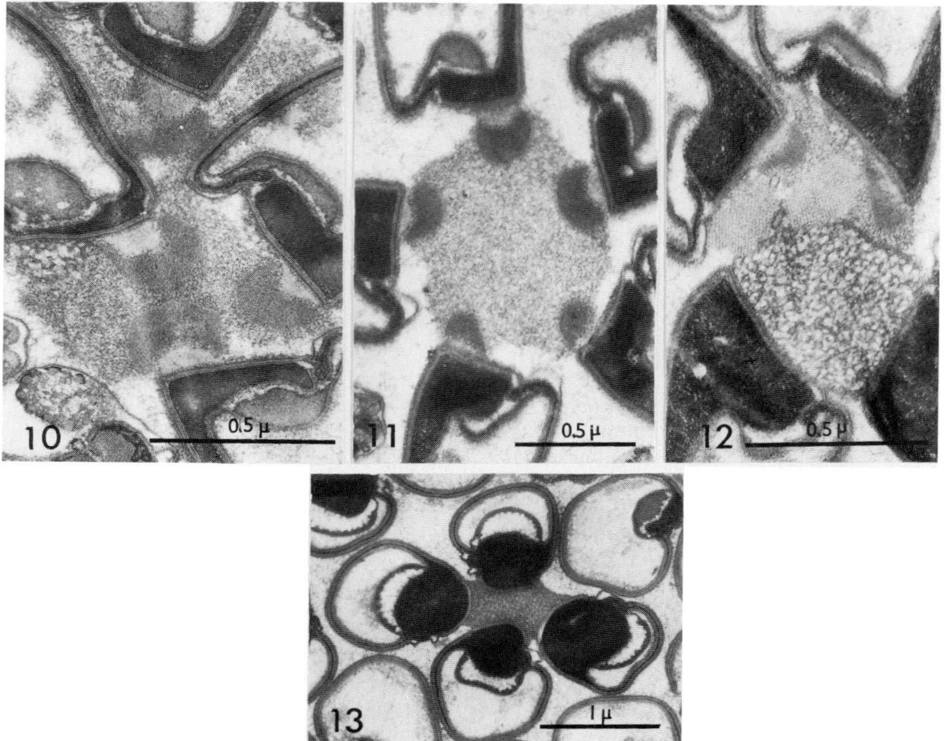


from three to six spermatozoa lying together by chance (fig. 10). In the preliminary junction the crescents are still visible (fig. 11).When the crystalline material is added as the last component the cement is transformed to a uniformly spongyly appearing material (fig. 12, 13). It extends over the whole length of the heads except the foremost part with the acrosomes which are surrounded by a loosely attached fibrillar network which is seen in less dense distribution all over the lumen of the duct.

REFERENCES

1. Ballowitz E., 1895. Die Doppelspermatozoen der Dyticiden. Z. Zool. 60, 458-499
2. Bawa SR., 1975. Joined spermatozoa. In Afzelius The functional anatomy of the spermatozoon. 259-266, Pergamon Press, Oxford
3. Mackie JB. Walker MH., 1974. A Study of the conjugate sperm of the dytiscid water beetles Dytiscus marginalis and Colymbetes fuscus. Cell Tiss. Res. 148, 505-519
4. Werner G., Entwicklung und Bau der Doppelspermien bei den Dytisciden Acilius sulcatus L., Dytiscus marginalis L. und Hydaticus transversalis Pont. (Coleoptera). Zoomorph. 83, 49-87.

# THE FATE OF SECRETED GLYCOPROTEINS IN THE MAMMALIAN EPIDIDYMIS ANALYSED BY ELECTRON MICROSCOPIC AUTORADIOGRAPHY

V. KOPEČNÝ[(1)], J-E. FLÉCHON[(2)], J. PIVKO[(3)]
(1) Institute of Histology and Embryology, J.E. Purkyně University, BRNO, Czechoslovakia. (2) Station de Physiologie Animale, INRA, 78350 Jouy-en-Josas, France. (3) Animal Production Research Institute, NITRA, Czecholosvakia

## SUMMARY

Autoradiography on semi-thin and thin sections was used to study the distribution of secretory products in the epididymis after the injection of precursors of glycoproteins, D-glucosamine 1-$^{3}$H and L-fucose 6-$^{3}$H. In guinea-pig labelling was associated with the sperm heads undergoing morphological changes in the acrosome and also with the sperm heads on spermatozoa aggregated in "rouleaux". In the mouse, the label was still present,mainly on the sperm heads,in cauda epididymis 5 days after injection ; in the epididymal epithelium, the clear cells were also labelled.

## INTRODUCTION

One of the sperm changes occuring during epididymal maturation is the coating of the cell surface by substances,mainly glycoproteins,elaborated by the epithelial cells. For a review of the biochemical and (immuno) cytochemical evidence, see Orgebin Crist et al., (1981) and Fléchon (1981). The aim of the present study using the autoradiographic technique, was to confirm and localize the association of labelled products of the epididymis with the sperm surface during epididymal transit. We used tritiated glycoprotein precursors among which fucose is the most specific (Corfield and Shauer, 1979). We chose to work with guinea-pigs in order to obtain a precise autoradiographic labelling at the electron microscopic level on the large sized acrosomal region.

## MATERIAL AND METHODS

A total dose of around 10 mCi of L-fucose 6-$^{3}$H (20 Ci/mmole) was injected into three adult male guinea-pigs and 8.5 mCi of D-glucosamine 1-$^{3}$H (11 Ci/mmole) into a fourth one. The epididymides were collected 24 hrs (or 3 days in one case) after injection. Five mCi of L-fucose 6-$^{3}$H were also injected into an adult male mouse and the material was dissected 5 days later. The epididymal ducts of the guinea-pig were immersed in fixative and divided into 7 segments according to Hoffer and Greenberg (1978) ; those of the mouse were fixed and divided into 8 segments. The fixatives used were :

- 5 % glutaraldehyde in cacodylate buffer with 1 % cetylpyridimium chloride added,
- 5 % glutaraldehyde in cacodylate buffer with 1 % tannic acid added.

The first fixative was either used alone or it was used first befcre the second one. The material was washed in buffer in the presence of cold precursor and embedded in paraffin (first experiment) or in epon. Ilford K5 liquid emulsion and Ilford L4 nuclear liquid emulsion were used respectively for light and for electron microscopy. Grain counts on the electron micrographs were necessary to obtain a figure of the labelling on the small mouse spermatozoa.

## RESULTS

Labelling was first found on guinea-pig spermatozoa 24 hrs after the injection of tritiated glucosamine. The labelled sperm cells were located in segments where the shape of the acrosome was changing (fig. 1). A second wave of labelling occured 3 days after tritiated fucose injection on spermatozoa associated in "rouleaux" in segments IV to VII. Silver grains were mostly located in the intercellular space between the superposed apical parts of the acrosomes (fig. 2). Quantitative analysis of the autoradiograms showed that the density of the grains on the mouse spermatozoa was twice that of the epididymal lumen in different segments of the cauda (e.g. V, VII). The labelling of the epididymal epithelium was less dense, except for the clear cells in the corpus and proximal cauda.

## DISCUSSION

The epididymis produces androgen-dependent glycoproteins some of them have been found to bind to the sperm surface (Acott and Hoskins, 1981 ; Lea and French, 1981 ; Moore, 1981). Several autoradiographic studies have analysed the synthetic activity of the epididymal epithelium (review in Flickinger, 1981). However, although the previous study of one of us (Kopečný and Pech, 1977), has shown for the first time the association of labelled material elaborated by the epididymis with spermatozoa, ultrastructural localization of the labelling was not yet described. Our positive results can be explained by the high radioactive dose used and by the addition of cetylpyridimium chloride to the fixative, which is known to preserve the surface coat (Shea, 1972). No improvement was obtained after further fixation with tannic acid.

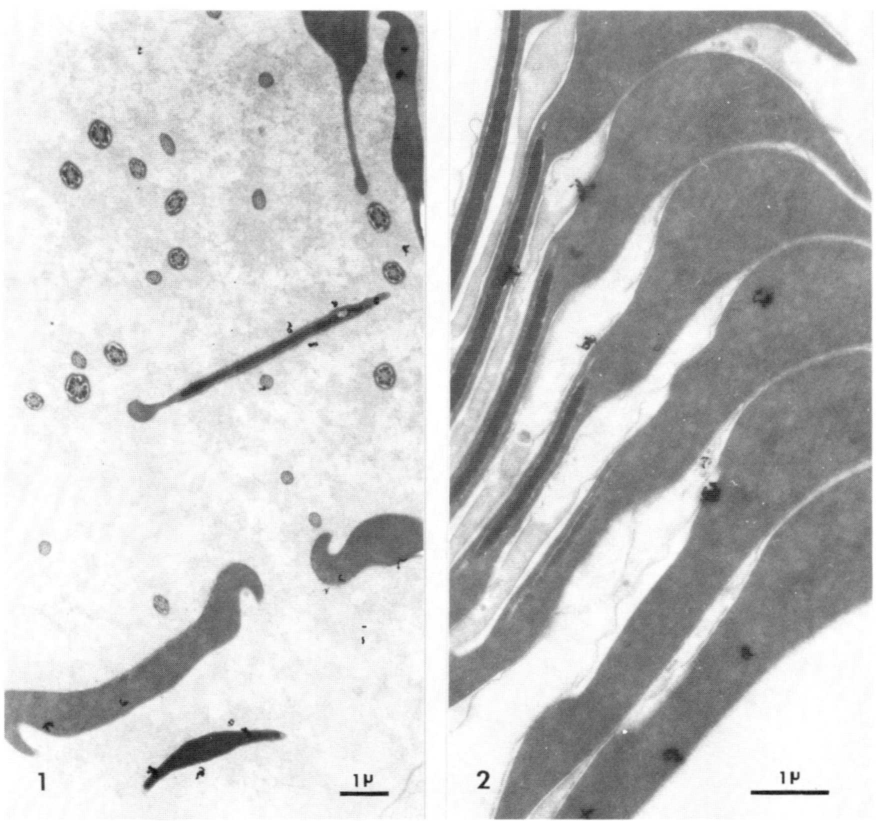


FIGURE 1. Labelling of guinea pig sperm heads in segment II of the epididymis 24 hrs after glucosamine-$^3H$ injection
FIGURE 2. Surface labelling of guinea pig sperm acrosomal regions associated in rouleaux (segment VII) 3 days after fucose-$^3H$ injection

Glycoproteins secreted by the guinea-pig epididymis associated mainly with the sperm head surface during at least two phases, i.e. the stage of the remodeling of acrosome shape and the stage of piling into "rouleaux". In the latter case, coating material (Bearer and Friend, 1982) may be responsible for "rouleaux" formation as well as sperm autoantigens (Tung et al., 1980). The association of labelled glycoproteins with the sperm surface also occurs in a species like the mouse where no acrosomal change or mutual binding occurs during maturation. The occurrence of labelling in the clear cells in the mouse 5 days after tritiated fucose injection was interpreted as proof of the resorptive activity of these cells (Moore and Bedford, 1979).

REFERENCES


1. ACOTT T.S. and HOSKINS D.D. 1981. Biol. Reprod. 24 : 234.
2. BEARER E.L. and FRIEND D.S. 1981. J. Cell. Biol. 92 : 604.
3. CORFIELD P. and SCHAUER R. 1979. Biol. Cell. 36 : 213.
4. FLÉCHON J.-E. 1981. In : Progress in Reproductive Biology, Vol. 8. Epididymis and Fertility : Biology and Pathology. Eds Bollack C, Clavert A., S. Karger 1981, p. 90.
5. FLICKINGER C.J. 1981. Biol. Reprod. 25 : 871.
6. KOPEČNÝ V. and PECH V. 1977. Histochemistry 50 : 229.
7. LEA O.A. and FRENCH F.S. 1981. Bioch. Biophys. Acta. 668 : 370.
8. MOORE H.D.M. 1981. J. Exptl. Zool. 215 : 77.
9. MOORE H.D.M. and BEDFORD J.M. 1979. Anat. Rec. 193 : 313.
10. ORGEBIN-CRIST M.C., OLSON G.E. and DANZO B.J. 1981. In : Intragonadal Regulation of Reproduction, Eds. Franchimont P. and Channing C.P., Academic Press, p. 393.
11. SHEA C.S. 1971. J. Cell Biol. 51 : 611.
12. TUNG K.S.K., OKADA A, and YANAGIMACHI R. 1980. Biol. Reprod. 23 : 877.

# EPIDIDYMAL PROTEIN SECRETION

D.E. BROOKS
Department of Animal Physiology, Waite Agricultural Research Institute, University of Adelaide, Glen Osmond, South Australia 5064.

The fluid within the lumen of the epididymis contains a variety of proteins. Some of these proteins are derived from the rete testis fluid, others such as albumin and transferrin traverse from the blood stream to the epididymal lumen. However, from a functional point of view, the most important proteins are likely to be those epididymal-specific proteins which are synthesized by the epididymal epithelium itself and then secreted into the epididymal lumen. Within the lumen these proteins may become associated with the sperm surface or in some other way modify the luminal milieu and hence contribute to the process of sperm maturation which occurs during passage through the epididymal duct. Characterization of these proteins and elucidation of the mechanisms regulating their secretion would therefore seem to be of particular importance in understanding the phenomenon of epididymal sperm maturation.

An excellent method of studying protein synthesis and secretion is to use radioactive precursors to label proteins synthesized and secreted by epididymal tissue _in vitro_. The resultant radioactive proteins are separated by analytical polyacrylamide-gel electrophoresis and then visualized and quantitated by exposing the polyacrylamide gel to x-ray film (1).

The epididymis has previously been subdivided into various segments based on morphological criteria (2), but this does not necessarily correlate with subdivision based on functional criteria. For the present study, the epididymis was subdivided into ten arbitrary segments (Fig. 1a) which were incubated separately with radioactive amino acids. The radioactive proteins formed within the tissue cytoplasm were very similar in all epididymal segments (Fig. 1b) but those which were secreted (Fig. 1c) varied markedly from segment to segment.

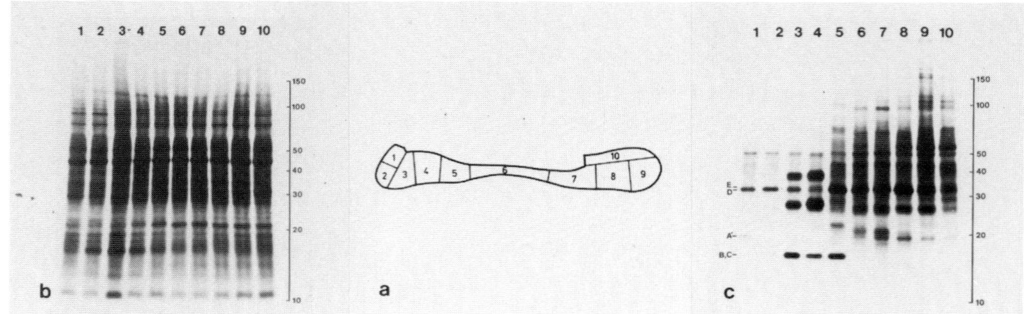

**FIGURE 1**. Regionality of synthesis and secretion of proteins from radioactive amino acids by the rat epididymis _in vitro_. a) Subdivision of epididymis into 10 segments. b) Segmental profile of radioactive cytoplasmic proteins. c) Segmental profile of radioactive secreted proteins.

Radioactive sugars can also be used to label those secretory proteins which are glycosylated. The most useful sugar precursors in this regard are [$^3$H]mannose and [$^3$H]fucose. One mannose-containing glycoprotein has been characterised previously and has been called AEG (3). This protein corresponds to protein D described by others (1,4). Protein D and a closely-related protein (protein E) have recently been purified in our laboratory and antisera have been raised against each separately (5). In all cases antiserum against protein D cross-reacts wtih that against protein E and vice versa. Antiserum has been used to specifically isolate radioactive proteins D and E from incubations of epididymal tissue with radioactive methionine _in vitro_. Protein D is synthesized throughout the epididymis with the exception of the initial segments (segments 1 and 2 in Fig. 1a); protein E is synthesized in segments 6 - 8.

Tunicamycin is an antibiotic which inhibits the incorporation of carbohydrate residues into N-linked glycoproteins. In some tissues, inhibition of glycosylation can prevent protein secretion. Evidently this is not the case for proteins D and E in the epididymis, for tunicamycin prevents glycosylation but not secretion of the non-glycosylated forms of these proteins (see proteins D' and E' migrating with lower apparent molecular weight in Fig. 2). Note that tunicamycin does not affect secretion or the molecular size of proteins B and C which are not glycoproteins.

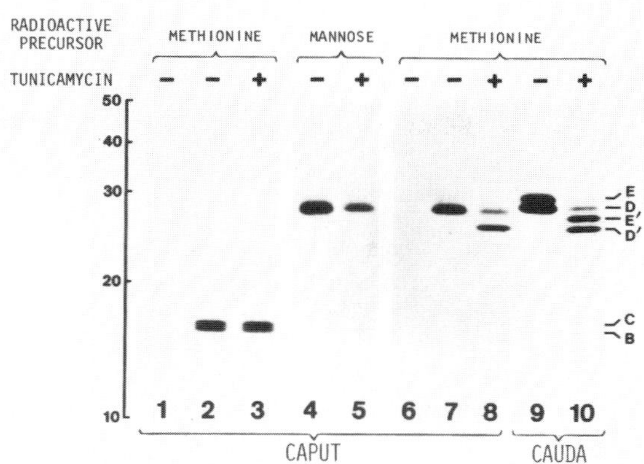


FIGURE 2. Effect of tunicamycin on secretion of specific rat epididymal proteins. Antiserum directed against proteins B and C or against proteins D and E was used to immunoprecipitate radioactive proteins secreted by epididymal tissue after incubation with radioactive precursor in vitro. Lanes 1 and 6 contained preimmune serum instead of immune serum. Note that tunicamycin caused proteins D and E to be secreted with a lower apparent molecular weight (D' and E' in lanes 8 and 10) and in the caput, band D' did not incorporate mannose (lane 5).

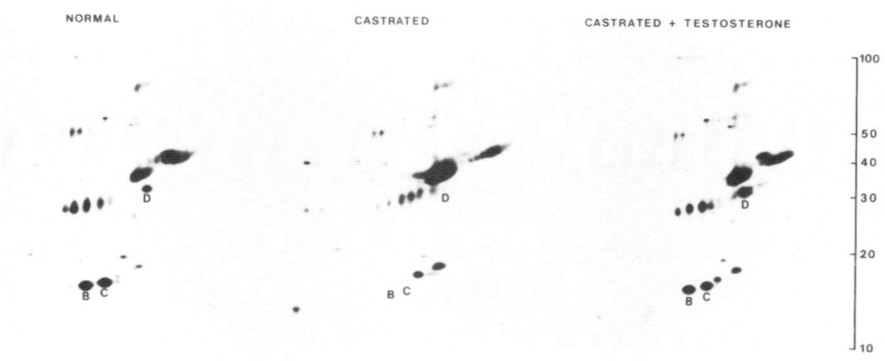


FIGURE 3. Effect of androgen status on secretion of radioactive proteins by the rat caput epididymidis following incubation with radioactive methionine in vitro. Two dimensional separation employed isoelectric focusing in the first dimension (left to right) and SDS-polyacrylamide gel electrophoresis in the second dimension (top to bottom). The position of androgen-dependent proteins B, C and D is indicated. Because equal amounts (cpm) of radioactive protein were loaded onto each gel, the disappearance of proteins B, C and D after castration has meant that the relative intensity of some other proteins has increased.

FIGURE 4. Effect of procaine on secretion of radioactive proteins by the rat caput epididymidis following incubation with radioactive methionine *in vitro*. Lanes 1-5 contain an equal amount of incubation medium. Lanes 1-6 contain an equal amount (cpm) of radioactive protein.

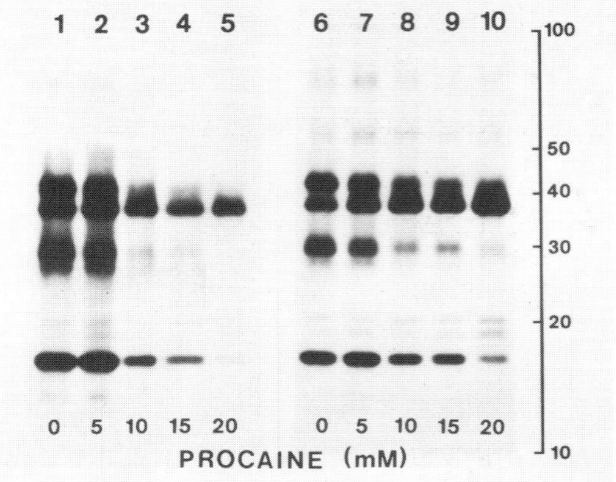


Many epididymal functions are regulated by androgens (6). Androgens also regulate epididymal protein synthesis and secretion, but not to the extent which might have been expected (1,7). Examination of secreted radioactive proteins separated by two-dimensional gel electrophoresis (Fig. 3) shows that androgens only regulate the synthesis and secretion of a few specific proteins (e.g. proteins B, C and D). However, androgens may have a greater effect on glycosylation reactions since incorporation of radioactive mannose and fucose into secretory glycoproteins is reduced to 30% of normal following castration.

Epididymal protein secretion can also be modified by agents which interact with cell membranes, such as various local anaesthetics. For instance, increasing concentrations of procaine cause increased suppression of the synthesis and secretion of some proteins, but not of others, and this leads to an altered profile of secreted proteins (Fig. 4). The precise mechanism of action of local anaesthetics on epididymal protein secretion is still under investigation.

REFERENCES

1. Brooks DE, Higgins SJ. 1980. J. Reprod. Fert. 59, 363.
2. Reid BL, Cleland KW. 1957. Aust. J. Zool. 5, 223.
3. Lea OA, Petrusz P, French FS. 1978. Int. J. Androl. Suppl. 2, 592.
4. Cameo MS, Blaquier JA. 1976. J. Endocr. 69, 47.
5. Brooks DE. 1982. Int. J. Androl., in press.
6. Brooks DE. 1981. Physiol. Rev. 61, 515.
7. Brooks DE. 1981. Biol. Reprod. 25, 1099.

# CHAPTER 3 NUCLEUS

## . SOME ASPECTS OF CHROMATIN ORGANIZATION IN SPERM NUCLEI

Ph. CHEVAILLIER
Laboratoire de Biologie Cellulaire, Université Paris-Val-de-Marne,
Avenue du Général de Gaulle, 94010 Créteil, France.

### 1. INTRODUCTION

During sperm differentiation, the nuclear chromatin is generally affected by dramatic modifications of its organization inside the nucleus, of the protein complement associated with DNA as well as of its biological properties. From an anatomical point of view, the general shaping of the nucleus and the different phases of chromatin modelling along spermiogenesis follow different schedules despite the fact that the ultimate step corresponds almost in all cases in highly compacted chromatin. Taking into account the results obtained with cytochemical tests, four main kinds of nuclear basic protein evolution were described (Bloch, 1969, 1976) : 1) Sperm nuclear proteins are histone-like proteins ; 2) Somatic histones are replaced by protamines ; 3) Sperm nuclear proteins have properties which seem intermediate between histones and protamines ; 4) No basic protein can be detected inside mature sperm nuclei. No relationship can be clearly established between the nuclear basic protein content, the degree of chromatin compaction and sperm nuclear morphology, except in case 4) where sperm nuclei are irregularly shaped and chromatin appears poorly condensed, such as in some Crustaceans. As the structure of the sperm histones and of the so-called intermediate basic proteins are discussed elsewhere during this Symposium, we shall discuss in this review only the known properties of protamines and their interactions with DNA in sperm chromatin.

### 2. STRUCTURE OF SPERM PROTAMINES

The first protamines whose sequences have been determined were proteins extracted from some fishes all belonging to species of Teleosts which are neighbours from a systematic point of view (for a review, see Ando, 1973). They were called monoprotamines as they have high amounts of only one basic amino acid which is arginine. These protamines can be described as small highly basic proteins with a few number of different amino acids (five to seven) and containing from 30 to 34 residues (fig. 1) ; in each of these species the protamine fraction appears heterogeneous being constituted of three or four different polypeptide chains with a high homology of amino acid sequence. Moreover, recent investigations by protamine gene cloning have revealed that the heterogeneity in a protamine family extracted from one species such as the rainbow *Salmo gairdnerii* is somewhat higher: at least twelve different gene structures or polypeptide sequences can be coined in this species (Jenkins, 1979 ; Jenkins *et al.*, 1979 ; Sakai *et al.*, 1978, 1981 ; Gedamu *et al.*, 1981).

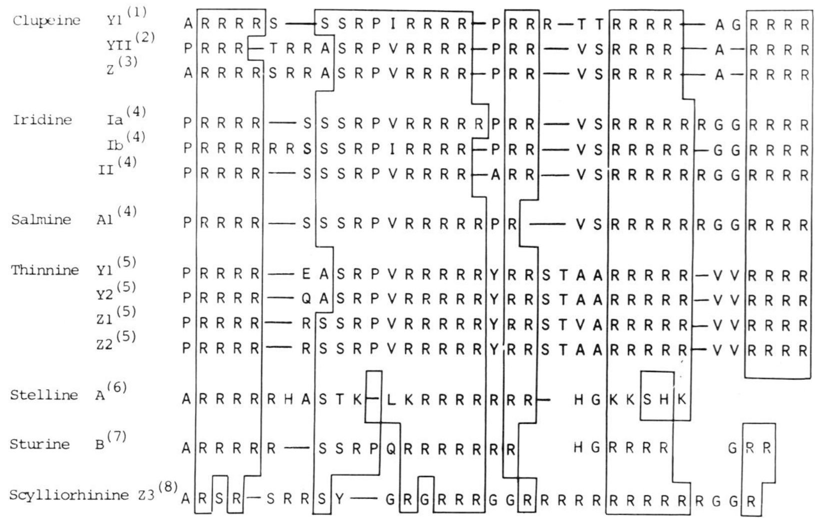


Fig. 1 - COMPARISON OF AMINO ACID SEQUENCES OF VARIOUS FISH PROTAMINES

(1) : Ando and Suzuki, 1967 ; (2) : Ando and Suzuki, 1966 ; (3) : Ando *et al.*, 1962 ; (4) : Ando and Watanabe, 1969 ; (5) : Bretzel, 1972, 1973 ; (6) : Yulikova *et al.*, 1979 ; (7) : Yulikova *et al.*, 1976 ; (8) Sautière *et al.*, 1981).

In all the species cited above, the primary structure of the different polypeptide chains shows little variation. However, this situation does not seem to be general. From the mature sperm of the dog-fish *Scylliorhinus caniculus*, we have extracted and characterized four different protamines which have been called scylliorhinines Z1, Z2, Z3 and S4 with increasing electrophoretic mobility (Gusse and Chevaillier, 1978, 1980a). Their electrophoretic mobility and their amino acid composition were significatively different (fig. 2 and 3). Scylliorhinine Z3, the first

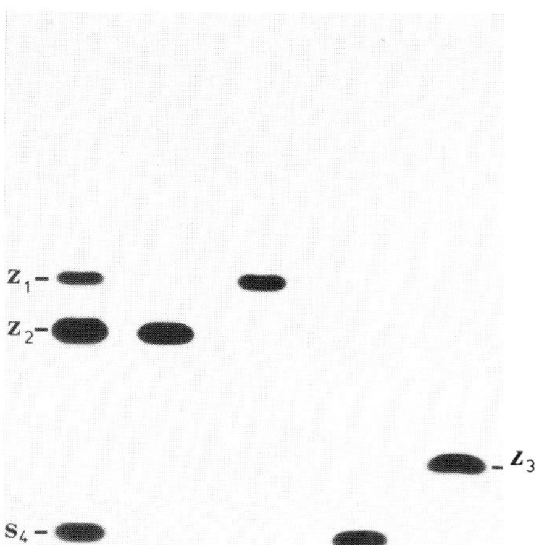


Fig. 2. The four protamines (= scylliorhinines) extracted and purified from dog-fish sperm nuclei.

Fig. 3 - AMINO ACID COMPOSITION OF SCYLLIORHININES

(P. Sautière, M. Gusse, G. Briand and Ph. Chevaillier, unpublished results).

| Amino acids \ Protamines | Scylliorhinines | | | |
|---|---|---|---|---|
| | Z1 [a] | Z2 [a] | Z3 | S4 |
| asp | 2 | - | - | - |
| thr | 2 | 1 | - | - |
| ser | 2 | 2 | 3 | - |
| glu | 1 | 1 | - | - |
| pro | 3 | 1 | - | 2 |
| gly | 1 | 1 | 6 | 1 |
| ala | 1 | 4 | 1 | 3 |
| cys | 5 | 5 | - | 5 |
| val | 2 | 1 | - | 1 |
| met | - | 1 | - | - |
| ile | - | 1 | - | - |
| leu | 4 | 1 | - | - |
| tyr | 1 | - | 1 | - |
| phe | 1 | - | - | - |
| his | 2 | 2 | - | - |
| lys | 9 | 5 | - | 14 |
| arg | 16-17 | 13 | 2 | 7 |
| Total number of residues | 52-53 | 39 | 31 | 33 |
| Amino acid diversity | 15 | 14 | 5 | 7 |
| Arginine % | ≈ 31 | 33 | 64.5 | 21.2 |

[a] The assumed number of amino acids per molecule of scylliorhinines Z1 and Z2 were calculated from amino acid analyses of protein after total acid hydrolyses.

which has been sequenced, is composed of only five different amino acids and contains 31 residues, arginine being the most represented (64.5 % of total residues). The polypeptide chain shows two different parts (fig. 4): the N-terminal one contains the three serine residues and the tyrosine residue present in the molecule ; the C-terminal part consists only of glycine and arginine residues. The longest sequence of arginine (11 residues) so far observed among protamines occurs in this very basic domain of that protein. Protamine Z3 is acid-soluble as the fish protamines previously described. By contrast, scylliorhinines Z1, Z2 and S4 although being acid-soluble when they appear in the testis during spermiogenesis are no more directly acid soluble in mature sperm nuclei. They can be

1 5 10
$_{(H)}$ala - arg - ser - arg - ser - arg - arg - ser - tyr - gly - arg - gly -
15 20
-arg - arg - arg - gly - gly - arg - arg - arg - arg - arg - arg - arg -
25 30
-arg - arg - arg - arg - gly - gly - arg$_{(OH)}$

Fig. 4 - AMINO ACID SEQUENCE OF SCYLLIORHININE 3

solubilized only after reduction of disulfide bonds which appear late during sperm maturation. Scylliorhinine S4 contains only seven different amino acids and 33 residues per molecule. Lysine (14 residues) and arginine (7 residues) are the two most abundant amino acids. Basic amino acids do not occur as long clusters as in other protamines and no part of the molecule seems to have any specific structure and properties as was observed for scylliorhinine Z3. Protamine S4 contains no amino acid with a hydrophilic lateral chain. A sequence of nine residues is duplicated and the two nonapeptides occur tandemly (Sautière P., Gusse M., Briand G. and Ph. Chevaillier, unpublished results). This protamine does not share any structural character with other protamines whose sequences have been determined previously, including the more recently sequenced sturine B and stelline A from two sturgeon species *Acipenser guldenstadti* (Yulikova *et al.*, 1976) and *Acipenser stellatus* (Yulikova *et al.*, 1979). Scylliorhinine S4 does not show any similitude with scylliorhinine Z3. By its highly basic character, the small size of the molecule and the few number of different amino acids, scylliorhinine S4 can be considered as a true protamine, although it contains a high amount of cysteine (5 residues per molecule). Scylliorhinines Z1 and Z2 are formed of about 40 and 50 residues respectively and are also highly basic proteins containing both arginine, lysine and histidine (fig. 3). Arginine remains the most abundant residue although fifteen or fourteen different amino acids are represented in scylliorhinines Z1 and Z2 respectively. Arginine and lysine represent almost half of the amino acid residues in each molecule. Cysteine is present in both proteins (five residues per molecule) as well as amino acids with a hydroxylated lateral chain although tyrosine is absent in scylliorhinine Z2. Protamine Z1 lacks methionine and isoleucine ; scylliorhinine Z2 lacks aspartic acid, tyrosine and phenylalanine. The amino acid compositions of all four scylliorhinines, the primary structures which have been determined for two of them and the primary structure of the N-terminal part of the other two lead to the conclusion that they are unrelated proteins. So it is difficult to imagine in the case of the dog-fish that the protamine genes emerge from the duplication of a common ancestral gene such as what was supposed for clupeine genes (Black and Dixon, 1967).

As these protamines are cross-linked by disulfide bonds in mature sperm chromatin, scylliorhinines Z1, Z2 and S4 are acid insoluble and the whole protamine complement of the dog-fish sperm could be classified-from the point of view of a cytochemist using the standard Alfert and Geschwind method (1953) - as intermediate proteins. This does not seem correct since scylliorhinines Z3 and S4 are true protamines. It seems to us highly probable that in some other species the situation would be the same and that cross-linking of sperm protamines rendering them acid insoluble it can be assumed that some proteins previously identified as intermediate proteins are in fact protamines (Bols and Kazinsky, 1976 ; Bols *et al.*, 1980). Two other examples can be given. Protamines from sperm sturgeons have been classified as intermediate basic proteins (Bloch, 1976). In fact, the determination of the primary structure of stelline A and sturine B (Yulikova *et al.*, 1976, 1979) have clearly shown that they are true protamines (fig. 1). The same observation can be made for the *Mytilus* protamines which are typical so-called intermediate proteins (*Mytilus* type according to Bloch, 1976). In fact, the basic proteins extracted from mollusc sperm nuclei appear highly heterogeneous even in a single species and contain proteins which migrate like histones or even slower, like protamines or with an intermediate mobility when submitted to polyacrylamide gel electrophoresis (Subirana *et al.*, 1973 ; Colom and Subirana, 1979 ; Zalensky and Zalenskaya, 1980). For all these reasons, the present classification of sperm nuclear proteins other than histones and the nuclear protein transitions occuring during spermiogenesis must be considered with great care as it lies mainly on cytochemical data concerning the whole nuclear protein complement. Neither the amino acid composition nor the properties of solubility of whole unpurified sperm nuclear proteins enable the prediction of the nature of individual fractions.

Outside of the group of fishes, few protamine sequences have been published. In some Mammals such as in the bull, only one protamine is present in mature sperm nuclei. In the mouse, two protamines with slightly different electrophoretic mobility were detected (Bellvé *et al.*, 1975 ; Bellvé and Carraway, 1978). In human sperm the situation is more confusing since from two to seven or more protamine fractions have been detected especially by polyacrylamide gel electrophoresis (Kolk and Samuel, 1975 ; Pongsawasdi and Svasti, 1976 ; Svasti and Talupphet, 1979).

The primary structure of the unique sperm protamine of the bull (Coelingh *et al.*, 1972) which contains 47 residues ($M_r$ = 6258) and 13 different amino acids reveals a highly basic central domain from residue 13 to residue 36 containing 20 of the 24 arginine residues of the molecule and arranged in three clusters of 6 to 7 residues. This central part of the molecule must represent the preferential binding site of the protein to DNA. It contains two identical basic octapeptides -arg-cys$(arg)_6$- which occur tandemly. The N-terminal and the C-terminal regions have common structural properties and especially both contain the seven amino acids with a hydroxylated lateral chain.

Partial primary sequences of protamines have been published in some other Mammals including rat, boar, ram, stallion and man (Monfoort *et al.*, 1973 ; Kistler *et al.*, 1976 ; Gaastra *et al.*, 1978). All these proteins contain about 50 residues and can be characterized as arginine- and cysteine-rich protamines : the arginine content corresponds to about half of the amino acid residues and six to eight cysteine residues are present per molecule. The amino acid diversity ranges from 8 to 13 (Bellvé, 1979). Lysine is present in rat and mouse protamines and in some human protamines ; mouse protamine II and human protamines contain also histidine. Proline which is a very important residue in the determination of the polypeptide chain configuration exists only in the rat and in some human protamines. Three to ten amino acids with a hydroxylated lateral chain are represented per molecule, tyrosine being present in all mammalian protamines except in the mouse protamine II (Bellvé, 1979). The N-terminal amino acid sequences of rat protamine and of human protamine I show homologies with bull sperm protamine. By contrast, the N-terminal sequence of human protamine II appears different from the other known sequences of the same region. Bull, boar, stallion and ram protamines have also some common structural characters in their C-terminal region, the rat protamine and human protamine I being rather different.

The primary structure of only one protamine extracted from Birds is presently known (Nakano *et al.*, 1976). Galline prepared from fowl sperm is a larger molecule of 65 residues ; its M.W. is 8433. Eight different amino acids are represented and arginine corresponds to 58.5 % of total residues. No other basic amino acid is present. Amino acids with a hydroxylated lateral chain are well represented in the molecule (24.5 %) especially serine with eleven residues per molecule. By contrast to mammalian protamines, galline contains no cysteine. The amino acid sequence reveals no specific domain, arginine residues being represented in all parts of the molecule. Arginine and amino acids with a hydroxylated lateral chain alternate in a regular way. Several internal homologies of sequence can be noticed such as the existence of two identical hexapeptides -(arg)$_3$-ser-pro-arg- (residues 16 to 21 and 24 to 29) or the similarity of the amino acid sequences around the four tyrosine residues (fig. 5).

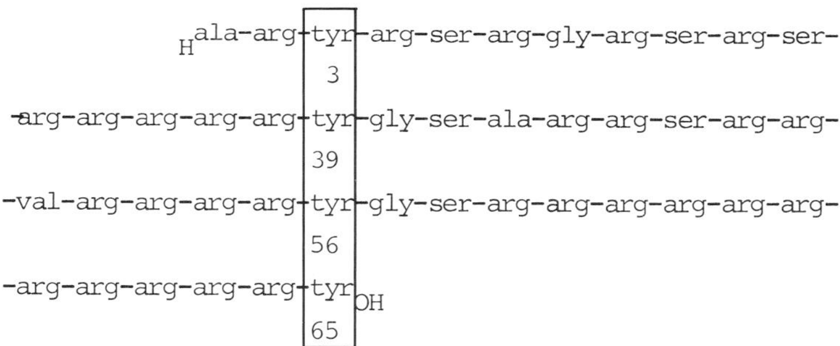


Fig. 5 - INTERNAL AMINO-ACID SEQUENCE HOMOLOGIES IN GALLINE AROUND THE FOUR TYROSINE RESIDUES
(Sequence determined by Nakano *et al.*, 1976)
The C-terminal amino acid sequence has been written twice.

The N-terminal amino acid of most protamines so far studied is either alanine (clupeines YI and Z, sturine B, stelline A, scylliorhinine Z3, galline, human protamine I) or proline (clupeine YII, salmine AI, iridines, thinnines). Glycine is present at the N-terminal end of scylliorhinines Z1 and S4 and of human protamine II ; methionine was found at the same position in scylliorhinine Z2. The C-terminal residue is often arginine but lysine, glutamine, cysteine and tyrosine have also been found at this position.

Nuclear basic proteins with protamine-like properties have been detected in some other Vertebrates such as in some Amphibians (Picheral, 1970 ;

Bols *et al.*, 1976) and also in some Invertebrates and these proteins have been partially characterized in various species of Molluscs (Subirana *et al.*, 1973 ; Phelan *et al.*, 1974 ; Suau and Subirana, 1977 ; Balhorn *et al.*, 1979 ; Colom and Subirana, 1979 ; Zalensky and Zalenskaya, 1980; Ausio and Subirana, 1982). The molecular size of the basic sperm proteins migrating faster than histones in polyacrylamide gel electrophoresis seems to range from about 45 to about 100 residues per molecule. Some of these protamines are either arginine-rich or lysine-rich or contain high levels of both basic amino acids. Serine and threonine are generally well represented. Proline is more variable and cysteine is absent. Proteins with a high molecular weight with 200 to 300 residues have also been described in Mollusc sperm (Colom and Subirana, 1979).

Protamine-like proteins with a greater electrophoretic mobility than histones have also been detected in some Insects such as in the cricket where the protamine fraction appears heterogeneous (Mc Master-Kaye and Kaye, 1976 ; Kaye and Mc Master-Kaye, 1982), but no structural data is available for these proteins.

Therefore, in conclusion, the diversity of structure of sperm protamines has been confirmed these last past years. All are rather small highly basic nuclear proteins which can replace partially or completely histones. Each contains from around 30 to around 60 residues ; the basic amino acids arginine and lysine are the most abundant residues, but almost all the other types of amino acids can be more or less represented depending on the origin of the protamines. The amino acid diversity is generally limited but ranges from four (in the squid *Loligo vulgaris*, Suau and Subirana, 1977) to fifteen such as in scylliorhinine Z1. No other general rule on protamine structure can be drawn neither about the amino acid sequences which have been determined nor on their spatial organization. More protamine sequences are needed to try to understand the common structural and biological properties of all these specific proteins.

## 3. DNA-PROTAMINE INTERACTIONS

In somatic cell nuclei, nucleohistones are organized in nucleosomes which appear as the now classical picture of beads-on-a-string when observed by electron microscopy (for recent reviews, cf. Chambon, 1977 ; Kornberg, 1977 ; Mc Ghee and Felsenfeld, 1980).

A lot of physical studies have been realized to describe the interactions between DNA and protamines such as by U.V. or I.R. spectroscopy, circular dichroism, X-ray diffraction or N.M.R. (Liquier *et al.*, 1975 ; Herskovits and Brahms, 1976 ; Suau and Subirana, 1977 ; Warrant and Kim, 1978 ; Bonora *et al.*, 1979 ; Toniolo *et al.*, 1979 ; Cozzone *et al.*, 1980). The general conclusions of all these studies is that DNA in DNA-protamine associations is stabilized in a B-form and that the polypeptide chains seem to be extended and are bound to DNA in the narrow groove of the double helix. It appears also that protamines whatever their amino acid composition have the same effect on DNA (Suau and Subirana, 1977). However the protamines studied by these authors share common structural properties, all being rich in arginine and containing few amounts or no lysine and no cysteine. By contrast, some experimental evidence is accumulating which suggests that individual fractions of protamines could have different properties and different roles in the structural organization of sperm chromatin (Bonora *et al.*, 1979 ; Toniolo *et al.*, 1979 ; Rybin and Yulikova, 1981).

Sperm chromatin is organized in a highly compacted form especially when sperm nuclei contain nucleoprotamines. This condensed state is obtained progressively after a characteristic series of evolutionary steps of the nuclear content and this evolution along spermiogenesis has been followed by electron microscopy in a lot of species.

The fine structure of sperm nucleoproteins have been studied on intact nuclei or on dispersed chromatin by a variety of methods and controversial data have been accumulated. The organization of DNA in sperm nuclei containing nucleohistones looks like the nucleosome organization of somatic cell nuclei ; however the amount and arrangement of DNA in sperm chromatin subunits can be slightly different as revealed by a kinetic analysis of nuclease digestion and the determination of mean DNA lengths of fragments generated after micrococcal nuclease or Dnase I. In sea urchin sperm, the DNA length in subunits of sperm chromatin is greater (240-260 base pairs), nuclease digestion is slower and the cutting site frequencies by Dnase I give a different pattern than in canonical nucleosomes (Spadafora and Geraci, 1975 ; Spadafora *et al.*, 1976 ; Keichline and Wassarman, 1977, 1979 ; Simpson and Bergman, 1980). Similar observations were made on the sea cucumber *Holothuria tubulosa* where 275 base pair chromatin fragments are released by micrococcal nuclease (Cornudella and Rocha, 1979). This original organization of sperm nucleosomes may be related to sperm specific histone variants. By contrast, the sperm chromatin of the gold-fish *Carassius auratus* contains histones which present no significant difference with somatic histones ; the mean DNA repeat length of gold-fish sperm nucleosomes is similar to somatic cell nucleosomes, being 205 base pairs, and a beads-on-a-string configuration of dispersed sperm chromatin has been observed by electron microscopy (Muñoz-Guerra *et al.*, 1982). In sea urchin sperm chromatin, nucleosomes are packed in globular aggregates of 20 to 26 nucleosomes corresponding to 4.8 to 6.2 Kb. DNA (Zentgraf *et al.*, 1980).

In sperm nuclei where histones have been replaced by protamines different chromatin organizations have been described. The persistence of a beaded chromatin structure and the existence of subunits in whole sperm chromatin or part of it have been reported in different Mammals (Wagner *et al.*, 1978 ; Wagner and Yun, 1979, 1981 ; Young and Sweeney, 1979 ; Kvist *et al.*, 1980 ; Tsanev and Avramova, 1981). Globular aggregates as well as lamellar aggregates have also been detected either by electron microscopy or after nuclease digestion (Evenson *et al.*, 1978 ; Delgado *et al.*, 1980 ; Wagner and Yun, 1981). By contrast, some authors have described a smooth chromatin organization and the absence of nucleosome-like subunits in mature sperm chromatin although a beaded structure could be observed in the early steps of spermiogenesis (Koehler, 1966 ; Plattner, 1971 ; Lung , 1972 ; Honda *et al.*, 1974 ; Kierszenbaum and Tres, 1975, 1978 ; Cech *et al.*, 1977 ; Loir and Courtens, 1979 ; Mc Master-Kaye and Kaye, 1979, 1980). Isolated protamine molecules or reconstituted nucleoprotamines have been shown to adopt a globular configuration (Ottensmeyer *et al.*, 1975 ; Bazett-Jones and Ottensmeyer, 1979 ; Cid and Arellano, 1982). However several secondary or tertiary structures have been proposed for these molecules and free or DNA-bound protamines may have different configurations (Warrant and Kim, 1978).

Electron microscopic observation of dispersed sperm chromatin of the dog-fish reveals a more or less regular organization of nucleoprotamine molecules with beads interspersed by linear segments of chromatin (fig. 6). This picture is reminiscent of somatic chromatin organization and was obtained whether the four sperm protamines were present or when protamine-Z3 was previously solubilized (fig. 6a). U.V. thermal denaturation and C.D. measurements both reveal two domains of different stabilities for total sperm chromatin and for protamine Z3-depleted chromatin ; no free

DNA was detected (Aubert J.P., Gusse M., Loucheux-Lefebvre M.-H. and Ph. Chevaillier, unpublished results). Therefore, the fibro-granular architecture of the dog-fish sperm chromatin is the result of DNA-protamine associations ; scylliorhinine Z3 does not seem to play a significant role in this organization and could be present in one domain or in both. This protamine is probably involved in the super-organization of nucleoprotamine molecules inside the sperm nucleus since its extraction is followed by the disappearance of nuclear anisotropy.

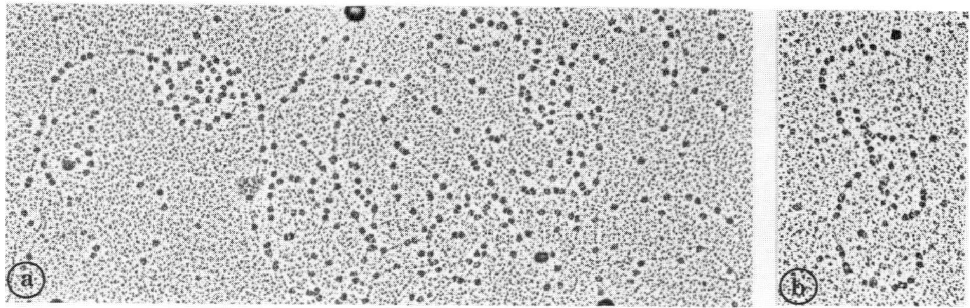


Fig. 6 - DISPERSED CHROMATIN FROM SPERM NUCLEI OF THE DOG-FISH. a) : Total sperm chromatin ; b) : Z3-depleted chromatin. X 75.000.

A similar beads-on-a-string aspect of dispersed chromatin has been also described in some other species (Gusse M. and Ph. Chevaillier, 1980b).

Micrococcal nuclease attack of sperm chromatin or sperm nuclei of the dog-fish gives rise to small amounts of soluble chromatin fragments which have been characterized by ultracentrifugation and by polyacrylamide gel electrophoresis (J. Altério and Ph. Chevaillier, unpublished results). The soluble chromatin subunits have lower S values compared to those obtained from somatic cell nuclei or from undifferentiated or differentiating germinal cells. The mean DNA length of these chromatin particles is about 123 base pairs when all four protamines were present. Therefore, the DNA sequence associated with the dog-fish sperm chromatin subunits is significantly lower than in somatic cell nuclei or in earlier steps of sperm differentiation and these subunits are also smaller as shown by ultracentrifugation and electron microscope measurements (Gusse M. and Ph. Chevaillier, 1980). The extraction of protamine Z3 enhances the accessibility of micrococcal nuclease to DNA and reduces both the S value and the mean DNA length of chromatin subunits solubilized by the enzyme. If micrococcal nuclease digestions are in keeping with previous electron microscope studies demonstrating the persistence of a beads-on-a-string structure in the dog-fish sperm chromatin, a rather different DNA organization in sperm chromatin compared to that of nucleosomes of somatic cell nuclei can be detected with an other nuclease, Dnase I. This enzyme is known to produce DNA fragments which contain multiples of 10 base pairs from somatic cell chromatin (Noll, 1974). A similar result was obtained from undifferentiated and differentiating germinal cells of the dog-fish testis but no regular pattern of DNA fragments was observed when mature sperm chromatin was digested by Dnase I (J. Altério and Ph. Chevaillier, unpublished results). Therefore, the periodic accessibility of the DNA double helix on the core-particle of nucleosomes is lost in nucleoprotamine associations.

In conclusion, sperm nuclear chromatin containing protamines instead of histones can be organized, at least in some species, in a repeated globular configuration with original properties when compared to nucleosomes of somatic cell nuclei.

## 4. NUCLEAR PROTEIN TRANSITIONS DURING SPERM DIFFERENTIATION

We shall discuss here only the transitions which affect the basic proteins of spermatid nuclei although the non-histone protein pattern show also dramatic modifications and testis-specific histone variants may be synthesized during spermatogenesis.

Where proteins of mature sperm nuclei are different from histones, their replacement is a very important event occuring during the second half of spermiogenesis. Several mechanisms may be implicated during this protein transition but no definitive conclusion can be drawn presently from the data which have been accumulated in several species. Histone acetylation (Bouvier and Chevaillier, 1976 ; Bouvier, 1977), protamine phosphorylation and dephosphorylation (Louie and Dixon, 1972 ; Marushige and Marushige, 1978), histone proteolysis (Marushige and Marushige, 1975 ; Chauvière, 1977) are some of the possible mechanisms which have been postulated to explain these nuclear basic transitions (for a review, see Dixon, 1972 ; Dixon *et al.*,

1975). In some species such as the trout, protamines replace histones directly, newly synthesized protamines being phosphorylated and then dephosphorylated when they become bound to DNA. In other cases, histones are displaced by more basic proteins called intermediate proteins and in a second transition which occur during the ultimate steps of spermiogenesis appear the definitive protamines which constitute the whole sperm nuclear protein complement or the main part of it.

These protein transitions have been thoroughly investigated in a number of species by a cytochemical or a biochemical approach and attempts to correlate molecular events to nuclear morphology have been made in some of them, for example, in the cricket (Kaye and Mc Master-Kaye, 1982), in the trout (Marushige and Dixon, 1971), in the newt *Pleurodeles waltlii* (Picheral, 1970, 1971 ; Picheral and Bassez, 1971), in rodents (Grimes *et al.*, 1975 ; Platz *et al.*, 1975 ; Bouvier, 1977 ; Goldberg *et al.*, 1977 ; Grimes *et al.*, 1977) and in the ram (Loir and Lanneau, 1975, 1978).

The complement of intermediate basic proteins or spermatid-specific proteins differs from species to species. A rat spermatid-specific protein has been sequenced (Kistler *et al.*, 1975) and on 54 residues, 21 are lysine and arginine. This protein contains 1 proline residue and 12 residues with a hydroxylated lateral chain but does not contain cysteine. By its structure, this intermediate protein is very different from mammalian sperm protamines. A similar testis-specific protein has been purified from human testis and the determination of its amino acid composition reveals that it is extremely similar to the rat spermatid-specific protein (Kistler *et al.*, 1975). In ram spermatids, Loir and Lanneau (1975, 1978) have shown that a dozen of such proteins are synthesized transiently, the main part of them containing cysteine.

During dog-fish spermiogenesis, two intermediate nuclear basic proteins appear transiently at the beginning of nuclear elongation (Gusse and Chevaillier, 1978, 1981). At the time of the presence of these proteins in spermatid nuclei, histones are progressively displaced. As nuclear elongation proceeds the four sperm protamines appear at the same time and the two intermediate proteins disappear progressively, the main component S1 faster than the minor species. The purification of these two proteins and the determination of their amino acid composition clearly demonstrate that they are two un-related proteins with no homology neither with histones nor with protamines (Chauvière, Laine, Sautière and Chevaillier, unpublished results).

Basic nuclear protein transitions have been followed in a number of species by cytochemical methods. The basic nuclear protein content shows different properties of solubility along dog-fish spermiogenesis. At the beginning of spermatid differentiation, the nuclear basic proteins are acid-insoluble according to the Alfert and Geschwind method as are histones. Later during sperm nuclear elongation, they are acid soluble as are protamines as salmon protamines and in sperm nuclei they are again acid-insoluble as are mammalian protamines (Gusse and Chevaillier, 1978). A biochemical approach of this evolution demonstrates that the disappearance of the nuclear stainability for basic proteins corresponds to the presence of protamines in elongating spermatids and that the acid stability of the nuclear protein content is not due to the binding of new proteins to DNA but instead is the result of cross-linking of preexisting protamines which then become acid insoluble. Therefore, the evolution which occurs along spermiogenesis is not exactly described by cytochemical methods : the replacement of histones by intermediate proteins is not detected as these proteins are acid insoluble as histones ; the replacement of the intermediate proteins by the definitive protamines is detected but the second transition is not explained by these methods used alone : this second transition is not due to the appearance of a new class of nuclear proteins but to the formation of disulfide cross-links. The late formation of disulfide bonds during the ultimate steps of nuclear protein evolution is well documented in Mammals (Calvin and Bedford, 1971 ; Bedford and Calvin, 1974).

Transient phosphorylation of basic nuclear proteins have been described such as in mammalian or trout spermiogenesis (Louie and Dixon, 1972 ; Marushige and Marushige, 1978). Phosphorylation of lateral chains of hydroxylated amino acids results in a decrease of DNA-protein binding and may facilitate nuclear protein transitions. During dog-fish spermiogenesis, histones, the intermediate protein S1 and protamines appear as phosphorylated molecules during spermiogenesis : histones H2A,B and H3 appear phosphorylated during the first steps of spermiogenesis whereas protamines Z2 and Z3 are phosphorylated during the second half of spermiogenesis ; sperm protamines are non-phosphorylated (Gusse, Martinage, Sautière and Chevaillier, unpublished results). Therefore, phosphorylation and dephosphorylation reactions seem to be classical modifications of basic nuclear proteins during the evolution of the protein complement which occur during spermatid differentiation.

## 5. CONCLUSIONS

Contrary to histones which have been highly conserved during evolution, the basic protein complement of sperm nuclei differs widely from species to species on several points : number of protein fractions, protein diversity in a single species, amino acid composition and primary structure of each protein fraction, occurence of post-synthetic modifications, conformation, protein interactions and DNA-protein complex organization in sperm chromatin. Nevertheless, protamines can be defined as highly basic proteins associated to DNA in sperm nuclei and which stabilize this DNA in a highly compacted configuration whatever their structural characteristics. More biochemical data on purified protamine fractions are needed to draw general conclusions on their structure, on their functions, on their origin.

## ACKNOWLEDGMENTS

The results reported on the structure of sperm chromatin of the dog-fish include research contributions made by Dr. M. Gusse, M. Chauvière and J. Altério of this Laboratory. The amino acid composition and primary structure of the dog-fish protamines, the purification of intermediary proteins and the determination of protein phosphorylation have been achieved in collaboration with Dr. P. Sautière, G. Briand, B. Laine and A. Martinage (Institut de Recherches sur le Cancer, Lille) ; physico-chemical studies on sperm chromatin have been realized in collaboration with Dr. Loucheux M.-H. and Aubert J.P. of the same Institution. The manuscript has been prepared by D. Tesson. Financial support was obtained from the Centre National de la Recherche Scientifique (E.R.A. CNRS n° 400) and from the Délégation à la Recherche Scientifique et Technique (Action : Programme et erreurs du développement embryonnaire).

## REFERENCES

1. Alfert M. and Geschwind I., 1953. *Proc. Nat. Acad. Sci. USA*, 39, 991-999.
2. Ando T. and Suzuki K., 1966. *Biochim. Biophys. Acta*, 121, 427-429.
3. Ando T. and Suzuki K., 1967. *Biochim. Biophys. Acta*, 140, 375-377.
4. Ando T. and Watanabe S., 1969. *Int. J. Protein Res.*, 1, 221-224.
5. Ando T., Yamasaki M. and Suzuki K., 1973. In *Molecular Biology Biochemistry and Biophysics*, 12, pp. 1-114, Springer-Verlag, Berlin.

6. Ando T., Iwaī K., Ishii S., Azegami M. and Nakahara C., 1962. *Biochim. Biophys. Acta*, 56, 628-630.
7. Ausio J. and Subirana J.A., 1982. *J. Biol. Chem.*, 257, 2802-2805.
8. Balhorn R., Lake S. and Gledhill B.L., 1979. *Exp. Cell Res.*, 123, 414-417.
9. Bazett-Jones D.P. and Ottensmeyer F.P., 1979. *J. Ultr. Res.*, 67, 255-266.
10. Bedford J.M. and Calvin H.I., 1974. *J. Exp. Zool.*, 188, 137-156.
11. Bellvé A.R., 1979. In *Oxford Reviews of Reproductive Biology*, Finn C.A. ed., Clarendon Press, Oxford, 1, 159-261.
12. Bellvé A.R. and Carraway R., 1978. *J. Cell Biol.*, 79, 177a.
13. Bellvé A.R., Anderson E. and Hanley-Bowdoin L.H., 1975. *Develop. Biol.*, 47, 349-365.
14. Black J.A. and Dixon G.H., 1967. *Nature*, 216, 152-154.
15. Bloch D.P., 1969. *Genetics* suppl. 61, 93-111.
16. Bloch D.P., 1976. In *Handbook of Genetics*, 5, 139-167, R.C. King edit., Plenum Press, New York.
17. Bols N.C. and Kasinsky H.E., 1976. *J. Exp. Zool.*, 198, 109-113.
18. Bols N.C., Byrd E.W. and Kasinsky H.E., 1976. *Differentiation*, 7, 31-38.
19. Bols N.C., Boliska S.A., Rainville J.B. and Kasinsky H.E., 1980. *J. Exp. Zool.*, 212, 423-433.
20. Bonora G.M., Ferrara L., Paolillo L., Toniolo C. and Trivellone E., 1979. *Eur. J. Biochem.*, 93, 13-21.
21. Bouvier D., 1977. *Cytobiologie*, 15, 420-437.
22. Bouvier D. and Chevaillier Ph., 1976. *Cytobiologie*, 12, 287-304.
23. Bretzel G., 1972. *Hoppe-Seylers Z. Physiol. Chem.*, 353, 933-943.
24. Bretzel G., 1972. *Hoppe-Seylers Z. Physiol. Chem.*, 353, 1362-1364.
25. Bretzel G., 1973. *Hoppe-Seylers Z. Physiol. Chem.*, 354, 312-320.
26. Bretzel G., 1973. *Hoppe-Seylers Z. Physiol. Chem.*, 354, 543-549.
27. Calvin H.I. and Bedford J.M., 1971. *J. Reprod. Fert.*, suppl. 13, 65-75.
28. Cech T., Potter D. and Pardue M.L., 1977. *Biochemistry*, 16, 5313-5321.
29. Chambon P., 1977. *Cold Spring Harbor Symp. Quant. Biol.*, 42, 1209-1234.
30. Chauvière M., 1977. *Exp. Cell Res.*, 108, 127-138.
31. Cid H. and Arellano A., 1982. *Int. J. Biol. Macromol.*, 4, 3-8.
32. Coelingh J.P., Monfoort C.H., Rozijn T.H., Gevers Leuven J.A., Schiphof R., Steyn-Parvé E.P., Braunitzer G., Schrank B. and Ruhfus A., 1972. *Biochim. Biophys. Acta*, 285, 1-14.
33. Colom J. and Subirana J.A., 1979. *Biochim. Biophys. Acta*, 581, 217-227.
34. Cornudella L. and Rocha E., 1979. *Biochemistry*, 18, 3724-3732.
35. Cozzone P., Toniolo C. and Jardetzky O., 1980. *Febs Lett.*, 110, 21-24.
36. Delgado N.M., Huacuja L., Merchant H., Reyes R. and Rosado A., 1980. *Arch. Androl.*, 4, 305-313.
37. Dixon G.H., 1972. *Karolinska Symposia on Research Methods in Reproductive Endocrinology*, 130-154.
38. Dixon G.H., Candido E.P.M., Honda B.M., Louie A.J., MacLeod A.R. and Sung M.T., 1975. In *"The Structure and Function of Chromatin"*, Ciba Foundation Symposium 28, Elsevier Amsterdam, 229-258.
39. Evenson D.P., Witkin S.S., de Harven E. and Bendich A., 1978. *J. Ultr. Res.*, 63, 178-187.
40. Gaastra W., Lukkes-Hofstra J. and Kolk A.H.J., 1978. *Biochem. Genet.*, 16, 525-529.
41. Gedamu L., Wosnick M.A., Connor W., Watson D.C. and Dixon G.H., 1981. *Nucl. Acids Res.*, 9, 1463-1482.
42. Goldberg R.B., Geremia R. and Bruce W.R., 1977. *Differentiation*, 7, 167-180.

43. Grimes S.R., Meistrich M., Platz R.D. and Hnilica L.S., 1977. *Exp. Cell Res.*, 110, 31-39.
44. Grimes S.R., Platz R.D., Meistrich M.L. and Hnilica L.S., 1975. *Biochem. Biophys. Res. Commun.*, 67, 182-189.
45. Gusse M. and Chevaillier Ph., 1978. *Cytobiologie*, 16, 421-443.
46. Gusse M. and Chevaillier Ph., 1980a. *Chromosoma*, 77, 57-68.
47. Gusse M. and Chevaillier Ph., 1980b. *J. Cell Biol.*, 87, 280-284.
48. Gusse M. and Chevaillier Ph., 1981. *Exptl. Cell Res.*, 136, 391-397.
49. Herskovits T.T. and Brahms J., 1976. *Biopolymers*, 15, 687-706.
50. Honda B.M., Baillie D.L. and Candido E.P.M., 1974. *Febs Lett.*, 48, 157-160.
51. Jenkins J.R., 1979. *Nature*, 279, 809-811.
52. Jenkins J.R., Bishop J.O. and Butterworth P.H.W., 1979. *Nucl. Acids Res.*, 6, 3805-3819.
53. Kaye J.S. and Mc Master-Kaye R., 1982. *Biochim. Biophys. Acta*, 696, 44-51.
54. Keichline L.D. and Wassarman P.M., 1977. *Biochim. Biophys. Acta*, 475, 139-151.
55. Keichline L.D. and Wassarman P.M., 1979. *Biochemistry*, 18, 214-219.
56. Kierszenbaum A.L. and Tres L.L., 1975. *J. Cell Biol.*, 65, 258-270.
57. Kierszenbaum A.L. and Tres L.L., 1978. *J. Cell Sci.*, 33, 265-283.
58. Kistler W.S., Geroch M.E. and Williams-Ashman H.G., 1975. *Invest. Urol.*, 12, 346-350.
59. Kistler W.S., Keim P.S. and Heinrikson R.L., 1976. *Biochim. Biophys. Acta*, 427, 752-757.
60. Kistler W.S., Noyes C., Hsu R. and Heinrikson R.L., 1975. *J. Biol. Chem.*, 250, 1847-1853.
61. Koehler J.K., 1966. *J. Ultr. Res.*, 16, 359-375.
62. Kolk A.H.J. and Samuel T., 1975. *Biochim. Biophys. Acta*, 393, 307-319.
63. Kornberg R.D., 1977. *Ann. Rev. Biochem.*, 46, 931-954.
64. Kvist U., Afzelius B.A. and Nilsson L., 1980. *Develop. Growth and Differ.*, 22, 543-554.
65. Liquier J., Pinot-Lafaix M., Taillandier E. and Brahms J., 1975. *Biochemistry*, 14, 4191-4197.
66. Loir M. and Courtens J.L., 1979. *J. Ultr. Res.*, 67, 309-324.
67. Loir M. and Lanneau M., 1975. *Exp. Cell Res.*, 92, 509-512.
68. Loir M. and Lanneau M., 1978. *Biochem. Biophys. Res. Commun.*, 80, 975-982.
69. Loir M. and Lanneau M., 1978. *Exptl. Cell Res.*, 115, 231-244.
70. Louie A.J. and Dixon G.H., 1972. *J. Biol. Chem.*, 247, 5498-5505.
71. Louie A.J. and Dixon G.H., 1972. *J. Biol. Chem.*, 247, 7962-7968.
72. Lung B., 1972. *J. Cell Biol.*, 52, 179-186.
73. Marushige K. and Dixon G.H., 1971. *J. Biol. Chem.*, 246, 5799-5805.
74. Marushige Y. and Marushige K., 1975. *Biochim. Biophys. Acta*, 403, 180-191.
75. Marushige Y. and Marushige K., 1978. *Biochim. Biophys. Acta*, 518, 440-449.
76. Marushige K., Ling V. and Dixon G.H., 1969. *J. Biol. Chem.*, 244, 5953-5958.
77. Mc Ghee J.D. and Felsenfeld G., 1980. *Ann. Rev. Biochem.*, 49, 1115-1156.
78. McMaster-Kaye R. and Kaye J.S., 1976. *Exptl. Cell Res.*, 97, 378-386.
79. McMaster-Kaye R. and Kaye J.S., 1979. *J. Cell Biol.*, 83, 225a.
80. McMaster-Kaye R. and Kaye J.S., 1980. *Chromosoma*, 77, 41-56.

81. Monfoort C.H., Schiphof R., Rozijn T.H. and Steyn-Parvé E.P.; 1973. *Biochim. Biophys. Acta*, 322, 173-177.
82. Muñoz-Guerra S., Azorin F., Casas M.-T., Marcet X., Maristany M.A., Roca J. and Subirana J.A., 1982. *Exptl. Cell Res.*, 137, 47-53.
83. Nakano M., Tobita T. and Ando T., 1976. *Int. J. Peptide Protein Res.*, 8, 565-578.
84. Noll M., 1974. *Nucl. Acids Res.*, 11, 1573-1578.
85. Ottensmeyer F.P., Whiting R.F. and Korn A.P., 1975. *Proc. Nat. Acad. Sci.*, 72, 4953-4955.
86. Phelan J.J., Colom J., Cozcolluela C., Subirana J.A. and Cole R.D., 1974. *J. Biol. Chem.*, 249, 1099-1102.
87. Picheral B., 1970. *Histochemie*, 23, 189-206.
88. Picheral B., 1971. *J. Micr.*, 12, 107-132.
89. Picheral B. and Bassez T., 1971. *J. Micr.*, 3, 441-452.
90. Plattner H., 1971. *J. Submicr. Cytol.*, 3, 19-32.
91. Platz R.D., Grimes S.R., Meistrich M. and Hnilica L.S., 1975. *J. Biol. Chem.*, 250, 5791-5800.
92. Pongsawasdi P. and Svasti J., 1976. *Biochim. Biophys. Acta*, 434, 462-473.
93. Rybin V.K. and Yulikova E.P., 1981. *Biokhimiya*, 46, 276-279.
94. Sakai M., Fujii-Kuriyama Y. and Muramatsu M., 1978. *Biochemistry*, 17, 5510-5515.
95. Sakai M., Fujii-Kuriyama Y., Saito T. and Muramatsu M., 1981. *J. Biochem.*, 89, 1863-1868.
96. Sautière P., Briand G., Gusse M. and Ph. Chevaillier, 1981. *Eur. J. Biochem.*, 119, 251-255.
97. Simpson R.T. and Bergman L.W., 1980. *J. Biol. Chem.*, 255, 10702-10709.
98. Spadafora C. and Geraci G., 1975. *Febs Lett.*, 57, 79-82.
99. Spadafora C., Noviello L. and Geraci G., 1976. *Cell Differentiation*, 5, 225-231.
100. Spadafora C., Bellard M., Compton J.L. and Chambon P., 1976. *Febs Lett.*, 69, 281-285.
101. Suau P. and Subirana J.A., 1977. *J. Mol. Biol.*, 117, 909-926.
102. Subirana J.A., Cozcolluela C., Palau J. and Unzeta M., 1973. *Biochim. Biophys. Acta*, 317, 364-379.
103. Svasti J. and Talupphet N., 1979. *Biochim. Biophys. Acta*, 577, 221-225.
104. Toniolo C., Bonora G.M., Marchiori F., Borin G. and Filippi B., 1979. *Biochim. Biophys. Acta*, 576, 429-439.
105. Tsanev R. and Avramova Z., 1981. *Eur. J. Cell Biol.*, 24, 139-145.
106. Wagner T.E. and Yun J.S., 1979. *Arch. Androl.*, 2, 291-294.
107. Wagner T.E. and Yun J.S., 1981. *Arch. Androl.*, 6, 47-51.
108. Wagner T.E. and Yun J.S., 1981. *Archiv. Androl.*, 7, 251-257.
109. Wagner T.E., Sliwinski J.E. and Shewmaker D.B., 1978. *Arch. Androl.*, 1, 31-41.
110. Warrant R.W. and Kim S.-H., 1978. *Nature*, 271, 130-135.
111. Young R.J. and Sweeney K., 1979. *Gamete Res.*, 2, 265-282.
112. Yulikova E.P., Evseenko L.K., Baratova L.A., Belyanova L.P., Rybin V.K. and Silaev A.B., 1976. *Bioorg. Khim.*, 2, 1613-1617.
113. Yulikova E.P., Rybin V.K. and Silaev A.B., 1979. *Bioorg. Khim.*, 5, 5-10.
114. Zalensky A.O. and Zalenskaya J.A., 1980. *Comp. Biochem. Physiol.*, 66B, 415-419.
115. Zentgraf H., Müller U. and Franke W.W., 1980. *Eur. J. Cell Biol.*, 20, 254-264.

# NUCLEAR PROTEINS IN SPERMATOZOA AND THEIR INTERACTIONS WITH DNA

JUAN A.SUBIRANA

Unidad de Quimica Macromolecular del CSIC, Escuela Ingenieros Industriales, Diagonal 999, Barcelona (28), Spain.

## INTRODUCTION

In this paper we will review the different types of proteins found in animal sperm nuclei, as well as the structure of the corresponding DNA-protein complexes. We will neither discuss the biological role of these proteins (Subirana, 1975), nor consider the specific protein components which appear at intermediate stages of spermiogenesis.

The proteins associated with DNA in spermatozoa were classified by Bloch (1969) on the basis of a cytochemical study. Five major classes were recognized, which are scattered throughout the animal kingdom:

- typical protamines, in which arginine predominates
- *Mytilus* type, which contains both lysine and arginine
- Cystine containing protamines
- histones
- absence of nuclear proteins

Since this classification was established, considerable work has been carried out in the detailed characterization of some of the proteins belonging to each class. As a result, it is becoming apparent that the classification suggested by Bloch is essentially correct, since no species has been described thus far, which could not be fitted to one of these classes. However the situation is complicated by the fact that many species contain several protein components, with different characteristics. For example, in the dog-fish *Scylliorhinus caniculus*, protamines with and without cystine are found (Gusse and Chevaillier, 1978). In *Mytilus*, which typifies a class according to Bloch, two minor components are present which have not been found in other related species (Ausió and Subirana, 1982a).

An additional complication is due to the fact that some protamines may be in the borderline between the classes suggested by Bloch. For example, the marine snails, which contain about 5-10 % lysine, could be placed in the borderline between the typical and the Mytilus type protamines. In fact, from a chemical point of view, the major basic proteins present in the spermatozoa of the first three classes suggested by Bloch share common characteristics and can all be studied together and called protamines.

Finally it should be pointed out that it is usually assumed that the basic proteins extracted from spermatozoa are associated with DNA "in vivo". This is probably correct in most cases, but it can not be excluded that some of the minor components described have a different cellular location. For example in the worm Urechis caupo basic proteins have been found in the acrosome (Das et al, 1967) and might well occur in the acrosome of other species.

Protamines

Typical protamines are characterized by a very high content of arginine, usually in the range 60 to 80 %, and a very low content or absence of acidic and hydrophobic amino acids. Serine also occurs in a relatively high amount (10-15 %). Some typical examples are shown in Table 1. A more extensive list can be found in the review by Bloch (1976). Proteins belonging to this class have been characterized in some groups of fishes (Ando et al, 1973; Bretzel, 1973; Sautiere et al, 1981), in the rooster (Nakano et al, 1976) and in some molluscs (Subirana et al, 1973).

In the bivalve molluscs, belonging to Bloch's Mytilus class, the major nuclear proteins are similar to the conventional protamines, but lysine and arginine occur in about similar amounts (Subirana et al, 1973; Colom and Subirana, 1979). These proteins also have a very high serine content, as it can be appreciated in the two examples given in Table 1, the surf clam and M.edulis Ø1.

Another particular class of protamines are those which contain cysteine. They have been characterized in mammals (Coelingh et al, 1972; Calvin, 1976; Bellvé et al, 1975), in an elasmobranch fish (Sautiere et al, 1981) and in the cephalopod Eledone cirrhosa (Subirana et al, 1973), although they occur in many other biological groups as reviewed by Bloch (1969; 1976). The amount of arginine in this group of protamines is slightly smaller than in fish. They also contain many different nonbasic residues.

All these different types of protamines share two common features, which can be equivalent to a definition of what a protamine is:

(Lys + Arg) = 45 - 80 %
(Ser + Thr) = 10 - 25 %

In spite of these common features, there are significant differences among the various groups. Marine snails contain some lysine (5-10 %), the tuna fish has a high valine percentage, the squid contains a high tyrosine percentage, etc. In spite of this variation in individual species, the acidic (Glx, Asx) and bulky hydrophobic residues (Met, Ile, Leu, Phe) seldom occur in protamines. Tyrosine appears in some species, particularly in the most evolved ones (squid, fowl, mammals). Histidine is usually absent, but it is present in some fish species (Yulikova et al, 1979) and is very common in mammals (Bellvé et al, 1975).

The molecular size of these proteins varies considerably, as shown in Table 1. In the most thoroughly studied fish species, these proteins are rather small, 27 to 34 amino acids long, but in other zoological groups they are much larger, up to 300 residues (Subirana and Colom, 1979; Ausió and Subirana 1982b, c).

The sequence of several fish (Ando et al, 1973; Bretzel, 1973; Yulikova et al, 1976, 1979; Sautiere et al, 1981), rooster (Nakano et al, 1976) and bull (Coelingh et al, 1972) protamines are presently available. Since all these are vertebrate species the results obtained can not be generalized. Nevertheless it is

| | Clustered basic residues | |
|---|---|---|
| Salmine A1 | $P-R_4-SSS-R-PV-R_5-P-R-VS-R_6-GG-R_4$ | |
| Galline | A-R-Y-R-S-R-G-R-S-R-S- | $-R_2-T-R_4-SP-R-SG-R_3-$<br>$-SP-R_4-S-R_5-YGSA-R_2-S-$<br>$-R_2-SGGV-R_4-YGS-R_6-Y$ |
| Bull | A-R-Y-R-CCLT-H-SGS-R-C-<br>→ VCYTVI-R-CT-R-Q | $-R_7-C-R_6-FG-R_6$ → |

FIGURE 1.Amino acid sequence homologies of vertebrate protamines

worth analyzing their general features. A few examples are shown on figure 1. A more detailed comparison is given by Chevaillier in another chapter of this book. The most striking feature is the tendency of the basic and neutral residues to be clustered. In the salmonid fish (Ando et al, 1973) 4 clusters of 4-6 arginines are present in the molecule. In galline most of the arginines appear clustered, except for the N-terminal twelve amino acids, in which an strictly alternating sequence of arginine and neutral residues is found (Nakano et al, 1976). The bull protamine also contains clusters of 6-7 arginines in its central region (Coelingh et al, 1972), but in this case the two ends of the molecule are very rich in uncharged residues.

In summary, the teleost fish protamines all have a clustered sequence. The same is true for the other species studied, but in these other cases the protamines contain fragments in one or both ends of the molecule with a different type of organization. The eventual structural significance of these observations will be discussed below. The biological role of these proteins is probably determined not only by its high charge but also by the sequence of its amino acid residues.

## Histones

Many animals of quite different phylogeny contain histones in their spermatozoa. Usually these proteins are accompanied by sperm specific components which can be related in composition to either histones or protamines. In fact, the somatic type histones may be just a small percentage of the total basic proteins in the sperm, as it is the case in bivalve molluscs (Subirana et al, 1973; Ausió and Subirana, 1982a,b). Therefore it is not quite clear how much variation can be accepted in the sperm protein components to call them histones as a whole. Bloch (1969) used a cytochemical criterion, but now that chromatin can be probed by nuclease digestion, perhaps it would be more adequate to consider that a spermatozoon belongs to the histone class when it gives a well defined set of degradation products as proof that it contains nucleosomes, as it has been shown for example in some echinoderm spermatozoa (Spadafora et al, 1976; Cornudella and Rocha, 1979). Unfortunately this test has not been carried out in most species, so that at the present moment it can not be applied in practice. In the following we will consider some examples which most likely satisfy this criterion. There appear to be four types of nuclear composition:

a) No detectable change in the H1 family
b) Slight changes in the H1 family
c) Additional sperm specific basic proteins
d) Considerable changes in histones H1 and H2b

The goldfish (Muñoz-Guerra et al, 1982a) is the only organism studied thus far in which the sperm histones appear to be identical with the somatic histones. It is the only example available of type a.

As an example of type b, the frog (Rana) has H1 subfractions which are unique in spermatozoa (Alder and Gorovsky, 1975), but the amino acid composition of the H1 specific component is very similar to other somatic H1 subcomponents (Roca, 1975). In the carp similar changes have been detected (Roca, 1975).

The third case is the appearance of sperm specific proteins, while the somatic histones are conserved in spermatozoa. The amino acid composition and relative amount of some of these additional components vary substantially in different cases, as shown in Table 2. However, they show the general trend of being more basic than somatic histones. In the table they have been grouped as H1 related, when they contain 20-30 % Ala and the amount of basic residues is close to 40 %, and H2b related. In the latter case, the amount of basic residues is 25-30 %, serine plus threonine amount to 10-20 %, and the other residues vary considerably in different cases. Most of these proteins are similar in size to histones, but $\phi_0$ from Holothuria is smaller, 78 amino acids long (Jordán, 1981), whereas flounder fraction A is a family of giant proteins (Kennedy and Davies, 1980). In the case of the sea cucumber, besides the presence of $\phi_0$, a new family of H1 histones is found in the spermatozoa (Phelan et al, 1972). Other species which belong to this third type are the horseshoe crab (Muñoz-Guerra et al, 1982b) and the shrimp (Sellos and Le Gal, 1981). Many amphibia also belong to this class, but no amino acid compositions have been reported thus far (Bols and Kasinsky, 1973; Kasinsky et al, 1978).

The fourth type is found in sea urchins and starfishes in which histones H2a, H3 and H4 are practically identical in sperm and somatic tissues, but H1 and H2b are substituted by sperm specific components which are much richer in arginine than their equivalent somatic fractions (Subirana and Palau, 1968; Palau et al, 1969; Vanhoutte-Durand et al, 1977; Zalenskaya and Zalensky, 1980; Zalenskaya et al, 1980). The sequence of some of these components has been determined by Von Holt et al (1979). In spite of the drastic changes in composition found in H2b, the chromatin from sea urchins has very clear nucleosomal structure (Spadafora et al, 1976).

In most of the cases we have described in this section, the sperm histone complement is more basic than in somatic cells and this fact will most likely influence the higher order arrangement of nucleosomes in sperm chromatin, but it is not clear what specific advantage for the sperm cell may result from the small changes found in some cases, as in Limmulus or Holothuria. It is even possible that some of these minor proteins are associated with a specific fraction of chromatin and play a role for example in sperm nuclear shaping or in fertilization.

Microheterogeneity of sperm proteins

Many of the protamines occur as a mixture of closely related species, as found in most teleost fishes (Ando et al, 1973; Bretzel, 1973) and in the squid (Subirana et al, 1973).

In other cases mixtures of protamines which differ in size and composition are found, as in the dogfish (Gusse and Chevaillier, 1978) or in the molluscs Cryptochiton stellerii and Donax trunculus (Colom and Subirana, 1979). The large variety of proteins which accompany the histones in different species of the genus Xenopus are a further example of this heterogeneity (Risley et al, 1982). In the latter case very large differences in electrophoretic mobility have been described, but it would not be surprising that the amino acid compositions were rather similar in the different cases.

In the case of sperm specific histones, microheterogeneity has also been found, which has been described in great detail for the H2b component of sea urchin sperm (Zalenskaya and Zalensky, 1980; Von Holt et al, 1979). It has been found that populations from different sites, as well as individual animals differ in the relative amount of the three H2b subfractions present (Strickland et al, 1981).

On the other hand in other cases it is clear that a single protamine species is present in the spermatozoa as found for example in the rooster (Nakano et al, 1976) and in the bull (Coelingh et al, 1972). It appears therefore that the presence of either one or several subcomponents does not have any specific physiological advantage, although it may have some evolutive implications. Another possibility which may apply in some cases is that protamines were synthesized as a large precursor protein which could be specifically degraded into the various protamines observed in the cell (Elsevier, 1982).

The structure of nuclei containing typical protamines

The nuclei of sperm cells which only contain typical protamines as their basic proteins do not appear to contain any other protein component. In the herring, a detailed study (Kawashima and Ando, 1978) has shown that only a small amount of non-histone proteins was present. In fact they could be due to contaminations from other sperm components. On the other hand, from a structural point of view, the X-ray diffraction pattern of whole purified nuclei and of reconstituted complexes of purified DNA and protamine are practically identical (Feughelman et al, 1955; Suau and Subirana, 1977), implying

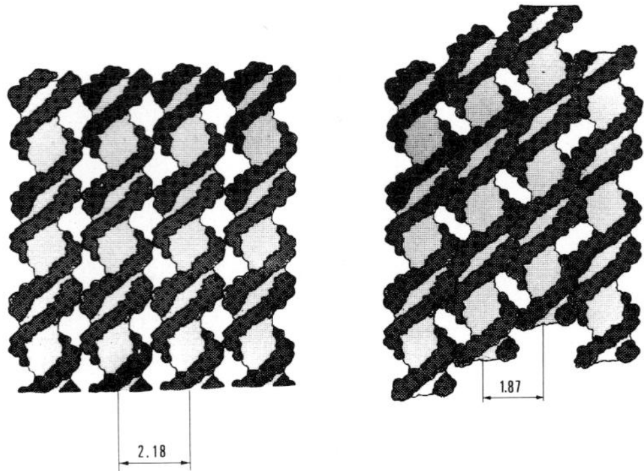


FIGURE 2. The relative arrangement of DNA molecules in complexes with protamine (left) and polyarginine (right). (Adapted from Fita, 1981). The numbers indicate the distance in nm between neighbouring molecules in the absence of water (0% relative humidity).

that the nucleus does not contain any other structural component with an influence on the average conformation of chromatin in these cells.

The structure of the nucleoprotamine complex has been studied by X-ray diffraction, as described by Feughelman et al (1955) and Suau and Subirana (1977). The DNA molecules tend to arrange themselves in a parallel fashion with protamines neutralizing the phosphate groups. The double helices are packed as a group of intertwinned screws in the hexagonal system. Surprisingly, this packing arrangement is different from that found in DNA-polyarginine complexes, which pack in the orthorhombic system. Both types of packing are shown in Fig. 2.

In order to understand the significance of the packing arrangements shown in the figure, it is important to discuss first the mode of interaction of the polypeptide chain with DNA. The orthorhombic packing found in the polyarginine complex (Fig. 2) shows that the polypeptide chain must be located in the wide groove of DNA, since there is no room available in the narrow groove. An analysis of the X-ray diffraction pattern confirms this observation (Fita, 1981).

The fact that polyarginine is associated with DNA on its wide groove is in contrast with the suggestion that protamine is associated with the narrow groove of DNA (Feughelman et al, 1955; Suau and Subirana, 1977). That suggestion was based on the X-ray diffraction pattern of nucleoprotamine complexes, since it is the simplest interpretation of the observation that the first layer line in the diffraction pattern is much stronger than in pure DNA. On the other hand, it is obvious from Fig. 2 that in the hexagonal packing there is room available both in the wide and in the narrow groove. From these observations we have concluded (Campos, Fita, Puigjaner and Subirana, in preparation) that protamine is in fact associated with the wide groove of DNA. The differences in intensity observed in the diffraction patterns can be atributed to the presence of neutral residues in nucleoprotamine, which will protrude from the wide groove and fit in the narrow groove of neighbouring DNA molecules as it can be concluded from the packing arrangement shown in Fig. 2. This suggestion is also supported by the chemical reactivity of nucleoprotamine (Mirzabekov et al, 1977).

In summary, the sperm heads which only contain protamine can be considered as small microcrystals in which the DNA molecules are packed in a parallel fashion as shown in Fig. 2.

## The structure of nuclei containing cystine protamines

The nucleoprotamine complex in this case is very tightly packed and difficult to dissociate due to the presence of cystine bridges. On the other hand, the amino acid sequence of bull protamine (Fig. 1) indicates that the central region of this molecule is rather similar to conventional protamines, so that the basic features of protamine-DNA interactions should be the same. In fact, recent X-ray diffraction experiments carried out in our laboratory (Fornells, 1982) show that DNA molecules are packed in bundles, with a structure similar to that presented in Fig. 2, but with a lower degree of order. It appears that the bundles are contorted and only show a regular order locally, so that DNA molecules at a certain distance apart are no longer parallel. The two terminal regions of the molecule with its lower charge, are probably instrumental in the lower degree of order found in these nucleoprotamine complexes.

Recently Balhorn (1982) has presented a detailed model of nucleoprotamine structure which appears to be incompatible with the results we have obtained (Fornells, 1982), since he assumes a way of packing of nucleoprotamine in which there would be five base pairs of DNA covered by three peptide chains followed by another five base pairs covered by only one peptide chain. This alternating thickening and narrowing of the nucleoprotamine complex does not appear to be compatible with the packing of DNA in regular bundles that we have shown to be present in this type of sperm. Furthermore Balhorn (1982) assumes that protamine is associated with the narrow groove of DNA, whereas in fact it appears to be bound to the wide groove as discussed in the previous section.

The structure of nuclei containing histones

The structure of sperm chromatin containing histones has been probed by nuclease digestion only in three cases. In all of them a typical nucleosome structure is present, in spite of a quite different histone composition. The repeat length of the DNA has been found to be 241 base pairs in sea urchin (Spadafora et al, 1976), 227 in sea cucumber (Cornudella and Rocha, 1979) and 205 in the gold fish (Muñoz-Guerra et al, 1982a). These values are slightly longer than in typical somatic nuclei, but the overall pattern of digestion is practically identical in all cases.

From the ultrastructural point of view, the chromatin of these nuclei is less condensed than in the other cases described above. Nevertheless it appears as a compact mass with no void spaces. In many cases clear 30 nm chromatin fibers can be detected (Longo and Anderson, 1969; Pladellorens and Subirana 1975; Fahrenbach, 1973; Muñoz-Guerra et al, 1982a), but in other species chromatin has a granular appearance (Casas et al, 1981). A detailed analysis of the chromatin fibers from sea cucumber spermatozoa has shown that they have a superbead structure (Subirana et al, 1981).

X-ray diffraction studies also show that nucleohistones

are similar in structure in different species (Subirana et al, 1975). However an interesting feature is observed in sea urchin sperm chromatin since it contains a much higher percentage of the fibrous phase, that is, of parallel DNA molecules ordered in bundles (Subirana et al, 1977). The higher basicity and relative amount of histone H1 present in this material gives to chromatin a characteristic feature of nucleoprotamine complexes, whereas the somatic-like histones present maintain the nucleosomal structure. Thus, sea urchin sperm chromatin, although belonging to the histone class, already shows some features of nucleoprotamine structure.

In fact, the significance of the various changes in the histones during spermiogenesis in different species, is not clear, since the overall properties of chromatin are rather similar. Perhaps the lower solubility of chromatin which has been found in some cases (Subirana et al, 1981) indicates that these changes may be instrumental in the stabilization of the sperm nucleus.

The structure of Mytilus type nuclei

In the original classification made by Bloch (1969), in this class were included those species which contained sperm proteins intermediate in composition between histones and protamines. As Bloch himself recognized, this definition is not very precise, since we have seen that protamines which are lysine rich can be found in some spermatozoa, whereas other species have arginine rich histones.

In fact in this class are included those species which do not belong to any of the other three classes. Therefore we may define it by including those species which do not fulfill any of the following criteria:

- Presence of Cys in their sperm proteins
- Crystalline arrangement of DNA in the sperm
- Nucleosome structure of the chromatin

At the present time the detailed molecular structure of any sperm belonging to this class is not known, in fact there could be different types of structure in different species. Furthermore in some cases there may be uncertainities with regard to the classification of a given sperm type, being in the borderline with either typical protamines or histones.

A clear example is *Mytilus*, which was taken as the prototype of this class by Bloch (1969). The sperm of this species contains a major protamine like component Ø1, which constitutes about 75 % of the total basic protein complement, and is rich in both arginine and lysine, as shown in Table 1. This protein is accompanied by two minor components, Ø3 and Ø2b, each of which amounts to about 10 % of the total sperm proteins (Ausió and Subirana, 1982a). A small amount of somatic histones is also found accompanying the sperm specific proteins (Subirana et al, 1973; Ausió and Subirana, 1982a).

The composition of Ø3 was included on table 1 because of its high basicity, but in fact it is unique in composition due to its high percentage of lysine and alanine. Its composition is more reminiscent of the C-terminal half of histone H1 than of a protamine, so that its inclusion on table 1 is very tentative. The other component, Ø2b, is a sperm specific histone similar to those found in some histone containing spermatozoa, as shown on Table 2.

In summary, the protein composition of *Mytilus* spermatozoa is complex, given the heterogeneity of its three components. In fact the major component Ø1 is a typical protamine and the two minor components are histone related.

From the ultrastructural point of view, the chromatin of these spermatozoa has a coarse structure with granules of about 40-70 nm in diameter (Longo and Dornfeld, 1967). Nuclease digestion does not show any nucleosomal structure, and X-ray diffraction indicates a tendency of DNA to be packed in regular bundles (Ausió and Subirana, 1982a), which are not so regular as in typical nucleoprotamines. In fact the diffraction pattern is similar to that found in nuclei containing cystine protamines which we have discussed above. However the Ø1 component alone does form typical nucleoprotamine structures when associated with DNA (Subirana and Puigjaner, 1973). It appears therefore that in this species the basic structure of the chromatin is nucleoprotamine-like, but its regularity is somehow modulated by the presence of the other two histone-like components.

The results for Mytilus can not be generalized to other species, since there may be strong compositional differences, even in closely related species. For example, another mollusc, the snail Gibbula divaricata, also contains a Ø2b component (shown in Table 2) and a protamine (shown in Table 1), but it does not contain any component related to Ø3. In general, the heterogeneity of the proteins present in the sperm of most molluscs (Subirana et al, 1973; Colom and Subirana, 1979; Zalensky and Zalenskaya 1980) does not allow to draw any clear general conclusions.

In conclusion, the structure and physiological significance of the nucleoproteins present in the sperm of the species belonging to this group is not clear and requires further study. Most likely its peculiarities reside more in the heterogeneity of the sperm proteins present than in the relative content of arginine and lysine.

Concluding remarks.

We have seen that from the chemical point of view there is a gradual change in the proteins found in the sperm from standard histones (goldfish) through strongly modified histones (sea urchin) and complex intermediate proteins (Mytilus), towards the simpler protamines (salmonid fish, squid). From the structural point of view this transition in the proteins is accompanied by a loss of the nucleosomal structure and the appearance of a well ordered parallel array of DNA molecules, but the structural rules which determine the spatial arrangement of DNA and protein in the intermediate cases are not yet understood.

The peculiar composition of Mytilus sperm points out our general lack of knowledge on the role played by sperm proteins. In fact the mature sperm represents a snapshot of a complex developmental process which starts at meiosis, continues at spermiogenesis and finishes in fertilization and early embryogenesis. Until all the nuclear transitions which occur in this process are understood, our picture of sperm structure will be incomplete. Furthermore, it would not be surprising if in the future additional minor components are discovered in the sperm nucleus, either as genome markers (Uschewa et al, 1982) or as structural elements involved in the overall morphogenesis of the nucleus.

## REFERENCES


Alder D and Gorovsky MA. 1975. J.Cell Biol. 64, 389-397

Ando T, Yamasaki M and Suzuki K. 1973. Protamines. Springer Verlag, Berlin, Heildelberg, New York

Ausió J. 1980. Ph.D.Thesis, Faculty of Biology, University of Barcelona

Ausió J and Subirana JA. 1982a. Exptl.Cell Res., in press

Ausió J and Subirana JA. 1982b. J.Biol.Chem. 257, 2802-2805

Ausió J and Subirana JA. 1982c. Biochemistry, in press

Balhorn R. 1982. J.Cell Biol. 93, 298-305

Bellvé AR, Anderson E and Hanley-Bowdoin L. 1975. Develop. Biol. 47, 349-365

Bloch DP. 1969. Genetics Suppl. 61, 93-110

Bloch DP. 1976. Handbook of Genetics Vol. 5. Molecular Genetics. R.C.King, ed. Plenum Press

Bols NC and Kasinsky HE. 1973. Can.J.Zool. 51, 203-208

Brandt WF, Strickland WN, Strickland M, Carlisle L, Woods D and Von Holt C. 1979. Eur.J.Biochem. 94, 1

Bretzel, G. 1973. Hoppe-Seyler's Z.Physiol.Chem. 354, 543-549

Calvin HO. 1976. Biochim.Biophys.Acta 434, 377-389

Casas MT, Muñoz-Guerra S, Subirana JA. 1981. Biol.Cell 40, 87-92

Coelingh JP, Monfoort CH, Rozijn TH, Gevers Leuven JA, Schiphof R, Steyn-Parvé, EP, Braunitzer G, Schrank B and Ruhfus A. 1972. Biochim.Biophys.Acta 285, 1-14

Colom J and Subirana JA. 1979. Biochim.Biophys.Acta 581, 217

Colom J and Subirana JA. 1981. Exptl.Cell Res. 131, 462-466

Cornudella L and Rocha E. 1979. Biochemistry 18, 3724-3732

Das NK, Micow -Eastwood J and Alfert M. 1967. J.Cell Biol. 35 455-458

Elsevier SM. 1982. Develop.Biol. 90, 1-12

Fahrenbach WH. 1973. J.Morphology 140, 31-52

Feughelman M, Landgridge R, Seeds WE, Stokes AR, Wilson HR, Hooper CW, Wilkins MHF, Barclay LK and Hamilton LD. 1955. Nature (London) 175, 834-838

Fita I. 1981. Ph.D.Thesis. Faculty of Biology. Universidad Autónoma, Barcelona

Fornells M. 1982. Ph.D.Thesis. Faculty of Biology. University of Barcelona

Gusse M, Chevaillier PH. 1978. Cytobiologie 16, 421-443

Jordán A. 1981. Ph.D.Thesis. Faculty of Chemistry. University of Barcelona

Kasinsky HE, Huang SY, Kwauk S, Mann M, Sweeney MAJ and Yee B. 1978. J.Exptl.Zool. 203, 109-126

Kawashima S and Ando T.1978. J.Biochem. 83, 1117-1123

Kennedy BP, Davies PL. 1980. J.Biol.Chem. 255, 2533-2539

Longo FJ and Anderson E. 1969. J.Ultrast.Res. 27, 486-509

Longo FJ and Dornfeld EJ. 1967. J.Ultrast.Res. 20, 462

Mirzabekov AD, San'ko DF, Kolchinsky AM and Melnikova AF. 1977

Eur.J.Biochem. 75, 379-389
Muñoz-Guerra S, Azorín F, Casas MT, Marcet X, Maristany MA, Roca J and Subirana JA. 1982a. Exptl.Cell Res. 137, 47-53
Muñoz-Guerra S, Colom J, Ausió J and Subirana JA. 1982b. Biochim.Biophys.Acta, in press
Nakano M, Tobita T and Ando T. 1976. Int.J.Pept.Prot.Res. 8, 565-578
Palau J, Ruiz-Carrillo A and Subirana JA. 1969. Eur.J.Biochem. 7, 209-213
Phelan JJ, Subirana JA and Cole RD. 1972. Eur.J.Biochem. 31, 63-68
Phelan JJ, Colom J, Cozcolluela C, Subirana JA and Cole RD. 1974. J.Biol.Chem. 249, 1099-1102
Pladellorens M and Subirana JA. 1975. J.Ultrast.Res. 52, 235-242
Risley MS, Eckhardt RA, Mann M and Kasinsky HE. 1982. Chromosoma in press
Roca J. 1975. Ph.D.Thesis. Faculty of Chemistry. University of Barcelona
Sautiere P, Briand G, Gusse M and Chevaillier Ph. 1981. Eur. J.Biochem. 119, 251-255
Sellos D and Le Gal Y. 1981. Cell Diff. 10, 69-77
Spadafora C, Bellard M, Compton JL and Chambon P. 1976. FEBS Let. 69, 281-285
Strickland M, Strickland WN and Von Holt C. 1981. FEBS Let. 135, 86-88
Suau P and Subirana JA. 1977. J.Mol.Biol. 117, 909-926
Subirana JA. 1975. in "The Biology of the Male Gamete" ed. J.G.Duckett and P.A.Racey (Suppl. 1 to the Biological Journal of the Linnean Society, Vol. 7) pp 239
Subirana JA and Palau J. 1968. Exptl.Cell Res. 53, 471-477
Subirana JA, Cozcolluela C, Palau J and Unzeta M. 1973. Biochim. Biophys.Acta 317, 364-379
Subirana JA and Puigjaner LC. 1973. In "Conformation of Biological Molecules and Polymers" E.D.Bergman & B. Pullman eds. The Jerusalem Symposia on Quantum Chemistry and Biochemistry V. The Israel Academy of Sciences and Humanities. Jerusalem
Subirana JA, Puigjaner LC, Roca J, Llopis R and Suau P. 1975. in "The Structure and Function of Chromatin" Ciba Foundation Symposium 28 (Associated Scientific Publishers, Amsterdam) pp 157
Subirana JA, Azorín F, Roca J, Lloveras J, Llopis R and Cortadas J. 1977. In "The Molecular Biology of the Mammalian Genetic Apparatus" P.Ts'O Ed. (Elsevier/North Holland Biomedical Press) pp 71
Subirana JA, Muñoz-Guerra S, Martínez AB, Pérez-Grau L, Marcet X and Fita I. 1981. Chromosoma 83, 455-471
Uschewa A, Avramova Z and Tsanev R. 1982. FEBS Let. 138, 50-54
VanHoutte-Durand G, Mizon J, Sautiere P and Biserte G. 1977. Comp.Biochem.Physiol. 57B, 121-126
Von Holt, C, Strickland WN, Brandt WF and Strickland MS. 1979 FEBS Let. 100, 201-218
Yulikova EP, Evseenko LK, Baratova LA, Belyanova LP, Rybin VK, and Silaev AB. 1976. Bioorg.Chem. 2, 1613-1618
Yulikova EP, Rybin VK and Silaev AB. 1979. Bioorg.Chem. 5 5-10

Zalenskaya IA, Zalensky AO. 1980. Comp.Biochem.Physiol. 65B 369-373
Zalenskaya IA, Zalenskaya EO and Zalensky AO. 1980. Comp. Biochem.Physiol. 65B, 375-378
Zalensky AO. and Zalenskaya IA. Comp.Biochem.Physiol. 1980. 66B, 415-419

TABLE 1. Amino acid composition (%) of some protamines

| Amino acid | Molluscs | | | | | Vertebrates | | |
|---|---|---|---|---|---|---|---|---|
| | Squid | Snail Gibbula | Mussel Ø1 | Mussel Ø3 | Surf clam | Tuna fish | Fowl | Bull |
| | a,c | a,c | b,c | | e | f | g | h |
| Lys | - | 5.8 | 21.7 | 50.6 | 24.8 | - | - | - |
| His | - | - | - | - | - | - | - | 2.1 |
| Arg | 78.0 | 56.3 | 28.7 | 4.9 | 23.1 | 64.7 | 58.4 | 51.1 |
| Asp | - | - | - | - | 0.6 | - | - | - |
| Thr | - | 2.0 | 3.5 | 2.1 | 4.3 | 2.9 | 1.5 | 6.4 |
| Ser | 12.2 | 17.1 | 16.9 | 9.4 | 21.7 | 8.8 | 16.9 | 4.2 |
| Glu | - | - | - | 0.9 | 0.6 | - | - | 2.1 |
| Pro | 2.4 | - | 6.3 | 9.5 | 2.4 | 5.9 | 3.1 | - |
| Gly | - | 4.9 | 6.3 | - | 3.0 | - | 9.2 | 4.2 |
| Ala | - | 9.6 | 13.6 | 20.2 | 14.2 | 5.9 | 3.1 | 2.1 |
| Cys | - | - | - | - | - | - | - | 12.8 |
| Val | - | 3.5 | 1.0 | 2.0 | 2.3 | 8.8 | 1.5 | 4.2 |
| Met | - | - | - | - | 0.4 | - | - | - |
| Ile | - | - | - | - | 0.5 | - | - | 2.1 |
| Leu | - | - | - | - | 1.7 | - | - | 2.1 |
| Tyr | 7.3 | - | - | - | 0.3 | 2.9 | 6.2 | 4.2 |
| Phe | - | - | - | - | 0.3 | - | - | 2.1 |
| Number of residues | 70 | 91 | 100 | 83 | 297 | 34 | 65 | 47 |

References

a Subirana *et al*, 1973

b Ausió, 1980

c Ausió & Subirana, 1982c

d Phelan *et al*, 1974

e Ausió & Subirana, 1982b (This protein also contains 0.3% Trp)

f Bretzel, 1973

g Nakano *et al*, 1976

h Coelingh *et al*, 1972

Table 2. Amino acid composition (%) of histone-related proteins from spermatozoa

| | H1-related | | H2b-related | | | | | Arginine rich |
|---|---|---|---|---|---|---|---|---|
| Amino acid | Holothuria Ø0 | Sea urchin Ø1 | Sea urchin H2b | Gibbula Ø2b | Mytilus Ø2b | Limmulus H1c | Shrimp Hsp | Flounder fraction A |
| | a | b | c | d | e | f | g | h |
| Lys | 17.0 | 24.6 | 11.1 | 13.4 | 20.0 | 20.0 | 15.3 | 14.9 |
| His | - | 1.2 | 1.4 | 2.9 | 0.8 | - | 2.2 | 0.5 |
| Arg | 25.9 | 11.7 | 14.6 | 9.1 | 7.9 | 6.7 | 9.8 | 23.7 |
| Asp | - | 2.9 | 3.5 | 8.4 | 5.3 | 6.1 | 6.1 | 1.1 |
| Thr | 2.1 | 2.9 | 8.3 | 3.7 | 3.3 | 6.0 | 3.3 | 5.1 |
| Ser | 9.4 | 7.4 | 10.4 | 8.0 | 12.2 | 9.4 | 7.5 | 22.8 |
| Glu | 2.9 | 2.6 | 7.6 | 8.6 | 3.6 | 10.1 | 7.1 | 3.0 |
| Pro | 6.5 | 8.0 | 4.9 | 2.7 | 6.7 | 8.7 | 3.4 | 13.7 |
| Gly | - | 4.9 | 9.7 | 9.5 | 9.4 | 5.8 | 10.8 | 1.1 |
| Ala | 29.5 | 23.3 | 6.9 | 7.7 | 13.9 | 10.0 | 10.6 | 2.8 |
| Val | 5.0 | 3.3 | 8.3 | 4.5 | 4.2 | 4.6 | 6.8 | 3.0 |
| Met | - | n.d. | 1.4 | 1.4 | 1.9 | 1.2 | 0.3 | 4.9 |
| Ile | 1.3 | 3.4 | 2.8 | 4.0 | 3.0 | 3.8 | 4.9 | 1.1 |
| Leu | - | 1.9 | 4.2 | 7.2 | 4.7 | 5.2 | 8.8 | 1.1 |
| Tyr | - | 1.1 | 3.5 | 5.6 | 0.7 | 1.5 | 2.0 | 0.5 |
| Phe | - | 0.6 | 1.4 | 3.3 | 1.6 | 1.3 | 1.3 | 0.5 |
| % of basic proteins | 4 | 35 | 15 | 15 | 10 | 3 | 18 | 25 |

References

a Jordán, 1980
b Palau *et al*, 1969
c Brandt *et al*, 1979
d Colom and Subirana, 1981
e Ausió and Subirana, 1982a
f Muñoz-Guerra *et al*, 1982b
g Sellos and Le Gal, 1981
h Kennedy and Davies, 1980

# STRUCTURAL ELEMENTS OF THE MAMMALIAN SPERM NUCLEUS

ANTHONY R. BELLVÉ AND STUART B. MOSS

Laboratory of Human Reproduction and Reproductive Biology, Harvard Medical School, Boston, Massachusetts 02115, U.S.A.

## 1. INTRODUCTION

During spermiogenesis the spherical nucleus of the germ cell condenses and assumes a configuration that is distinct for each mammalian species. This nuclear transformation, whether yielding a falciform, spatulate or discoid shape, serves to protect the haploid genome and to facilitate penetration of the ovum by the motile sperm cell. Remodeling of the nucleus is complex. The process involves the removal of histones, the selective elimination of most nonhistone chromosomal proteins, the deposition of protamines, and the insertion of novel nuclear proteins (Bellvé, 1979; Bellvé & O'Brien, 1982). A microtubular array, the manchette, exists transiently in association with the condensing nucleus, acting as an external scaffold rather than imposing a direct mechanical force (Myles & Hepler, 1982). Consequently, sperm nuclear shape may be defined intrinsically by nuclear proteins, as suggested previously by Fawcett et al. (1971).

## 2. PROTAMINES AS DETERMINANTS OF SPERM NUCLEAR SHAPE

A comparison among mammalian species having sperm with falciform, spatulate or discoid nuclei failed to reveal any correlation between nuclear morphology and the complement of protamines. The mouse sperm nucleus, which is *falciform* in shape, contains two distinct protamines differing substantially in their primary sequences, as shown by amino acid and C-terminal analyses (Table 1). The type 1 protein has a high arginine content, while the type 2 protein also contains histidine (Bellvé, 1979). In addition, protamine 2 has a peculiar C-terminal sequence of 25 basic amino acids that is interrupted by only three cysteinyl groups. A similar complement of types 1 and 2 protamine exists in the human sperm nucleus and yet this species is *discoid* in shape (Gaastra et al., 1978). Conversely, rat and guinea pig sperm contain only type 1 protamine, even though the nucleus of the former is *falciform* and the latter is *spatulate* (Bellvé, 1979). Based on these species, the protamine complement does not appear to influence the shape of the sperm nucleus.

TABLE 1

Partial Sequence Analysis of the C-termini of Mouse Protamines 1 and 2*

| | |
|---|---|
| Protamine 1 | ---------------------------------------Lys-Cys-Arg-Arg-Lys-$Tyr^{55}$ |
| Protamine 2 | ----($Arg_{11}$-$His_2$-Lys-$Cys_2$)-(His-$Arg_4$)-Lys-Cys-His-Arg-Arg-Arg-$His^{63}$ |

*Sperm nuclei were dissociated in 6.0 M guanidine hydrochloride, 50 mM DTT, 1 mM PMSF, 0.5 M Tris-HCl (pH 8.6). The proteins were alkylated by adding ethylenimine to 0.25 M, and then were purified by chromatography on Bio-Rex 70, Bio-Gel P-10 and Amberlite IRC-50 (Bellvé et al., 1975). The proteins were digested (0-60 min) using carboxypeptidases A, B and Y. Amino acids released were quantitated using a Beckman 121M amino acid analyzer. The sequence of amino acids within brackets was not determined.

Furthermore, during rat and mouse spermiogenesis synthesis of the protamines occurs after the primary phase of nuclear transformation (Grimes et al., 1977; Mayer et al., 1981). These observations suggest that protamines are not the primary determinants of sperm nuclear shape.

## 3. A STRUCTURAL ROLE OF THE PERINUCLEAR THECA

Conformation of the sperm nucleus may be defined by a skeletal matrix analogous to that delineating somatic nuclei (Franke et al., 1981). This structure could be assembled from the nonprotamine nuclear proteins that are synthesized during spermiogenesis coincident with nuclear transformation (O'Brien and Bellvé, 1980a,b).

The existence of a sperm nuclear matrix was tested by developing a procedure for selectively removing the protamines and DNA. Mouse sperm nuclei, in which the protamines were prelabeled with [$^3$H]arginine (Bellvé et al., 1975) were incubated with DTT to reduce intra- and intermolecular disulfide bonds. These nuclei then were exposed to selected dissociating reagents. Following a 30 min incubation, the samples were centrifuged at 50,000 g for 13 hr at 5$^o$C, and each supernatant and pellet was quantified for soluble and DNA-associated [$^3$H]protamine, respectively. Most reagents tested failed to displace the protamines. These included 1) heparin (0.2 mg/ml) and dextran sulfate (2 mg/ml), which are known to displace histones from metaphase chromosomes (Paulson & Laemmli, 1977); 2) sodium dodecyl sulfate (1%) and sarkosyl (1%), which promote swelling of sperm nuclei (Bedford and Calvin, 1974; Evenson et al., 1978); and 3) the monovalent cations NaCl (1.0 M) and KCl (1.0 M). However, the divalent cations $Ca^{2+}$· $Mg^{2+}$ (3:2 ratio) displaced protamine in a dose-dependent manner, with maximal displacement occurring at 225-250 mM total concentration (Bellvé, 1982). These nuclei retained their normal configura-

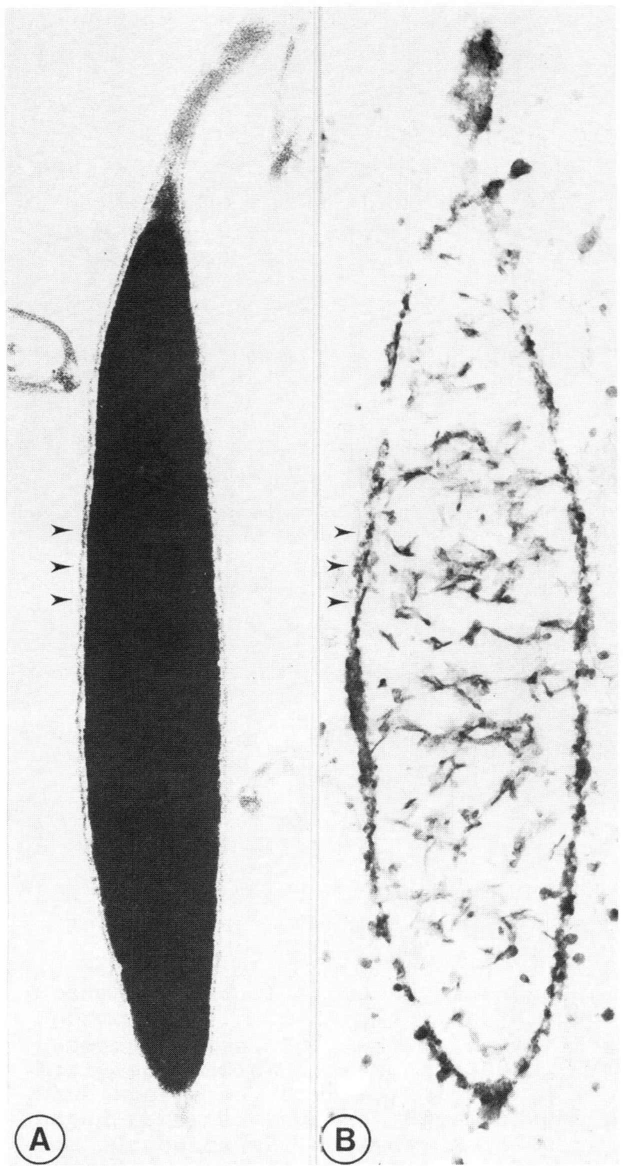


FIGURE 1. Electron micrographs of a mouse sperm nucleus and its associated perinuclear theca and matrix. X 46,000.

A) Nucleus with the apical perforatorium and the perinuclear theca (arrowheads) surrounding the dense chromatin. The cell's acrosome, nuclear envelope and tail were removed previously by incubation in 1% SDS.

B) Perinuclear theca and internal matrix fibers delineate the original shape of the nucleus after the removal of protamine and DNA with ~ 190 mM $Ca^{2+}$. $Mg^{2+}$ (3:2 ratio) and DNase I (50 μg/ml).

tion, even though a marked ethidium bromide-positive DNA halo surrounded the nucleus. Significantly, when the nuclei were exposed to DNase I (∿ 50 μg/ml), either concurrently or following incubation with 190 mM $Ca^{2+}\cdot Mg^{2+}$, ∿ 98% of the DNA was hydrolyzed. Removal of both protamine and DNA left the perinuclear theca conforming to the original shape of the sperm nucleus (Fig. 1). This skeletal structure is composed of the perforatorium, the perinuclear material and the postacrosomal dense lamina (see Lalli and Clermont, 1981), and contains polypeptides ranging from $M_r$ ∿ 8,000 to 80,000 daltons (Bellvé, 1982).

The molecular organization of the sperm perinuclear theca is being resolved using xenogeneic monoclonal antibodies. Mouse sperm nuclei were used to immunize two female Wistar rats and the spleen cells were fused with P3/NS1/1 mouse myeloma cells. Hybrid cells were selected using HAT medium and those producing antibody were identified using sperm nuclei in a solid phase immunoabsorbent assay. Nine positive hybridomas were selected initially, grown in quantity and stored in liquid $N_2$. After cloning in soft agar, three cell lines, SpN-1, SpN-2 and SpN-3, demonstrated specificity for sperm nuclear constituents. These monoclonal antibodies showed quantitative differences in their binding to intact sperm nuclei and to nuclei that were selectively depleted of protamines, certain nonprotamine chromosomal proteins or constituents of the perinuclear theca. Moreover, these antibodies also exhibited distinct patterns on binding to sperm nuclei, suggesting that certain antigens are localized preferentially to discrete regions of the nucleus.

REFERENCES

1. Bedford JM, Calvin HI. 1974. J Exp Zool. 188:137-156.
2. Bellvé AR. 1979. In: Finn CA, ed. Oxford Reviews of Reproductive Biology, pp. 159-261. Oxford, Oxford University Press.
3. Bellvé AR. 1982, in press. In: Amann RP, Seidel GE Jr, eds. Prospects for Sexing Mammalian Sperm. Boulder, Colorado Assoc. University Press.
4. Bellvé AR, Anderson E, Hanley-Bowdoin L. 1975. Develop Biol. 47:349-365.
5. Bellvé AR, O'Brien DA. 1982, in press. In: Hartman JF, ed. Mechanisms and Control of Fertilization. New York, Academic Press, Inc.
6. Evenson DP, Witkin SS, deHarven E, Bendich A. 1978. J Ultrastruct Res. 63:178-187.
7. Fawcett DW, Anderson WA, Phillips DM. 1971. Develop Biol. 26:220-251.
8. Franke WW, Scheer U, Krohne G, Jarasch E-D. 1981. J Cell Biol. 91:39s-50s.
9. Gaastra W, Lukkes-Hofstra J, Kolk AJH. 1978. Biochem Genet. 16:525-529.
10. Grimes SR Jr, Meistrich ML, Platz RD Hnlica LS. 1977. Exp Cell Res. 110:31-39.
11. Lalli M, Clermont Y. 1981. Am J Anat. 160:419-434.
12. Mayer JF, Chang TSK, Zirkin BR. 1981. Biol Reprod. 25:1041-1051.
13. Myles DG, Hepler PK. 1982. Develop Biol. 90:238-252.
14. O'Brien DA, Bellvé AR. 1980a. Develop Biol. 75:386-404.
15. O'Brien DA, Bellvé AR. 1980b. Develop Biol. 75:405-418.
16. Paulson JR, Laemmli UK. 1977. Cell 12:817-828.

# TRANSFORMATION AND STABILIZATION OF CHROMATIN IN RAM SPERMATIDS.

M. LOIR[a], D. BOUVIER[b], M. FORNELLS[c], M. LANNEAU[a] & J.A. SUBIRANA[c].

a) I.N.R.A., Physiologie de la Reproduction, 37380 Nouzilly, France.
b) Pathologie cellulaire, Institut biomédical des Cordeliers, 75270 Paris.
c) Unidad de Quimica Macromolecular del C.S.I.C., Diagonal 999, Barcelona (28), Spain.

## 1. INTRODUCTION

At the time mammalian spermatids differentiate, histones are replaced by one or more protamines and the chromatin becomes tightly packaged and resistant to various agents. What relationships exist between the biochemical and the structural nuclear changes? Which proteins amongst the various ones which alternate in the nuclei are responsible for the progressive chromatin reorganization and stabilization? We report the results of recent studies partly answering these questions.

## 2. MATERIAL

Homogeneous populations of round (steps 1-8,fig. 1), elongating (steps 9-12) and elongated (steps 13-15) ram spermatid nuclei were prepared by elutriation and Triton X-100 (1), by the EDTA method and by sonication (2). The nuclei were purified through 1.5 M sucrose.

## 3. BIOCHEMICAL CHANGES (Fig. 1)

When they leave chromatin in elongating spermatids, the core histones, namely H4, are hyperacetylated (3). In spermatid nuclei which begin to condense, the $H_1^{\circ}$ histone is present, while it is absent in spermatocytes, in round spermatids and in elongated spermatids.

During the replacement of histones by the protamine, four spermatid-specific, cystein-containing, small (6,600 to 18,000 daltons) basic proteins are present (P1, 3, 7, T). Their complete extraction needs a reducing treatment (2). Two of them (7, T) are soluble in 0.75 M perchloric acid (PCA), but they cannot be extracted with 0.35 M NaCl.

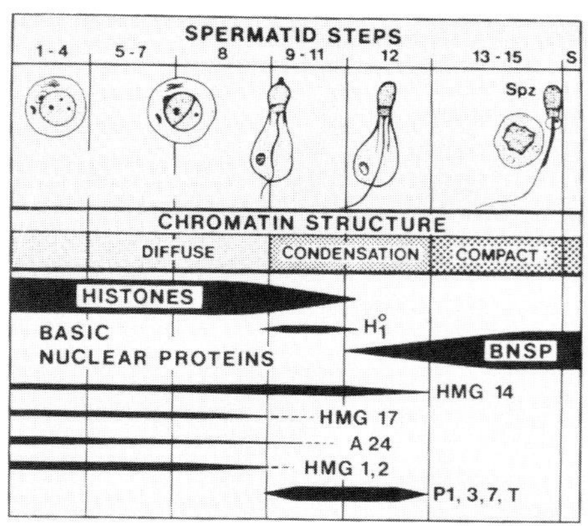


Figure 1. Summary of changes in the nucleus throughout ram spermiogenesis.

HMG proteins (Fig. 2) have been identified by their extractability, by their migration on two-dimensional slab gels, and by reference with calf thymus HMG proteins. The two large HMG proteins 1 and 2 are present in spermatocyte and round spermatid chromatin. They disappear when chromatin becomes inactive and begins to condense. This is true also for the small HMG

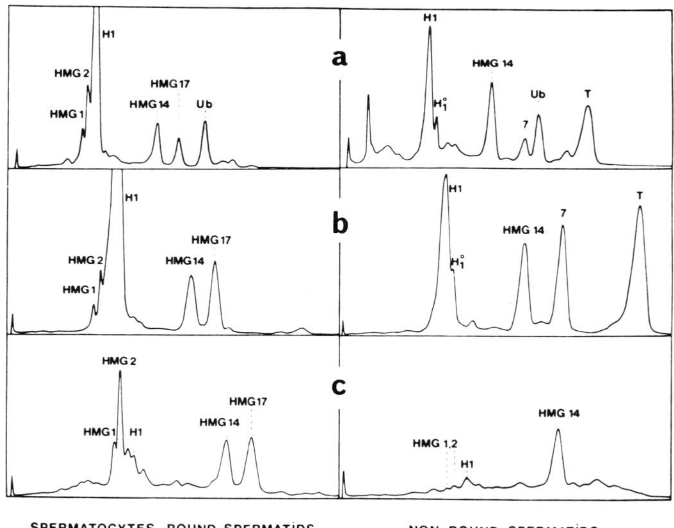


Figure 2. Proteins in round germ cells and in non-round spermatids.
a) Cellular proteins extracted by 0.75 M PCA (4). Ub: ubiquitin. b) Nuclear proteins extracted similarly. c) Nuclear proteins extracted by 0.35 M NaCl and soluble in 3% TCA (5) (HMG proteins).

protein 17. By contrast, the amount of HMG 14 increases at the time histones are removed, then it disappears. It is important to note that free ubiquitin (HMG 20) is never present in the nuclei of spermatocytes and spermatids although it is present in their cytoplasm.

As soon as the protamine (BNSP) appears in the nuclei, its extraction needs a reducing pretreatment. This indicates that soon after it is deposited on DNA, SS bridges begin to cross-link the nucleoprotein molecules.

## 4. STRUCTURAL CHANGES

In round germ cell nuclei, cations are necessary for the chromatin stabilization, but in non-round spermatid nuclei the chromatin becomes resistant to EDTA treatment then to sonication, DNase and Trypsin (2). The chromatin can be solubilized by Heparin in round nuclei, but in elongating spermatids it is only decondensed and is not changed in elongated spermatids. In contrast with epididymal and ejaculated spermatozoa, the decondensation of the non-round spermatid nuclei needs only a reducing treatment. Proteolysis or a detergent are not necessary but they do speed up the decondensing effect of DTT or 2-mercaptoethanol. SDS alone (30 mM) has no decondensing effect on the elongated spermatid nuclei.

At the time histones are removed, the chromatin loses its beaded appearance and is constituted of smooth fibers which aggregate in parallel.

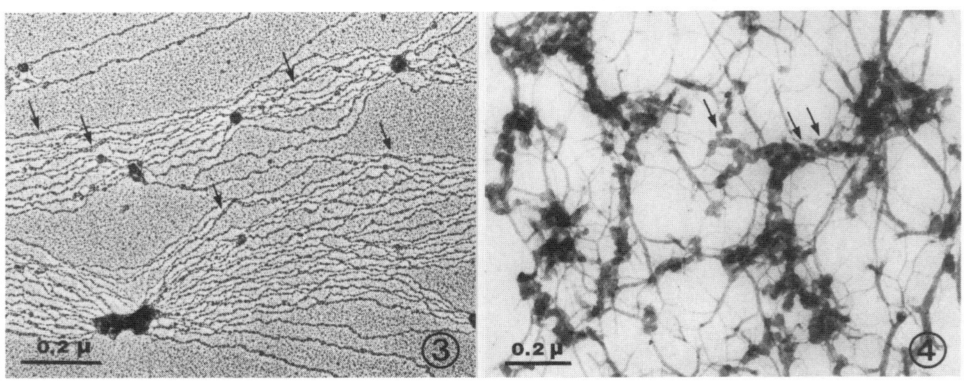


Figure 3. Spermatid step 12 chromatin. Whole-mount preparation. Note the parallel orientation of fibers and some side-by-side pairings (arrows).

Figure 4. Spermatid step 13 nuclei decondensed by a SDS-DTT treatment. Note the coiled portions of large fibers (arrows).

This is confirmed by the study of whole-mount preparations (6) of unextracted and undecondensed elongating nuclei (Fig. 3). When the electron-opaque elongated nuclei are decondensed by a short SDS-DTT treatment, large smooth branched and coiled fibers are observed on sections (Fig. 4), while on whole-mount preparations only thin fibers are visible. However, such a treatment may extract partly the protamine or at least disturb the structural organization of chromatin. So, X-ray diffraction studies have been carried out on sperm nuclei. The results suggest that the DNA is organized in bundles of parallel DNA molecules which may be bent in a smooth manner. No higher order structures of a size between 30 and 200 A are present.

## 5. CONCLUSIONS

HMG 14 is known to inhibit histone deacetylase (7). The increase in this protein could be responsible for the core histone hyperacetylation, which should initiate their removal from the DNA.

The start of chromatin condensation could be promoted by two events: a) the disappearance of HMG 1, 2 and 17 which usually are associated with transcriptionnally active chromatin (8), and b) the appearance of $H_1^{\circ}$, already supposed to play a role in chromatin condensation (9).

The four spermatid-specific proteins, distributed evenly along the DNA fibers and forming interfiber cystine bridges, should be responsible for the first step of chromatin compaction. They should lead the chromatin to a moderately condensed state, allowing the protamine to play its extreme terminal packaging role.

The protamine should not induce nucleosome-like structures in elongated spermatid and spermatozoa chromatin.

## REFERENCES

1. Loir M, Lanneau M. 1982. Gamete research 6, in press.
2. Loir M, Lanneau M. 1978. Exptl. Cell. Res. 115, 231-243.
3. Loir M, Lanneau M, Le Gac-Jegou F. 1982. Proceeding Special FEBS Meeting, Athens, 214.
4. Sanders C. 1977. Biochem. Biophys. Res. Com. 78, 1034-1042.
5. Christensen M.E., Dixon G.H. 1981. J. Biol. Chem. 256, 7549-7556.
6. Miller O.L., Bakken A.H. 1972. Acta Endocr. Suppl. 168, 155-177.
7. Reeves R., Candido E.P.M. 1980. Nucl. Acids. Res. 8, 1947-1962.
8. Weisbrod S., Weintraub H. 1979. Proc. Natl. Acad. Sci. USA, 76, 630-634.
9. Gjerset R., Ibarrando F., Saragosti S., Eisen H. 1981. Biochem. Biophys. Res. Com. 99, 349-357.

# CHROMATIN AGGREGATION OF HUMAN SPERM (*)

E. BUSTOS-OBREGON AND S. LEIVA.
Dept. Cell Biology and Genetics, Faculty of Medicine, University of Chile, Casilla 6556, Santiago 7, Chile.

## 1. INTRODUCTION

Testicular spermatozoa in eutherian mammals have to undergo a series of morpho-physiological changes along the epididymal transit, in a process known as sperm maturation. Among all these changes, stabilization of the DNA/protein complex due to disulfide bond formation in the histone is an important event (1, 2, 3, 4).

The present work reports on cytochemically detectable differences in the chromatin of normal and teratospermic human ejaculated spermatozoa.

## 2. Material and Methods.

Ejaculates with high percentages of oval heads were considered as normospermic whereas high incidence of either round or elongated heads were contrasted to the former, as being teratospermic.

Three groups of techniques were used in order to characterize: a) the complex DNA/proteins; b) the organization and stabilization of the nuclear chromatin, and c) the morphologic pattern of distribution of sulfhydrilated proteins in the sperm nuclei.

Since a detailed analysis of the number of techniques used is beyond the aim of this paper, only some of them will be mentioned together with the results.

(*) Supported by OEA, OMS and U. of Chile.

## 3. Results and Discussion.

"In vitro" incubations of sperm samples with the anionic detergent SDS, followed by a thiol-reducing agent (DTT), were run to find optimal concentrations and timing for a differential swelling of normal or teratospermic sperm nuclei. Decondensation is atributed to cleavage of S-S- bonds due to the thiol-reducing substance, of which also sodium thyoglycolate was also used. The latter gives good results without using SDS and it was thus considered as a better method to get chromatin decondensation (Fig.1a, b, c).

Cytophluorometric DNA mensuration using a fluorescent Feulgen reaction (5) on sperm nuclei treated with S-S reducing agents (2- mercapto ethanol at different concentrations), show that the teratospermic spermatozoa have greater fluorescence at shorter acid hydrolisis periods than the normospermic. (Table I). Without reduction of S-S groups, DNA values are very similar in the three types of sperm nuclei studied, as it has been previously reported by conventional Feulgen cytophotometry (6).

Total proteins, identified by sulfaflavin reaction with the -NH2 free protein group, after extraction with TCA, give a greater reaction for teratospermic spermatozoa at shorter times of TCA extraction, thus giving another proof that the degree of packing of DNA and protein differ from that found in normospermic nuclei (Table II).

The ratio ADN/Total protein was 1:2,1 (normospermic); 1:2,0 (spheric teratospermic nuclei) and 1: 1.7 (elongated teratospermic nuclei), the latter being the lowest value observed. It implies a greater accesibility to the DNA, due to a loose binding to its associated proteins.

Histones were identified after TCA extraction of DNA, using a cytophluorometric method (Table III). This time, it is the spheric nuclei that reveal higher histone reactivity at rather short DNA extraction times, again suggesting a weaker binding of DNA to its basic proteins in this sperm type. It should be stressed that in all three nuclei types analyzed the ratio Total proteins/Histones was the same (1: 0.7).

A certain morphological heterogeneity has been described in the ultrastructure of the chromatin in the nuclei of human spermatozoa (7).

A method was deviced for specific staining of SH groups of proteins, after thyoglycolate reduction. The p-Cl-Mercurial benzoate (sodium salt) was selected for its high specificity and adequate electron contrast, which is reinforced by a somehow selective binding of osmium to the Hg-containing sites, by a mechanism yet to be determined.

Normal (oval) sperm nuclei, decondensed by thyoglycolate, and stained for SH groups, display a regular pattern of SH-rich granules in the chromatin that form a network all through the nucleus, with a higher concentration close to the nuclear membrane, and in particular, towards its cephalic portion (Fig. 2 a, b, c, d). Teratospermic spermatozoa tend to give a weaker SH- reaction and the reactive points are more irregularly distributed in the nucleus (Fig.3 a, b), only rarely assuming the regular granulo-fibrilar pattern of highly SH-reactive chromatin of normal nuclei.

This ultrastructural difference should be further explored to understand its physiological implications, analysis for which the ultracytochemical method illustrated here can be of relevance.

Table I. % decrease of intensity of fluorescence.
(by reabsorption and/or interference)
Feulgen reaction (4 N HCl + Mercapto ethanol 0.1 M)

| Hydrolisis time | 40' | 60' | 80' |
|---|---|---|---|
| Normospermic | 13,7 | 23,5 | 28,7 |
| Spherical teratospermic | 25,1 | 27,5 | 33,9 |
| Elongated teratospermic | 28,7 | 34,5 | 40,9 |

Table II. Cytophluorometric determination of Total proteins.
(In arbitrary units)

| DNA extraction | Normospermic | Spherical t. | Elongated t. |
|---|---|---|---|
| TCA 2.00 hrs. | 121.0 (± 3.9) | 127.2 (± 7.2) | 128.0 (± 4.6) |
| TCA 3.00 hrs. | 108.0 (+ 3.8) | 109.2 (± 3.5) | 110.8 (± 3.5) |
| TCA 3.5 hrs. | 144.6 (± 2.5) | 144.0 (± 3.6) | 141.6 (± 4.0) |

Table III. Cytophluorometric determination of Histones. (In arbitrary units)

| DNA extraction | Normospermic | Spherical t. | Elongated t. |
|---|---|---|---|
| TCA 2.00 hrs. | 62.2 (± 5.4) | 70.5 (± 4.3) | 67.4 (± 3.5) |
| TCA 3.00 hrs. | 40.9 (± 4.7) | 45.4 (± 4.7) | 43.3 (± 6.0) |
| TCA 3.5 hrs. | 101.2 (± 5.5) | 105.3 (± 4.2) | 103.6 (± 5.3) |

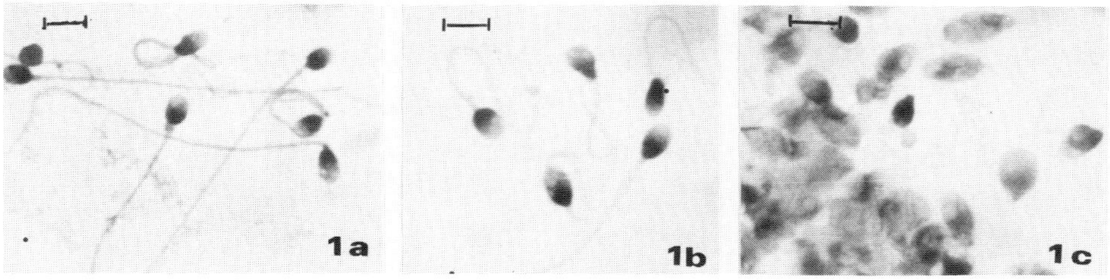

1a 1b 1c

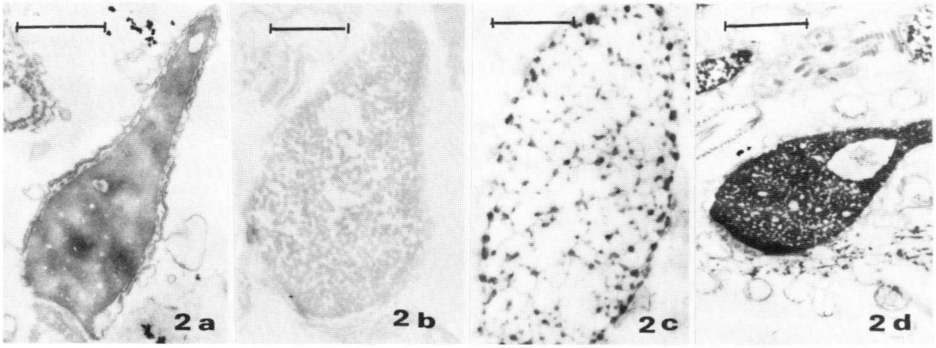

2a 2b 2c 2d

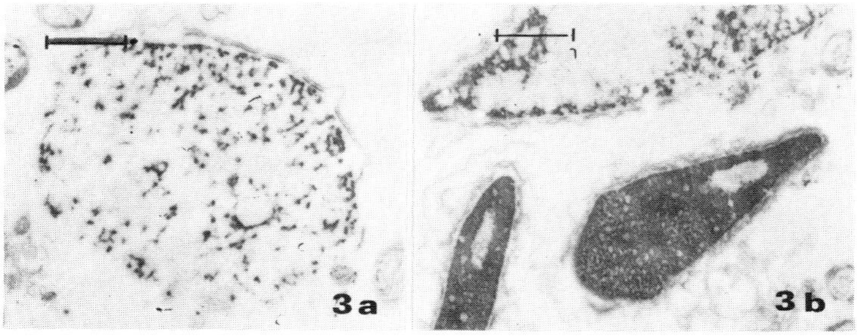

3a 3b

Figs.1a, b, c: Control, 4 and 10 min., incubation in Na thioglycolate ( bar = 1000 nm).
Figs. 2a, b (control), c, d. Chromatin decondensation (normospermic cells) by Na thioglycolate and -SH electronstaining ( bar | = 10 nm).
Figs. 3a, b. As in Figs. 2 but teratospermic cells ( bar = 10 nm). See text for detailed explanations of figures.

REFERENCES

1. Bedford J.M., Calvin H.I., Cooper G.W. (1973a). The maturation of spermatozoa in the human epididymis.J.Reprod. Fert. Supp.18: 199-213.
2. Bedford J.M., Bent J.M., Calvin H. (1973b). Variation in the structural character and stability of the nuclear chromatin in morphologically normal semen spermatozoa. J. Reprod. Fert. 33: 19-29.
3. Bedford J.M., Calvin H. 1974. The ocurrence and possible functional significance of S-S crosslink in sperm head with particular reference to eutherian mammals. J.Exp.Zool. 88: 137-156.
4. Bedford J.M. 1975 Handbook of Physiology.Endocrinology 5, D.W. Hamilton and R.O. Greep (Eds.) Phys.Soc., Washington D.C.
5. Prenna G., Leiva S., Mazzini G. 1974. Quantitation of DNA by cytofluorometry of the conventional Feulgen reaction. Histochem.J. 6: 467-489.
6. Bustos-Obregón E., Leiva S., Guadarrama A. 1978. Estudio citofotométrico de teratospermia en semen humano. Rev. de Micr.Elect. 5: 146-147.
7. Pedersen H. 1974. The human spermatozoa. Danish.Med.Bull. 21, Suppl. I: 1-36.

# PRESUMED IMPAIRMENT OF THE X INACTIVATION DURING MEIOTIC PROPHASE : A COMMON PHENOMENON CAUSING SPERMATOGENIC BREAKDOWN IN VARIOUS CYTOGENETIC CONDITIONS.

J.M. LUCIANI and M.R. GUICHAOUA - Department of Embryology and Cytogenetics- Faculty of Medicine - Bd P. Dramard, Marseilles, France.

## INTRODUCTION

Balanced autosomal translocations are more frequently observed in men attending an infertility clinics that in general population (1). The sterilizing effects of these autosomal translocations appear to disturb gamete production in male heterozygotes , meanwhile no effects have been reported in females carrying the same translocation (1). The mechanism of spermatogenic breakdown is unclear : some translocations are associated with severe oligospermia, azoospermia and subsequent sterility; others have no effect on gamete production. Moreover, both sterility and fertility were observed in translocation carriers from the same family (2). Various tentative explanations have been proposed in the past (3) : the position effect of the genes or a mutation associated with one or both of the break points leading to pairing disturbance and a high frequency of univalents.

A more satisfactory hypothesis could offer an explanation for the spermatogenic breakdown occuring in autosomal translocation carriers. This hypothesis was first postulated in sterile Drosophila with reciprocal translocation X-autosome (4) : the inactivation of the single X chromosome in the primary spermatocytes of species with heterogametic males may represent a basic mechanism controlling spermatogenesis. In man as well as in other heterogametic species, normal inactivation of the single chromosome X occurs at the end of zygotene stage with the formation of the sex vesicle and is efficient during pachytene stage on. Anything which interferes with the normal inactivation of the sex chromosomes during male meiotic prophase might disturb biochemistry of the cell and cause failure of the germ cell development.

Impairment of the X chromosome inactivation might occur in man with various cytogenetic conditions causing anomalies in sex vesicle formation:

X-autosomal translocations; Y-autosomal translocations; dissociation between X and Y chromosomes; purely autosomal translocations.

1 - X-autosomal translocations.

Males carrying a reciprocal translocation between the X chromosome and an autosome are kwnown to be sterile in the mouse, in Drosophila and also in man. In the latter, the only meiotic study performed showed either large anomalies of the sex vesicle formation or its absence (5).

2 - Y-autosomal translocations.

Y-autosomal translocations found in infertile men are usually balanced translocations. Meiotic studies have been performed in several cases (for review see 6) : pachytene analysis showed association between the Y-translocation segment and the sex vesicle. It was assumed that the absence of integrity of the Y chromosome induces an abnormal sex vesicle.

3 - Dissociation between X and Y chromosomes.

Failure of sex chromosome pairing has been found in association with spermatogenic breakdown in both mouse (7) and man (8). In one personal case, specific silver staining of the sex chromosomes failed to show abnormal pairing at pachytene stage (unpublished data). But these observations made at the light microscope level do not exclude a fine anomaly of the X-Y link.

4 - Purely autosomal translocations.

It is not obvious how reciprocal or robertsonian translocations could interfere with inactivation of the X chromosome, but in some male sterile translocations of the mouse, a non-random association was found between the X chromosome and the translocation configuration at diakinesis and metaphase I (9, 10, 11). The non-random association involves the centromeric heterochromatin (C-bands) of the X chromosome and of the translocation configuration. The non-random association is confined to early and late diakinesis. The frequency of C-bands contact decreases rapidly from early diakinesis on. Its non randomness disappears at early metaphase I. This observation suggested that the association could occur before early diakinesis. Recent observations of the synaptonemal complexes visualized by silver staining techniques displayed in sterile mice a high frequency (about 60%) of pachytene spermatocytes with non-random association between autosomal translocation and the XY pair (12). Thus, the non-random contact is already present at the pachytene stage when the X-inactivation takes place. This phenomenon is not restricted to the mouse. Recently, in the human, we reported in a sterile male with a 13-14 robertsonian translocation, a non-random association between the trivalent configuration and the sex vesicle at the pachytene stage (13 and fig. 1).

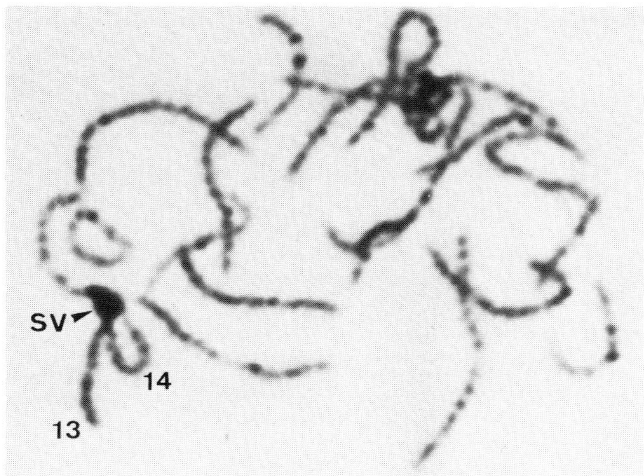


Fig.1 - Human pachytene nucleus showing the 13-14 trivalent, identified by its chromomere pattern, associated with the sex vesicle (S.V.) (x1300).

This observation was possible thanks to the availability of a precise identification of all autosomal bivalents at pachytene stage in human spermatocyte nuclei (14). The association between the 13-14 trivalent and the sex vesicle was found in a large number of cells (36 over 59 judged informative for analyze). The percentage frequency (61%) is highly significant compared to the rate of association between the acrocentric bivalents and the sex vesicle (from 7.1% to 12.2%) observed in a sample of men presenting excretory infertility, normal testis histology, normal somatic karyotype and referred as control for these reasons. As already reported in the mouse (9), the tight association between the translocation configuration and the sex vesicle always concerned the centromeric heterochromatin region of the 13-14 trivalent.

Non-random association between rearranged autosomes and XY pair observed both in mouse and man may represent a widespread phenomenon. This association, already present at pachytene stage, would interfere with the normal X chromosome inactivation disturbing the orderly biochemical machinery of the developing germ cells, mainly during spermiogenesis. Early storage of information in spermatocytes, necessary for acrosome development in ram spermatids, was recently reported (15).

In conclusion, we would like to emphasize as a working hypothesis, that any process interfering with the X chromosome inactivation at the pachytene stage during male meiosis, such as X-autosome translocations, Y-autosome translocations and XY dissociations, would disturb the genetic control on germ cell differentiation, thus causing the death of the germinal cells and subsequent infertility of the individual.


1 - Chandley, A.C., 1979 Brit. Med. Bull., 35, 181-186.

2 - Luciani, J.M., Stahl, A., 1978, Ann. Biol. anim. Bioch. Biophys., 18, 377-382.

3 - Chandley, A.C., Christie, S., Fletcher, J. Frackiewicz, A. and Jacobs P.A. 1972, Cytogenetics, 11, 516-533.

4 - Lifschytz, E. and Lindsley, D.L., 1972, Proc. Nat. Acad. Sci. U.S.A., 69, 182-186.

5 - Dutrillaux, B., Couturier, J., Rotman, J., Salat, J. and Lejeune J., 1972 - C.R. Acad. Sci. (Paris), 274, 3324-3327.

6 - Gonzales, J., Lesourd, S. and Dutrillaux, B., 1981, Hum. Genet., 57, 111-114.

7 - Beechey, C.V., 1973, Cytogenet Cell Genet, 12, 60-67.

8 - Chandley, A.C., 1973, Heredity, 30, 262.

9 - Forejt J., 1974, Cytogenet Cell Genet., 13, 369-383.

10 - Forejt, J. and Gregorova, S., 1977, Cytogenet Cell Genet., 19, 159-179.

11 - Forejt, J., 1979 Cytogenet Cell Genet., 23, 163-170.

12 - Forejt, J., Gregorova, S. and Goetz, P., 1981, Chromosoma, 82, 41-53.

13 - Guichaoua, M.R. and Luciani, J.M. (in press).

14 - Luciani, J.M., Morazzani, M.R. and Stahl, A., 1975, Chromosoma, 52, 275-282.

15 - Courtens, J.L. and Courot, M., 1980, Anat. Rec., 197, 143-152


Aknowledgements. We wish to thank Mrs Marie Régine Morazzani for technical and photographic assistance. This investigation was supported by a grant from the C.N.A.M.T.S.

# CHROMATIN DECONDENSATION ABILITY OF THE HUMAN SPERMATOZOON

U. KVIST
Department of Physiology, Karolinska Institutet,
104 01 Stockholm, Sweden.

## INTRODUCTION

Studies on human ejaculated spermatozoa indicate that the human spermatozoon has an inbuilt capacity to decondense its nuclear chromatin(5). The intrinsic mechanism is triggered to work by two necessary but independent, pre-requisites. One is zinc-removal (effectuated by sperm exposure to albumin or EDTA) and the other is fulfilled by sperm exposure to the detergent sodium dodecyl sulphate (SDS). As studied by light microscopy, scanning and transmission electron microscopy decondensation induced by activation of the intrinsic mechanism for NCD (7) resembles that observed in spermatozoa entering the ovum (1,12) and results in separation of individual chromosomal fibers, as reported for in vitro decondensation accomplished by exogenous thiols (4,8).

The aims of the present work were: 1. To study how sperm NCD-ability is influenced upon sperm storage and 2. To further evaluate the role of zinc for preservation of sperm NCD-ability.

## MATERIAL AND METHODS

Semen samples from 16 healthy donors were used in three series of experiments. Washed spermatozoa were centrifuged (400 g,15 min), supernatant removed and the sperm pellet dispersed in a buffered salt solution (BSS; mM: NaCl 123, KCl 5, HCl 24, TRIS 37, pH=8.0). This wash procedure was repeated twice.

<u>In the first series of experiments</u> two aliquots of whole semen were withdrawn from each of 6 semen samples 5 min after ejaculation. One aliquot was exposed to SDS (1% sodium dodecyl sulphate in 0.05 M borate buffer, pH=9.0) and the other to SDS containing 6 mM EDTA (SDS-EDTA).

In the second series of experiments 5 semen samples were stored for 24h at 22°C. One aliquot from each sample was withdrawn and exposed to SDS-EDTA at 20 min, 1h,2h,5h and 24h after ejaculation, respectively.

In the third series of experiments each of 5 semen samples were divided in two parts after liquefication. Parts one and two were washed; parts one twice in BSS, and parts two once in BSS and once in BSS containing 6 mM EDTA. Before the last centrifugation all sperm suspensions were further divided in two halves. Halves one were resuspended in pure BSS, and halves two in BSS containing 1.5 mM zinc(II). Thus, from each original semen sample spermatozoa treated in four different ways were obtained (BSS washed,BSS washed and zinc supplemented,BSS-EDTA washed, BSS-EDTA washed and zinc supplemented). Sperm suspensions were stored at 22° C for 24h. Aliquots were withdrawn and exposed to SDS-EDTA at 0h and 24h of storage.

The NCD ability was assessed from the degree of sperm head swelling after 60 min exposure to SDS or SDS-EDTA. One volume of sperm suspension was mixed with 9 volumes of SDS/SDS-EDTA and the reaction was stopped by adding the equal amount of 2.5% glutaraldehyde in 0.05 M sodium borate buffer (pH=9.0). Nuclear swelling was assessed with phase contrast microscopy (500x) and the spermatozoa were classified as stable (=not swollen) and unstable (=moderately and grossly swollen) (cf 5).

Analytical statistics was performed with the Randomization Test for Matched Pairs (11). Mean and range are given for the frequency of stable spermatozoa.

## RESULTS

### I. NCD ability close to ejaculation

A high proportion (87%,77%-99%) of spermatozoa remained stable when exposed to SDS alone 5 min after ejaculation. In contrast, a majority of spermatozoa exposed to SDS-EDTA reacted with nuclear chromatin decondensation,NCD, (90%, 71%-97%), Fig.1,Group A).

### II. Loss of NCD ability upon semen storage

The percentage spermatozoa decondensing in SDS-EDTA promptly decreased upon semen storage (Fig.1,Group B).

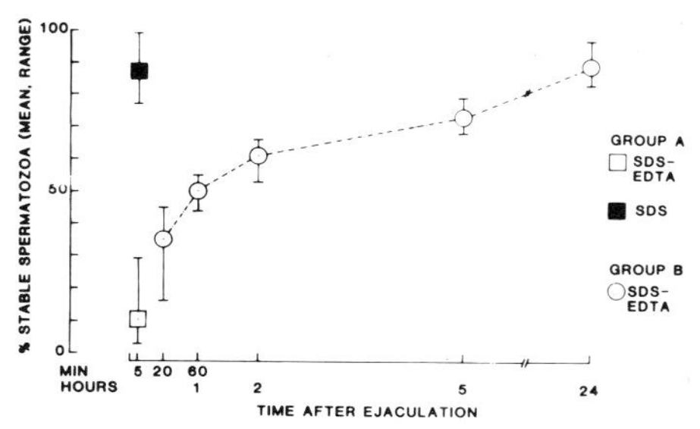


FIGURE 1. Loss of sperm NCD ability upon semen storage. Group A) Spermatozoa exposed to SDS and SDS-EDTA 5 min after ejaculation (N=6). Group B) Spermatozoa stored in seminal plasma for 24 h. Exposed to SDS-EDTA at given intervals (N=5).

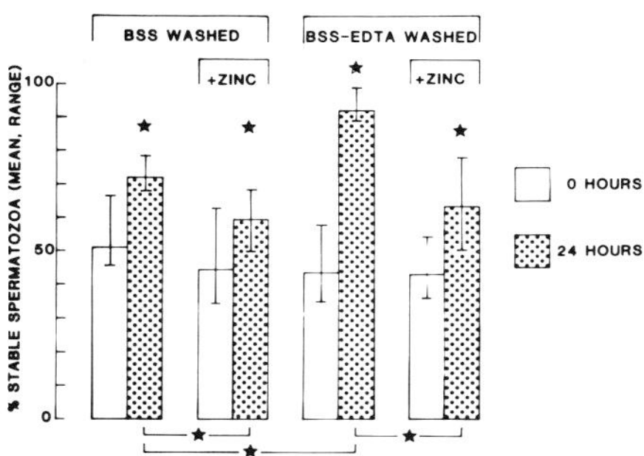


FIGURE 2. Enhanced loss of sperm NCD ability upon sperm saline storage, induced by EDTA pre-treatment, restored by zinc(II) supplementation. (Significance for all differences, p 0.05, N=5)

III. Zinc preserved NCD ability

Upon 24 h of saline storage all four sperm treatment groups (ie BSS washed; BSS washed and zinc supplemented; BSS-EDTA washed; BSS-EDTA washed and zinc supplemented) increased their proportions of non-decondensing spermatozoa (Fig. 2:dotted bars vs open bars). Almost 100% (93,88-100) of spermatozoa pre-washed in BSS-EDTA lost their NCD reacting capacity upon storage. The corresponding percentage for BSS washed controls was 72 (67-88). Thus, EDTA pre-treatment enhanced the loss of NCD capacity (Fig. 2: BSS-EDTA washed controls(dotted) vs BSS washed controls(dotted)). When zinc was added to the storage medium zinc not only counteracted the EDTA-enhanced loss of NCD ability (Fig. 2: BSS-EDTA washed (dotted): zinc vs controls) but also reduced the "native" loss of NCD ability revealed by BSS washed controls (Fig. 2: BSS washed (dotted):zinc vs controls).

## DISCUSSION

Fertilization can not be completed until the tightly condensed chromatin fibers of the sperm nucleus have loosened and the individual chromosomal fibers unravelled. Chromatin decondensation of the eutherian sperm nucleus and separation of individual chromosomal fibers have earliar been reproducably accomplished in vitro only by the use of a combination of detergents and exogenous disulphide-disrupting compounds (4,8). Chromatin disulphide bridges are evidently instrumental in the stabilization of eutherian sperm chromatin and have to be broken for a decondensation to take place ( 2,9,13). It has been suggested that the ooplasm provides the spermatozoon the thiols necessary for a reductive cleavage of the crucial disulphide bridges allowing the chromatin to decondense ( 2,9,13).

However, it was recently suggested that the human spermatozoon itself provides the thiols necessary for reductive cleavage of chromatin S-S crosslinks (5). It was demonstrated that the human spermatozoon exposed to a detergent (SDS), spontaneously reacted with NCD, if also released of zinc,shortly before,during or after SDS exposure (5). This inbuilt capacity was blocked by various SH-inactivating compounds which led to the conclusion that the sperm NCD ability is dependent upon the action of intrinsic thiols (6). Whenever studied, a certain proportion of ejaculated spermatozoa fails to react with NCD and as demonstrated in this paper, this proportion increased upon sperm storage. For several reasons oxidative "destruction" of thiols,which become comitted into surplus S-S crosslinks seems to be a plausible explanation for this phenomenon. Thus, oxidizing compounds eg iodosobenzoate and copper(II) ions totally abolish the NCD ability (6), and has been demonstrated to decrease the spermatozoal content of free SH-groups (3). Moreover, zinc (a) inhibits copper induced oxidation of sperm SH-groups into S-S bridges (3), (b) reversibly inhibits the NCD ability (5) and (c) as here reported,zinc also can serve to protect the NCD ability from destruction upon sperm storage. Along these lines the dual effects of EDTA on sperm NCD ability may also be explained. Thus, by EDTA-induced chromatin zinc removal (10) normally zinc-masked thiols are liberated and become potential breakers of chromatin S-S crosslinks. However, once unmasked the thiols become susceptible to oxidative destruction which enhances the loss of NCD ability upon semen storage.

In conclusion, the present and previous studies (5,6,7,10) on human spermatozoa support the idea that the spermatozoon has an intrinsic thiol-mechanism for NCD, that is temporary inhibited and preserved by zinc derived from the concomitantly expulsed prostatic fluid upon ejaculation. Such an effect of zinc may be of importance to extend the functional life span of spermatozoa where copulation and ovulation are not closely related events.

REFERENCES


1. Bedford JM. 1972. An electron microscopic study of sperm penetration into rabbit egg after natural mating. Am J Anat 133:165-178.
2. Calvin HI, Bedford JM. 1971. Formation of disulphide bonds in the nucleus and accessory structures of mammalian spermatozoa during maturation in the epididymis. J Reprod Fert, Suppl 13:65-75.
3. Calvin HI, Yu CC, Bedford JM. 1973. Effects of epididymal maturation. Zn(II) and Cu(II) on the reactive sulfhydryl elements in rat spermatozoa. J Exp Cell Res 81:333-341.
4. Evenson DP, Witkin SS, Harven E de, Bendich A. 1978. Ultrastructure of partially decondensed human spermatozoal chromatin. J Ultrastruct Res 63: 178-187.
5. Kvist U. 1980. Sperm nuclear chromatin decondensation ability. An in vitro study on ejaculated human spermatozoa. Acta Physiol Scand 109, Suppl 486:1-24.
6. Kvist U. 1982. Spermatozoal thiol-disulphide interaction: A possible event underlying physiological sperm nuclear chromatin decondensation. Acta Physiol Scand (in press).
7. Kvist U, Nilsson L, Afzelius BA. 1980. The intrinsic mechanism of chromatin decondensation and its activation in human spermatozoa. Develop Growth and Differ 22: 543-554.
8. Lung B. 1972. Ultrastructure and chromatin disaggregation of human sperm head with thioglycolate treatment. J Cell Biol 52: 179-186.
9. Mahi CA, Yanagimachi R. 1975. Induction of nuclear decondensation of mammalian spermatozoa in vitro. J reprod Fert 44: 225-229.
10. Roomans GF, Lundevall E, Björndahl L, Kvist U. 1982. Removal of zinc from subcellular regions of human spermatozoa by EDTA treatment studied by X-ray microanalysis. Int J Androl (in press)
11. Siegel S. 1956. Nonparametric statistics for the behavioral sciences. McGraw-Hill Kogakusha Ltd, Tokyo.
12. Soupart P, Strong PA. 1974. Ultrastructural observations on human oocytes fertilized in vitro. Fert Steril 25:11-44.
13. Young RJ. 1979. Biol Reprod 20: 1001-1004.

# MICROTUBULES AND SPERMIOGENESIS

J.L. COURTENS, B. DELALEU, M. LOIR
I.N.R.A. Station de Physiologie de la Reproduction Nouzilly 37380 MONNAIE

## 1. INTRODUCTION

Numerous works on spermiogenesis did not focus on the occurence of microtubules. They have been conducted before the first use of glutaraldehyde as fixative agent or in very archaic phylla such as spongia (54) or in species with so scarce microtubules that they remain hidden to the observer sagacity. On the other hand, most recent ultrastructural investigations of spermiogenesis have considered or evaluated the importance of microtubules, especially those of the manchette, for the shaping of the sperm heads.

However, other microtubules are also linked to the nuclei of spermatids. They are "cytoplasmic" microtubules and flagellar microtubules. In the following pages, we intend to show that at least one microtubular system is generally attached to the nucleus of spermatids in a broad spectrum of species and that the presence of this system may often be related to special features of the adjacent chromatin. These considerations led us to propose new roles for the microtubules during spermiogenesis, based on both phyllogenetic considerations and on some recent experimental findings in mammalian spermatids.

## 2. RELATIONS OF MICROTUBULES WITH THE SPERMATID NUCLEUS

### 2.1. The Manchette

The manchette have been described as component of spermatids in many species. Its location near the elongating nucleus has led numerous observers to hypothesize it could have a morphogenetic role. This is due in part because:

- It appears just before, and disappears just after nuclear morphogenesis in insects (58, 24, 25, 21, 49), birds (36), and mammals (62).
- The microtubules of the manchette are often linked together (9) and

to the nuclear envelope by numerous fibers (37, 49) suggesting a strong rigidity for the microtubular system.

- they often fit the shape of nuclei (see the helicoïdal nuclei described by (36, 25, 18) just as do an extranuclear mold.

However, several results argue against this thesis.

- The nuclear morphogenesis of spermatids takes place in the absence of any typical manchette in numerous species. This is the case of hydra (38, 59) and of crustacea (28, 17) whose spermatids only contain few microtubules. In peudo scorpion the microtubules are present for too a short time to ensure a role of mold (39). In the squallus (48) and the raja (3) microtubules are replaced by microfilaments, while both are absent in most teleostei (47).

2.2. Other microtubular systems

2.2.1. Scarce cytoplasmic microtubules. The relations between the nucleus of spermatids and microtubules appears in diblastic organisms as rudimentary as hydra: microtubules are few, located near the diplosoma and they form a protocytoskeletton, umbrella-shaped (38, 59) near the nucleus. The chromatin condenses first in this region. In most crustacea (17), spermatid do only contain some microtubules linked to the nuclear envelope, often in the pore-rich area, which is destroyed during fertilization (28). Primitive crustacea, the pycnogonidae (56) possess a very simple "proto manchette" made of microtubules, unlinked together, but attached to the nuclear envelope.

2.2. . Flagellar microtubules and mixed systems. Privileged relations between the nuclei and the flagella, appear in the species which are devoided of, or which are poor in, free microtubules. The flagellum could be deeply engaged in a nuclear fossa or in a central nuclear channel. The chromatin condenses first in the channel area. This situation is described in numerous telestei such as mullidae (5) and poecilidae (40). The channel is lateral in dactylopterus (cephalacantidae) (4) and oligocottus (Teleostei) (47). It do exist in acanthocephala (32), in nematodes (22), and in gasteropodia (57, 19). Sometimes, the intranuclear channel contains other microtubules than those of the flagellum. This is described in nematodes (22) and in the fish upeneus (mullidae) (5).

The spermatids of molluscs often possess two microtubular systems associated with the nucleus: ie in nucella (gasteropodia), an external manchette coexist with an internal flagellum located in a deep nuclear fossa; the spiralization of the flagellum is accompagned by spiralization of the nucleus. A double system made of cytoplasmic microtubules and of a flagellum both engaged in a lateral nuclear fossa has been described in a columbiform bird (34).

The nuclear channel may be replaced by the simple lateral apposition of the flagellum against the nucleus. This situation is met in some chaetognaths (56) with the flagellum applyed all along the nucleus; or in a more distal position in Diplosoma listerianum (53). In this later ascidian, the flagellum migrates backwards when the manchette appears revealing a trend for the replacement of priviliged contact with the flagellum toward privileged contacts with the manchette.

The reverse seems true in arachnids. For instance, the spermatid nucleus of Mastigoproctus giganteus (Uropygia) is first surrounded by a manchette, but, at the time of manchette removal, it becomes so coiled that the 9 + 3 type flagellum takes the place of the manchette (39). This type of nuclear morphogenesis could be the rule in numerous spiders (39).

The scorpion spermatids do not possess a typical manchette (23). The chromatin, attached to the nuclear envelope condenses first in face of the proximal controle and of the centriolar adjunct.

In the fish albula vulpes (albulidae), a second pseudo flagella, joining the proximal centriole is applyed against the spermatid nucleus (33).

### 3. POSSIBLE ROLES OF MICROTUBULES DURING SPERMIOGENESIS

Many hypothesis have been raised in the species provided with a typical manchette. In annelid worms the manchette could arise from one centriole (1, 42, 41). It plays an important, functionnal role in the binding of the nuclear enveloppe to the chromatin in the Tubifex and the Lymnodrilus. The manchette is linked to both the nucleus and the endoplasmic reticulum in Lumbricus (29) and play and "inductive role upon the condensation of the chromatin" (31).

The manchette of insects is well developped and is linked to parts, or to most parts, of the nuclear envelope by filaments (24). Sometimes, the manchette surrounds the acrosome: in the Gerris (50) and the bee (21). The microtubules synthetised at, or near, the centriolar adjunct (49) play a morphogenetical action whether by modeling the nuclear envelope and favourizing the release of excess nuclear envelope in the dragonfly (24) or by the transport of substances in the Gerris (50) or by "favourizing" the chromatin condensation in Sarcophaga (58).

This last function seems effective in Drosophila. In this species (51, 49) a clear topographical relation between the condensing chromatin and the presence of extranuclear microtubules has been demonstrated. Shoup (46) demonstrated that in the manchette-devoided spermatids of the T (I, 2h) 25 (20) yl 25/ FM6 heterozygotes the nuclear histones are not replaced by nucleoprotamines. This should also be the case in the SD (segregation distorter) and the ms 165 mutants (26, 20).

Many other examples could be cited in invertebrates and vertebrates (3, 4, 5, 33, 34, 36) indicating the occurence of at least a topographical relation between extranuclear microtubules and intranuclear related condensation of the chromatin (1, 41, 25, 18, 49, 2). Among species, Loligo (2) and monotreme mammals (6) are specially demonstrative in that respect.

In mammals, a constant contact exists between the centriolar adjunct, and some microtubules of the manchette. Even if no functional link can be drawn from this observation, its remains true that the centriolar adjunct and the manchette appear and disappear in the same times in most mammals. Only Homo could escape this rule because a short centriolar adjunct has been described in human spermatozoa (61).

Have the microtubules of the manchette and those of the flagellum a common origin? If this was the case, the manchette should be considered as a specialized structure issued from the flagellum or from one centriole. It could assume the role originaly played by the flagellum itself when inserted in a deep intranuclear fossa. This type of trend, from a centraly to a fully extermaly located systems of microtubules, ressemble the evolution of the central mitotic apparatus of protists to the external one of more specialized cells.

This asks the question of what role could play a mitotic apparatus, or part of it, whether archaic or developped, in spermïogenesis.

## 4. THE MANCHETTE OF MAMMALS AS HALF A MITOTIC APPARATUS: RELATION OF MICROTUBULES WITH THE INTRA-NUCLEAR COMPARTMENT OF SPERMATIDS

Relations of the manchette with the nuclear envelope are fully documented in rat spermatids (37). The numerous links which attach the manchette to the nuclear envelope are also present in ram, boar stallion, goat and bull spermatids (10, 11, 13) and they appear in numerous microphotographs published in papers on other species, including in vertebrates.

In domestic mammals, numerous fibers originating in the chromatin passes through the nuclear envelope of spermatids from the time of nuclear elongation to the fully formed spermatozoa. A lot of fibers are linked to the manchette microtubules (14). Such fibers linking chromosomes to the manchette could facilitate the anaphase-like chromosomes movements, already described during amphibian and bird spermiogenesis (35, 45). The manchette should then be similar to half the mitotic apparatus of protozoans, because only one "pole" is present, and because the nuclear envelope persists.

This hypothesis rises several questions: Is the anaphase-like movement of chromosomes a general feature of spermiogenesis? Is the chromatin linked to the flagellum in species without manchette? Is the intranuclear part of the flagellum equivalent to the mitotic apparatus of some protozoans?

## 5. THE MANCHETTE AS PART OF AN APPARATUS FOR NUCLEOCYTOPLASMIC EXCHANGES

In numerous species, the manchette is called "caudal sheat". This reflects a difference in shape with that of mitotic spindles and the lack of a conventional "mitotic pole" for the manchette. The open shape of the manchette could favour nucleocytoplasmic exchanges. Several facts argue for this thesis.

- Microtubules are generaly present only near the zones of nuclear envelope that are rich in nuclear pores (43, 44, 16, 7),
- The manchette of rat (8) of ram, boar and stallion (10) is internally lined with a network of endoplasmic reticulum, in continuity with the nuclear envelope and the future redundant nuclear envelope,
- Cytochemical techniques reveal that these endoplasmic reticulum cisternae are filled with abundant material at the time nucleoproteins are lost from, and arrive in, the nuclei fo elongating spermatids (10),
- The temporary depolymerization of the manchette of rat spermatids by colcemid treatment, induces nuclear misshaping in the only cells the relations between the nucleus and the endoplasmic reticulum are not formed again when the manchette repolymerizes (11).

The nuclear protein changes which promote the changes of the nuclear shape in mammalian spermatids (27, 30, 12) should then be controlled at least partly by the presence and action of the manchette. Such relations, between the endoplasmic reticulum and the nuclei remain, however, to be demonstrated in manchette devoided spermatids, as are the occurence (and roles) of the nucleocytoplasmic exchanges in these species.

## 6. CONCLUSION

To the two following opposite theses: - and intranuclear control of nuclear shape of spermatids by nucleoproteins (15) - and a control of nuclear shaping by external forces developped in part by the manchette microtubules (36), recent advances allow to propose a more integrated model, in which the manchette, or cytoplasmic, or flagellar microtubules, may act as half a primitive mitotic apparatus. The anaphase-like sliding of chromosomes along cytoplasmic microtubules during spermiogenesis should be, in some way, advantageous in preparing the nuclear events of fertilization and orientating the kinetochores towards the pole of the future sperm aster. It could also favour the correct packaging of D.N.A. fibers inside the nuclei of spermatids at the time of chromatin condensation.

The comparison of spermiogenesis with the last events of mitosis in unicellular organisms is also curious in several respects such as: the persistance of the nuclear envelope during remodeling of the nuclear contents; lost of nucleoli; exchanges of nucleoproteins related to chromatin condensation; relation of intranuclear material with extranuclear microtubules.

The comparison of spermiogenesis in different phylla reveals several degrees of organization that are also encountered in unicellular organisms. They are: intranuclear channel filled with microtubules; trend toward the formation of a flagellum; the "progressive" (?) replacement of an archaic model by a more sophisticated one, containing both a flagellum and free microtubules linked to the nucleus; the final differentiation of a double system of microtubules, one being devoted to cell locomotion, and the other to nuclear (help in) shaping.

## REFERENCES


1. Anderson WA, Ellis RA. 1968. Acromosome morphogenesis in Lumbricus terrestris. Z. Mikrosk. Anat. Forsch. 85: 398-407.

2. Bergstrom BH, Arnold JM. 1974. Nonkinetochore association of chromatin and microtubules. A preliminary note. J. Cell Biol. 62: 917-920.
3. Boisson C, Mattei X, Mattei C. 1968. La spermiogenèse de Rhinobatus cemiculus Geof. St-Hilaire (Selacien Rhinobatidae). Etude au microscope electronique. Bulletion de l'I.F.A.N., T. xxx, ser. A, n°2 659-690.
4. Boisson C, Mattei X, Mattei C. 1968. Le spermatozoïde de Dactylopterus volitans, Linné (Poisson Cephalacantidae), étudié au microscope électronique. C.R. Soc. Biol. 3: 820-823.
5. Boisson C, Mattei X, Mattei C. 1969. Mise en place et évolution du complexe centriolaire au cours de la spermiogenèse d'Upeneus prayensis c.v (Poisson mullidae). J. Microscopie 8: 103-113.
6. Carrick FN, Hugues RL. 1981. Aspects of the structure and development of monotreme spermatozoa and their relevance to the evolution of mammalian sperm morphology. Cell Tissue Res. 222: 127-141.
7. Chemes HB, Fawcett DW, Dym M. 1978. Unusual features of the nuclear envelope in the human spermatogenic cells. Anat. Rec. 192: 493-512.
8. Clermont Y, Rambourg A. 1978. Evolution of the endoplasmic reticulum during rat spermiogenesis. Am. J. Anat. 151: 191-212.
9. Courot M, Flechon JE. 1966. Ultrastructure de la manchette de la spermatide chez le Bélier et le Taureau. Ann. Biol. anim. Bioch. Biophys., 6: 479-482.
10. Courtens JL. 1982a. Some nucleocytoplasmic exchanges during spermiogenesis of boar, ram and stallion. Gamete Res. 5: 137-152.
11. Courtens JL. 1982b. Rôles indirects des microtubules dans la morphogenèse nucléaire des spermatides. Reprod. Nutr. Develop. sous presse.
12. Courtens JL, Amir D, Durand J. 1980. Abnormal spermiogenesis in bulls treated with ethylene dibromide: an ultrastructural and ultracytochemical study. J. Ultrastr. Res. 71: 103-115.
13. Courtens JL., Loir M. 1981a. A cytochemical study of nuclear changes in boar, goat, mouse, rat and stallion spermatids. J. Ultrastr. Res., 74: 327-340.
14. Courtens JL, Loir M. 1981b. The spermatid manchette of mammals: formation and relations with the nuclear envelope and the chromatin. Reprod. Nutr. Develop. 21: 467-477.
15. Fawcett DW, Anderson W, Phillips DM. 1971. Morphogenetic factors influencing the shape of the sperm-head. Developmental Biol. 26: 220-251.
16. Fawcett DW, Chemes HE. 1979. Changes in distribution of nuclear pores during differentiation of the male germ cells. Tissue and Cell. 11: 147-162.
17. Fahrenbach WH. 1973. Spermiogenesis in the Horseshoe Crab Limulus polyphemus. J. morphol. 140: 31-52.
18. Ferraguti M, Lanzavecchia G. 1971. Morphogenetic effects of microtubules I spermiogenesis in annelida Tubificidae. J. Submicr. cytol. 3: 121-137.
19. Garreau de Loubresse N. 1971. Spermiogenèse d'un gastéropode prosobranche : Nerita senegalensis; évolution du canal intranucléaire. J. Microscopie, 12: 425-440.
20. Habliston DL, Stanley HP, Bowman JT. 1977. Genetic control of spermiogenesis in Drosophila melanogaster: the effects of abnormal association of centrosome and nucleus in mutant ms (I) 65[1]. J. Ultrastr. Res. 60: 221-234.
21. Hoage TR, Kessel RG. 1968. An electron microscope study of the process of differentiation during spermatogenesis in the drone honey bee (Apis

mellifera L.) with special reference to the centriole replication and elimination. J. Ultrastr. Res. 24: 6-32.
22. Jamuar MP. 1966. Studies of spermiogenesis in a nematode, Nippostrongylus brasiliensis. J. Cell Biol. 31: 381-396.
23. Jespersen A, Hartwick R. 1973. Fine structure of spermiogenesis in scorpions from the family Vejovidae. J. Ultrastr. Res. 45: 366-383.
24. Kessel RG. 1966. The association between microtubules and nuclei during spermiogenesis in the dragonfly. J. Ultrastr. Res. 16: 293-304.
25. Kessel RG. 1970. Spermiogenesis in the Dragonfly with special reference to a consideration of the mechanism involved in the development of cellular assymetry. In "Comparative spermatology" Baccetti Ed. pp 531-552. Academic Press N.Y.
26. Kettaneh NP, Hartl DL. 1976. Histone transition during spermiogenesis is absent in segregation distorter male of Drosophila melanogaster. Science, 193, 1020-1021.
27. Kierszenbaum AL, Tres LL. 1975. Structural and transcriptional features of the mouse spermatid genome. J. Cell Biol. 65: 258-270.
28. Langreth SG. 1969. Spermiogenesis in Cancer Crabs. J. Cell Biol. 43: 574-603.
29. Lanzavecchia G, Lora Lamia Donin C. 1972. Morphogenetic effects of microtubules II Spermiogenesis in Lumbricus terrestris. J. Submicrosc. Cytol. 4: 247-260.
30. Loir M, Courtens JL. 1979. Nuclear reorganization in ram spermatids. J. Ultrastruct. Res. 67: 309-324.
31. Malecha J. 1975. Etude ultrastructurale de la spermiogenèse de Piscicola geometra L. (Hirudinee Rhynchobdelle). J. Ultrastr. Res. 51 188-203.
32. Marchand B, Mattei X. 1976. La spermiogenèse des Acanthocéphales. L'appareil centriolaire et flagellaire au cours de la spermiogenèse d'Illiosentis urcatus var africana Golvan, 1956. (Paleacanthocephala, Rhadinorhynchidae). J. Ultrastr. 54: 347-358.
33. Mattei C, Mattei X. 1973. La spermiogenèse d'Albula vulpes (L. 1758) (Poisson Albulidae). Etude ultrastructurale. Z.Zellforsch., 142: 171-192.
34. Mattei C, Mattei X, Manfredi JL. 1972. Electron microscope study of the spermiogenesis of Streptopelia roseogrisea. J. Submicros. Cytol. 4: 57-73.
35. MacGregor HC, Walker MH. 1973. The arrangement of chromosomes in nuclei of sperm from phethodontic salamanders. Chromosoma. 40: 243-262.
36. MacIntosh JR, Porter KR. 1967. Microtubules in the spermatids of the domestic fowl. J. Cell Biol. 35: 153-173.
37. MacKinnon EA, Abraham PJ. 1972. The manchette in stage 14 rat spermatids: a possible structural relationship with the redundant nuclear envelope. Z. Zellforsch.., 124 1-11.
38. Moore GPM, Dixon KE. 1972. A light and electron microscopical study of spermatogenesis in Hydra cauliculata. J. morphol. 137: 483-502.
39. Phillips DM. 1976. Nuclear shaping in the whip scorpion. J. Ultrastr. Res. 54: 397-405.
40. Porte A, Follenius E. 1960. La spermatogenèse chez Lebistes reticulatus. Etude au microscope électronique. Bull. Soc. Zool. Fr. 85: 82-89.
41. Postwald HE. 1967. An electron microscope study of spermiogenesis in spirobis (Lacospira) morchi Levinsen (polychaeta) Z. Zellforsch. 83: 231-248.

42. Reger JF. 1967. A study on the fine structure of developing spermatozoa from the oligochaete Enchytraeus albidus. Z. Mikrosk. Anat. Forsch. 82: 257-269.

43. Sandoz D. 1970. Evolution des ultrastructures au cours de la formation de l'acrosome du spermatozoïde chez la Souris. J. Microscopie. 9: 535-558.

44. Sandoz D. 1974. Modifications in the nuclear envelope during spermiogenesis of Discoglossus pictus (anuran Amphibia). J. Submicr. Cytol. 6: 399-419.

45. Schmid M, Krone W. 1976. The relationship of a specific chromosomal region to the development of the acrosome. Chromosoma. 56: 327-347.

46. Shoup JR. 1967. Spermiogenesis in wild type and in a male sterile mutant of Drosophila melanogaster. J. Cell Biol. 32: 663-675.

47. Stanley HP. 1969. An electron microscope study of spermiogenesis in the teleost fish. Oligocottus maculosus. J. Ultrastr. Res. 27: 230-243.

48. Stanley HP. 1971. Fine structure of spermiogenesis in the elasmobranch fish Squalus suckleyi I. Acrosome formation, nuclear elongation and differention of the midpiece axis. J. Ultrastr. Res. 36: 86-102.

49. Stanley HP, Bowman JT, Romrell LJ, Reed SC, Wilkinson RF. 1972. Fine structure of normal spermatid differentiation in Drosophila melanogaster. J. Ultrastr. Res. 41: 433-466.

50. Tandler B, Moriber LG. 1966. Microtubular structures associated with the acrosome during spermiogenesis in the water strider Gerris remigis (say). J. Ultrastr. Res. 14: 391-404.

51. Tokuyasu KT. 1974. Dynamics of spermiogenesis in Drosophila melanogaster. IV Nuclear transformation. J. Ultrastr. Res. 48: 284-303.

53. Tuzet O, Bogoraze D, Lafargue F. 1972. Recherches ultrastructurales sur la spermiogenèse de Diplosoma listerianum (MILNE-EDWARD, 1841) et Lissoclinum pseudoleptoclinum (VON DRASCHE, 1883) (Ascidies composées, aplousobranches). Ann. Sci. Nat. Zool. Paris. 12e ser. Tome XIV pp. 177-190.

54. Tuzet O, Garonne R, Pavans de Ceccatty M. 1970. Observations ultrastructurales sur la spermatogenèse chez la démosponge Aplysilla rosea (SCHULZE) (dendroceratide) : une metaplasie exemplaire. Ann. Sci. Nat. Zool. Paris 12e ser. Tome XII pp. 27-50.

56. Van Deurs B. 1974. Spermatology of some pycnogonida (arthropoda), with special reference to a microtubule-nuclear envelope complexe. Acta. Zool. Stockholm, 55: 151-162.

57. WALKER M, MacGrecor HC. 1968. Spermatogenesis and the structure of the mature sperm in Nucella lapillus (L). J. Cell. Sci. 3: 95-104.

58. Warner FD. 1971. Spermatid differentiation in the Blowfly Sarcophaga bullata with particular reference to flagellar morphogenesis. J. Ultrastr. Res. 35: 210-232.

59. Weissman A, Lentz TL, Barnett R. 1969. Fine structural observations on nuclear maturation during spermiogenesis in hydra littoralis. J. Morphol. 128: 229-240.

60. Wilkinson RF, Stanley HP, Bowman JT. 1974. Genetic control of spermiogenesis in Drosophila melanogaster: the effects of abnormal cytoplasmic microtubules population in mutant ms (3) IO R and its colcemid - induced phenocopy. J. Ultrastr. Res. 48: 242-258.

61. Zamboni L, Stefanini M. 1971. The fine structure of the neck of mammalian spermatozoa. Anat. Rec. 1969, 155-172.

62. Zirkin BR. 1971. Some ultrastructural aspects of nuclear morphology in developing spermatids and mature spermatozoa of the bull. Mikroskopie, 27: 10-16.

# ARE MICROTUBULES INDISPENSABLE FOR SPERM MORPHOGENESIS ? A COMPARATIVE STRUCTURAL STUDY

S.R. BAWA and G. WERNER
Department of Biophysics, Panjab University, Chandigarh, India and Department of Medical Biology, Institute of Biophysics, University of Saarland, Homburg, West Germany.

In the early spermatid of the insect *Callosobruchus maculatus* the cytoplasmic microtubules are dispersed around the nucleus except its furrowed-in surface where the nuclear envelope is distinctly electron-opaque (Fig. 1). In late spermatids microtubules are disposed uniformly around the electron-opaque elongating nuclear head (Fig. 2). However, they disappear in the mature spermatozoa which lie compacted together in the female spermatheca. In the primate *Loris tardigradus* "manchette" microtubules terminate in the nuclear ring (Fig. 3) and probably after eliminating unwanted cytoplasm in late spermatids these tubules disappear without affecting the shape of the sperm head. Contractile actin-like filaments of the Sertoli cells may be involved in the shaping and elongation of the nuclear head (Fig. 4). *Loris* sperm head obtains a definitive form during its transport through the epididymis without any participation of the microtubules (Fig. 5). In the crow *Corvus splendens* microtubules are intimately associated with the morphogenetic changes of the nucleus and acrosome. In early spermatids tubules appear in a bundle which is helically arranged around the cylindrical sperm head (Figs. 6 & 7) ; the pitch of the helix appears quite short. Later the microtubules are located preferentially in one of the furrows of the nuclear and acrosomal surface. At the same time the helix pitch increases thereby transforming the nucleus and the acrosome into a corkscrew-shaped head (Figs. 8, 9 & 10). The mature spermatozoa in the male duct maintain their corkscrew-shaped head in the absence of such microtubules. The giant acrosome with its lateral decorations and rod-like electron-opaque nucleus in the pseudo-scorpion *Diplotemnus* spermatid acquires such a shape without any involvement of the cytoplasmic microtubules (Figs. 11 & 12). It is obvious from the foregoing that the time of appearance, spatial distribution and the relationship of the microtubules *vis a vis* nucleus and acrosome morphogenesis are variable. In *Callosobruchus* and *Loris* although the microtubules are identifiable in various phases of spermatid development, they do not affect the shape of the sperm head. In *Corvus* it is apparent that the microtubules are

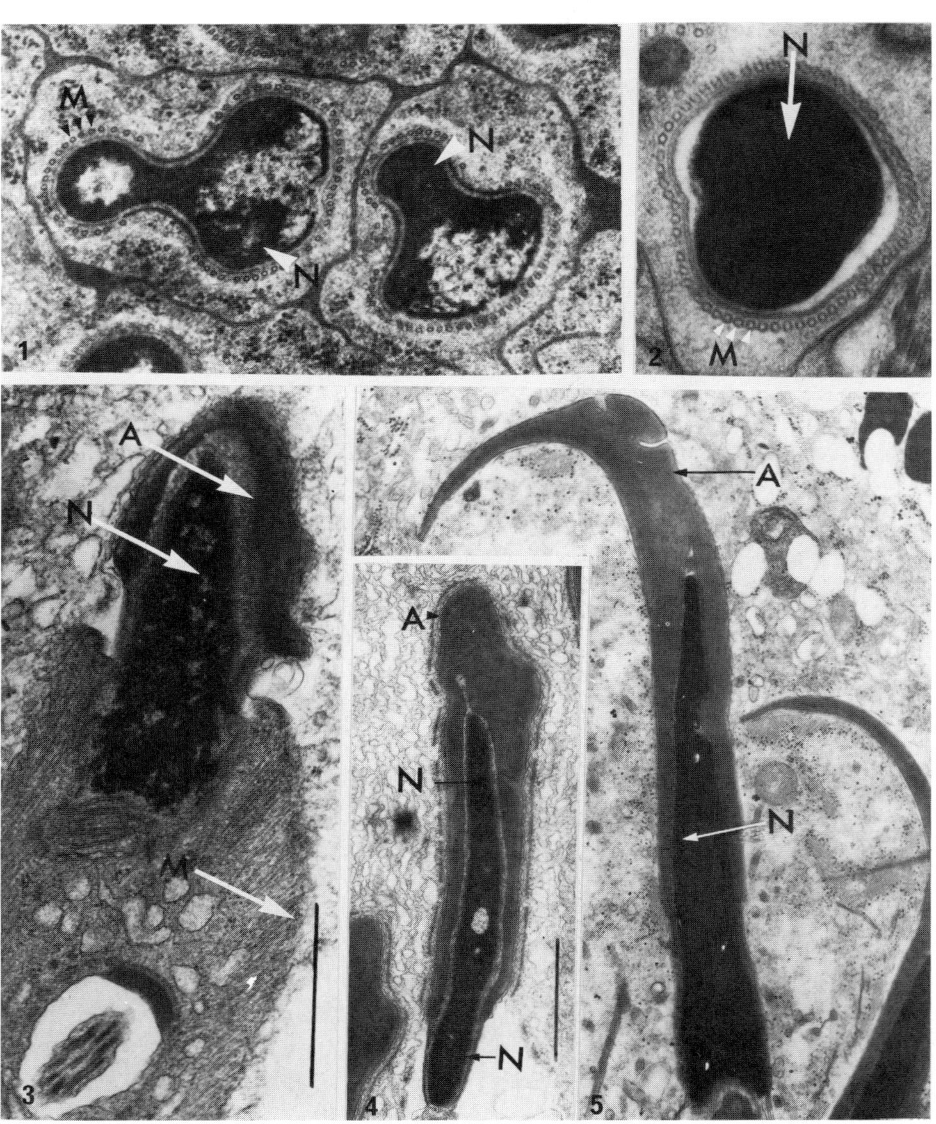

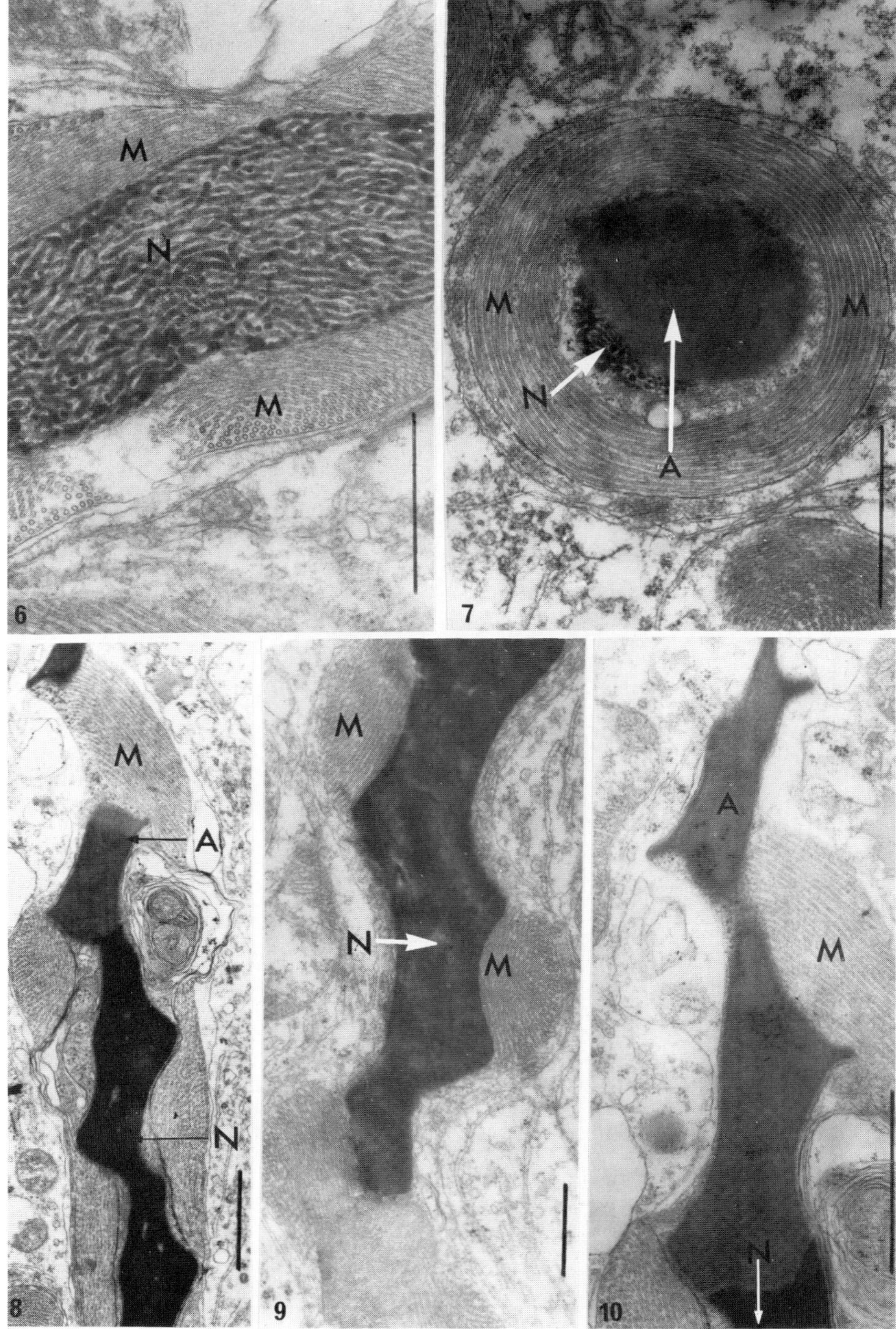
M
N
M
6
M
M
N
A
7
M
A
N
8
M
N
M
9
A
M
N
10

intimately involved in endowing a corkscrew shape to the sperm head. In *Diplotemnus* the acrosome and nucleus of the spermatozoa attain massive size and bizarre shape without any involvement of the microtubules.

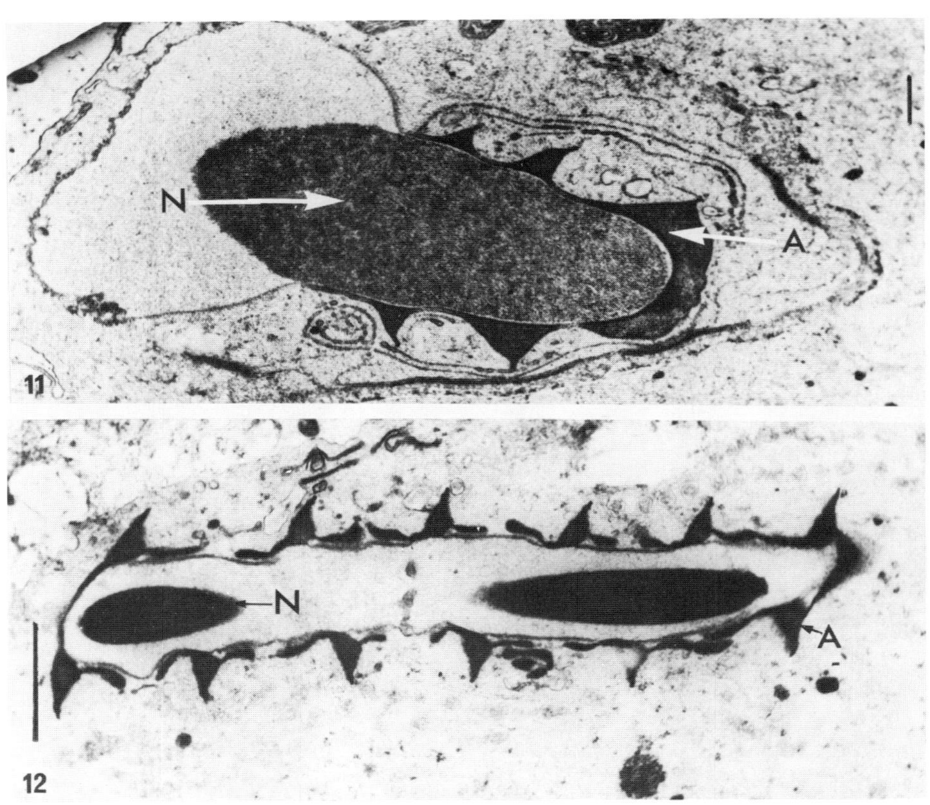


| | |
|---|---|
| Figs. 1 & 2 | Spermatids. *Callosobruchus maculatus*. X 40000, 56000 |
| Figs. 3 & 4 | Spermatids. *Loris tardigradus*. X 30000, 20000 |
| Fig. 5 | Late Spermatid. Epididymis. *Loris*. X 20000 |
| Figs. 6 & 7 | Early spermatids *Corvus*. X 32000 |
| Fig. 8 | Late spermatid. *Corvus*. X 18000 |
| Fig. 9 | Late Spermatid. *Corvus*. Microtubules present in one furrow of the nucleus. X 16000 |
| Fig. 10 | Late Spermatid. *Corvus*. Microtubules present in one furrow of the acrosome. X 32000 |
| Fig. 11 | Early Spermatid. *Diplotemnus* sp. Acrosome with associated Golgi elements. X 8000 |
| Fig. 12 | Late spermatid. *Diplotemnus*. Acrosomal decoration and electron-opaque rod-like nucleus. X 20000 |

# ANILINE BLUE STAINING OF HUMAN SPERMATOZOON CHROMATIN. EVALUATION OF NUCLEAR MATURATION.

A. TERQUEM and J.P. DADOUNE+
Laboratoire d'Histologie, Université René Descartes,
75270 Paris, France.

## 1) INTRODUCTION

Cytochemical studies have shown that, in late spermatids, basic nucleoproteins, rich in lysine, are replaced by spermatozoon - specific nucleoproteins (3, 5). Acidic aniline blue is known to strongly stain the chromatin of early spermatids (7). Beyond these results, our observations demonstrate the variable ability of ejaculated human spermatozoa to stain with aniline blue.

## 2) MATERIAL AND METHODS

Samples were obtained from 56 donors. In each specimen, the motility, numeration and vitality were normal. After semen fluidification, centrifugation and two successive washings in a Spinner Salt solution, without calcium, and 0.2 M phosphate buffer, resuspended cells were spread on glass slides. Smears were fixed for 30 min. in different fixatives (10% neutral formol, 3% glutaraldehyde in 0.2 M phosphate buffer, Bouin fluid, alcohol-ether, methanol-acetone). Staining was done at different times (5 to 30 min.) in aqueous aniline blue with 4% acetic acid used at different concentrations (0.01 to 5%) and pH (2.5 to 7.5). The optical density of nuclei was measured using a Leitz MPV II single cell cytophotometer. The percentages of stained sperm heads, normal and abnormal, were calculated from 100 spermatozoa per preparation.

After washing and fixation in 3% buffered glutaraldehyde (3h.) pellets from the same samples were successively stained "en bloc" with acidic aniline blue (4h.) and alcoholic phosphotungstic acid - a.PTA - (2) then prepared for electron microscopy.

+ *We wish to thank Mrs M.F. ALFONSI for her skilful technical assistance.*

Controls included 1) pretreatment before staining by acid hydrolysis (Cf. réf. 5), acetylation, nitration, n- ethyl maleimide + malonaldehyde (Cf. réf. 2) 2) characterization of basic nucleoproteins by other cytochemical techniques (1,5,6,8).

## RESULTS AND DISCUSSION

Except after fixation in alcohol-ether and methanol-acetone, all the fixatives tested give similar results. After fixation of cells in 3% buffered glutaraldehyde, an optimal intensity of the reaction is obtained in smears stained for 5 min. in 5% aniline blue at pH 3.5 (Fig. 1). At alkaline pH, the staining specificity disappears. In general, chromatin is not stained; however, certai sperm heads are completely or partly stained. Optical density ($\lambda$= 580 nm), measured at random in 400 sperm heads, gradually increases in about 25% of stained heads. In partly stained heads, chromatin staining varies in intensity and localization: scattere patches, the apex and the base of the head.

A comparison of the percentages by $\chi_2$ test shows significant differences between stained normal ($9.71 \pm 0.01$) and abnormal ($36.24 \pm 0.02$) sperm heads ($P < 0.001$).

Pretreatment by acid hydrolysis in TCA or picric acid (5) and blocking of thiol and guanidyl groups by N-ethyl-maleimide + malonaldehyde have no effect on aniline blue staining. On the other hand, blocking of primary amino groups by acetylation and deamination of epsilon amino groups of lysine by nitration (2) suppresses the staining. The characterization of basic nucleoproteins by the alkaline fast green method (Cf. réf. 5), of arginin-rich proteins by a modified Sakaguchi technique (8) and possibly by PTA hematoxylin staining (6) shows a positive reaction of most of sperm heads observed. The ammoniacal silver staining, specific for lysine-rich and arginin-rich nucleoproteins (1), reveals about 20% positive-lysine heads, which is near to the percentage of nuclei stained with aniline blue. The hypothesis that lysine-rich proteins are visualized by this latter method, as suggested by these results, is however restricted by the fact that smear staining involves different cells, even from the same sample.

Combined "en bloc" staining with aniline blue and a.PTA specific for lysine (2), allows one to examine concomitantly the same cells in light and electron microscopy. In this case, blue-stained nuclei in semi-thin sections correspond to PTA-positive nuclei in ultra-thin sections (Fig. 2). As described in other mammals (3), nuclei of mature spermatozoa are generally non-stained. Nevertheless, in certain heads, the chromatin presents an aspect exactly identical to that seen in spermatids (steps 4-5) stained under the same conditions (4) (Fig. 3). The results of the effects of blocking guanidyl and primary amino groups on aniline blue + a.PTA staining are in agreement with those obtained on blue-stained smears.

In conclusion, the present observations show that changes in basic nucleoprotein composition of ejaculated human spermatozoa are revealed by aniline blue staining. This technique gives a positive reaction which is related to the a.PTA-positive reaction, specific for lysine.

REFERENCES

1. Black M.M. and Ansley H.R. 1966. Histone specificity revealed in ammoniacal silver staining. J. Histochem. Cytochem., 14, 177-181.
2. Courtens J.L. and Loir M. 1981a. Ultrastructural detection of basic nucleoproteins: alcoholic phosphotungstic acid does not bind to arginine residues. J. Ultrastruct. Res., 74, 322-326.
3. Courtens J.L. and Loir M. 1981b. A cytochemical study of nuclear changes in boar, bull, goat, mouse, rat and stallion spermatids. J. Ultrastruct. Res., 74, 327-340.
4. Dadoune J.P., Alfonsi M.F. and Terquem A. 1982. PTA cytochemical characterization of lysine residues in the nucleus of spermatids and ejaculated spermatozoa in man. First European Congress on Cell Biology, Paris 1982.
5. Gusse M. et Chevaillier Ph. 1978. Etude ultrastructurale et chimique de la chromatine au cours de la spermiogenèse de la roussette Scyliorhinus caniculus (L.). Cytobiologie, 16, 421-443.
6. Issidorides M.R. and Katsorchis T. 1981. Dispersed and compact chromatin demonstrated with a new EM method: phosphotungstic acid hematoxylin block-staining. Histochemistry, 73, 21-31.
7. Mc Kay R.B. 1962. An investigation of the anomalous staining of chromatin by the acid dyes, methyl blue and aniline blue. Quart. J. micr. Sci., 103, 519-530.
8. Notenboom C.D. Van de Veerdonk F.C.G. and Van de Kamer J.C. 1967. A fluorescent modification of the Sakaguchi reaction on arginine. Histochemie, 9, 117-121.

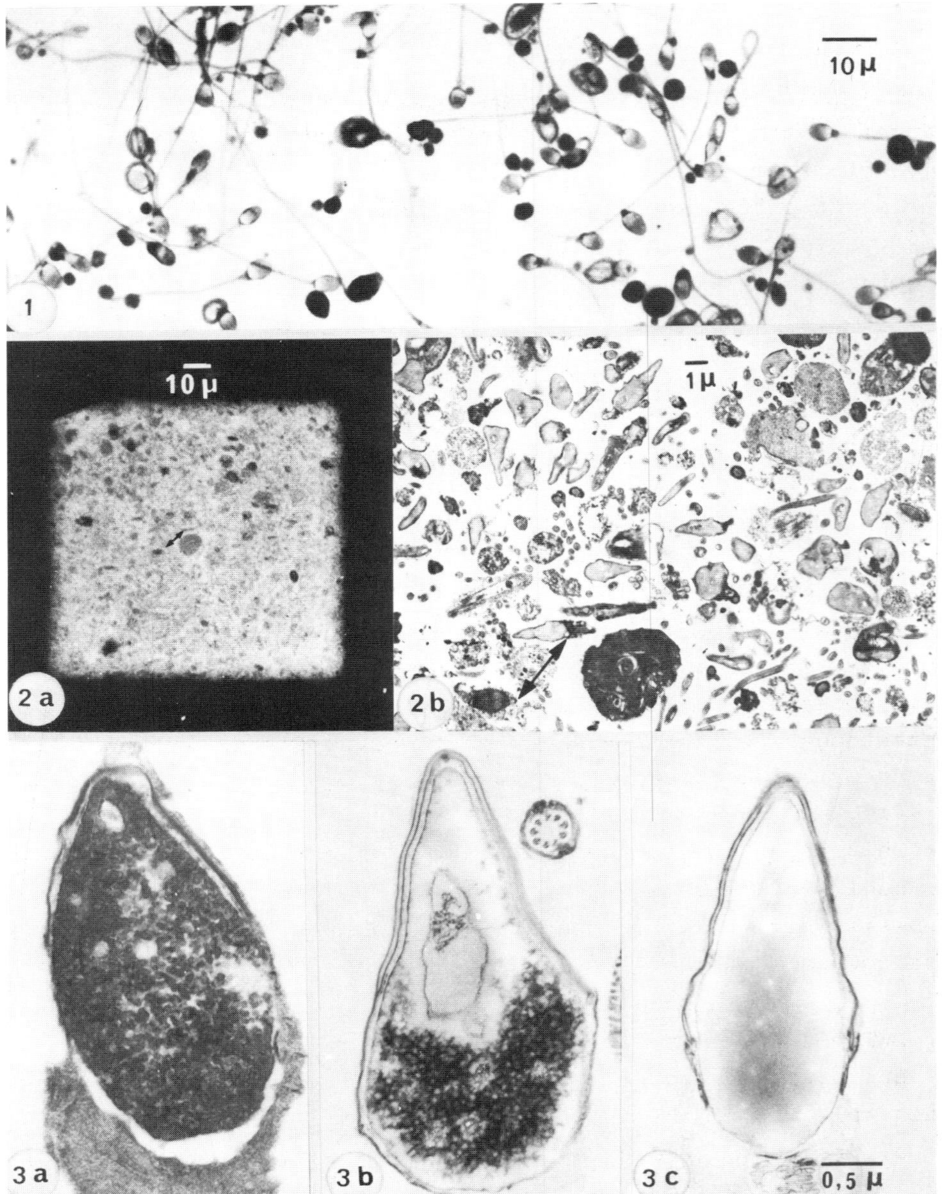


FIGURE LEGENDS: Fig. 1: Smear of spermatozoa stained with aniline blue; Fig. 2: 0.2 µm thick section of the same sperm heads, observed in light (a) and electron (b) microscopy after aniline blue a.PTA block-staining; Fig. 3: Ultrastructural aspects of immature (a, b) and mature (c) sperm heads after aniline blue a.PTA block-staining.

# CYTOPLASMIC CONTROL OF THE TRANSFORMATION OF SPERM NUCLEUS INTO MALE PRONUCLEUS IN THE MOUSE: AN EXPERIMENTAL APPROACH

A.K. TARKOWSKI
Department of Embryology, Zoological Institute, University of Warsaw, Poland

## 1. INTRODUCTION

Under physiological conditions the oocytes of the majority of mammals are fertilized at the stage of metaphase of the 2nd meiotic division. As a result of sperm penetration meiosis is completed and the maternal chromosomes form the female pronucleus. In the meantime the sperm nucleus decondenses, swells and transforms into the male pronucleus. It is conceivable that development of both male and female pronuclei is controlled by the ooplasm and that conditions optimal for pronuclear growth are transient. However, it is difficult, and perhaps even impossible, to prove this point by simple observation of the fertilization process in intact eggs. Direct evidence for the cytoplasmic control of the transformation of sperm nucleus into a pronucleus can only be provided by experimental studies.

This paper summarizes that part of the work of our laboratory on nucleo-cytoplasmic interactions in oogenesis and early embryogenesis which directly concerns the behaviour of sperm nucleus in the oocyte under various experimental conditions.

## 2. METHODS

The studies which are described below have been carried out with the help of the two following techniques :

a. Bisection of oocytes at various stages of maturation into nucleate and anucleate fragments (for description of the technique see 1).

b. *In vitro* fertilization of zona-free whole eggs and egg fragments.

3. RESULTS AND DISCUSSION : Conditions required for transformation of sperm nucleus into functional male pronucleus

3.1. Mixing of germinal vesicle karyoplasm with ooplasm. Fully grown mammalian oocytes with the intact germinal vesicle (GV) liberated from ovarian follicles resume meiosis when cultured *in vitro* and proceed to metaphase II (for references see 2, 3). The final and direct determinant of germinal vesicle breakdown (GVBD) and condensation of bivalents is autonomous change in the physiological state of the ooplasm (it occurs also in anucleate oocyte fragments) soon after explantation of the egg (4). As a result of GVBD and mixing of karyoplasm with ooplasm, all extra-chromosomal nuclear constituents become accessible to the spermatozoon when it penetrates the egg (this, of course, also applies to maternal chromosomes when they transform into a female pronucleus). Are these nuclear constituents really essential for the formation of pronuclei ?

To answer this question two experiments were carried out. The aim of the first was to discover whether both nucleate and anucleate fragments produced by bisection of ovulated oocytes in metaphase II can incorporate spermatozoa and initiate embryonic development. Since the result of this experiment was positive (5), we extended this approach to follicular oocytes (6). Oocytes were liberated from follicles and bisected into nucleate and anucleate fragments : (a) at the germinal vesicle stage, (b) immediately after GVBD, and (c) in metaphase I. Pairs of sister fragments were cultured until the nucleate ones extruded the 1st polar body and were mixed with sperm. Several hours later the fragments were examined for the fate of sperm nuclei. The results were clear cut. In nucleate halves the sperm nuclei were transformed into large pronuclei, regardless of the maturation stage at which the oocytes had been bisected. In contrast, the behaviour of sperm nuclei in anucleate halves depended on whether they were isolated before or after GVBD. In the first case (before) the sperm nuclei underwent only slight decondensation, whereas in fragments obtained after GVBD they transformed into large pronuclei. It can be thus concluded that materials in the nuclear sap are essential for the complete transformation of the sperm nucleus, although they are not required for the first stages of decondensation (6).

While the first part of the above conclusion appears to be beyond question and is in full agreement with results of similar studies on amphibian and invertebrate eggs (for references see 6), the second part requires further comments. According to several studies the sperm nucleus does not decondense at all in oocytes before GVBD or, at best, decondenses only partly and with a considerable delay (7-12). These observations were meant to imply that some constituents of the GV karyoplasm are essential for decondensation of the sperm nucleus. The discrepancy between our results and those by other authors may be due to the fact that anucleate fragments have initiated cytoplasmic maturation (sister nucleate halves have resumed meiosis ; see also 4), while GV oocytes with incorporated spermatozoa have not (maintenance of intact GV testifies for cytoplasmic immaturity of the oocyte). I suggest therefore that a prerequisite of decondensation of sperm nucleus (in contrast to transformation into a pronucleus) is the initiation of cytoplasmic maturation rather than a direct or indirect effect of the germinal vesicle karyoplasm.

3.2. Male pronucleus growth factor (MPGF). Thibault and co-workers (13-15) have observed that rabbit and cow oocytes which have undergone apparently normal maturation *in vitro*, can incorporate sperm and become activated, but only the female pronucleus develops while the sperm nucleus remains in the condensed state. These observations led the above authors to postulate that the oocyte undergoing maturation *in vitro* outside the follicle does not produce or acquire a special "male pro-nucleus growth factor" which is essential for transformation of sperm nucleus into male pronucleus.

In contrast to rabbit oocytes, mouse oocytes which have undergone maturation *in vitro* can support transformation of sperm nucleus into functional male pronucleus (16, 17, 19). However, since development of these eggs beyond birth or even to advanced postimplantation embryos is very rare (16, 17), their developmental potential is clearly inferior to eggs undergoing maturation in the ovary. On several occasions we did observe, however, aberrant development of the sperm nucleus in the mouse oocytes, one similar to that described in the rabbit. For instance : in 10 out of 105 penetrated nucleate oocyte fragments and in 6 out of 85 non-bisected oocytes that have reached metaphase II *in vitro*, the incorporated spermatozoa underwent only slight decondensation while the female pronucleus developed normally (6). Thus the difference between the *in vitro* maturing oocytes of the mouse and the rabbit appears to be quantitative rather than qualitative. In the mouse at least transformation of sperm nucleus into a pro-nucleus appears to depend largely on culture conditions during the period of oocyte maturation. It seems also that the factor in question is synthesized (or activated) by the oocyte itself rather than absorbed from the follicle, as suggested for the rabbit (14).

In view of the observations described above, one has to conclude that both in the mouse and the rabbit, mixing of GV karyoplasm with ooplasm is not sufficient by itself to promote the growth of male pronucleus, and that the cytoplasmic conditions required for the development of the male and the female pronucleus are not identical.

3.3. Activation of the egg. Once GVBD occurs and maturation is initiated, mouse oocytes readily incorporate spermatozoa and the sperm nuclei undergo decondensation. However, the decondensed sperm nuclei do not transform into pronuclei unless penetration occurs after the extrusion of the 1st polar body (18, 19). It is at that late period of maturation that the oocyte acquires potential for activation (fertilizability). Spermatozoa that have penetrated earlier do not produce this effect and the sperm nuclei do not develop beyond decondensation. In addition, it has been shown by Uehara and Yanagimachi (20, 21) that the fate of sperm heads injected into ovulated hamster oocytes in metaphase II depends on whether the oocyte has or has not been activated by the injection procedure : in the first case the male pronucleus develops normally, in the second case the sperm nucleus decondenses but does not transform into a pronucleus. These observations taken together with the results of our studies on virus mediated fusion of oocytes with interphase cells (22) demonstrate that activation of the egg, whether physiological (spermatozoon) or artificial, brings about a profound change in the physiological state of the cytoplasm : rather than maintaining egg chromosomes in the condensed state, it now promotes formation and maintenance of interphase nuclei (pronuclei).

## 4. CONCLUSIONS

Transformation of sperm nucleus into functional male pronucleus requires "preconditioning" of the oocyte cytoplasm with GV karyoplasm, appearance during maturation of the male pronucleus growth factor (MPGF) and activation of the egg.

## REFERENCES


1. Tarkowski AK. 1977. J. Embryol. exp. Morph. 38, 187-202.
2. Donahue RP. 1968. J. exp. Zool. 169, 237-250.
3. Thibault C. 1977. J. Reprod. Fert. 51, 1-15.
4. Bałakier H, Czołowska R. 1977. Exp. Cell Res. 110, 466-469.
5. Tarkowski AK. 1980. Exp. Cell Res. 128, 73-77.
6. Bałakier H, Tarkowski AK. 1980. Exp. Cell Res. 128, 79-85.
7. Barros C, Munoz G. 1973. J. Exp. Zool. 186, 73-78.
8. Usui N, Yanagimachi R. 1976. J. Ultrastruct. Res. 57, 276-288.
9. Moore HDM, Bedford JM. 1978. J. Ultrastruct. Res. 62, 110-117.
10. Berrios M, Bedford JM. 1979. J. Cell Sci. 39, 1-12.
11. Thadani VM. 1979. J. Exp. Zool. 210, 161-168.
12. Barros C, Rhim A. 1980. Arch. Biol. Med. Exp. 13, 325-334.
13. Thibault C, Gérard M. 1970. C.R. Acad. Sc. Paris, 270, 2025-2026.
14. Thibault C, Gérard M. 1973. Ann. Biol. anim. Bioch. Biophys. 13, 145-156.
15. Thibault C, Gérard M., Menezo Y. 1975. Ann. Biol. anim. Bioch. Biophys. 15, 705-714.
16. Cross PC, Brinster RL. 1970. Biol. Reprod. 3, 298-307.
17. Mukherjee AB. 1972. Nature, Lond. 237, 397-398.
18. Iwamatsu T, Chang MC. 1972. J. Reprod. Fert. 31, 237-247.
19. Wojewódzka M. (personal communication).
20. Uehara T, Yanagimachi R. 1976. Biol. Reprod. 15, 467-470.
21. Uehara T, Yanagimachi R. 1977. Biol. Reprod. 16, 315-321.
22. Tarkowski AK, Bałakier H. 1980. J. Embryol. exp. Morph. 55, 319-330.

# CHAPTER 4 ACROSOME

## THE ACROSOME, ITS HYDROLASES, AND EGG PENETRATION

R.A.P. HARRISON
A.R.C. Institute of Animal Physiology, Animal Research Station, Cambridge CB3 0JQ, England.

### 1. INTRODUCTION

The acrosome of the spermatozoon has long been recognized to play an essential role in fertilization in the vast majority of animal species studied. Located in the anterior part of the sperm head, this membrane-bounded vesicle is believed to contain hydrolytic enzymes whose action is necessary for egg penetration. The enzymes are deployed following an exocytotic event known as the acrosome reaction which takes place in the close vicinity of the egg at the time of fertilization, and spermatozoa which have not undergone an acrosome reaction will penetrate neither zona-invested nor zona-free eggs (5, 35, 59, 97).

Because of its obvious importance, a great deal of attention has been paid to the acrosome, to its contents, and to the dynamic changes it undergoes at fertilization. But despite all this research effort, we remain ignorant of the details concerning the enzymes involved in egg penetration and the way in which they are deployed. Similarly the details of the process by which the acrosome reaction is caused to take place are equally unclear. This article reviews our current state of knowledge concerning the acrosome in mammalian species.

### 2. ACROSOME STRUCTURE AND ENZYME CONTENT

The size and shape of the acrosome varies greatly between species, but the organelle exhibits a constancy of general design (20). It overlies the anterior part of the condensed sperm nucleus in the form of a cap-like vesicle beneath the plasma membrane, and consists of a semisolid matrix bounded by a normal bilayer membrane structure. By virtue of its position in relation to the nucleus, the portion of the bounding membrane which apposes the nuclear membrane is known as the inner acrosomal membrane, while the portion which apposes the overlying plasma membrane is known as the outer acrosomal membrane. The molecular architecture of these two portions clearly differ, for the outer acrosomal membrane undergoes fusions with the overlying plasma membrane at the time of the acrosome reaction, whereas at a later stage when the sperm cell fuses with the oocyte the inner acrosomal membrane does not fuse with the oolemma but is dispersed within the egg (5, 35, 97). Moreover the plasma membrane is easily separated mechanically from the outer acrosomal membrane (43), but the inner acrosomal membrane always remains tightly attached to the underlying nuclear membrane.

The acrosome can be described as consisting of three main morphological regions. At its posterior end, the inner and outer acrosomal membranes are closely apposed and separated by only a thin layer of matrix: this region is known as the equatorial segment. The rest of the part of the acrosome which overlies the nucleus (and which constitutes the major portion of the acrosome in species such as the human, the rabbit, and the farm animals) is known as the principal segment. The final part, the apical segment, is the apical portion of the acrosome which projects beyond the tip of the nucleus. This region varies considerably in size and shape between species. It constitutes the major portion of the acrosome in the guinea pig, for example, extending way beyond the nucleus; but in the ram or bull it is only a slight swelling to one side of the tip of the nucleus. The apical segment is of a particular interest because it is in this segment that a regionalized structure of the acrosomal contents has been observed (20). Most obvious in the guinea pig, this morphological heterogeneity can also be seen under certain conditions in other species; the different regions of the matrix in the apical segment appear to undergo dispersion at different rates during the acrosome reaction (31, 87). Histochemical staining for glycoproteins reveals other, different, regionalization of the matrix throughout the acrosome (21, 40). At the present time, no specific association of identifiable components with any of these regionalizations has been proven, though in the guinea pig the distribution of histochemical staining for both $Ca^{++}$ and aryl sulphatase is related to some of them (29).

Although a large array of sperm hydrolases have been detected (64), unequivocal evidence for their localization within the acrosome is mostly lacking, and indeed is difficult to obtain. Certainly, the presence of enzyme activity in sperm extracts obtained by a procedure which disrupts the acrosome (so-called 'acrosomal extracts') does not prove an acrosomal location, for other regions of the sperm cell are almost inevitably disrupted. Only those procedures which can be shown to cause specifically the release of acrosomal contents are of value in this type of approach. Immunocytochemical studies at the light microscope level, although helpful in ascribing the enzyme's location to (for example) the anterior acrosomal region, are not definitive, because at this level an intra-acrosomal location cannot be distinguished from a peri-acrosomal location. Studies involving the use of specific antibodies with transmission electron microscopy are most likely to yield precise details of localization of individual enzymes. But unfortunately there are considerable technical difficulties involved in procuring adequate labelling of intracellular components while preserving sufficient morphological structure to allow identification of location: many antigens are very sensitive to fixation.

Knowledge of the intracellular localization of the enzymes thought to be involved in egg penetration is obviously important, because it provides clues as to when and how they are likely to be deployed. By no means all sperm hydrolases are localized in the acrosome. The cytoplasmic droplet for example contains a considerable proportion of the total (testicular) sperm hydrolase content (65). The lysosomal enzyme marker, acid phosphatase, is located in the post-acrosomal region and in the peri-acrosomal cytoplasm; it has not been detected within the acrosome (34, 73, 93). Another 'lysosomal' hydrolase, aryl sulphatase, though apparently localized within the acrosome in the human (34) and the rat (86), has been shown to be largely peri-acrosomal in the guinea pig (29).

At the present time only acrosin and hyaluronidase can be shown with reasonable certainty to have an intra-acrosomal origin in all species so far studied. Acrosin is found within intact spermatozoa entirely as its zymogen form proacrosin (10, 61, 74). Activation of the zymogen occurs following induction of the acrosome reaction (32), and it is released at this time together with hyaluronidase, while cytoplasmic enzymes remain within the cell (32, 87). Located in the anterior acrosomal region by immunocytochemical methods (65), acrosin was earlier thought to be bound to the inner acrosomal membrane. However, it has now been shown that the enzyme disperses following zymogen activation (38). The activation and dispersal is correlated with dissolution and dispersal of the acrosomal matrix. Since we have recently succeeded in labelling the acrosomal matrix in ram spermatozoa with anti-ram acrosin antibodies which cross-react specifically with proacrosin, we conclude that proacrosin is associated with (or is a component of) the acrosomal matrix material (43).

Hyaluronidase is also specifically and rapidly released from spermatozoa undergoing the acrosome reaction (87, 94). At the light microscope level, it has been localized immunocytochemically in the anterior acrosomal region of several sperm species (65). It has also been detected at the electron microscope level within the acrosome of ram spermatids (80). However, despite a wide range of evidence that (the bulk of the) sperm hyaluronidase is located within the acrosome, some recently published results (50) imply that at least some of the hyaluronidase activity in rat spermatozoa is definitely accessible to external substrates prior to an acrosome reaction! Other evidence for externally bound hyaluronidase activity in spermatozoa also exists (see 36). If confirmed, is this activity due to a true exoenzyme or does it result from adsorption of seminal plasma hyaluronidase to the sperm surface?

As well as being uncertain as to the location of many of the enzymes which have been assumed to be acrosomal, we are also entirely ignorant of many of the entities known to be within the acrosome: the glycoproteins, for example, and the components which make up the visible matrix. An intriguing question is how the activation of proacrosin is prevented within the intact acrosome. At one stage, it was postulated that activation of proacrosin required the active participation of a second protease, acrolysin (57). However it has since been shown that the activation process is an autocatalytic intramolecular process, requiring neither an auxiliary enzyme nor even active acrosin (37, 46). However there is a protein acrosin inhibitor within spermatozoa (11) and Harrison & Brown (37) have postulated that if it were localized within the acrosome, this inhibitor would be sufficiently concentrated to prevent zymogen activation while the acrosome was intact; release and dilution of the inhibitor following the onset of the acrosome reaction would allow initiation of proacrosin activation. It is known that the essentially irreversible inhibitor of acrosin, p-nitrophenylguanidinobenzoate, reversibly inhibits proacrosin activation (84), thus the zymogen contains an 'active' site similar to but not identical to that in the active enzyme. On the other hand, Kennedy et al. (47) have recently isolated a protein which is not an acrosin inhibitor but which prevents proacrosin activation; unfortunately this protein seems to be more firmly bound to the sperm head than proacrosin, thus its mode of function is difficult to visualize. An important aspect of acrosomal physiology which undoubtedly has a bearing on zymogen stasis is its acid internal environment. The contents of the intact acrosome are apparently maintained below pH 5 by the action of a proton-translocating $Mg^{++}$-dependent membrane ATPase (60, 95). This environment would mediate greatly against proacrosin activation, which is very low below pH 6.5 (10). Obviously there would be a very marked rise in acrosomal pH following the onset of the acrosome reaction.

## 3. SPERM ENZYMES AND EGG PENETRATION

Interest in the acrosome and its contents has been stimulated very largely because of its role in egg penetration. Yet despite some 17 years' effort, our knowledge of the mechanisms and enzymes involved directly in egg penetration is still very scanty. The confusion that currently exists can be illustrated by data regarding the actions of hyaluronidase and acrosin.

The freshly ovulated mammalian oocyte presents three barriers to the fertilizing spermatozoon: the oocyte plasmalemma or vitelline membrane, the heterogeneous oocyte coating layer known as the zona pellucida, and a mass of surrounding follicular cells (the cumulus mass) more or less firmly attached to each other and to the zona surface by intercellular matrix material; the innermost layers of follicular cells are relatively regular and orientated towards each other and are termed the corona radiata. The first two barriers are always present at the time of fertilization, but the persistence of the cumulus mass varies between species; in the sheep and cow, for example, the mass is rapidly dispersed and is usually absent at the time of fertilization.

Early studies ascribed a cumulus-dispersing role to hyaluronidase, following observations of the effects of hyaluronidase-containing sperm extracts and of commercial hyaluronidase preparations (58). Yet this role does not equate with an enzyme localized within the acrosome, for it is believed that the acrosome reaction of the fertilizing spermatozoon only takes place close to the zona surface, after passage through any cumulus cells that might be present. Of course, it could be argued that the acrosomal contents of prematurely reacted spermatozoa might act to aid the passage of non-reacted colleagues. But recent attempts to demonstrate a specific effect of the hyaluronidase inhibitor thioauromalate have provided inconsistent results. In the mouse (76) the compound blocked fertilization *in vitro* only if cumulus cells were present, consistent with its proposed role. On the other hand, in the hamster (69), fertilization of both cumulus-surrounded and cumulus-free eggs was strongly inhibited. Moreover Lorton & First (52) have reported that commercial hyaluronidase preparations will not disperse the cumulus of bovine follicular oocytes, although the enzyme is used routinely to remove cumulus cells from the ovulated oocytes of laboratory animals. Unfortunately such data are difficult to correlate. The purity of enzyme preparations are rarely defined, and Brown (9) has shown that acrosin (which can be considered a likely contaminant) will itself readily disperse cumulus masses. The possibility also exists that the dispersibility of the cumulus alters following ovulation.

Early work showed that dissolution of the zona pellucida by sperm extracts could be prevented by inhibitors of trypsin-like enzymes. The same inhibitors could also block fertilization, both *in vivo* and *in vitro*. The trypsin-like esterolytic activity in sperm extracts was isolated as acrosin (58). But, so far, efforts to prove a direct specific role for pure acrosin in zona dissolution or penetration have been to no avail. Pure ram acrosin, for example, dissolves the zona from mouse eggs, but is quite ineffective on sheep, pig and gerbil eggs (9). Other roles for acrosin in fertilization have been postulated, in particular in the acrosome reaction (59). But involvement of acrosin in the actual fusion process has now been largely discounted (27, 32, 87), and although dispersal of the matrix depends on proacrosin activation (27, 32, 87), I personally find it hard at present, from a philosophical standpoint, to accept this latter as the role for acrosin in fertilization.

Evidence (81) for the involvement of a trypsin-like enzyme in sperm-zona binding in the mouse is difficult to link with acrosin, because it seems that such binding necessarily takes place before the acrosome reaction has occurred (82): at this juncture, acrosin would be sequestered within the acrosome in the form of proacrosin. On the other hand, Fraser (27) has recently demonstrated that in a highly defined mouse *in vitro* fertilization system *p*-aminobenzamidine, an inhibitor of acrosin and proacrosin activation, prevents zona penetration while fusion with zona-free eggs is almost unaffected. The fusion events of the acrosome reaction occur normally, but acrosomal matrix dispersion is inhibited. It is, of course, very difficult to correlate data from the effects of inhibitors on the action of whole cells or crude cell extracts with data obtained from the use of pure enzymes. In the case of studies on acrosin, for example, more than one trypsin-like enzyme might have been active in the less defined systems. None of the inhibitors used are specific for acrosin, and no detailed cataloguing of the multiple protease activities present in spermatozoa (58, 90) has yet been attempted. Moreover, the possibility also exists that compounds such as p-aminobenzamidine are exerting non-specific effects: controls using similar compounds are not always performed.

Other sperm enzyme activities have been invoked in mammalian egg penetration, but few have survived scrutiny. For example, the contentions of the esterase reputed to remove the corona radiata from rabbit eggs (58) have diminished following a demonstration that dispersal of the corona radiata occurs spontaneously around pH 8.0 (8); such conditions prevail in the rabbit oviduct at ovulation (7). Amid this confusion, it would be naïve to assume that processes such as zona dissolution can be ascribed to the action of individual enzymes. It is far more likely that several enzymes are involved at each stage, either acting in concert or in a specific sequence. Farooqui & Srivastava (19) have reported that seminal N-acetyl hexosaminidase will cause the rabbit zona pellucida to swell only after pretreatment of the egg with aryl sulphatase.

The problem of the involvement of acrosomal enzymes in penetration of the egg investments is compounded not only by the point in time at which the acrosome reaction occurs in the fertilizing spermatozoon, but also by the speed with which the acrosomal contents, both soluble and matrix-associated, are voided. While these facets have been long recognized and discussed, we are no closer to resolving them. On the contrary, Yanagimachi and his colleagues (24, 42) have recently demonstrated elegantly that in the guinea pig fully acrosome-reacted spermatozoa were perfectly capable of fertilizing zona-invested eggs, when seemingly all their diffusible acrosomal contents had been lost! Even in species in which the acrosome reaction is thought to take place close to or at the zona surface, the acrosomal contents are apparently released well before zona penetration is complete (49, 92). It is clear that further searches must be made for enzyme activities firmly attached to the inner acrosomal membrane. Neuraminidase may be one of these (89).

Involvement of acrosomal enzymes in sperm-oocyte fusion is questionable. As mentioned above, the inner acrosomal membrane does not fuse with the oocyte plasmalemma but is endocytosed and dispersed within the oocyte cytoplasm. Some discussion has surrounded the relevance of the delay in fusion of the acrosomal and plasma membranes in the equatorial segment during the acrosome reaction, and the possible action of enzymes contained within this portion of the acrosome. But studies have shown there is no correlation between this latter fusion event and oocyte penetration; also, fusion between the sperm and egg membranes is initiated in the post-acrosomal region just posterior to the equatorial segment (see 5). In any case, few data are available as to the enzymic events involved in fusion, though out of a wide range of hydrolases, only phospholipase C was shown to affect the vitelline membrane so as to prevent fertilization of zona-free eggs in the hamster (39).

A final point of great interest is the apparent requirement for 'whip-lash' or 'activated' motility in achieving zona penetration. Normally, in successful in vitro fertilization systems, as well as undergoing the acrosome reaction, spermatozoa of laboratory species exhibit a special sort of motility pattern in which the flexure of the flagellum is greatly enhanced (25, 96). Using dibutyryl cyclic AMP in the absence of glucose, Fraser (26) induced acrosome reactions in motile mouse spermatozoa without inducing whip-lash motility. These spermatozoa were unable to fertilize zona-invested eggs although penetration of zona-free eggs was obtained. When glucose was included in the system, whip-lash motility was induced and fertilization was normal. Fleming & Yanagimachi (24) have obtained a close correlation between the degree of whip-lash motility and the ability of acrosome-reacted guinea pig spermatozoa to fertilize zona-invested eggs; once whip-lash motility had declined, the weakly motile cells were only able to penetrate zona-free eggs. This requirement for whip-lash motility may indicate the need for considerable mechanical force as well as enzyme action to cut a path through the zona, and could complicate considerably attempts to identify the enzymes involved.

## 4. THE BIOCHEMISTRY OF THE ACROSOME REACTION

The acrosome reaction is the exocytotic event which results in the release or exposure of the acrosomal contents to the external environment of the sperm cell. The morphology of the reaction has been described in detail with respect to a number of sperm species (see 5, 35). It involves point fusions of the outer acrosomal membrane with the overlying plasma membrane at sites all over the dorsal and ventral faces of the anterior acrosomal region; as described above, fusions also occur at later stages in the equatorial region. The fusion sites develop into fenestrations, through which the acrosomal contents escape. Release of soluble material is rapid, whereas the matrix material disperses more slowly, leaving eventually remnants associated with the fenestrated plasma membrane/outer acrosomal membrane complex. Finally this entire complex is sloughed off, leaving the inner acrosomal membrane as the bounding cell membrane over the anterior portion of the sperm head. The fusion process appears to be the vital stage; enzyme release and matrix dispersal appear to follow automatically (see previous sections for discussion of matrix dispersal). Beyond the fact that the reaction is triggered by $Ca^{++}$ entry, the mechanism of the fusion process remains largely a mystery. However it appears that phospholipid modifications, cyclic nucleotides and energy metabolism are all involved.

The acrosome reaction can be induced in a wide variety of sperm species by treatment with the divalent cation ionophore A23187 in the presence of $Ca^{++}$ (31, 87, 94). In the absence of $Ca^{++}$, no reaction ensues, and it can be shown that enhanced uptake of $Ca^{++}$ occurs as the reaction is induced to take place (94). The fact that the acrosome reaction cannot normally be induced by simple treatment with $Ca^{++}$, and that $Ca^{++}$ entry is usually very low (e.g. 70), implies that the spermatozoon must become more permeable to $Ca^{++}$ in order that the acrosome reaction be triggered. *In vivo*, spermatozoa require a period of residence in the female tract before they will undergo the acrosome reaction and fertilize eggs. The changes that take place in the sperm cell during this time are collectively known as 'capacitation' and involve removal or modification of cell coating material and alterations in the phospholipid and protein architecture of the plasma membrane, especially over the acrosomal region (4, 14, 18, 48, 68). To induce the acrosome reaction *in vitro* without the use of ionophore appears to require similar pretreatment and modification of the spermatozoon (77). One may deduce that a major result of capacitation is increased permeability to $Ca^{++}$ (88), due either to membrane 'leakiness' or to unmasking of a $Ca^{++}$ 'gate'. In the case of the former, raised levels of external $Ca^{++}$ would be sufficient to trigger the acrosome reaction (as has been demonstrated in the guinea pig (98)). In the case of the latter, a hormonal agonist might be required to 'open' the $Ca^{++}$ gate (63). It is thus of considerable interest that catecholamines, which are well known as such agonists, induce the acrosome reaction in suitably treated hamster spermatozoa (62).

Studies on membrane fusion in other systems have shown that $Ca^{++}$ promotes the fusion process. But the phospholipid composition of the membranes is also very important in enhancing or suppressing fusibility (53, 91). It is clear that this finding relates to the acrosome reaction as well. Fleming & Yanagimachi (23) have shown that pretreatment of guinea pig spermatozoa with certain lysophospholipids causes the acrosome reaction to occur very rapidly following addition of $Ca^{++}$. In the continuous presence of $Ca^{++}$, it is glycerylmono-oleate which provokes the acrosome reaction. Meizel's group have detected release of $Ca^{++}$-stimulated phospholipase A following the acrosome reaction in hamster spermatozoa (51), and have shown that compounds which inhibit phospholipase activity inhibit the acrosome reaction (56). And treatment of guinea pig spermatozoa with a compound ($A_2C$) which increases the fluidity of membrane lipids will permit the induction of the acrosome reaction by subsequent addition of $Ca^{++}$ (22). There is thus considerable evidence to suggest that phospholipid modifications within the membranes of the acrosomal region are involved in the onset of the fusion events. Some of these modifications are probably brought about through $Ca^{++}$-stimulated phospholipase action.

Much attention has been paid to the involvement of cyclic nucleotides in the acrosome reaction. In mouse and hamster, treatments which might be expected to cause a rise in the intracellular cyclic AMP level enhance the rate at which the acrosome reaction can be induced (26, 66). In the guinea pig, however, cyclic AMP appears inhibitory and it is cyclic GMP which is stimulatory (83). Cyclic nucleotides cannot replace $Ca^{++}$, and, as yet, it has not been shown that they have a direct role in the mechanism of the acrosome reaction.

An important prerequisite for successful induction of the acrosome reaction is a suitable energy source. In the mouse, a supply of glycolysable sugar is essential (28), in the hamster glucose and lactate (16), and in the guinea pig, pyruvate (79); in the latter species, glucose is inhibitory, possibly because it suppresses respiration via the Crabtree effect (78). From these observations, it must be concluded that ATP is required in the fusion process.

A final intriguing clue is the apparent specific requirement for serum albumin in systems modelled on physiological conditions (54). Serum albumin is present in considerable quantities in the female tract, but particularly in follicular fluid (6, 7, 17). The potency of commercial preparations of this protein in stimulating the induction of acrosome reactions is enhanced by prior removal of any bound lipids and is reduced by prior saturation with fatty acids (55). It is of interest that this protein has been shown to interact with fatty acids and lysolecithin from erythrocyte membranes, thereby potentiating phospholipase action (15).

A hypothetical schematic mechanism for the acrosome reaction might thus run as follows. During capacitation, the surface coating, membrane surface proteins, and membrane lipids are modified, with the result either that the plasma membrane over the acrosomal region becomes more intrinsically permeable to $Ca^{++}$ or that $Ca^{++}$ gates and their controlling receptor sites are unmasked. $Ca^{++}$ entry into the acrosomal region is enhanced, either as a result of increased permeability and the relatively high $Ca^{++}$ levels found in the vicinity of the egg (7), or following stimulation of the $Ca^{++}$ gate by agonists such as adrenalin acting at unmasked receptor sites on the membrane surface (c.f. 63). The level of free $Ca^{++}$ rises in the cytoplasmic space between the plasma membrane and the outer acrosomal membrane, as increased influx overcomes the efflux mediated by external $Ca^{++}$ pumps (3). Phospholipases are thereby stimulated, and breakdown of phospholipids in the plasma and outer acrosomal membranes produces fusogenic lysophospholipids and diacylglycerols; the latter may be converted to the highly fusogenic phosphatidic acids (91) with the aid of ATP (63). Continued phospholipase action and enhanced fusibility is ensured through the external presence of serum albumin which combines with and removes free fatty acids and other breakdown products from the membranes (15). Soon, the combined presence of the fusogenic phospholipids and free internal $Ca^{++}$ results in multiple fusions between the plasma membrane and the outer acrosomal membrane.

The increased free internal $Ca^{++}$ may also stimulate adenylate cyclase and inhibit cyclic nucleotide phosphodiesterase, via a calmodulin-$Ca^{++}$ complex (44, 45). The resultant rise in cyclic AMP would stimulate energy metabolism to produce more ATP, and might also act directly on motility (41). A rise in local ATP levels could also enhance the production of phosphatidic acid from diacylglycerol. It is noteworthy that both $Ca^{++}$ binding sites and adenylate cyclase have been localized in some concentration on the inner face of the plasma membrane over the acrosomal region (71, 72).

## 5. THE ACROSOME AS A SECRETORY GRANULE

At a time of great interest in lysosomes, the discovery of a wide range of hydrolases in spermatozoa and the correlation of some properties of the acrosomal region with properties of lysosomes naturally led to the proposal that the acrosome could be considered as a specialized lysosome (1). This concept found favour with many reproductive biologists and it is still a generally accepted view. However, in the interim, it has become increasingly clear that the acrosome is better viewed as a secretory granule. This latter concept has already been put forward by several workers (29, 33, 67), but as yet does not appear to have been widely considered.

Because they share a common origin (the Golgi complex), lysosomes and secretory granules have many similarities, especially in their contents of hydrolytic enzymes. But the two classes of organelle differ fundamentally in their functions. Lysosomes are essentially involved in intracellular digestive processes and are associated with endocytotic events (2). Secretory granules, on the other hand, are involved in extracellular release of active components via exocytotic events (12). Clearly, on this basis, the acrosome is a secretory granule rather than a lysosome.

In fact, close parallels can be drawn between the acrosome and the secretory zymogen granule of the pancreas (12, 75, 85). Both are formed via the Golgi complex and pass through an intermediate condensation stage. Both contain considerable quantities of inactive protease zymogen as well as other active hydrolases. Both contain a solid matrix of condensed material in an acid environment probably mediated via the action of proton-translocating ATPases. And the zymogen granule contains the pancreatic secretory trypsin inhibitor, which may act to prevent activation of zymogen (c.f. Section 2). Exocytosis of the contents of both organelles proceeds via fusion events between the cell plasma membrane and the bounding membrane of the organelle, triggered by $Ca^{++}$ and involving phospholipid modifications, cyclic nucleotides, and ATP. Moreover Mrsny & Meizel (67) have recently reported a requirement for $K^+$ influx and $Na^+/K^+$-ATPase activity in the hamster sperm acrosome reaction; similar requirements have been demonstrated for zymogen granule discharge.

Such close similarity between the organelles persuades searches for other parallels. The almost universal requirement for an external hormonal trigger in secretory granule exocytosis has led to the well-known 'stimulus-secretion coupling' concept, with agonists acting at receptor sites on the cell surface. Are there such sites on the plasma membrane of spermatozoa? Although catecholamines will stimulate the acrosome reaction in hamster spermatozoa (62), no direct evidence has yet been found for the existence of β-adrenergic receptor sites in the acrosomal region (13), though the capacitation status of the spermatozoon might well affect their detection (see Section 3). Also, it appears that in all secretory granules so far investigated the original $Ca^{++}$ release which triggers exocytosis can be shown to be from intracellular $Ca^{++}$ stores (12, 85). So far, it has been generally assumed that external $Ca^{++}$ is essential for the sperm acrosome reaction. However, both Friend (29) and Gravis (30) have detected localized sites containing high concentrations of $Ca^{++}$ in the sperm head cytoplasm of guinea pig and Syrian hamster respectively.

## 6. CONCLUDING REMARKS

Despite a great deal of research during the past decade, we are only just beginning to unravel the mysteries of the acrosome. In the past we have suffered not only from shortcomings of available techniques but also as a result of making assumptions which were not justified and of drawing erroneous conclusions from (equivocal) data. However new methodology (particularly the establishing of highly defined *in vitro* fertilization systems), advances in related fields, and a more rationalized approach will doubtless force the acrosome to yield up most if not all of its secrets in the next decade.

## REFERENCES

1. Allison AC, Hartree EF. 1970. J. Reprod. Fertil. 21, 501-515.
2. Bainton DF. 1981. J. Cell Biol 91, 665-765.
3. Barritt GJ. 1981. Trends Biochem. Sci. 6, 322-325.
4. Bearer EL, Friend DS. 1982. J. Cell Biol. 92, 604-615.
5. Bedford JM, Cooper GW. 1978. In Membrane Fusion, eds. Poste G, Nicolson GL. Elsevier, Amsterdam. pp. 65-125.
6. Beier HM. 1974. J. Reprod. Fertil. 37, 221-237.
7. Brackett BG, Mastroianni L. 1974. In The Oviduct and Its Functions, eds. Johnson AD, Foley CW. Academic Press, NY. pp. 133-159.
8. Bradford MM, McRorie RA, Williams WL. 1976. Biol. Reprod. 15, 102-106.
9. Brown CR. 1982. J. Reprod. Fertil. 64, 457-462.
10. Brown CR, Harrison RAP. 1978. Biochim. Biophys. Acta 526, 202-217.
11. Brown CR, Hartree EF. 1975. Hoppe-Seyl. Z. Physiol. Chem. 356, 1909-1913.
12. Case RM. 1978. Biol. Rev. 53, 211-354.
13. Cornett LE, Meizel S. 1980. J. Histochem. Cytochem. 28, 462-464.
14. Davis BK. 1981. Proc. Natl. Acad. Sci. USA. 78, 7560-7564.
15. Drainas D, Harvey E, Lawrence AJ, Thomas A. 1981. Eur. J. Biochem. 114, 239-245.

16. Dravland E, Meizel S. 1981. Gamete Res. 4, 515-523.
17. Edwards RG. 1974. J. Reprod. Fertil. 37, 189-219.
18. Eng LA, Oliphant G. 1978. Biol. Reprod. 19, 1083-1094.
19. Farooqui AA, Srivastava PN. 1980. Biochem. J. 191, 827-834.
20. Fawcett DW. 1975. Devel. Biol. 44, 394-436.
21. Fléchon JE. 1979. Gamete Res. 2, 43-51.
22. Fleming AD, Kosower NS, Yanagimachi R. 1982. Gamete Res. 5, 19-33.
23. Fleming AD, Yanagimachi R. 1981. Gamete Res. 4, 253-273.
24. Fleming AD, Yanagimachi R. 1982. J. Exp. Zool. 220, 109-115.
25. Fraser LR. 1977. J. Exp. Zool. 202, 439-444.
26. Fraser LR. 1981. J. Reprod. Fertil. 62, 63-72.
27. Fraser LR. 1982. J. Reprod. Fertil. 65, 185-194.
28. Fraser LR, Quinn PJ. 1981. J. Reprod. Fertil. 61, 25-35.
29. Friend DS. 1977. In Immunobiology of Gametes, eds. Edidin M, Johnson MH. Cambridge University Press, Cambridge, UK. pp. 5-30.
30. Gravis CJ. 1979. Am. J. Anat. 154, 245-266.
31. Green DPL. 1978. J. Cell Sci. 32, 137-151.
32. Green DPL. 1978. J. Cell Sci. 32, 153-164.
33. Green DPL. 1978. In Development in Mammals, vol. 3, Ed. Johnson MH, North-Holland, Amsterdam. pp. 65-81.
34. Guedenet JC. 1978. C.R. Soc. Biol. 172, 523-528.
35. Gwatkin RBL. 1976. In The Cell Surface in Animal Embryogenesis and Development, eds. Poste G, Nicolson GL. Elsevier, Amsterdam. pp. 1-54.
36. Harrison RAP. 1977. In Frontiers in Reproduction and Fertility Control, eds. Greep RO, Koblinsky MA. MIT Press, Cambridge, USA. pp. 379-401.
37. Harrison RAP, Brown CR. 1979. Gamete Res. 2, 75-87.
38. Harrison RAP, Fléchon JE, Brown CR. 1982. J. Reprod. Fertil. 66, in press.
39. Hirao Y, Yanagimachi R. 1978. Gamete Res. 1, 3-12.
40. Holt WV. 1979. J. Ultrastruct. Res. 68, 58-71.
41. Hoskins DD, Casillas ER. 1975. In Handbook of Physiology, Sect. 7, Endocrinology, Vol. 5. Eds. Hamilton DW, Greep RO. American Physiological Society, Washington DC. pp. 453-460.
42. Huang TTF, Fleming AD, Yanagimachi R. 1981. J. Exp. Zool. 217, 287-290.
43. Huneau D, Harrison RAP, Fléchon JE. These Proceedings.
44. Hyne RV, Garbers DL. 1970. Biol. Reprod. 21, 1135-1142.
45. Jones HP, Lenz RW, Palevitz BA, Cormier MJ. 1980. Proc. Natl. Acad. Sci, USA. 77, 2772-2776.
46. Kennedy WP, Polakoski KL. 1981. Biochemistry 20, 2240-2245.
47. Kennedy WP, Swift AM, Parrish RF, Polakoski KL. 1982. J.Biol.Chem. 257, 3095-3099.
48. Koehler JK. 1981. Arch. Androl. 6, 197-218.
49. Kopecny V, Fléchon JE. 1981. Biol. Reprod. 24, 201-216.
50. Lewin LM, Nevo Z, Gabsu A, Weissenberg R. 1982. Int. J. Androl. 5, 37-44.
51. Llanos MN, Lui CW, Meizel S. 1981. Fedn. Proc. 40, 1617.
52. Lorton SP, First NL. 1979. Biol. Reprod. 21, 301-308.
53. Lucy JA. 1978. In Membrane Fusion, eds. Poste G, Nicolson GL. Elsevier, Amsterdam. pp. 267-304.
54. Lui CW, Cornett LE, Meizel S. 1977. Biol. Reprod. 17, 34-41.
55. Lui CW, Meizel S. 1977. Differentiation, 9, 59-66.
56. Lui CW, Meizel S. 1979. J. Exp. Zool. 207, 173-185.
57. McRorie RA, Turner RB, Bradford MM, Williams WL. 1976. Biochem. Biophys. Res. Commun. 71, 492-498.
58. McRorie RA, Williams WL. 1974. Annu. Rev. Biochem. 43, 777-803.

59. Meizel S. 1978. In Development in Mammals, Vol. 3. Ed. Johnson MH. North Holland, Amsterdam. pp. 1-64.
60. Meizel S, Deamer DW. 1978. J. Histochem. Cytochem. 26, 98-105.
61. Meizel S, Mukerji SK. 1975. Biol. Reprod. 13, 83-93.
62. Meizel S, Working PK. 1980. Biol. Reprod. 22, 211-216.
63. Michell RH. 1979. Trends Biochem. Sci. 4, 128-131.
64. Morton DB. 1976. In Lysosomes in Biology and Pathology, Vol. 5. Eds Dingle JT, Dean RT. North-Holland, Amsterdam. pp. 203-255.
65. Morton DB. 1977. In Immunobiology of Gametes, eds. Edidin M, Johnson MH. Cambridge University Press, Cambridge, UK. pp. 115-156.
66. Mrsny RJ, Meizel S. 1980. J. Exp. Zool. 211, 153-157.
67. Mrsny RJ, Meizel S. 1981. J. Cell Biol. 91, 77-82.
68. O'Rand MG. 1979. In The Spermatozoon. Maturation, Motility, Surface Properties and Comparative Aspects, eds. Fawcett DW, Bedford JM. Urban and Schwarzenberg, Baltimore. pp. 195-204.
69. Perreault S, Zaneveld LJD, Rogers BJ. 1980. J. Reprod. Fertil. 60, 461-467.
70. Peterson RN, Freund M. 1976. Fertil. Steril. 27, 1301-1307.
71. Peterson RN, Russell L, Bundman D, Freund M. 1979. Biol. Reprod. 21, 583-588.
72. Peterson RN, Russell L, Hook L, Bundman D, Freund M. 1980. J. Cell Sci. 43, 93-102.
73. Poirier GR. 1975. J. Reprod. Fertil. 43, 495-499.
74. Polakoski KL, Parrish RF. 1977. J. Biol. Chem. 252, 1888-1894.
75. Pollard HB, Pazoles CJ, Creutz CE, Zinder O. 1979. Int. Rev. Cytol. 58, 159-197.
76. Reddy JM, Joyce C, Zaneveld LJD. 1980. J. Androl. 1, 28-32.
77. Rogers BJ. 1978. Gamete Res. 1, 165-223.
78. Rogers BJ, Chang L, Yanagimachi R. 1979. J. Exp. Zool. 207, 107-112.
79. Rogers BJ, Yanagimachi R. 1975. Biol. Reprod. 13, 568-575.
80. Sakai Y, Yamamoto N, Yasuda K. 1979. Acta Histochem. Cytochem. 12, 141-150.
81. Saling PM. 1981. Proc. Natl. Acad. Sci. USA, 78, 6231-6235.
82. Saling PM, Storey BT. 1979. J. Cell Biol. 83, 544-555.
83. Santos-Sacchi J, Gordon M. 1980. J. Cell Biol. 85, 798-803.
84. Schleuning WD, Hell R, Fritz H. 1976. Hoppe-Seylers Z. Physiol. Chem. 357, 207-212.
85. Schulz I, Stolze HH. 1980. Annu. Rev. Physiol. 42, 127-156.
86. Seiguer AC, Castro AE. 1972. Biol. Reprod. 7, 31-42.
87. Shams-Borhan G, Harrison RAP. 1981. Gamete Res. 4, 407-432.
88. Singh JP, Babcock DF, Lardy HA. 1978. Biochem. J. 172, 549-556.
89. Srivastava PN, Abou-Issa H. 1977. Biochem. J. 161, 193-200.
90. Srivastava PN, Akruk SR, Williams WL. 1979. J. Exp. Zool. 207, 521-529.
91. Sundler R, Papahadjopoulos D. 1981. Biochim. Biophys. Acta, 649, 743-750
92. Szöllösi D, Hunter RHF. 1978. J. Anat. 127, 33-41.
93. Teichman RJ, Bernstein MH. 1971. J. Reprod. Fertil. 27, 243-248.
94. Triana LR, Babcock DF, Lorton SP, First NL, Lardy HA. 1980. Biol. Reprod. 23, 47-59.
95. Working PK, Meizel S. 1982. Biochem. Biophys. Res. Commun. 104, 1060-1065.
96. Yanagimachi R. 1970. J. Reprod. Fertil. 23, 193-196.
97. Yanagimachi R. 1977. In Immunobiology of Gametes, eds. Edidin M, Johnson MH. Cambridge University Press, Cambridge, UK. pp. 255-295.
98 Yanagimachi R, Usui N. 1974. Exp. Cell Res. 89, 161-174.

# INDUCTION OF ACROSOME REACTION IN CIONA INTESTINALIS

R. DE SANTIS[1], M. HOSHI[1,2], F. COTELLI[1,3] and M.R. PINTO[1]

[1]Stazione Zoologica, Napoli, Italy; [2]Department of Biology, Nagoya University, Japan; [3]Dipartimento di Biologia, Università di Milano, Italy.

## 1. INTRODUCTION

The acrosome of *Ciona intestinalis* spermatozoon is a small vesicle(s) at the flute-beak-shaped tip of the anterior part of the head (1). In this region the fusion of the outer and inner membranes gives rise to an apparent thickening of the nuclear envelope called dense-plate (1). The first step of fertilization consist of the binding of the spermatozoa to the vitelline coat (v.c.) (2). This is a fibrillar structure which contains the specific sperm receptors (3,4). Only a few of the spermatozoa which bind to the v.c. undergo acrosome reaction and penetrate through the v.c. itself (3). Vesiculation of the tip of the sperm head (3) is the first change noticed upon the occurrence of the acrosome reaction. A sliding of the mitochondrion towards the tail during the process of sperm penetration has also been described in *Ascidia ceratodes* (5). On the other hand, when the spermatozoa bind to the v.c. of glycerol treated eggs they fail to undergo acrosome reaction and to penetrate the v.c. (3). Induction of acrosome reaction by $Ca^{2+}$ ionophores has been described in a number of animals; accordingly we have investigated the effect of two ionophores (A 23187 and BrX537A) on spermatozoa of *Ciona intestinalis* with the specific aim of finding out whether or not the events leading to the acrosome reaction and mitochondrion sliding (sperm activation), are related. In addition we report on some effects of a v.c.-soluble fraction on sperm activation.

## 2. MATERIALS AND METHODS

Spermatozoa of *Ciona intestinalis* were collected from the spermiduct with a pasteur pipette and stored in a test tube. Their motility was checked and their concentration was estimated as previously described (6). They were incubated for 10' in $Ca^{2+}$ ionophores A 23187 and BrX537A either in artificial sea water (ASW) or in $Ca^{2+}$- free sea water (CFSW) and fixed in 1% final concentration of gluteraldheyde in sea water. For negative staining, after fixation, a drop of sperm suspention was dried on a grid, rinsed with distilled water, stained with 0.5 % PTA and observed with a Philips 400 electron microscope.

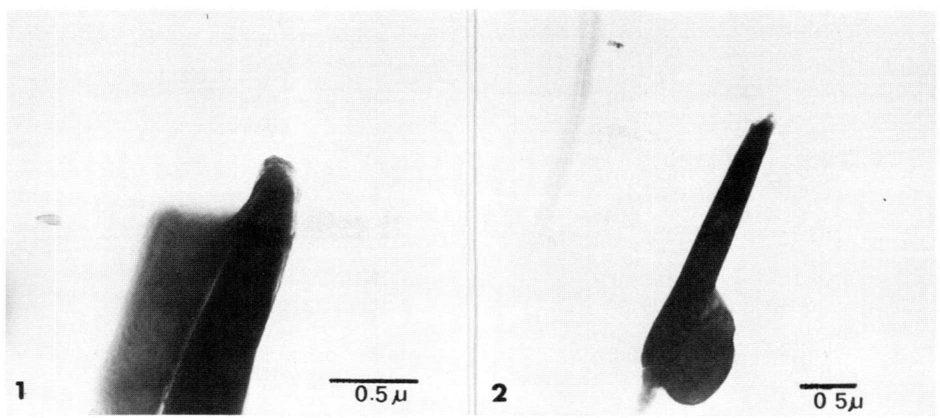


3. RESULTS AND CONCLUSIONS

Since ionophore BrX537A gave variable results, ionophore A 23187 was used. Among the concentration tested, 2 µM was the most suitable. Under these conditions three classes of changes involving both the location of the mitochondrion and the tip of the head were recognizable. The mitochondrion which in the untreated sperm runs parallel and partially surrounds the nucleus almost reaching its tip (fig. 1), slides towards the tail. In the spermatozoa in which the sliding of the mitochondrion has just started, small blebs can be distinguished in the plasma membrane of the sperm tip (fig. 2). Also the outline of the proximal region of the tail is not so clearly delineated as in the controls. The sliding process of the mitochondrion is associated with its acquiring a spherical shape.This further displacement is concurrent with a change in the dense plate which becomes concave, while at the same time the structures of the sperm tip become fuzzy. The position of the mitochondrion with respect to the tail is different in different spermatozoa and in the sperm in which the mitochondrion has travelled along the tail, material with variable elec-

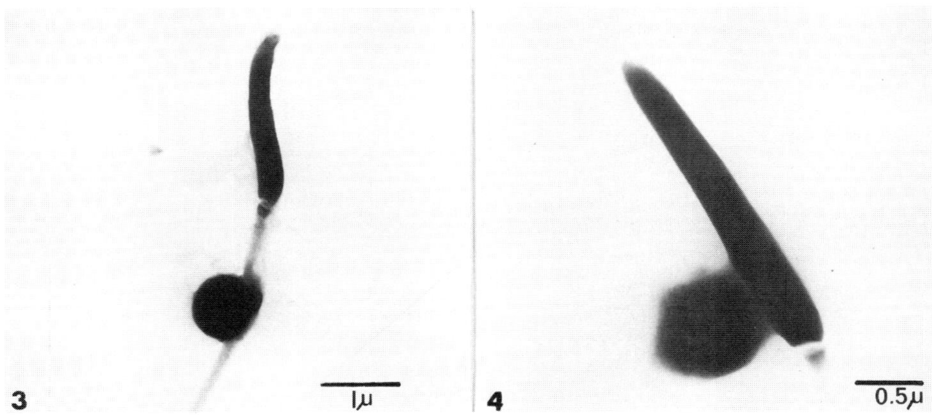

trondensity is seen to protrude from the sperm tip (fig. 3). Some spermatozoa even lack the mitochondrion.

Treatment of the spermatozoa with a v.c.-soluble fraction produces a similar results (fig. 4).

It has been shown that $Ca^{2+}$ ionophore A 23187 (in *Ascidia ceratodes*) and pH change (in *Ciona intestinalis*) elicit shedding of the mitochondrion (5). Preliminary experiments using $Ca^{2+}$ ionophore in CFSW suggest that these phenomena are $Ca^{2+}$ dependent also in *Ciona intestinalis*.

Our experiments thus indicate that the changes in the acrosomal region of the spermatozoon (acrosome reaction) and the shedding of the mitochondrion are somehow related and hence possibly a part of the process of sperm activation.

REFERENCES


1. Cotelli F, De Santis R, Rosati F and Monroy A.1980. Develop. Growth and Differ. 22, 561 - 569.
2. Rosati F and De Santis R. 1978. Exptl. Cell Res. 112 , 111 - 119.
3. De Santis R, Jamunno G and Rosati F. 1980. Dev. Biol. 74, 490 - 499.
4. Pinto MR, De Santis R, D'Alessio G and Rosati F. 1981. Exptl. Cell Res. 132, 289 - 295.
5. Lambert CC and Epel D.1979. Dev. Biol. 69, 296 - 304.

# FREEZE-FRACTURE OBSERVATIONS ON THE IONOPHORE-INDUCED ACROSOMAL REACTION IN RAM SPERMATOZOA.

J.-E. FLÉCHON[(1)], B. FLÉCHON[(1)], J. ESCAIG[(2)], R.A.P. HARRISON[(3)]
(1) Station de Physiologie Animale, I.N.R.A., 78350 Jouy-en-Josas, France, (2) C.N.R.S. Bd. Raspail, Paris, (3) A.R.C. Institute of Animal Physiology, Cambridge, England.

Substances are acquired by the sperm surface during epididymal maturation and ejaculation. One of their possible roles is to protect the sperm cell, and especially its labile acrosomal complex, until the counterpart phenomenon of capacitation (OLIPHANT and SINGHAS, 1979). We observed previously by freeze-fracture that the pattern of the intramembranous particle distribution of the plasma membrane was essentially unchanged throughout the course of the sperm cell surface modifications undergone in the epididymis (FLÉCHON, 1981). In this preliminary study, we used freeze-fracture techniques in an attempt to reveal the ultrastructural changes induced in the plasma membrane of the ram sperm head during the acrosomal reaction induced with an ionophore.

## MATERIAL AND METHODS

The method used to induce the acrosomal reaction (AR) was developed by SHAMS-BORHAN and HARRISON (1981). Ejaculated ram semen was diluted (3 % V/V) in buffered saline containing 3 mM $Ca^{2+}$. Incubation took place _in vitro_ at 37°C in the presence or absence (control) of 1 μM calcium ionophore A 23187. Aliquots were fixed in a glutaraldehyde-formaldehyde mixture at different times up to 1 h. The percentage of AR was estimated by phase-contrast microscopy. For freeze-fracture, the spermatozoa, equilibrated with glycerol, were frozen in nitrogen slush and cleaved in the "cryfract" apparatus (ESCAIG and NICOLAS, 1976).

## RESULTS

Spermatozoa showed no significant changes under the phase-contrast microscope until 10 min of ionophore treatment, after which the AR increased with time ; only a few acrosomes were still present after 1 h. The observations on replicas of freeze-fracture reported here were limited to the plasma membrane (PM) of the acrosomal and postacrosomal regions of the sperm head. Before treatment the PM on both the cytoplasmic (P) and the exterior (E) faces contained few intramembranous particles larger than 10 nm over the acrosomal region (fig. 1) compared with the postacrosomal

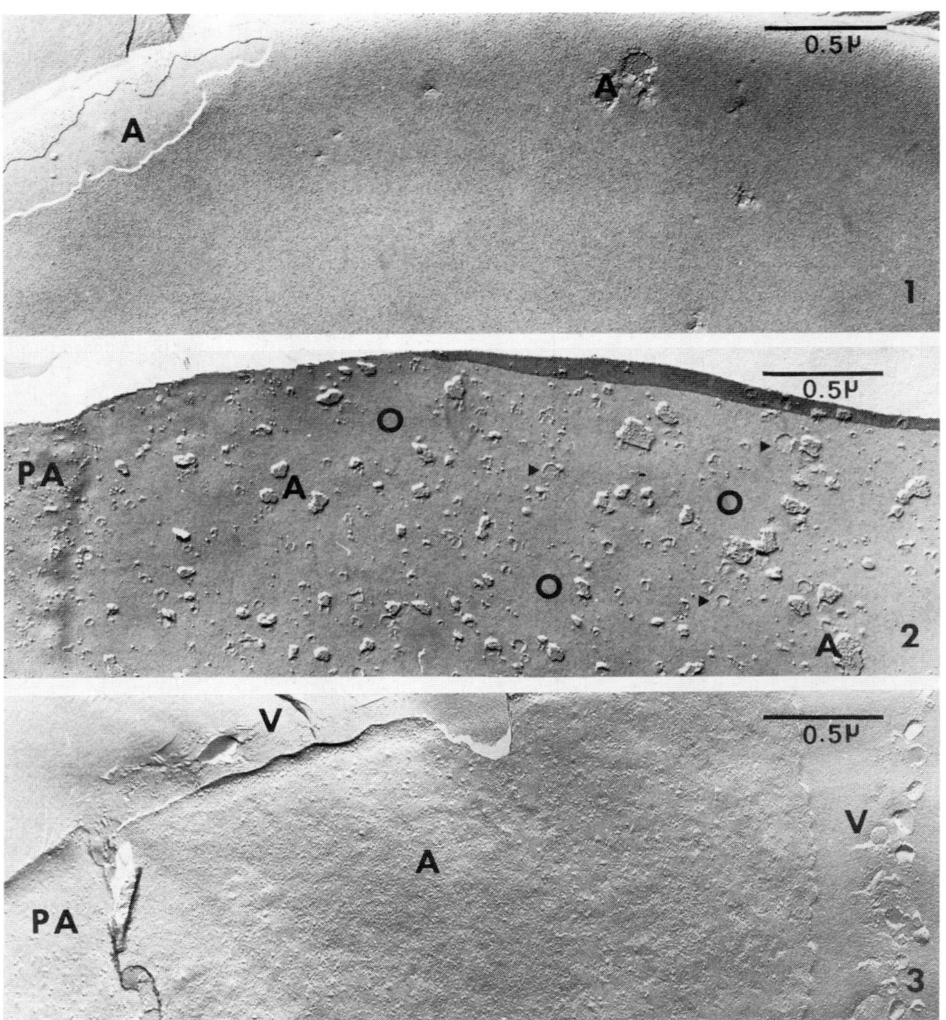


FIGURE 1. In a control sperm cell before treatment, the plasma membrane (PM) of the acrosomal region shows relatively few large intramembranous particles (IMP) scattered over the cytoplasmic face (P face). In some areas the fracture passes through the outer acrosomal membrane (AM) : A (E face).
FIGURE 2. IMP accumulated in the PM (E face) at the anterior border of the postacrosomal region (PA) after 10 min treatment ; in the acrosomal region large IMP-free areas (O) are found together with small craters (▶). Strips (A) of the outer AM (P face)
FIGURE 3. After the occurrence of the acrosomal reaction (40-min treatment), the inner AM (E face) replaces the PM in the acrosomal region (A). The fused PM and outer AM appear as a fenestrated network (V). Same IMP distribution in the PM (E face) of the postacrosomal region (PA) as in fig. 2.

region. This differential particle distribution was mostly unaffected after treatment. However, in treated spermatozoa many particles became apparent along the anterior limit of the postacrosomal region (fig. 2) and in the posterior-central domain of the equatorial segment. Possible modifications preceding the AR, e.g. segregation of particles around several particle-free areas, were observed in the PM over the acrosomal region. These features were accompanied by small holes or invaginations on the P face that corresponded to craters on the E face (fig. 2). All the changes described in the PM occurred in almost all the sperm cells treated for at least 10 min, whereas they were less constant before 10 min. Some of these changes, but not all, occurred in controls, especially after long incubation.

The AR expressed itself as multiple fusion points between the PM and the outer acrosomal membrane (AM), resulting in the formation of a uniformly fenestrated acrosomal cap (fig. 3). The mixed membrane resulting from fusion showed a relatively large particle density. Fusion extended to the equatorial segment, with the frequent exception of the posterior-central domain, where the PM contained only particle-free areas surrounded by aggregated particles. The PM of the postacrosomal region and the inner AM of the equatorial segment fused together.

DISCUSSION

The present results provide some insight into the events occurring in the head PM during the AR ; they suggest that some destabilization of the PM was induced before the AR, but they don't entirely reveal the relative contribution of $Ca^{2+}$ and of the ionophore. Finely do the changes occurring in this model system (in which an AR was obtained by bypassing capacitation) reflect the actual modifications of spermatozoa undergoing capacitation ?

Our observations, confirming those of SHAMS-BORHAN and HARRISON (1981), indicate that : (1) the AR was significantly obtained only in the presence of ionophore and $Ca^{2+}$ and (2) the spermatozoa reacted after some delay and with relative asynchrony. This means that membrane fusion depended on the accumulation of intracellular $Ca^{2+}$, but that the sperm populations were not homogeneous. Were the holes or invaginations we observed for the first time, signs of fusion of the PM with the outer AM ? They would correspond nicely with the early "vesiculation" stage observed on thin sections.

As in other cases of exocytosis, particle-free areas in the PM, already described in guinea-pig spermatozoa incubated _in vitro_ for capacitation, have been interpreted as potential domains of fusion (FRIEND et al., 1977). Preexisting particle-free areas have also been described in fresh guinea-pig spermatozoa (BEARER and FRIEND, 1982). In any case, the size of particle-free areas might well be exaggerated by relatively slow chemical fixation vs rapid freezing (PLATTNER, 1981). Nevertheless it seems clear that the whole range of concomitant PM changes we observed were induced by incubation in the presence of ionophore and $Ca^{2+}$.

Some of these modifications may be induced at a lower rate by $Ca^{2+}$ only. This is likely, as few "spontaneous" AR probably occur in the absence of ionophore : in fact the percentage of AR tended to increase with time, although not significantly in these conditions.

After ionophore-induced AR, the equatorial segment was generally disrupted, as frequently occurs after _in vitro_ capacitation (FRIEND et al., 1977). In both cases, the induction of the fusion process may be exaggerated (excess of intracellular $Ca^{2+}$ ?) or acrosin release from a large number of reacted spermatozoa may induce lysis of the trypsin-sensitive content (RUSSEL et al., 1977) of the equatorial segment. In fact a complete loss of the equatorial segment was frequently observed only after 1-h treatment. Acrosin sensitivity however is not required for membrane fusion over the anterior segment (SHAMS-BORHAN and HARRISON, 1981).

REFERENCES

1. BEARER E.L. and FRIEND D.S., 1982. J. Cell Biol. 92, 604.
2. ESCAIG J. and NICOLAS, 1976. C.R. Acad. Sci. Paris. Serie D 283,1254.
3. FLÉCHON J.E., 1981. Progr. Reprod. Biol. 8, 90.
4. FRIEND D.S., ORCI L., PERRELET A. and YANAGIMACHI R., 1977. J. Cell Biol. 74, 561.
5. OLIPHANT G. and SINGHAS C.A., 1979. Biol. Reprod. 21, 937.
6. PLATTNER H., 1981. Cell Biol. Intern. Rep. 5, 435.
7. RUSSELL L., PETERSON R.N. and FREUND M., 1980. Anat. Rec. 198, 449.
8. SHAMS-BORHAN G. and HARRISON R.A.P., 1981. Gamete Res. 4, 407.

# ACROSOMAL ACTIVATION OF MAMMALIAN SPERM

G. BERRUTI, Dept. Biology, Univ.Milano, MILANO, ITALY
R.A.McRORIE, Dept.Biochemistry, Univ.Georgia, Athens,GA 30602 USA
E. MARTEGANI, Dept.Physiology & Biochemistry,Univ.Milano,MILANO.

The current model proposed for proacrosin activation (1)is essentially that proposed originally (2) with the addition of inhibitor :

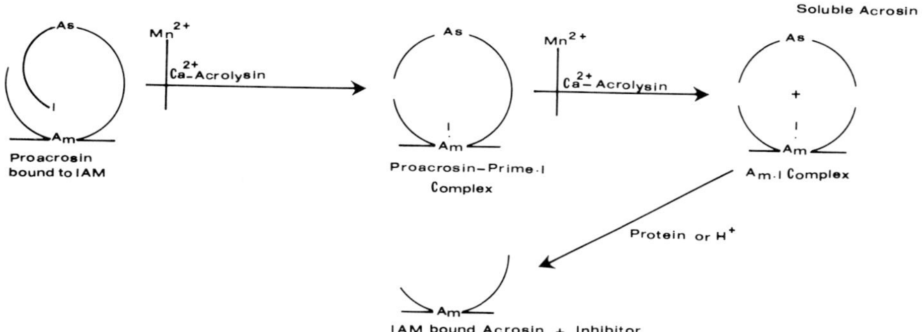


The acrosin/inhibitor (A:I) ratio of 2:1 in rabbit testis proacrosin after activation by thermolysin is shown in Fig.1. BzArgOEt activity is released rapidly and total activity doubles after disruption of [AI] (pH 3) which forms in solution after dissociation from proacrosin-prime•I and $A_m$• I.

The role of acrolysin (3) in the normal activation of rabbit sperm proacrosin is shown by activity (4) on polyacrylamide gels of ejaculated rabbit sperm after membrane disruption and extraction (freeze-thaw) in the presence of $Mn^{++}$, an acrosin activator (5) and an acrolysin inhibitor. Inhibition of acrolysin is apparent and formation of $A_s$ and $A_m$ obvious (Fig.2). Sperm with damaged membranes always show proacrosin activation.

Similar activation and inhibition have been observed during proacrosin "auto-activation" in ejaculated boar sperm extracts (6). Extracts of 700 rabbit ejaculates pooled over weekly periods showed 47.6% [AI] and 52.4% free A. Thus, all inhibitor in rabbit sperm acrosomes must be in proacrosin. Additional acrosomal inhibitor would alter the 50:50 ratio observed.

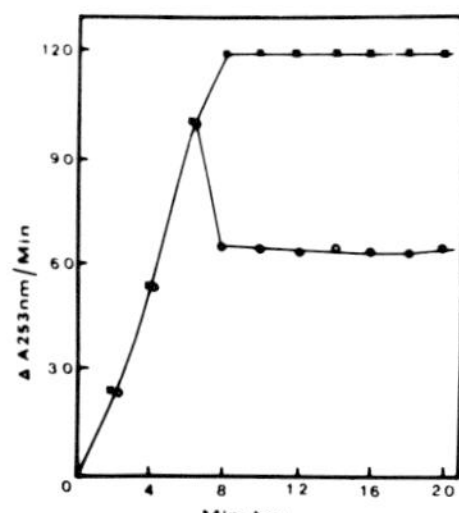


Fig. 1. BzArgOEt hydrolysis before (-o-) and after (-■-) treatment at pH 3.The incubations were performed in 0.05 M Tris-HCl, pH 8.0,with 0.05 M $Ca^{++}$, (5).

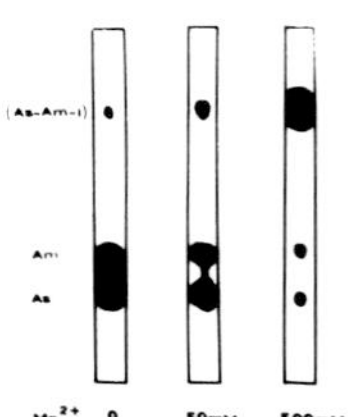


Fig. 2. Zymographic analyses performed on polyacrylamide gels stained for BzArgNab amidohydrolase activity (4) of rabbit sperm proacrosin extracted in the presence of different $Mn^{++}$ concentrations.

Proacrosin-prime and its weak association with inhibitor was first detected in rabbit acrosomal extracts chromatographed on proflavin Sepharose columns (7). The proacrosin-prime•I complex is weakly held by the acrosin affinity column eluting with 0.05 M NaCl (Fig.3). Inhibitor dissociates in the presence of the protein substrate azocoll or after acid treatment at pH 3 with a 10-fold activity increase.

Chromatography of the combined proacrosin-prime fractions on Sephadex G-75 at pH 3 is shown in Fig.4. During processing proacrosin-prime (PeaK 1,$M_r$ = 66,000) was partially degraded to $A_m$ (Peak 2,$M_r$ = 45,000) and $A_s$ (Peak 3,$M_r$ = 32,000). Total activities of $A_s$ and $A_m$ were essentially equal. The broader $A_s$ peak indicates partial binding to the column. I is released at pH 3 and elutes much later ($M_r$ = 8-10,000). The apparent A:I ratio is low due to low proacrosin-prime activity(1/8 - 1/10).

$A_m$, in solution, degrades to a molecule that does not separate from $A_s$ during acid polyacrylamide gel electrophoresis suggesting that the molecules are very similar if not identical. The degree to which this degradation occurs on the inner acrosomal membrane (IAM) during normal fertilization is not known. Proof of the existence of $A_m$ over the entire IAM of rabbit and hamster sperm has been presented (1) using the fluorescent active site directed inhibitor, dansylalanyllysyl-$CH_2$-Cl (DALCK). Fluorescence fades rapidly but similar results have been obtained for human sperm (Fig.5a). BSA was used to dissociate $A_m \cdot I$ for the rabbit and hamster, acid was used for human sperm. DALCK treatment also provided valuable information on $A_s$ using human sperm treated with $Ca^{++}$ + ionophore A23187 (Fig.5b).

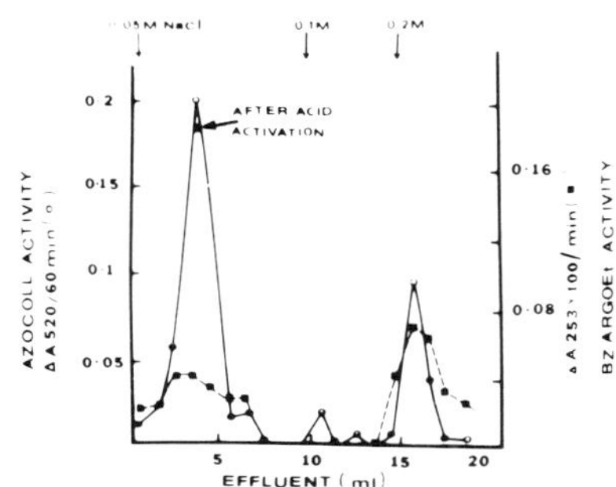


Fig. 3. Elution profiles of rabbit proacrosin-prime after affinity chromatography on a proflavin Sepharose column (7).

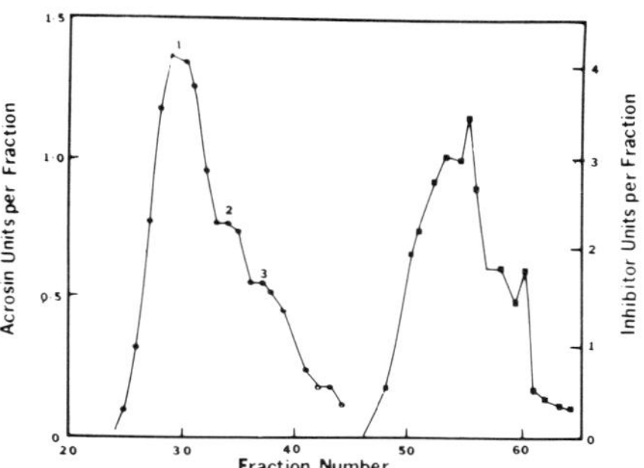


Fig.4. Elution profile of proacrosin-prime after chromatography on Sephadex G-75 at pH 3. Acrosin units (-o-) = Inhibitor units (-■-) = BzArgOET units.

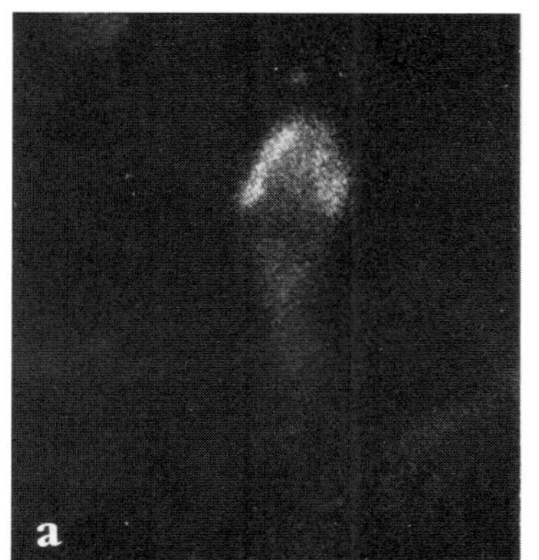


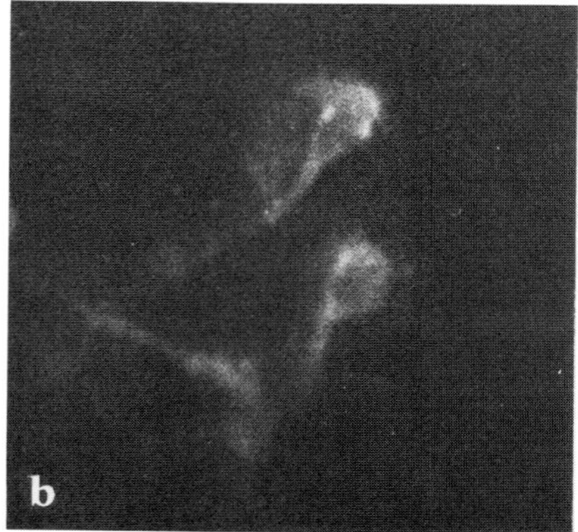


Fig.5. Detection of human $A_m$ and $A_s$ with DALCK. a) Triton-treated spermatozoon; b) $Ca^{++}$ + A23187 treated spermatozoa.

Fluorescence is seen initially at the anterior of the acrosome and as the acrosome reaction proceeds $A_s$ fluorescence can be seen as "towers" of enzyme emerging through small membrane openings. These results, coupled with the data that acrosin inhibitors markedly reduce the acrosome reaction (8), suggest that digestion of membrane proteins by $A_s$ initiates production of membrane orifices in the acrosome reaction. Since cytoplasmic droplet membranes have the same origin as acrosomal membranes and the soluble enzyme content is equivalent (9), proacrosin occurs only in sperm acrosomes.

When rabbit sperm and cytoplasmic droplets were treated with $Ca^{++}$ + A23187 to introduce $Ca^{++}$ into the acrosome, activate proacrosin to $A_m$ + $A_s$ and initiate the acrosome reaction, the following results were obtained using arylsulfatase to measure droplet enzyme release :

| Treatment | Acrosin (units) | | Arylsulfatase (units) | |
|---|---|---|---|---|
| | Droplets | Sperm | Droplets | Sperm |
| None | 0 | 0 | 0 | 0 |
| $Ca^{++}$,A23187 | 0 | $A_s$ 8.0 | 0 | 0.19 |
| Followed by extraction* | 0 | $A_m$ 8.8 | 0.10 | 0.46 |

*Dialysis for droplets, Triton for sperm.

Proacrosin activation occurred with expected release of $A_s \simeq A_m$ and concomitant release of arylsulfatase from sperm. No activity was released from treated droplets. Subsequent extraction of droplets indicated no destruction of droplets arylsulfatase in the experiment. Acrosin activity has previously been observed in small units of acrosomal membranes (10).

The multiple described "boar proacrosin", cf,review (11), may be an artifact of proacrosin isolation resulting from proteolytic cleavage within the $A_s$ moiety of the proacrosin molecule. Activation sequences reported (12) :

Z $\rightarrow$ $\alpha$-acrosin $\rightarrow$ $\beta$-acrosin $\rightarrow$ $\gamma$-acrosin may correspond to :

Artifact $\rightarrow$ $A_m$ $\rightarrow$ $A_s$ $\rightarrow$ Degradation product in solution.

ACKNOWLEDGMENTS- We are indebted to colleagues MM Bradford,JH Branter, PN Srivastava and JK Thakkar for major contributions to this work.

REFERENCES

1. Bradford MM, Dudkiewicz AB, Penny GS, Dyckes BD, Burleigh BD, Wooley RE, McRorie RA. 1981. Am J Vet Res 42, 1082.
2. Mukerij SK, Meizel S. 1975. Arch Biochem Biophys 168, 720.
3. McRorie RA, Turner RB, Bradford MM, Williams WL. 1976. Biochem Biophys Res Commun 71, 492.
4. Garner DL, Salisbury GW, Graves CN. 1971. Biol Reprod 4, 93.
5. Polakoski KL, McRorie RA. 1973. J Biol Chem 248, 8183
6. Schleuning WD, Hell R, Fritz H.1976. H-S Z Physiol Chem 357, 207.
7. Branter JH, Medicus RG, McRorie RA. 1976. J Chromatog 129, 97.
8. Lui CW, Meizel S. 1979. J Exp Zool 207, 173.
9. Srivastava PN. 1981. Biol Reprod 24, Supp 1, 82A.
10. Stambaugh R. 1978. Gamete Res 1, 65.
11. Parrish RF, Polakoski KL. 1979. Int J Biochem 10, 391.
12. Polakoski KL, Parrish RF. 1977. J Biol Chem 252, 1888.

# ULTRASTRUCTURAL LOCALIZATION OF PROACROSIN AND ACROSIN DURING THE ACROSOME REACTION IN RAM SPERMATOZOA

D. HUNEAU, R.A.P. HARRISON°, J.-E. FLECHON
Station Centrale de Physiologie Animale, I.N.R.A., 78350 Jouy-en-Josas, France, and ° A.R.C. Institute of Animal Physiology, Animal Research Station, Cambridge CB3 0JQ, England.

## 1. INTRODUCTION

The sperm-specific serine proteinase, acrosin, plays an essential though as yet undefined role in mammalian fertilization (1). Located within the acrosome as its zymogen form proacrosin, active acrosin is produced following the acrosome reaction (2) or other disruption of the sperm head membranes (3). Early evidence suggested that acrosin played an essential part in zona penetration (4). In particular the enzyme was believed to be located on the inner acrosomal membrane, where it remained to act after the rest of the acrosomal contents had been released following the acrosome reaction. However we have recently obtained indirect biochemical and immunocytochemical evidence to suggest that proacrosin is localized in the acrosomal matrix and not on the inner acrosomal membrane (5).

In the present work, we relate morphological changes in the acrosome to the proacrosin activation process and we show the ultrastructural localization of (pro)acrosin at different stages of the ionophore-induced acrosome reaction in ram spermatozoa.

## 2. PROCEDURE

2.1. Ejaculated ram spermatozoa were washed and disrupted, and the heads were isolated in sucrose/MES media, buffered to pH 6.0-6.5 in the presence of 0.5 mM p-amino-benzamidine (pAB) to prevent proacrosin activation (procedures as described in (3)). The heads were then incubated at 20°C in NaCl/HEPES pH 7.8 in the presence or absence of pAB. Aliquots of the head suspensions were removed at 10-min intervals and either acidified to determine acrosin activity (3) or fixed for electron microscopy.

2.2. Acrosome reactions were induced in ejaculated ram spermatozoa using 1 µM A23187 (divalent cation ionophore) and 3 mM $Ca^{2+}$ (6). Cells were incubated with the reagents for 20 min at 37°C and then fixed with 0.5 % ($^{w}/v$) 1-ethyl-3 (3-dimethylaminopropyl)-carbodiimide in 0.2M sucrose/0.1M cacodylate pH 7.3/0.5 mM pAB. An anti-acrosin serum which reacted with several acrosin forms including proacrosin was raised in rabbits (7), and the IgG fraction from this serum was used in the first stage of an immunoperoxidase localization procedure with non-immune IgG as control. Peroxidase labelled anti-rabbit IgG Fab fragments were used in the second stage of the procedure.

3. RESULTS

3.1. Although the plasma membrane was removed from the sperm heads during the isolation procedure, the acrosome remained _in situ_. In the presence of pAB, in which no activation of proacrosin took place, the acrosomal matrix remained essentially undispersed within the acrosomal membranes (fig. 1a), although its granular aspect and the appearance of a vacuole in the apical region suggested some loss of (soluble ?) components as compared with the intact organelle. After 10 min incubation at pH 7.8 in the absence of pAB, when 60 % proacrosin activation had occured, the contents of the anterior segment of the acrosome had already largely dispersed, leaving only a layer of material associated with the outer acrosomal membrane (fig. 1b). By 60 min, when proacrosin activation was essentially complete, the outer acrosomal membrane had ballooned away from the head surface and the equatorial segment had emptied of material.

3.2. After 20 min incubation in the presence of ionophore and $Ca^{2+}$, 75 % of the spermatozoa had undergone an acrosome reaction. The response being non-synchronous (6), different stages could be observed. When matrix disaggregation had hardly begun, the contents of the anterior segment of the acrosome were homogeneously labelled by anti-acrosin (fig. 2a). In later stages there was labelling of the matrix material remaining associated with the membrane vesicles and also labelling of the (by now) vesiculated equatorial segment, whereas no labelling of the inner acrosomal membrane was seen (fig. 2b).

4. DISCUSSION AND CONCLUSION

The studies with the isolated heads demonstrate that the acrosomal matrix remains in place while proacrosin activation is prevented, whereas it disperses rapidly once activation has begun. One may therefore deduce that in an acrosome-reacted spermatozoon in which the acrosomal matrix is still compact, much of the acrosin remains in the zymogen form. Using an anti-acrosin serum which reacts strongly with proacrosin (7), we obtained

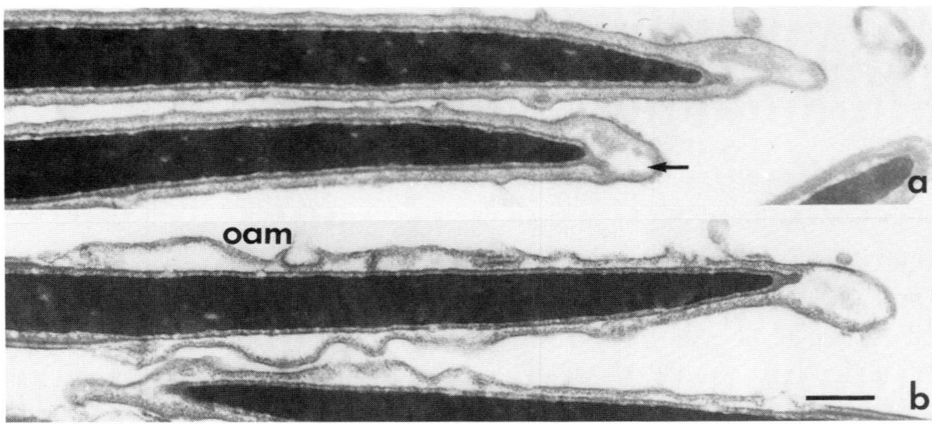


FIGURE 1 - Heads isolated from disrupted ram spermatozoa and incubated for 10 min in saline at pH 7.8 in the presence (a) or the absence (b) of 0,5 mM p-amino-benzamidine. a) Acrosomal matrix has remained mostly *in situ*. Note vacuole in the apical ridge (→). b) After acrosomal disaggregation had occured, only a layer of material was left under the weaving outer acrosomal membrane (oam). The bar represents 0.25 µm.

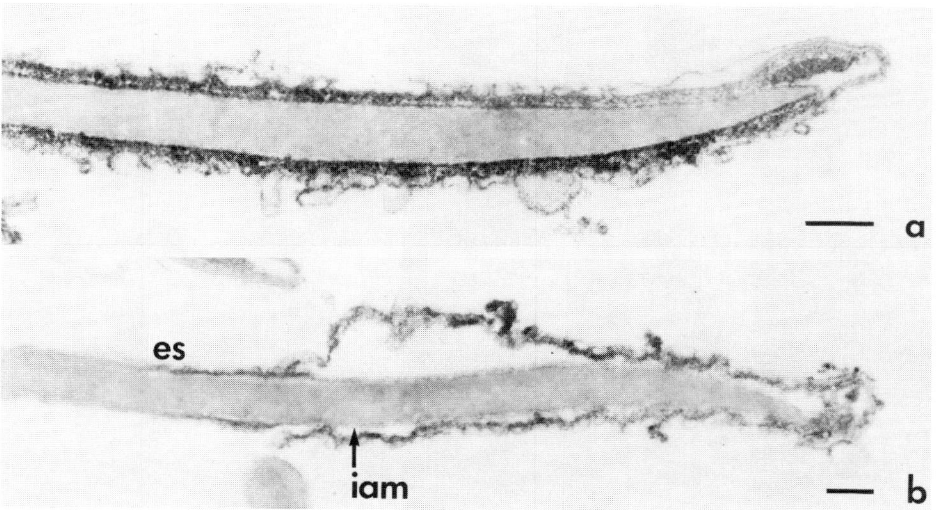


FIGURE 2 - Ram spermatozoa after ionophore induced acrosome reaction. Immunoperoxidase method with anti-acrosin ; unstained sections. a) Early stage of acrosome reaction ; over-all labelling of the still compact acrosomal matrix. b) Remnants of acrosomal matrix associated with membrane vesicles and the anterior part of equatorial segment(es) are labelled whereas the inner acrosomal membrane (iam) is not. The bar represents 0,25 µm.

labelling throughout the matrix. We therefore conclude that at this stage proacrosin is distributed throughout the anterior acrosomal matrix. Moreover (pro)acrosin remains associated with matrix material throughout its dispersion.

Thus proacrosin appears to be a fundamental constituent of the acrosomal matrix in the ram ; its activation is necessary for the dispersion of the matrix following the acrosome reaction (see also (2)). Whether or not this latter can be interpreted as a "role" for acrosin in the fertilization process, (pro)acrosin localization within the dispersable matrix rather than tightly bound to the inner acrosomal membrane is difficult to reconcile with the enzyme's originally proposed role as a zona lysin.

REFERENCES


1. Harrison R.A.P. 1982. The acrosome, its hydrolases, and egg penetration (see this volume).
2. Green D.P.L. 1978. J. Cell Science 32, 153-164.
3. Brown C.R., Harrison R.A.P. 1978. Biochim. Biophys. Acta 526, 202-217.
4. Mc Rorie R.A., Williams W.L. 1974. Annu. Rev. Biochem. 43, 777-803.
5. Harrison R.A.P., Fléchon J.-E., Brown C.R. 1982. J. Reprod. Fert. 66, (in press).
6. Shams-Borhan G., Harrison R.A.P. 1981. Gamete Res. 4, 407-432.
7. Harrison R.A.P. 1982. J. Reprod. Immunol. 4 (in press).

# CHAPTER 5 SPERM MOTILITY

## CONTROL MECHANISMS IN SPERM FLAGELLA

CHARLES J. BROKAW

Division of Biology, California Institute of Technology, Pasadena, California 91125, U.S.A.

### 1. INTRODUCTION

My title should read "Some Control Mechanisms in Sperm Flagella," because I may neglect someone's favorite control mechanism, and there may well be control mechanisms in spermatozoa that we do not yet suspect. Certainly the biochemical complexity of even simple flagellar axonemes, evidenced by the studies of David Luck and his colleagues on Chlamydomonas flagella (1) should inhibit any of us from taking a simplistic view of spermatozoa. The control mechanisms I will discuss are those with which I have some direct experience, and I will mainly discuss invertebrate sperm flagella.

### 2. CONTROL OF THE ACTIVE SLIDING PROCESS

Flagellar bending is caused by an active sliding process operating between the microtubular doublets of the 9+2 axoneme. There is some reason to think that the force-generating capability of a 9+2 axoneme may need to be augmented by other mechanisms in order to generate the large bending waves seen on some mammalian spermatozoa (cf. 2), but attempts to identify such mechanisms have been unrewarding. For simple sperm flagella bending in a plane, the relationship between bending and sliding can be stated quite precisely if we assume that the doublets have no longitudinal compliance, that no sliding occurs at the base of the flagellum, and that the axoneme does not twist: with these restrictions, the rate of bending, $d\kappa/dt$, is a simple function of the rate of sliding, $d\sigma/dt$, where $\sigma$ is an angular measure of shear between doublets:

$$d\kappa/dt = d\,(d\sigma/dt)/ds \qquad (1)$$

Bending is caused by differences in sliding rate at different positions along the length of the flagellum.

Evidence from experiments of Sale and Satir (3) indicates that the active sliding process operates in only one direction. The dynein arms on the A-tubule of a doublet interact with the B-tubule of the adjacent doublet and push it towards the tip, or distal, end of the flagellum—at least in Tetrahymena cilia, which normally propagate bends towards the tip. This means that both active and passive sliding must occur in flagella. At any location on a flagellum where sliding ($d\sigma/dt$) is occurring, there must be sliding between doublets on both sides of the axoneme. The direction of sliding will be appropriate for active sliding on only one side of the axoneme; on the other side, passive sliding must occur. There must be a control mechanism that activates sliding between some doublets in selected parts of the flagellum and allows passive sliding to occur in other parts of the flagellum.

Such a control mechanism is presumed to be responsible for oscillation and bend propagation by sperm flagella. In principal, any desired pattern of flagellar bending can be generated by specifying the rates of active sliding throughout the flagellum. However, flagella do not appear to work this way, because the bending pattern, and the rates of sliding, can be altered by external factors such as viscosity. Largely because of observations of the effects of viscosity, a different view of the generation of flagellar bending waves has evolved. In this view, flagellar bending results from the interaction between active forces generated by the active sliding process, and resistive forces resulting from the resistance of the fluid medium surrounding the sperm flagellum (these are viscous forces) and the bending resistances of flagellar structures (these are usually approximated as simple elastic resistances). Since bending is what we observe, it is most straightforward to deal with these forces as bending moments. Then, the movement of the flagellum is the solution to an equation that balances active, viscous, and elastic bending moments

$$M_A + M_V + M_E = 0 \qquad (2)$$

at every point along the length of the flagellum.

Computer simulation of the movement of a sperm flagellum involves successive solutions of this equation, and is useful to test our ideas about mechanisms that control $M_A$ by controlling the active sliding process, or other control mechanisms that might enter this equation by controlling the bending resistance of flagellar structures.

If a bend propagating along a flagellum is a region of constant curvature—i.e., a circular arc—then equation (1) indicates that the rate of sliding must be constant throughout the bend. The rate of sliding will change in the interbend regions, where the curvature also changes. In the simplest cases, the transition from a bend of positive curvature to a bend of negative curvature in the interbend region will be associated with a change in the direction of sliding within the bends. A natural extension of this is the suggestion that the curvature of the flagellum controls the active sliding mechanism (4). Computer simulations, such as the one illustrated in Fig. 1, have been used to show that control mechanisms of this type are sufficient for initiation and propagation of bends on a flagellum (5,6). However, confirmatory evidence for this control mechanism has not appeared. We do know that flagellar ATPase activity is tightly coupled to bending (7,8),

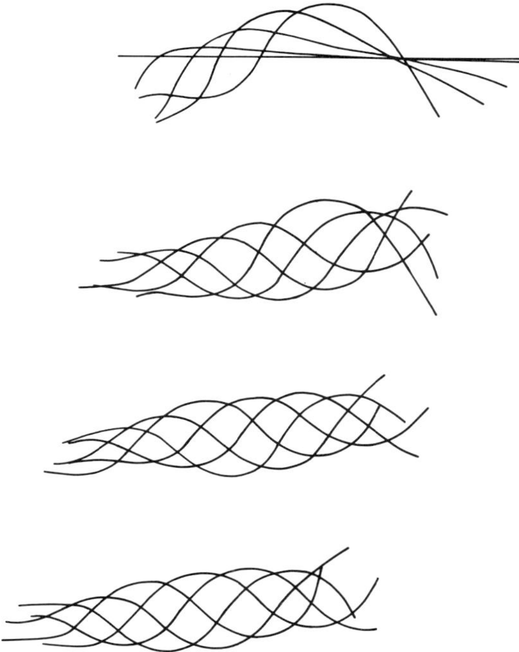

FIGURE 1. Computer simulation of the movement of a flagellar model with a curvature-controlled active sliding process. This model contains a 4-state model for the dynein cross-bridges (6). The model was started from a straight rest position by activating cross-bridges on one side of the axoneme. These cross-bridges remained active until the local curvature of the flagellum became less than -0.16 rad/μm; at this point, they were deactivated and the cross-bridges on the other side of the axonemal segment were activated. This state was maintained in that segment until the local curvature became greater than 0.16 rad/μm, and then the activation and deactivation of cross-bridges on the two sides of the axonemal segment was reversed. This model has a length corresponding to 40 μm, and was broken into 24 segments of equal length for computation; 500 time steps were used for computing each of the 4 bending cycles shown in this figure. Each of the 4 bending cycles contains images separated by time intervals corresponding to 8.33 msec. The oscillation of the model appears to stabilize at a frequency of 30 Hz. The horizontal displacement of the images accurately represents the forward swimming of this flagellar model; the vertical displacement of each group of images is arbitrary.

but control of the active sliding process by bending is not a unique explanation for this observation.

Recent experiments by Kamimura and Takahashi (9), using apparatus designed for direct measurement of the force generated by the active sliding process, have provided evidence for oscillatory sliding between flagellar doublets in the absence of bending. A theoretical mechanism for oscillatory sliding was developed previously (10-12) but computer simulations of flagella containing this type of oscillatory mechanism have so far been less successful than the simulations involving control of active sliding by curvature (13; Brokaw, unpublished observations).

I think we have to conclude that we still do not know how active sliding in sperm flagella is controlled to produce oscillation and bend propagation.

## 3. CONTROL OF BENDING WAVE PARAMETERS

When Triton-demembranated sea urchin sperm flagella are reactivated with $MgATP^{2-}$, their frequency of oscillation can be varied from less than 1 Hz to more than 30 Hz by varying the $MgATP^{2-}$ concentration. Throughout this range, the amplitude of the bending waves, as measured by the mean bend angle, remains nearly constant (14). There is also some variation in wavelength, and at low $MgATP^{2-}$ concentration, only the wavelength decreases when the viscosity of the medium is increased, and again the bend angle remains nearly constant (14). These observations suggest that the axoneme has a control mechanism that regulates the bend angle, independently of changes in sliding velocity or curvature of the bends.

When these reactivated sperm flagella are exposed to $CO_2$ or to antitubulin antibodies, the bend angle is decreased, while the frequency of oscillation remains constant, at least initially (15-18). These observations suggest that the axoneme has a control mechanism that regulates the frequency, independently of changes in bend angle, curvature, and sliding velocity. Under other conditions, this control mechanism appears to behave abnormally, and abrupt transitions between different frequencies of oscillation can be observed (19,20).

When these reactivated sperm flagella are exposed to digestion by elastase, or changes in the $Mg^{2+}$ concentration, reciprocal changes in bend angle and frequency occur, such that the rate of sliding between tubules remains nearly constant (21,22). Under these conditions, a control mechanism that determines the velocity of active sliding appears to be dominant.

Some of the control mechanisms needed to explain these observations may be an integral part of the control of active sliding that is required for oscillation and bend propagation. They are not contained in the simple curvature-controlled models that I have examined by computer simulation—these models do not behave properly when their parameters are varied in an attempt to reproduce the experiments. Perhaps the best hope for progress here is more detailed knowledge of quantitative properties of the active sliding process, obtained from direct and indirect measurements on the sliding disintegration of axonemes, such as those being carried out in the laboratories of Miki-Nomoura, Takahashi, and Warner (23,24).

In addition to these parameters, there are other important properties of flagellar bending that must be controlled. I have been discussing observations on sea urchin sperm flagella, which produce planar bending waves. A flagellar axoneme would appear to have relatively little resistance to twisting (25,26). What keeps a planar wave planar? How is sliding organized to generate planar bends? Interesting observations on patterns of sliding disintegration that indicate that such organization exists have been reported by Olson and Linck (27), Mohri and Yano (28), and Miki-Nomoura (personal communication). Why do some spermatozoa produce nonplanar bends?

At the other extreme, there are subtle features of sperm flagellar bending that remain to be explained. For instance, in sea urchin sperm flagella, new bends form at the basal end of the flagellum with relatively small radius of curvature. Then the radius of curvature increases as the bends elongate and propagate away from the base (29,30). This change in curvature does not occur in our simple computer models, nor in the symmetrical reverse beating mode of flagella of _Chlamydomonas_ gametes. It appears to be a special adaptation found in sea urchin spermatozoa and some other spermatozoa, but not in all spermatozoa. The simplest explanation—a greater flexibility of the flagellum in the basal region—appears to be excluded by measurements of flagellar bending resistance (31,32).

## 4. CONTROL OF SPERM FLAGELLAR BENDING BY $Ca^{2+}$

Spermatozoa of many species show chemotaxis. In the vicinity of conspecific eggs, the spermatozoa make turns in their swimming paths that increase the probability that they will encounter an egg. Sperm chemotaxis has now been demonstrated in many species of Cnidaria (33-35), Mollusca (36), Echinodermata (37), and Chordata (38), and in several other scattered species. The most complete information about the turning behavior has been obtained from spermatozoa of a Cnidarian, _Tubularia_ (39). Turns are generated by brief periods of highly asymmetric flagellar bending, which separate longer periods of symmetric bending wave generation. The frequency of oscillation remains nearly constant. Chemotactic response in some Cnidarian spermatozoa requires the presence of $Ca^{2+}$ (35,40). Asymmetric beating can be induced by $Ca^{2+}$ in demembranated _Tubularia_ spermatozoa (Brokaw; Miller; unpublished observations). However, more complete information about the relationship between flagellar beat asymmetry and $Ca^{2+}$ has been obtained with sea urchin spermatozoa, although chemotaxis has not been demonstrated in this group.

Triton-demembranation of sea urchin sperm flagella in the presence of a high $Ca^{2+}/Mg^{2+}$ ratio produces "potentially symmetric" axonemes that beat symmetrically at low (e.g., $10^{-9}$ M) $Ca^{2+}$ concentration and show a gradual increase in asymmetry as the $Ca^{2+}$ concentration is raised to $10^{-3}$ M (41,42). This change in asymmetry involves an increase in bend angle and curvature of bends in one direction (principal bends) at the expense of bends in the other direction (reverse bends) with little or no change in mean bend angle, wavelength, or frequency (43,44). If the oscillation of these demembranated sperm flagella is inhibited with 10 μM of the dynein ATPase inhibitor, vanadate, changes in $Ca^{2+}$ concentration are still able to reversibly change the shape of the flagellum from a straight configuration to a bent configuration (45). This suggests that $Ca^{2+}$ may act on a control system that is independent of the dynein-ATPase driven active sliding process, to cause a change in flagellar configuration on which bending waves are superimposed by the normal mechanism for bend generation. Such a mechanism is attractive because it seems most easily to fit the observation that frequency, mean bend angle, and wavelength remain constant when the symmetry changes. However, there appear to be some features of the bending waves that are difficult to reconcile with this control mechanism (44).

A calcium regulated transition between symmetric and asymmetric bending waves is also exhibited by flagella of Chlamydomonas, but in this case the normal, low-$Ca^{2+}$ mode is the asymmetric mode appropriate for forward swimming of these biflagellate cells, and $Ca^{2+}$ induces a symmetric, reverse mode (46). We have recently been examining these waveforms, using a mutation, uni-1, that produces uniflagellate cells that are easier to photograph. With these flagella, the transition from symmetric to asymmetric beating is accompanied by a 15-20% reduction in beat frequency and a corresponding increase in wavelength, while the shear amplitude remains constant (Brokaw and Luck, in preparation). The asymmetric mode differs from that of sea urchin sperm flagella in having unequal rates of sliding in principal and reverse bends, and the lengths of principal and reverse bends are also unequal (47). The time of initiation of sliding in principal bends, relative to the time of initiation of sliding in reverse bends, is much earlier in the symmetric mode than in the asymmetric mode; this feature is important for reorientation of the bending pattern relative to the cell axis (Brokaw and Luck, in preparation). These features suggest that a simple mechanism involving superposition of normally generated bending waves on a $Ca^{2+}$-regulated axonemal configuration is inadequate to explain the transition between symmetric and asymmetric modes in these flagella.

When Triton-demembranation of sea urchin spermatozoa is carried out at a low $Ca^{2+}/Mg^{2+}$ ratio, the axonemes are "potentially asymmetric." They beat asymmetrically even when reactivated at low $Ca^{2+}$ concentration, and at high $Ca^{2+}$ concentration in the presence of 1 mM $MgATP^{2-}$ they enter a quiescent state in which movement is inhibited and the flagellum is bent into a characteristic "cane" shape (41). A similar quiescent state is seen during spontaneous arrests of live spermatozoa exposed to intense illumination or to the calcium ionophore A23187 in the presence of $Ca^{2+}$ (48). The quiescent state produced with potentially asymmetric axonemes is an unstable state which leads to disintegration by active sliding, indicating that quiescence does not involve a complete inhibition of the active sliding mechanism.

Calcium sensitivity of potentially asymmetric axonemes appears to be associated with a component that can be reversibly removed by exposure to $Ca^{2+}$ and Triton (42) and that is extremely sensitive to trypsin (2,20,41).

Calcium sensitivity of many cellular processes is mediated by a ubiquitous calcium-binding protein, calmodulin. Calmodulin has been identified in sea urchin spermatozoa (49) and in other flagella and cilia (50,51). Suggestive evidence for involvement of calmodulin in control of flagellar motility has been obtained in several studies (52,53). However, there is no evidence demonstrating that the effects of $Ca^{2+}$ on sea urchin sperm flagellar motility are mediated by calmodulin.

## 5. THE ACTIVATION OF SPERM FLAGELLAR MOTILITY

During storage in the male, and in some species at other times, spermatozoa conserve energy by remaining immotile; motility is then activated at an appropriate time.

What environmental signals control activation?

At what level is motility controlled?

What are the biochemical mechanisms?

Two environmental factors are particularly relevant: dilution and proximity to conspecific eggs. Dilution may coincide with the removal of inhibitory factors. Since the use of ATP by flagella is tightly coupled to flagellar bending (7,54), mechanical inhibition of bending when spermatozoa are tightly packed into storage ducts is a reasonable mechanism for inhibition, and will be relieved by dilution (55). Alternatively, dilution may cause activation by removal of a chemical inhibitor. Perhaps the best understood example of this is in salmonid fishes, where activation by dilution appears to result from removal of inhibition by $K^+$ ions (56,57); spermatozoa diluted into solutions containing 24 mM $K^+$ remained in a quiescent state. On the other hand, in sea urchin spermatozoa activation by dilution appears to require positive exposure to $Na^+$ ion; spermatozoa diluted into choline-sea water remain quiescent (58). $Na^+$ ions are also required for the action of "speract," a decapeptide obtained from sea urchin egg jelly that activates sea urchin sperm respiration and motility at pH 6.6 (59-61). In both cases, $Na^+$ appears to be required for a $Na^+$-$H^+$ exchange transport that causes $H^+$ efflux from the spermatozoa, and intracellular pH and $Ca^{2+}$ may be involved in regulation of motility (58,62). Regulation by internal pH might explain earlier observations suggesting that $CO_2$ inhibits spermatozoa during storage, but $CO_2$ can also inhibit the axoneme directly at higher pHs (16).

In contrast to sea urchin spermatozoa which become motile upon dilution into sea water at normal pH (8.2), spermatozoa of *Limulus* show only transient motility when diluted into sea water, and can then be activated by egg extracts (63-65). Activation by egg extracts has also been reported in other species, including the herring (66) and Ciona (38). In many of these cases, activation of nonmotile spermatozoa can be achieved with a variety of chemical agents—such as $Cu^{++}$ or $Ni^{++}$ ions in Limulus (64,67); and $NH_4^+$, histidine or theophylline in Ciona (68,69).

In some of these cases, it has been possible to distinguish between control at the metabolic level and at the level of axonemal motility. For instance, in Ciona spermatozoa, normal procedures for Triton-demembranation and ATP-reactivation do not work unless sperm motility is activated before demembranation (68). Similar results have been reported with trout spermatozoa (57) and starfish spermatozoa (Okuno, personal communication). On the other hand, activation of motility does not appear to be a prerequisite for demembranation and reactivation of spermatozoa of Limulus (Clapper, personal communication), sea urchins (70) and many other species.

In Ciona spermatozoa, this functional dissection has been taken one step further by demonstrating that demembranated spermatozoa prepared without prior activation of motility, so that they do not reactivate in ATP, do show active sliding disintegration following brief digestion with trypsin (68). Therefore, in this species, the activation of motility does not occur at the level of the energy supply or the active sliding process, but at higher level control mechanisms that are involved in converting active sliding into oscillation and bend propagation. The inactive state may therefore be closely related to the quiescent state that can be induced by $Ca^{2+}$.

In many cases, cAMP has been implicated in the activation of sperm motility. The cleanest results have been obtained with trout spermatozoa (57). In this case, demembranated spermatozoa prepared without prior activation of motility can be reactivated by $MgATP^{2-}$ if and only if the reactivation solution contains cAMP. Somewhat similar results were reported for Ciona spermatozoa, but the quality of the cAMP-stimulated motility was poor (68). Enhancement of the motility of reactivated spermatozoa by cAMP has been observed in mammalian spermatozoa (28,71). Increases in sperm cAMP content by motility-activating conditions have also been reported (cf. 72,73). These observations indicate that phosphorylation by the cAMP-dependent protein kinase is responsible for the activation of motility in some, if not all, species of spermatozoa, and the presence of this enzyme has been confirmed in several species.

The connection between protein phosphorylation and activation of axonemal bending is under active investigation. Many axonemal proteins of Chlamydomonas are phosphoproteins (74). Many peptides in both axonemal and detergent-solubilized fractions show cAMP-dependent phosphorylation in Ciona (Opresko and Brokaw, unpublished observations) and dog (75) spermatozoa. So far, the best correlations with activation of motility appear to involve cAMP-dependent phosphorylation of nonaxonemal proteins of bull (76) and sea urchin (77) spermatozoa. In the sea urchin system, there appears to be a soluble component that sensitizes the flagellum to $Ca^{2+}$-induced quiescence; this component appears to be inactivated by cAMP-dependent phosphorylation. If we are lucky, this work will not only provide us with an understanding of sperm activation, but also with an understanding of biochemical aspects of control mechanisms responsible for conversion of active sliding into oscillation and bend propagation in sperm flagella.

## 6. FINIS

Although this survey has been broken up into four sections, one of the more exciting conclusions is the interrelatedness of the control mechanisms that have been discussed. Control of sperm activation and oriented movement (chemotaxis) appear to interact with the basic mechanisms for oscillation and bend propagation. Many results appear to point to the use of ubiquitous biochemical processes—an ATP-driven active sliding process involving transient cross-bridging; and the use of $Ca^{2+}$ and cAMP as intracellular messengers. Does this reflect a real economy on the part of evolution, or just our limited abilities to detect new mechanisms? Only further work will tell. In spite of the complexity of the controls I have surveyed, only part of the story has been told. For instance, I do not know where to fit into this picture the evidence for a cholinergic control system (e.g., 78) or the various alterations in the motility of mammalian spermatozoa in different portions of the male and female reproductive tracts. In *Chlamydomonas*, we hope that genetic analysis and the analysis of the motility of mutant flagella will be a powerful tool for unraveling these problems. Genetic analysis of spermatozoan motility remains a challenge for the future.

## 7. ACKNOWLEDGEMENT

The author's work is supported by NIH research grants, GM 18711 and GM 21931.

## REFERENCES

1. Piperno, G, Huang, B, Luck, DJL. 1977. Two-dimensional analysis of flagellar proteins from wild-type and paralyzed mutants of Chlamydomonas reinhardii. Proc. Natl. Acad. Sci. USA 74: 1600-1604.
2. Brokaw, CJ, Gibbons, IR. 1975. Mechanisms of movement in flagella and cilia. In: Wu, TY-T, Brokaw, CJ, Brennan, C (eds.) "Swimming and Flying in Nature." New York, Plenum Publ. Corp. 89-126.
3. Sale, WS, Satir, P. 1977. The direction of active sliding of microtubules of *Tetrahymena* cilia. Proc. Natl. Acad. Sci. USA 74: 2045-2049.
4. Brokaw, CJ. 1971. Bend propagation by a sliding filament model for flagella. J. Exp. Biol. 55: 289-304.

5. Brokaw, CJ. 1972. Computer simulation of flagellar movement. I. Demonstration of stable bend propagation and bend initiation by the sliding filament model. Biophys. J. 12: 564-586.
6. Brokaw, CJ. 1982, in press. Models for oscillation and bend propagation by flagella. In: Amos, WB, Duckett, JG, Goldsmith, GJ (eds.) "Eukaryotic and Prokaryotic Flagella." Symp. Soc. Exp. Biol. 35.
7. Brokaw, CJ, Benedict, B. 1968. Mechanochemical coupling in flagella. I. Movement-dependent dephosphorylation of ATP by glycerinated spermatozoa. Arch. Biochem. Biophys. 125: 770-778.
8. Brokaw, CJ, Simonick, TF. 1977. Mechanochemical coupling in flagella. V. Effects of viscosity on movement and ATP dephosphorylation of Triton-demembranated sea urchin spermatozoa. J. Cell Sci. 23: 227-241.
9. Kamimura, S, Takahashi, K. 1981. Direct measurement of the force of microtubule sliding in flagella. Nature 293: 566-568.
10. Brokaw, CJ. 1975. Molecular mechanism for oscillation in flagella and muscle. Proc. Natl. Acad. Sci. USA 72: 3102-3106.
11. Brokaw, CJ. 1976. Computer simulation of flagellar movement. IV. Properties of an oscillatory two-state cross-bridge model. Biophys. J. 16: 1029-1041.
12. Brokaw, CJ, Rintala, D. 1977. Computer simulation of flagellar movement. V. Oscillation of cross-bridge models with an ATP-concentration-dependent rate function. J. Mechanochem. Cell Mot. 4: 205-232.
13. Hines, M, Blum, JJ. 1979. Bend propagation in flagella. II. Incorporation of dynein cross-bridge kinetics into the equations of motion. Biophys. J. 25: 421-442.
14. Brokaw, CJ. 1975. Effects of viscosity and ATP concentration on the movement of reactivated sea urchin sperm flagella. J. Exp. Biol. 62: 701-719.
15. Brokaw, CJ, Simonick, TF. 1976 $CO_2$ regulation of the amplitude of flagellar bending. In: Goldman, RD, Pollard, TD, Rosenbaum, J (eds.) "Cell Motility." Cold Spring Harbor Lab., Cold Spring Harbor, New York 933-940.
16. Brokaw, CJ. 1977. $CO_2$-inhibition of the amplitude of bending of Triton-demembranated sea urchin sperm flagella. J. Exp. Biol. 71: 229-240.
17. Asai, DJ, Brokaw, CJ. 1980. Effects of antibodies against tubulin on the movement of reactivated sea urchin sperm flagella. J. Cell Biol. 87: 114-123.
18. Asai, DJ, Brokaw, CJ, Harmon, RC, Wilson, L. 1982. Monoclonal antibodies to tubulin and their effects on the movement of reactivated sea urchin spermatozoa. Cell Mot. Suppl. 1: 175-180.
19. Brokaw, CJ. 1966. Effects of increased viscosity on the movements of some invertebrate spermatozoa. J. Exp. Biol. 45: 113-139.
20. Brokaw, CJ, Simonick, TF. 1977. Motility of Triton-demembranated sea urchin sperm flagella during digestion by trypsin. J. Cell Biol. 75: 650-665.
21. Okuno, M, Brokaw, CJ. 1979. Inhibition of movement of Triton-demembranated sea urchin sperm flagella by $Mg^{2+}$, $ATP^{4-}$, ADP and $P_i$. J. Cell Sci. 28: 105-123.
22. Brokaw, CJ. 1980. Elastase digestion of demembranated sperm flagella. Science 207: 1365-1367.
23. Yano, Y, Miki-Noumura, T. 1981. Recovery of sliding ability in arm-depleted flagellar axonemes after recombination with extracted dynein 1. J. Cell Sci. 48: 223-239.
24. Mitchell, DR, Warner, FD. 1980. Interactions of dynein arms with B subfibers of Tetrahymena cilia. Quantitation of the effects of magnesium and adenosine triphosphate. J. Cell Biol. 87: 84-97.
25. Gibbons, BH, Gibbons, IR. 1974. Properties of flagellar "rigor waves" formed by abrupt removal of adenosine triphosphate from actively swimming sea urchin sperm. J. Cell Biol. 63: 970-985.
26. Hines, M, Blum, JJ. 1982. Three-dimensional ciliary mechanics. Cell Mot. Suppl. 1: 153-158.

27. Olson, GE, Linck, RW. 1977. Observation of the structural components of flagellar axonemes and central pair microtubules from rat sperm. J. Ultrastruct. Res. 61: 21-43.
28. Mohri, H, Yano, Y. 1982. Reactivation and microtubule sliding in rodent spermatozoa. Cell Mot. Suppl. 1: 143-147.
29. Brokaw, CJ. 1965. Non-sinusoidal bending waves of sperm flagella. J. Exp. Biol. 43: 155-169.
30. Brokaw, CJ. 1970. Bending moments in free-swimming flagella. J. Exp. Biol. 53: 445-464.
31. Okuno, M, Hiramoto, Y. 1979. Direct measurements of the stiffness of echinoderm sperm flagella. J. Exp. Biol. 79: 235-243.
32. Okuno, M. 1980. Inhibition and relaxation of sea urchin sperm flagella by vanadate. J. Cell Biol. 85: 712-725.
33. Miller, RL. 1966. Chemotaxis during fertilization in the hydroid Campanularia. J. Exp. Zool. 162: 22-45.
34. Miller, RL. 1979. Sperm chemotaxis in the hydromedusae. I. Species specificity and sperm behavior. Mar. Biol. 53: 99-124.
35. Carre, D, Sardet, C. 1981. Sperm chemotaxis in siphonophores. Biol. Cell 40: 119-128.
36. Miller, RL. 1978. Chemotactic behavior of the sperm of chitons (Mollusca polyplacophora). J. Exp. Zool. 202: 203-212.
37. Miller, RL. 1981. Sperm chemotaxis occurs in echinoderms. Am. Zool. 21: 985.
38. Miller, RL. 1975. Chemotaxis of the spermatozoa of Ciona intestinalis. Nature 254: 244-245.
39. Miller, RL, Brokaw, CJ. 1970. Chemotactic turning behavior of Tubularia spermatozoa. J. Exp. Biol. 52: 699-706.
40. Miller, RL. 1975. Effect of calcium on Tubularia sperm chemotaxis. J. Cell Biol. 67: 285a.
41. Gibbons, BH, Gibbons, IR. 1980. Calcium-induced quiescence in reactivated sea urchin sperm. J. Cell Biol. 84: 13-27.
42. Okuno, M, Brokaw, CJ. 1981. Effects of Triton-extraction conditions on beat symmetry of sea urchin sperm flagella. Cell Mot. 1: 363-370.
43. Goldstein, SF. 1977. Asymmetric waveforms in echinoderm sperm flagella. J. Exp. Biol. 71: 157-170.
44. Brokaw, CJ. 1979. Calcium-induced asymmetrical beating of Triton-demembranated sea urchin sperm flagella. J. Cell Biol. 82: 401-411.
45. Okuno, M, Brokaw, CJ. 1981. Calcium-induced change in form of demembranated sea urchin sperm flagella immobilized by vanadate. Cell Mot. 1: 349-362.
46. Hyams, JS, Borisy, GG. 1978. Isolated flagellar apparatus of Chlamydomonas: Characterization of forward swimming and alteration of waveform and reversal of motion by calcium ions in vitro. J. Cell Sci. 33: 235-253.
47. Brokaw, CJ, Luck, DJL, Huang, B. 1982. Analysis of the movement of Chlamydomonas flagella: The function of the radial-spoke system is revealed by comparison of wild-type and mutant flagella. J. Cell Biol. 92: 722-732.
48. Gibbons, BH. 1980. Intermittent swimming in live sea urchin sperm. J. Cell Biol. 84: 1-12.
49. Garbers, DL, Hansbrough, JR, Radany, EW, Hyne, RV, Kopf, GS. 1980. Purification and characterization of calmodulin from sea urchin spermatozoa. J. Reprod. Fert. 59: 377-381.
50. Gitelman, SE, Witman, GB. 1980. Purification of calmodulin from Chlamydomonas: Calmodulin occurs in cell bodies and flagella. J. Cell Biol. 87: 764-770.
51. Jamieson, GA, Jr, Vanaman, TC, Blum, JJ. 1979. Presence of calmodulin in Tetrahymena. Proc. Natl. Acad. Sci. USA 76: 6471-6475.
52. Witman, GB, Minervi, N. 1982. Role of calmodulin in the flagellar axoneme: Effect of phenothiazines on reactivated axonemes of Chlamydomonas. Cell Mot. Suppl. 1: 199-204.

53. Reed, W, Satir, P. 1980. Trifluoperazine inhibits mussel gill lateral cell ciliary arrest. J. Cell Biol. 87: 39a.
54. Brokaw, CJ, Benedict, B. 1968. Mechanochemical coupling in flagella. II. Effects of viscosity and thiourea on motility and metabolism of Ciona spermatozoa. J. Gen. Physiol. 52: 283-299.
55. Gray, J. 1928. The effect of dilution on the activity of spermatozoa. Brit. J. Exp. Biol. 5: 337-344.
56. Benau, D, Terner, C. 1980. Initiation, prolongation, and reactivation of the motility of salmonid spermatozoa. Gamete Res. 3: 247-257.
57. Morisawa, M, Okuno, M. 1982. Cyclic AMP induces maturation of trout sperm axoneme to initiate motility. Nature 295: 703-704.
58. Lee, HC, Schuldiner, S, Johnson, C, Epel, D. 1980. Sperm motility initiation: Changes in intracellular pH, $Ca^{2+}$ and membrane potential. J. Cell Biol. 87: 39a.
59. Hansbrough, JR, Garbers, DL. 1981. Sodium-dependent activation of sea urchin spermatozoa by speract and monensin. J. Biol. Chem. 256: 2235-2241.
60. Suzuki, N, Nomura, K, Ohtake, H, Isaka, S. 1981. Purification and the primary structure of sperm-activating peptides from the jelly coat of sea urchin eggs. Biochem. Biophys. Res. Comm. 99: 1238-1244.
61. Garbers, DL, Watkins, HD, Hansbrough, JR, Smith, A, Misono, KS. 1982. The amino acid sequence and chemical synthesis of speract and of speract analogues. J. Biol. Chem. 257: 2734-2735.
62. Christen, R, Schackmann, RW, Shapiro, BM. 1981. Coupling of sperm respiration, motility, and intracellular pH ($pH_i$) in Strongylocentrotus purpuratus. J. Cell Biol. 91: 174a.
63. Clapper, DL, Brown, GG. 1980. Sperm motility in the horseshoe crab, Limulus polyphemus L. I. Sperm behavior near eggs and motility initiation by egg extracts. Devel. Biol. 76: 341-349.
64. Clapper, DL, Brown, GG. 1980. Sperm motility in the horseshoe crab, Limulus polyphemus L. II. Partial characterization of a motility initiating factor from eggs and the effects of inorganic cations on motility initiation. Devel. Biol. 76: 350-357.
65. Clapper, DL, Epel, D. 1981. Isolation and utilization of a sperm motility initiating peptide in the horseshoe crab, Limulus polyphemus: Evidence for involvement of $Ca^{2+}$ but not intracellular pH on membrane potential in motility initiation. J. Cell Biol. 91: 179a.
66. Yanagimachi, R. 1957. Some properties of the sperm-activating factors in the micropyle area of the herring egg. Anot. Jap. 30: 114-119.
67. Bishop, DW, Nidek, DJ. 1977. Reversible regulation of sperm motility. Fed. Proc. 36: 371.
68. Brokaw, CJ. 1982. Activation and reactivation of Ciona spermatozoa. Cell Mot. Suppl. 1: 185-189.
69. Woollacott, RM. 1977. Spermatozoa of Ciona intestinalis and analysis of ascidian fertilization. J. Morph. 152: 77-88.
70. Gibbons, BH, Gibbons, IR. 1972. Flagellar movement and adenosine triphosphatase activity in sea urchin sperm extracted with Triton X-100. J. Cell Biol. 54: 75-97.
71. Lindemann, CB. 1978. A cAMP-induced increase in the motility of demembranated bull sperm models. Cell 13: 9-18.
72. Kopf, GS, Tubb, DJ, Garbers, DL. 1979. Activation of sperm respiration by a low molecular weight egg factor and by 8-bromoguanosine 3',5'-monophosphate. J. Biol. Chem. 254: 8554-8560.
73. Garbers, DL, Kopf, GS. 1980. The regulation of spermatozoa by calcium and cyclic nucleotides. Adv. Cyclic Nucleotide Res. 13: 251-306.
74. Piperno, G, Huang, B, Ramanis, Z, Luck, DJL. 1981. Radial spokes of Chlamydomonas flagella: Polypeptide composition and phosphorylation of stalk components. J. Cell Biol. 88: 73-79.
75. Tash, JS, Means, AR. 1982, in press. Regulation of protein phosphorylation and motility of sperm by cAMP and calcium. Biol. Reprod.

76. Brandt, H, Hoskins, DD. 1980. A cAMP-dependent phosphorylated 'motility protein in bovine epididymal sperm. J. Biol. Chem. 255: 982-987.

77. Ishiguro, K, Murofushi, H, Sakai, H. 1982. Evidence that cAMP-dependent protein kinase and a protein factor are involved in reactivation of Triton X-100 models of sea urchin and starfish spermatozoa. J. Cell Biol. 92: 777-782.

78. Nelson, L. 1976. α-Bungarotoxin binding by cell membranes: Blockage of sperm motility. Exp. Cell Res. 101: 221-224.

# SPERM MOTILITY: MECHANISMS AND CONTROL

I.R. GIBBONS, Pacific Biomedical Research Center, University of Hawaii, Honolulu, Hawaii and Istituto di Zoologia, Università di Siena, Siena, Italy.

I was asked by the Organizing Committee to present a general review of the area of mechanisms and control of sperm flagellar motility in order to provide an introduction to the more detailed papers which follow. Since this is a large area and I could not possibly cover it comprehensively in the time available, I shall devote most of my attention to aspects that are developing particularly rapidly at the present time and will mention other aspects only briefly. For more general accounts reference may be made to the recent reviews of Blum and Hines (1) and Gibbons (2).

In spermatozoa of most animal species, motility is the result of the undulatory beating of a cylindrical axoneme, consisting of the well known 9+2 flagellar tubules and a fairly complex array of associated accessory structures, that runs the length of the flagella. Basic hydrodynamic theory indicates that stable wave propagation requires that energy must be fed in along most of the length of the flagellum (3); this energy is normally provided by ATP which is generated by mitochondria in the basal or mid regions and which passes distally by diffusion within the space delimited by the flagellar membrane. In sperm of many species, it has been possible to render the motile mechanism directly accessible to experimentation by solubilizing selectively the sperm membranes with a non-ionic detergent, and then reactivating normal flagellar movement by placing the demembranated spermatozoa in a saline medium containing exogenous $MgATP^{2-}$. In such preparations the rate of ATP hydrolysis can be correlated with their motility, and with sea urchin spermatozoa ATP hydrolysis diminishes to about 15-20% of its maximal level when the flagella are rendered non-motile by brief homogenization or by inhibition with $HCO_3^-$ (4,5).

Recent work has suggested that the efficiency of mechanochemical energy conversion may be increased by having an organic acid, such as acetate or glutamate, as the predominant anion in the reactivation medium (6) and this may be related to the fact that such anions tend to maintain tubulin in its polymerized microtubule form (7). The conditions for effective reactivation of mammalian sperm flagella with their 9+9+2 structure are somewhat more critical than those for most invertebrate sperm flagella with a 9+2 structure; however under the proper conditions bull spermatozoa reactivated at 25°C progress forward at about 50 $\mu m\ s^{-1}$ with a flagellar beat frequency of 10 Hz and a roll frequency of 4 Hz, approximately the same as the corresponding values for the live spermatozoa at the same temperature (8). The reactivation of such demembranated mammalian sperm flagella may help to clarify the nature of the changes in motility that occur during sperm maturation in the epididymis (9), as well as those occurring during capacitation in the female genital tract.

In the basic mechanism of flagellar motility, the primary event appears to be the generation of shear stress (i.e. a longitudinally directed force) between adjacent outer doublet tubules by an ATP-driven cross-bridge cycle in which the two rows of dynein arms on the A-tubule of each doublet detach transiently and reattach to successive binding sites along the B-tubule of the adjacent doublet. This longitudinal shear stress can be visualized most directly in the sliding disintegration of trypsin-treated flagellar axonemes when they are exposed to ATP (10). In such preparations, the rate of sliding between adjacent doublets is 10-15 $\mu m\ sec^{-1}$ in 1 mM ATP (11,12), and examination of the slid preparations by electron microscopy shows that the usual direction of sliding is such that arms on the A-tubule of each doublet move the adjacent doublet distally from the former's base toward its tip (13).

The detailed functional substructure of the dynein arms has become a subject of intensive investigation over the past 2-3 years. The outer dynein arms can be solubilized from sea urchin sperm flagella as 21S particles of molecular weight 1,250,000 composed of at least 9 distinct polypeptides and having 2 sites of ATPase activity associated with different high molecular weight polypeptides (14-16); these 21S outer arm particles are capable of rebinding in a functionally active manner, as indicated by their ability to increase the beat frequency of dynein-depleted sperm flagella (17). In Chlamydomonas, the solubilized outer arms appear to consist of a mixture of 18S and 12S ATPases containing a total of about 12 distinct polypeptides, and they are capable of rebinding only when both forms of ATPase are present (18-21). While the general form of the ATP-driven mechanochemical cycle of the dynein arms appears to resemble that of the myosin crossbridges in muscle (22-25), the much greater size and polypeptide complexity of the dynein arms suggests that this parallel will be, at best, a substantial simplification. The recent development of techniques for studying solubilized dynein arms by quantitative scanning transmission electron microscopy (26), for using quick freezing to preserve dynein arm structure for electron microscopy in unfixed axonemes in known chemical and physiological states (27,28), as well as the development of monoclonal antibodies directed against particular polypeptides in the dynein arms (29), suggest that more complete information regarding the mechanochemistry of the dynein arms may soon be available.

The longitudinal shear stress generated by the dynein arms is converted into the transverse bending moments required for the generation and propagation of undulatory waves by the resistances to sliding resulting from the various accessory structures that interconnect the doublet tubules. The factors most important in the conversion of sliding into bending appear to be the localized block to sliding at the flagellar base, due to the centriole equivalent, and the distributed resistance along the flagellar length that results from the elastic-like circumferential links ("nexin") between adjacent doublets (30,31). In addition, the flexible stalk structures recently described as one of the components of the dynein arm (26-28) may also constitute interconnections between doublets that, under some conditions, constitute a passive elastic resistance to sliding. If this proves to be the case, each arm, in addition to generating a shear stress when active, would also have built into it a passive elastic component that tended to resist the resultant shear strain and convert it into a bending moment.

The radial spokes that connect each of the doublets to one of the two halves of the central sheath were earlier considered as the primary resistive structures converting sliding into bending (32), but recent observations indicate that mutant Chlamydomonas flagella with defective radial spokes in which the doublets are not connected to the central sheath are nevertheless still capable of forming and propagating modified planar bending waves (33). This suggests that the normal function of the radial spokes and of the 2 central tubules is to strengthen the axonemal structure and to coordinate the activity of the dynein arms to produce particular patterns of beating. The occurrence of species with motile sperm flagella having reduced 9+0, 6+0, and 3+0 axonemal structures, lacking both the radial spokes and central tubules, indicates that these structures are not essential for symmetric helicoidal bending waves adequate for sperm propulsion (Table I; see also ref. 38). On the other hand, motile cilia which typically have a much longer functional life-span and have an asymmetric beat pattern that requires synchronous sliding along almost the full length of the individual doublet tubules, appear always to possess a complete 9+2 axonemal structure, including both the radial spokes and central tubules, even in those species in which the sperm flagellar structure is greatly reduced (39,40).

The function of the additional 9 peripheral dense fibers that surround the 9+2 axonemal core in mammalian sperm flagella has been the subject of some controversy. A variety of evidence favors the view that these fibers are passive structures whose function is to strengthen the flagellum and the attachment of the sperm head to the flagellum. The work of Phillips (41) has shown that the sperm flagella of mammalian species in which the peripheral fibers are thickest have a relatively low amplitude of bending in the proximal region, suggesting that these fibers are major factors in flagellar stiffness; the isolated thick fibers are composed

Table I. Motility in flagella with Reduced Structure

| Organism | Structure | Spokes | Dynein Arms outer | Dynein Arms inner | Beat Frequency (Hz)[a] | Form of Beat | Reference |
|---|---|---|---|---|---|---|---|
| Sea urchin | 9+2 | + | + | + | 46 | planar | 4 |
| " | 9+2 | + | + | + | 32[b] | planar | 4 |
| " | 9+2 | + | − | + | 17[b] | planar | 34 |
| Eel | 9+0 | − | − | + | 112 | helicoidal | 35 |
| Gregarine | 6+0 | − | − | + | 0.5 | helicoidal | 36 |
| Gregarine | 3+0 | − | ? | ? | 1.5 | helicoidal | 37 |
| Chlamydomonas wt | 9+2 | + | + | + | 65 | planar | 33 |
| Chlamydomonas pf 17, sup 3 | 9+2 | (−)[c] | + | + | 35 | planar | 33 |

a, mostly at 25°C

b, demembranated flagella reactivated in 1 mM ATP

c, portion of spoke structure remains, but it is relieved to be non-functional because spoke-head is absent.

Table II. Power output of sperm flagella

| Organism | Temperature (0°C) | Beat frequency (Hz) | Power output (per beat) | (per sec) |
|---|---|---|---|---|
| Sea Urchin | 16 | 30 | 1.4 | 42 |
| Ciona | 16 | 35 | 2.0 | 70 |
| Chaetopterus | 16 | 26 | 0.8 | 21 |
| Bull | 37 | 22 | 6.9 | 164 |

Power output is expressed in units of $10^{-17}$ J/sperm/beat and $10^{-17}$ J/sperm/sec.
For full details see references 44-47.

of a keratin-like protein with no apparent enzymatic activity (42,43); the greater power output of mammalian sperm flagella compared to that of sperm flagella with the simpler 9+2 pattern (Table II; see also ref. 40) could be accounted for by the 9+2 axoneme in mammalian sperm functioning at a higher temperature (37°C) without need for a supplemental motile mechanism associated with the thick fibers. The principal objection to attributing all the active force generation in mammalian sperm to dynein arms has been that such small structures located close to the central axis would be unable to generate a sufficient bending moment to account for the observed bending of the peripheral fibers, as well as that of the mitochondrial helix and the fibrous sheath. Although in the past this objection has usually been raised in qualitative terms, the recent development of procedures for measuring the actual force generated by the dynein arms (48,49) and for measuring directly the stiffness of sperm flagella under known physiological conditions (50-52) provide the basis for obtaining a quantitative answer to the question.

Small amounts of actin have been found associated with the inner dynein arm in algal flagellar axonemes (53). However, most of the actin and myosin reported to be present in spermatozoa appears localized in the acrosomal region of the head (54-59) or in the vicinity of the centriole, rather than in the flagellum. There is no evidence for either actin or myosin playing a significant role in sperm flagellar movement.

There appear to be two distinct control mechanisms that modulate the motility of sperm flagella in at least some species. A cAMP-dependent mechanism is involved with the initial activation of sperm motility and may also regulate the vigor of beating, whereas a $Ca^{2+}$-dependent mechanism modulates the asymmetry of flagellar beating and is involved in the mechanism of chemotaxis. The cAMP-dependent control mechanism involved in the initial activation of sperm flagellar motility is seen particularly clearly in fish sperm, where the natural trigger for activation is the change in the osmotic or ionic composition of the medium surrounding the sperm, from that of seminal plasma to that of the external environment, which occurs upon shedding (60). The involvement of cAMP in the activation process has been shown directly for trout sperm, in which sperm that have already been activated normally by dilution are immediately motile when demembranated with Triton X-100 and placed into a medium containing $MgATP^{2-}$, whereas sperm that have not been activated by dilution prior to demembranation remain non-motile under otherwise equivalent conditions but can be made to become motile by addition of cAMP to the demembranation or $MgATP^{2-}$-containing solution (61,62). Similar evidence for the involvement of cAMP in initial activation has been obtained for various salmonid fish by Benau and Terner (63), and for the invertebrate Ciona by Brokaw (64).

In mammalian spermatozoa, the process corresponding to the activation of motility in fish sperm appears to be the gradual maturation of the sperm that occurs during their transit of the epididymis. Sperm taken directly from the testis are non-motile and those from the caput of the epididymis show minimal motility in most species, whereas sperm from the cauda epididymis have about the same high capacity for motility as ejaculated sperm (although they are maintained non-motile during storage in the epididymis, presumably by the high $CO_2$ tension). The involvement of cAMP in the epididymal maturation process was first identified by the activating effects of dibutyryl cAMP and of such phosphodiesterase inhibitors as caffeine and theophylline on the motility of bovine sperm from the caput epididymis (65).

The action of cAMP involves activation of a cAMP-dependent protein phosphokinase (66,67). However, the maturational process of sperm in the epididymis is a complex process involving changes in their sensitivity to intracellular levels of $Ca^{2+}$ and the action of a "forward motility protein" which by increasing the roll frequency of sperm, decreases their tendency to swim in circles and results in much faster forward progression (68-70). At least part of the maturational process appears to involve changes in the properties of the plasma membrane, for the difference between the motility of testicular caput and cauda sperm is diminished when demembranated sperm reactivated with $MgATP^{2-}$ are utilized (9). However, in at least some species, the demembranated sperm show significant enhancement of motility when incubated with cAMP (71), and these changes in motility appear to parallel changes in the phosphorylation state of certain flagellar polypeptides (72).

Sperm of many invertebrate classes show chemotactic attraction toward the eggs of the same species (73). The basis of attraction is the occurrence of brief episodes of asymmetric flagellar beating which result in abrupt changes in the direction of forward progression so that the sperm move gradually closer to the egg and once in its vicinity remain there (74,75). The basis for the episodes of asymmetric beating appears to be a brief influx of $Ca^{2+}$ ions into the cell when it detects a decreasing concentration of the chemoattractant. Concentrations of $Ca^{2+}$ above $10^{-6}$ M cause increasing asymmetry in demembranated sperm reactivated with $MgATP^{2-}$ (76). The mechanism by which $Ca^{2+}$ increases the asymmetry of flagellar beating is not yet clear (77-79). While there are large amounts of calmodulin typically present in cilia and sperm flagella (81-83), there is little evidence linking it to the mechanism of asymmetry; and micromolar concentrations of $Ca^{2+}$ appear also to have a direct effect on tubulin subunit packing within the flagellar tubules (84).

---

Supported in part by NIH grant HD-06565.

## REFERENCES


1. Blum JJ, Hines M. 1979. Quart. Revs. Biophys. 12, 103-180.
2. Gibbons IR. 1981. J. Cell Biol. 91, 107s-124s.
3. Machin KE. 1958. J. Exp. Biol. 35, 796-806.
4. Gibbons BH, Gibbons IR. 1972. J. Cell Biol. 54, 75-97.
5. Brokaw CJ, Simonick TF. 1975. In Cell Motility, ed. R. Goldman, T. Polland, J. Rosenbaum. Cold Spring Harbor Laboratory, Cold Spring Harbor, N.Y.
6. Gibbons IR, Evans JA, Gibbons BH. 1982. Cell Motil. in press.
7. Hamel E, Lin CM. 1981. Biochim. Biophys. Acta 675, 226-231.
8. Lindemann CB, Gibbons IR. 1975. J. Cell Biol. 65, 147-162.
9. Mohri H, Yanagamachi R. 1980. Exp. Cell Res. 127, 191-196.
10. Summers KE, Gibbons IR. 1971. Proc. Natl. Acad. Sci. USA 68, 3092-3096.
11. Gibbons IR. 1975. In The Functional Anatomy of the Spermatozoon, ed. B.A. Afzelius, Pergamon Press, Oxford, pp. 127-140.
12. Yano Y, Miki-Noumura T. 1980. J. Cell Sci. 44, 169-186.
13. Sale WS, Satir P. 1977. Proc. Natl. Acad. Sci. USA 74, 2045-2049.
14. Gibbons IR, Fronk E. 1979. J. Biol. Chem. 254, 187-196.
15. Bell CW, Fronk E, Gibbons IR. 1979. J. Supramol. Struct. 11, 311-317.
16. Tang W-JY, Bell CW, Sale WS, Gibbons IR. 1982. J. Biol. Chem. 257, 508-515.
17. Gibbons BH, Gibbons IR. 1979. J. Biol. Chem. 254, 197-201.
18. Huang B, Piperno G, Luck DJL. 1979. J. Biol. Chem. 254, 3091-3099.
19. Piperno G, Luck DJL. 1979. J. Biol. Chem. 254, 3084-3090.
20. Fay RB, Witman GB. 1977. J. Cell Biol. 75, 286a.
21. Fay RB, Witman GB. 1982. Cell Motil. 2, suppl. (in press).
22. Sale WS, Gibbons IR. 1979. J. Cell Biol. 82, 291-298.
23. Satir P, Wais-Steider J, Lebduska S, Nasr A, Avolio J. 1981. Cell Motil. 1, 303-307.
24. Johnson KA, Porter ME. 1982. Biophys. J. 83, 345a.
25. Gibbons IR. 1982. Cell Motil. 2, suppl. (in press).
26. Johnson KA, Wall JS. 1982. Biophys. J. 37, 345a.
27. Heuser JE, Goodenough UW. 1981. J. Cell Biol. 91, 49a.
28. Goodenough UW, Heuser JE. 1982. J. Cell Biol. (in press).
29. Piperno G, Fleming J. 1983. J. Submicrosc. Cytol. (in press).
30. Gibbons IR. 1963. Proc. Natl. Acad. Sci. USA 50, 1002-1010.
31. Summers KE, Gibbons IR. 1973. J. Cell Biol. 58, 618-629.
32. Warner FD, Satir P. 1974. J. Cell Biol. 63, 35-63.
33. Brokaw CB, Luck DJL, Huang B. 1982. J. Cell Biol. 92, 722-732.
34. Gibbons BH, Gibbons IR. 1973. J. Cell Sci. 13, 337-357.
35. Gibbons BH, Gibbons IR, Baccetti B. 1983. J. Submicrosc. Cytol. (in press).
36. Schrevel J, Besse C. 1975. J. Cell Biol. 66, 492-507.
37. Prensier G, Vivier E, Goldstein S, Schrevel J. 1980. Science 207, 1493-1494.
38. Crowley PH, Benham CJ, Lenhart SM, Morgan JL. 1981. J. Theoret. Biol. 93, 769-784.

39. Silviera M, Porter KR. 1964. Protoplasma 59, 240-265.
40. Baccetti B, Burrini AG, Pallini V. 1981. J. Submicrosc. Cytol. 13, 479-481.
41. Phillips DM. 1972. J. Cell Biol. 53, 561-573.
42. Baccetti B, Pallini V, Burrini AG. 1973. J. Submicrosc. Cytol. 5, 237-256.
43. Calvin HI, Hwang FH-F, Wohlrab H. 1975. Biol. Reprod. 13, 228-239.
44. Brokaw CJ. 1965. J. Exp. Biol. 43, 155-169.
45. Brokaw CJ, Gibbons IR. 1975. In Swimming and Flying in Nature, ed. T.Y. Wu, C.J. Brokaw, C. Brennan, Plenum Publ. Corp. N.Y., pp. 89-126.
46. Rikmenspoel R, Sinton S, Janick JJ. 1969. J. Gen. Physiol. 54, 782-805.
47. Rikmenspoel R, Jacklet AC, Orris SE, Lindemann CB. 1973. J. Mechanochem. Cell Motil. 2, 7-24.
48. Kamimura S, Takahashi K. 1981. Nature 293, 566-568.
49. Takahashi K, Kamimura S. 1983. J. Submicrosc. Cytol. (in press).
50. Lindemann CB, Rudd WG, Rikmenspoel R. 1973. Biophys. J. 13, 437-448.
51. Okuno M, Hiramoto Y. 1979. J. Exp. Biol. 79, 235-243.
52. Okuno M. 1980. J. Cell Biol. 85, 712-725.
53. Piperno G, Luck DJL. 1979. J. Biol. Chem. 254, 2187-2190.
54. Tilney LG, Hatano S, Ishikawa H, Mooseker M. 1973. J. Cell Biol. 59, 109-126.
55. Tilney LG. 1975. J. Cell Biol. 64, 289-310.
56. Mabuchi I. 1976. J. Bichem. (Tokyo) 80, 413-415.
57. Clarke GN, Yanagamachi R. 1978. J. Exp. Zool. 205, 125-132.
58. Tamblyn TM. 1980. Biol. Reprod. 22, 727-734.
59. Tamblyn TM. 1981. Gamete Res. 4, 499-506.
60. Morisawa M, Suzuki K. 1980. Science 210, 1145-1147.
61. Morisawa M, Okuno M. 1982. Nature 295, 703-704.
62. Morisawa M, Okuno M, Suzuki K, Morisawa S, Ishida K. 1983. J. Submicrosc. Cytol. (in press).
63. Benau D, Terner C. 1980. Gamete Res. 3, 247-257.
64. Brokaw CJ. 1982. Cell Motil. 2, suppl. (in press).
65. Garbers DL, Lust WD, First NL, Lardy HA. 1971. Biochemistry 10, 1825-1831.
66. Hoskins DD, Casillas ER, Stephens DT. 1972. Biochem. Biophys. Res. Commun. 48, 1331-1338.
67. Garbers DL, First NL, Lardy HA. 1973. J. Biol. Chem. 248, 875-879.
68. Acott TS, Hoskins DD. 1981. Biol. Reprod. 24, 234-240.
69. Hoskins DD, Acott TS, Critchlow L, Vijayaraghavan S. 1983. J. Submicrosc. Cytol. (in press).
70. Acott TS, Hoskins DD. 1983. J. Submicrosc. Cytol. (in press).
71. Lindemann CB (1978). Cell 13, 9-18.
72. Tash JS, Means AR. 1981. J. Cell Biol. 91, 335a.
73. Miller RL. 1977. In Advances in Invertebrate Reproduction, Vol. I, ed. K.G. Adiyodi, R.G. Adiyodi. Peralam Kenoth Press, Kerala, India.
74. Miller RL, Brokaw CJ. 1970. J. Exp. Biol. 52, 699-706.
75. Gibbons BH. 1980. J. Cell Biol. 84, 1-12.

76. Brokaw CJ, Josslin R, Bobrow L. 1974. Biochem. Biophys. Res. Commun. 58, 795-800.
77. Brokaw CJ. 1979. J. Cell Biol. 82, 401-411.
78. Gibbons IR. 1982. In Eukaryotic and Prokaryotic Flagella, eds. W.B. Amos, J.G. Duckett, Symp. Soc. Exp. Biol. 35, 203-224.
79. Cosson MP, Carré D, Cosson J, Sardet C. 1983. J. Submicrosc. Cytol. (in press).
80. Blum JJ, Hayes A, Jamieson GA, Vanaman TC. 1980. J. Cell Biol. 87, 386-397.
81. Garbers DL, Hansbrough JR, Radany EW, Hyne RV, Kopf GS. 1980. J. Reprod. Fert. 59, 377-381.
82. Gitelman SE, Witman GB. 1980. J. Cell Biol. 98, 764-770.
83. Takasaki Y, Miki-Noumura T. 1982. J. Mol. Biol. 158, 317-324.

# A Study of the Motility of Limulus Polyphemus Spermatozoa Using Quasielastic Light Scattering and Cinematography

Tom Craig and F. R. Hallett
Physics Department
University of Guelph
Guelph, Ontario, Canada

## Introduction

Previous studies on the sperm of the Limulus have concentrated on morphology (André, 1963), immunology (Cooper and Brown, 1972), sperm-egg interactions (Shoger and Brown, 1970), fertilization (Brown and Krouse, 1973) and motility initiation (Clapper and Brown, 1980a and b). This study however examines the type and quality of motility after initiation using cinematography and quasi-elastic light scattering.

## Materials and Methods

Both sperm and eggs were collected from Limulus using electrostimulation (0.2 mA, ~10 V ac). All sperm were diluted in artificial sea water before study. Experiments were always performed within two hours of collection.

The cinematography was performed in phase contrast using a Locam high speed movie camera mounted on a Zeiss microscope and analyzed on a motion analyzing projector. The quasi-elastic light scattering was performed using a standard apparatus (Hallett et al., 1978) at a scattering angle of 15°. The correlation functions were gathered using a Langley-Ford Model 64 autocorrelator. No sophisticated analysis of the quasi-elastic light scattering data was undertaken; rather, attempts were made to relate the linewidths of the data to the overall swimming speed of the spermatozoa and to characterize the shapes of autocorrelation functions from motile and dead samples.

Two sets of experiments were performed at room temperature (~21°C). Firstly, motility was initiated by the presence of an egg. In these experiments only cinematography was performed with subsequent measurement of size parameters for the head of the spermatozoa. Here, an egg was placed in the center of a welled microscope slide then a sperm suspension was introduced around it. Secondly, experiments were performed where motility had been initiated by the sea water itself, presumably by

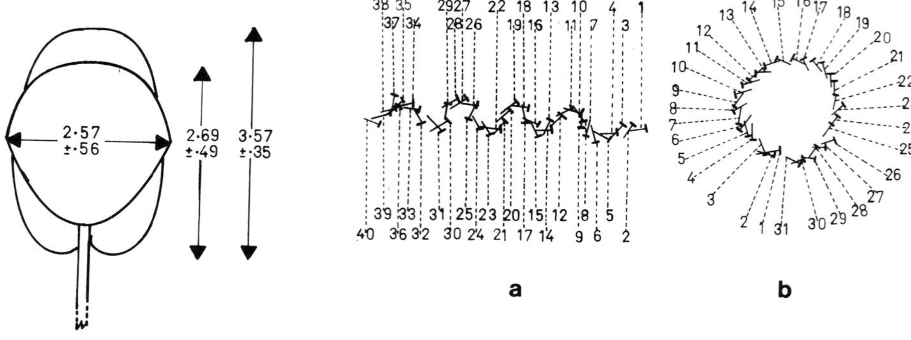


Fig. 1. Sketch of the spermatozoon head with the average measurements show.

Fig. 2. Tracings of head positions of a) normally and b) defectively swimming spermatozoa for successive frames at ~50 frames/s. The long line lies along the midline of the head while the cross-bar indicates the dividing line between the head and the tail. These were both taken using film speeds of ~50 frames/s.

divalent cations. More detailed measurements of speeds and trajectories were undertaken and light scattering was done at the same time as the cinematography.

Results

Cinematography

Egg Initiation

Initially hemocytometer counts were performed and it was determined that the raw semen had a sperm concentration of about $1.5 \times 10^9$ cells/ml. This indicated that a dilution of 0.1 ml semen in 5.0 ml sea water would be appropriate for doing the light scattering work so it was incorporated for this initial cinematography. Cinematography was difficult to do on this system due to depth of field problems posed by the introduction of the egg. However, it was possible to make some size measurements of the head of the sperm. It was found that although the head area of the sperm was axially symmetric it was not spherical. As shown in fig. 1 it had an overall length of $3.57 \pm 0.35$ μm, and a width of $2.57 \pm 0.49$ μm.

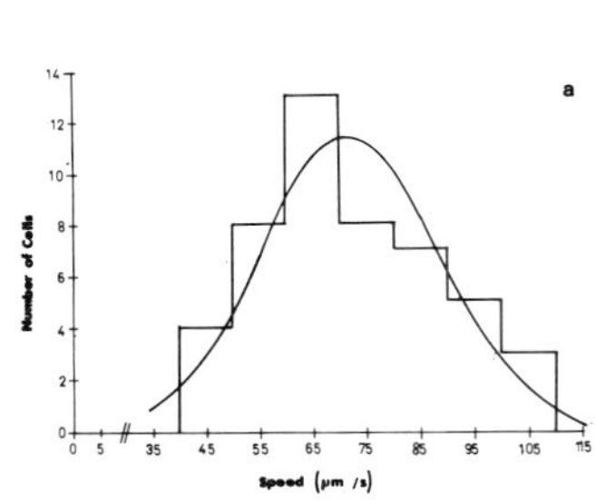


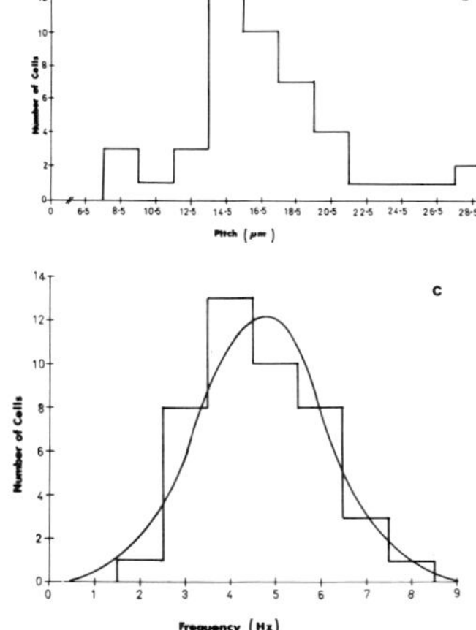


Fig. 3. Histograms of the parameters of the motion of normally swimming cells as taken from the cinematographic film for a) speed of the cells b) pitch of the helix c) frequency of the helix.

It was noticed that the average trajectory of the sperm was usually a straight line. The detailed trajectory appeared helical in nature. Due to the depth of field problems however trajectories could not be determined nor speeds measured with any confidence.

Contaminant Initiation

In the subsequent study it was found that the sperm were motile upon dilution in the artificial sea water. This allowed for the determination of dynamical information from the cinematography. Cinematography and coincident light scattering were performed on these samples although a further dilution by a factor of three was necessary for the light scattering.

From the cinematographic results it was noticed that two types of motility were present in the sample; namely, sperm which swam in straight lines along helical paths and those which swam in circles. Two examples of these tracks are shown in fig. 2. The straight line swimming sperm will be referred to as the normal cells while the circularly swimming cells will be referred to as defective.

In the normal cell case their linear speed, the pitch of the helix (i.e. the space-extent of the helix), the frequency of the helix (the time-extent) and the diameter of the helix were measured. The averages and

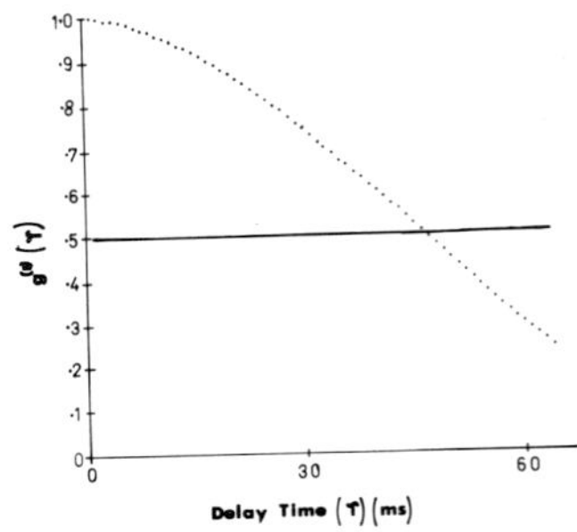


Fig. 4. An autocorrelation function taken from a sample of Limulus sperm. 40% of a non-motile sample has been subtracted.

standard deviations for these determinations were found to be 71.8 ± 16.4 μm/s, 16.7 ± 4.66 μm/s, 4.7 ± 1.4 Hz and 4.4 ± 1.1 μm respectively. Histograms of the values for individual cells were also plotted and are shown in fig. 3. Superimposed on the speed and frequency histograms are normal distributions with the same mean and standard deviations as those above and described by

$$P(x) = \frac{1}{2\pi\sigma_x} e^{-\frac{(x-\bar{x})^2}{2\sigma_x^2}} \tag{1}$$

where x represents either the speed or frequency. The agreement is not as good as one would like due to the apparent skewing in the histograms. This is probably due to an observer bias against low speeds and frequencies.

Unfortunately it was not possible to obtain data from enough defective cells to construct histograms. However, from the limited data available an average speed around the circle of 86.8 ± 30.8 μm/s was obtained, an average radius for the circle of 9.24 ± 0.42 μm and an average frequency of 1.49 ± 0.42 Hz. The diameter of this circle is very small compared to other species such as the bull or ram.

## Light Scattering

Unlike the spermatozoa of species such as the bull where the halfwidth of the correlation functions are primarily determined by orientational effects and the frequency of rotation of the cell (Craig et al., 1979) in the case of the sperm of Limulus this halfwidth is determined primarily by the speed because of the axial symmetry of the cell. As well, since the cell is not spherical, even though the swimming speeds are comparable with those of the bull one obtains light scattering data with halfwidths that are about 100 times larger than those from the bull. This is due to the fact that one preferentially observes those cells whose velocity components along the scattering vector are small (Chen and Hallett, 1982).

Figure 4 shows light scattering data taken at a scattering angle of 15°. This data was obtained from a mixture of motile and non-motile cells by subtracting a portion of the non-motile cell function to obtain the true scattering function from motile cells. This fraction of non-motile cells was estimated from the cinematographic work. The halfwidth of this function is directly related to the average speed of the spermatozoa in the sample and the shape of the curve is related to the shape of the speed distribution of the cells. Rayleigh-Gans-Debye models of the cell, however, are required in order to convert the halfwidths into speeds. After this is done it would be very easy to follow the motility during and immediately after initiation.

## Conclusions

It is interesting to note the similarities in type of motility of a so-called primitive spermatozoa (*Limulus*) and a more advanced one (bull). Both of these systems show normal (helical) swimming motion as well as defective (circular) swimming motion. In the future it would be interesting to study the type of motility obtained when initiation was undertaken using the methods of Clapper and Brown (1980a and b) and to examine the mechanisms for these different motility patterns using cine and quasi-elastic light scattering.

## Acknowledgements

We would like to thank Dr. G. H. Renninger for supplying the *Limulus* from which the semen was abstracted and A. Blaber for examining and measuring some of the cinematographic films.

## References

André, J. 1963. Ann. Fac. Sci. Univ. Clermont-Ferrand, 26, 27-38.

Brown, G. G. and J. R. Krouse. 1973. Biol. Bull., 144, 462-470.

Chen, S.-H. and F. R. Hallett. 1982. Quarterly Reviews of Biophysics 15, 131-222.

Clapper, D. L. and G. G. Brown. 1980a. Develop. Biol. 76, 341-349.
Clapper, D. L. and G. G. Brown. 1980b. Develop. Biol. 76, 350-357.
Cooper, C. D. and G. G. Brown. 1972. Biol. Bull. 142, 397-406.
Craig, T., F. R. Hallett and B. Nickel. 1979. Biophys. J. 38, 457-472.
Hallett, F. R., T. Craig and J. Marsh. 1978. Biophys. J. 21, 203-216.
Shoger, R. L. and G. G. Brown. 1970. J. Submicro. Cytol. 2, 167-179.

# STUDIES ON SPERM MOTILITY BY LASER-LIGHT SCATTERING

WYLIE I. LEE AND PENELOPE GADDUM-ROSSE, Center for Bioengineering and Department of Biological Structure, School of Medicine, University of Washington, Seattle, Washington, U.S.A.

## 1. INTRODUCTION

Measurement of sperm motility has received considerable attention of many investigators since the earliest observations of Leeuwenhoek. However, sperm motility has often been judged subjectively by an individual. Therefore, a suitable definition for the measurement of sperm motility continues to confuse the observer. Many methods of measurement have been developed in recent years (1,2). Essentially, these techniques can be divided into two categories depending upon the objectives of the study, which may be associated with the mechanism of flagellation of single cells or may be concerned with the evaluation of relevant statistical parameters which describe the quality of sperm motility. In the studies of effects of exogenous factors on sperm motility and clinical evaluations of semen quality, the microphotographic technique for studying single cells becomes cumbersome due to its inherent disadvantage of being time-consuming. This paper summarizes some of our recent studies of sperm motility by laser-light scattering. Results of these measurements show that this simple and objective method has considerable potential for both basic research and clinical use in biomedical studies.

## 2. PRINCIPLES AND METHODS

### 2.1. Laser-Light Scattering

The spectral study of light scattered from a suspension of microorganisms was first reported by Bergé *et al.* (3). The principle of this method is the scattering of a monochromatic laser-light by a population of motile microorganisms. According to the optical Doppler effect, the scattered light intensity is spread into frequencies giving a spectrum which depends on the parameters that characterize the "motility" of microorganisms. The theoretical foundation of this method was published by Nossal *et al.* (4). Briefly, the spectral study of light scattered from a suspension of microorganisms is achieved by measuring the intermediate scattering function

$$F_s(k,t) = \int d^3v\ P_s(v)p(k,\mu)e^{ikv\mu t}, \tag{1}$$

where k is the scattering vector. Its magnitude depends only on the scattering angle in any experimental set up. v is the speed of a microorganism. $P_s(v)$ is the swimming speed distribution function, $\mu$ is the cosine of the angle between two directions of k and v. $P(k,\mu)$ is the intensity distribution function of the microorganisms. This complicated methematical formula can be interpreted as the following: The measured quantity contains information about the swimming speed of microorganisms, the speed distribution function, and even their size and shape. For a specially designed experiment where the measurement is made at a very small forward scattering, the root-mean-square speed can be readily evaluated by the following equation (5):

$$V_r = \frac{1.33}{kt_{\frac{1}{2}}}, \tag{2}$$

where $t_{\frac{1}{2}}$ is the time at which the amplitude of measured function drops to half of its initial value. This is an important parameter for determining the quality of sperm motility because it is a statistical number that includes the swimming speed (defined as the rate of displacement of sperm head) and the percentage of microorganisms that swim at that speed.

## 2.2. Motility of Spermatozoa

Applications of laser-light scattering to studies of spermatozoa were first reported by Bergé *et al.* (6,7). This French group also described their technique at the Third Symposium of this series, that was held at Woods Hole, Massachusetts, U.S.A., May 2-5, 1978 (7). However, the method is not well understood by many investigators and is regarded as a research method even though many studies were reported by the French group and investigators elsewhere (8). It is notable that a few minutes of measurement by this method is equivalent to a microcinematographic study of hundreds of spermatozoa at a film speed of 500 to 2,000 frames per second. Furthermore, laser-light scattering is a non-invasive method, thus it becomes an ideal method for studies of the effect of exogenous factors on sperm motility. The deatils of the laser spectrometer can be found in the cited references. A newly developed spectrometer, which is dedicated for the motility measurements, will be described by us elsewhere.

2.3. Spermatozoa

Human spermatozoa were obtained from ejaculates of healthy male donors who had abstained from sexual activity for at least 48 hours. Rat spermatozoa were obtained from the cauda epididymis of mature rats. The suspending medium for human spermatozoa was a modified BWW medium containing 0.4% bovine serum albumin and HEPES. The medium for rat spermatozoa was a modified Krebs-Ringer bicarbonate solution.

2.4. Washing Procedures

Liquified semen samples were homogenized by gentle pipetting then were mixed with the modified BWW medium in a ratio of 1 part of semen to 4 parts medium. The diluted semen was then centrifuged for 8 minutes at specified centrifugal force, for example, 300xg. Most of the supernatant was aspirated, leaving 1ml above the pellet, and 4ml of fresh medium was added. After homogenizing was completed, the same washing procedure was repeated for two more times, but for less time: 6 minutes for the second wash and 3 minutes for the third wash.

2.5. Sperm Penetration Test

The procedure we used for this test was similar to the well-known Kremer's test, except two modifications were made. The capillary was a flat capillary tube; thus it provided easy observation under a microscope. The mucus filled inside the capillary was estrous cow cervical mucus. The details of the procedure have been reported by us previously (9).

2.6. In vitro Fertilization

The rat eggs in cumulus were collected from 3-4 week old Sprague-Dawley rats. Superovulation was induced in these rats by sequential treatment with PMSG and HCG. The collection was made at 14 hours after HCG. The eggs in cumulus were placed directly into the dilute suspension of rat epididymal spermatozoa. This dilute sperm suspension was kept under paraffin oil in a petri dish. The dishes containing the gametes were kept undisturbed in a $CO_2$ incubator at 37°C for at least 10 hours, after which the eggs were checked for fertilization using standard criteria (10), and the fertilization rate was calculated.

## 3. RESULTS

### 3.1. Effect of Temperature

The temperature dependence of root-mean-square (RMS) speed, $V_r$, of rat sperm was reported recently (11). The speed increases linearly with temperature in the range 21°C to 40°C, only when the sample contained sufficient oxygen. Otherwise, the RMS speed decreased rapidly as temperature rised above 34°C (Table 1). Only the slow twitching movement of spermatozoa was observed when temperature was in the range of 36°C to 40°C. Motility of these spermatozoa could be reactivated when sperm samples were exposed to air for 5 minutes.

Table 1. Effect of Temperature on RMS Speed* of Rat Spermatozoa

| | Temperature, °C | | | | | |
|---|---|---|---|---|---|---|
| Sperm sample | 21 | 26 | 30 | 35 | 37 | 40 |
| Concentrated | 126 (±28) | 119 (±40) | 133 (±52) | 198 (±36) | 118 (±66) | 37 (±30) |
| Diluted | 97 (±15) | -- | 166 (±19) | 187 (±34) | 215 (±56) | 254 (±56) |

* RMS speed is expressed in the unit of μm/sec

### 3.2. Effect of Washing

The RMS speed of human spermatozoa was found sensitive to the centrifugal force used in washing and the number of repeated washings (12). Washing at forces 600xg or more reduced sperm motility and also their ability to penetrate cervical mucus *in vitro* (Table 2).

Table 2. Effect of Washing on Sperm Penetration

| Centrifugal Force | RMS Speed | Mean Migration Rate ± SD (mm/min) 10-20 min | 30-40 min |
|---|---|---|---|
| 325xg | 66±15 | 1.7 ± 0.6 | 0.7 ± 0.2 |
| 800xg | 18±5 | 1.1 ± 0.4 | 0.4 ± 0.2 |

### 3.3. Effect of Caffeine

Motility of rat epididymal spermatozoa increases when 0.1mM to 1mM of caffeine is introduced in the suspension (13). However, our recent results show an inhibitory effect when the concentration of caffeine in suspension exceeds 2mM. Furthermore, low doses of caffeine have no stimulatory effect on capacitated rat spermatozoa but 1mM or higher doses of caffeine inhibits sperm motility.

Stimulation of sperm motility by caffeine also affects the in vitro fertilization rate. Compared to a rate of 64% in control preparations, a sperm suspension containing 0.5mM caffeine resulted in a rate of 22% fertilization. 2mM caffeine in suspension markedly increased the rate to 77%, but 6mM caffeine in suspension resulted in only 39%.

### 3.4. Effect of Gossypol

Gossypol, a yellowish phenolic compound isolated from seeds and stems of cotton plants, was reported by the Chinese to have a potent antifertility effect on men (14). For our in vitro study on mature rat spermatozoa, we prepared fresh gossypol solutions by dissolving gossypol acetic acid in pure ethanol immediately before use. The solutions were prepared such that the alcohol concentration in the final mixture of sperm suspension was less than 0.5% (v/v). This precaution was taken because excessive alcohol also affects sperm motility (15). Our results showed gossypol had a biphasic effect on sperm motility: a light stimulatory effect in the range 5-10μM followed by a strong inhibitory effect as the concentration increased beyond 10μM (Figure 1).

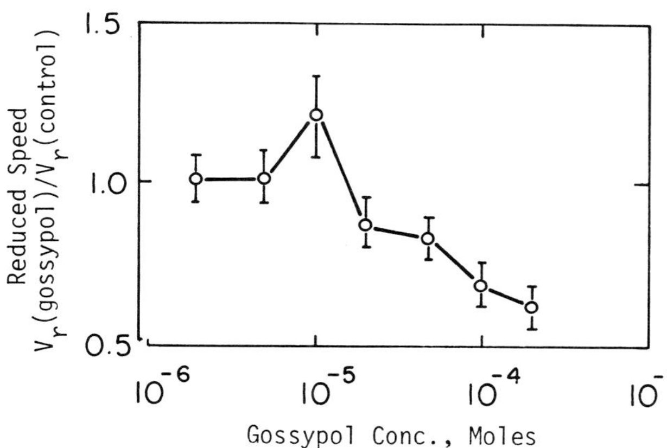


*Figure 1. Effect of gossypol on RMS speed of rat epididymal spermatozoa.*

### 4. DISCUSSION

Sperm motility is essential for reproduction. In mammals, sperm flagellation assists in the transport of spermatozoa from the site of deposition to the site of fertilization. In particular, sperm motility is necessary during their passage through the uterine cervix and uterotubal junction. It becomes more crucial in the penetration of cumulus cells and zona pellucida of the ovum. Therefore, studies of factors affecting sperm motility are important in both fertility and reproductive biology.

Our results of the temperature-effect study suggest the role of an enzyme, LDH-X (lactate dehydrogenase-X) in sperm respiration. Sperm metabolism seems to depend primarily on glycolysis at temperatures that are below body temperature. At body temperature and above, the respiration dominates and oxygen becomes a factor affecting sperm motility.

Evaluation of sperm motility is also influenced by the rheological properties of seminal plasma and the debris presented in the seminal fluid. Therefore, there is a need to separate spermatozoa from seminal plasma and then resuspend them in a certain harmless medium. Our results indicate that the centrifugal force used in centrifugation, a popular method for separation, should not exceed 600xg in order to avoid deleterious effects on sperm motility. It was reported recently that no significant changes in sperm motility were observed even up to 2,500xg (16). However, when a correction of viscosity is taken into account for the difference in dilution of our process (1:4) and of their process (1:1), the sedimentation velocity in both centrifugations becomes identical. In other words, the washing of 1:1 dilution at 2,500xg has the same effect on spermatozoa as the washing of 1:4 dilution at 400xg.

Our results obtained from the study of the effect of caffeine not only confirm the previous report of a stimulatory effect (17) but also present additional questions above the inhibitory effect at higher concentrations (>2mM), and about the insensitivity of capacitated spermatozoa to caffeine. It is clear that more study will be required before we understand the mechanism of the effect of caffeine on sperm motility.

The effect of gossypol on sperm motility looks like the effect of temperature on sperm respiration. In fact, gossypol is hypothesized to reduce sperm motility through its uncoupling effect on oxidative phosphorylation (18), and through its inhibitive effect on LDH-X (19). Our present study on mature rat spermatozoa shows the effective dosage of gossypol is only a factor of 2 more than the dosage reported previously.

REFERENCES

1. Mitchell JA, Nelson L, Hafez ESE. 1976. Motility of spermatozoa. In Hafez ESE (ed.): Human Semen and Fertility Regulation in Men. St. Louis, C.V. Mosby Co., pp. 83-89.
2. Atherton RW. 1977. Evaluation of sperm motility. In Hafez ESE (ed.): Techniques of Human Andrology. New York, Elsevier/North-Holland Biomedical Press, pp. 173-187.

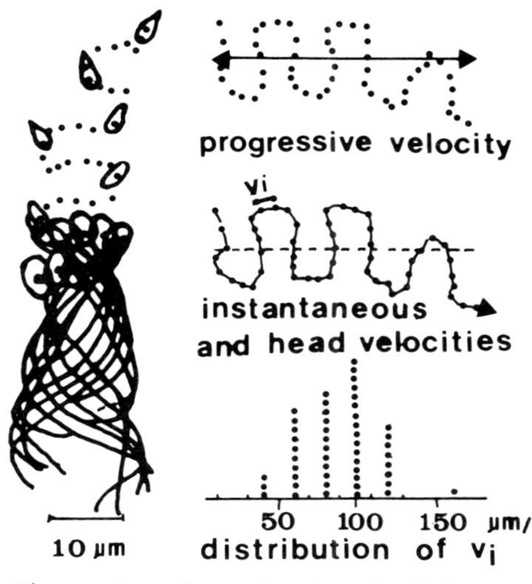


*Figure 1 : Sperm head velocities in seminal plasma, 1s, 50 frames. VP=34 $\mu ms^{-1}$, VH=86 $\mu ms^{-1}$, VP/VH=0.40*

the head lateral displacement and/or the beat frequency are growing.

There was a large heterogeneity of sperm head velocities in the ejaculate of two fertile men with normal sperm characteristics (table I, sample I and II).

1.1.2. Flagellar velocities

The flagellar wave may be defined by its propagation velocity (Vc) and its angular development velocity (Vd). These two velocities are measurable by MC (5). There was a large heterogeneity of Vc and Vd even when considering only the swiming straight spermatozoa (table II)

| | VP | VH | VP/VH |
|---|---|---|---|
| Sample I | 20.3±5.6 (9-37) | 42.3±12.8 (21-72) | 0.50±0.12 (0.24-0.77) |
| Sample II | 24.1±8.7 (8.4-44) | 49.1±22.2 (12.4-99) | 0.49±0.18 (0.22-0.89) |
| Sample III | 27.4±4.8 (14.6-40) | 29.3±5.4 (20.2-47) | 0.94±0.05 (0.72-0.99) |

Table I : Mean values ± sd, min. and max. Values (between brackets) of sperm head velocities ($\mu m\ s^{-1}$) in seminal plasma (sample I and II) and cervical mucus (sample III), VP = Progressive velocity, VH = head velocity.

| $\mu m.s^{-1}$ | 100 | 150 | 200 | 250 | 300 | 350 | 400 | | Mean± sd |
|---|---|---|---|---|---|---|---|---|---|
| Vc | 1.8 | 3.5 | 17.5 | 24.6 | 29.8 | 8.8 | 10.5 | 3.5 | 274.2 ± 113.0 |

| Degrees.$ms^{-1}$ | 0.5 | 1 | 1.5 | 2 | 2.5 | 3 | 3.5 | Mean ± sd |
|---|---|---|---|---|---|---|---|---|
| Vd | 10.5 | 31.6 | 33.3 | 19.3 | 3.5 | 1.8 | | 1.60 - 0.56 |

Table II : Frequency distribution (%) and mean values of sperm flagellar velocities in seminal plasma (57 progressive spermatozoa from 12 samples) (5). Vc = propagation velocity of the wave. Vd = angular velocity of wave development.

1.2. Velocities measured by laser Doppler Velocimetry

From the frequency spectrum of the light scattered by the motile sperm, it is possible to calculate the velocity distribution of the population (2) (Fig. 2). The comparison of the results obtained by LDV and MC analysis of a same semen sample shows that the velocities measured by LDV are close to Vi distribution except in the higher values (Fig. 2). The mode of the distribution calculated by LDV is called characteristic velocity of the popu-

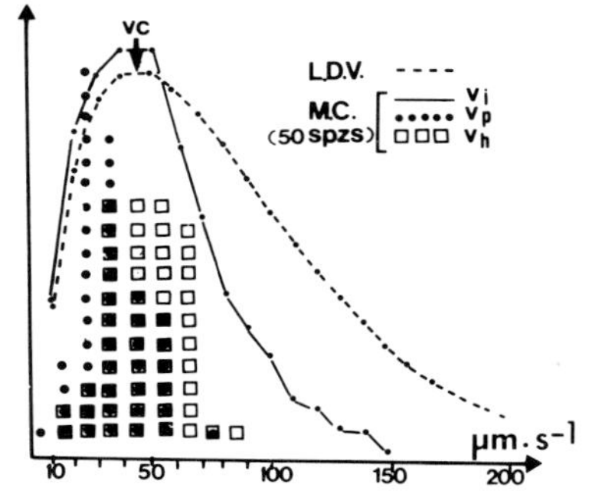


*Figure 2 : Spermatozoa velocities measured by LDV and MC in seminal plasma. VC= 43 $\mu ms^{-1}$, mean values of VH and VP were 45.9 $\mu ms^{-1}$ and 32.2 $\mu ms^{-1}$ respectively.*

lation (VC) and is in good agreement with the mean VH measured by MC.

2. EVALUATION OF SPERM VELOCITIES IN CERVICAL MUCUS

2.1. Velocities measured by Microcinematography

In CM, there is an important reduction of the lateral displacement of the head compared to sperm movemert in the seminal plasma ; VP and VH have very close values and $\frac{VP}{VH}$ is almost equal to one (Fig 3). There is a relative heterogeneity of VP and VH but the type of movement is very homogeneous (85 % of $\frac{VP}{VH}$ are between 0.90 and 0.99).

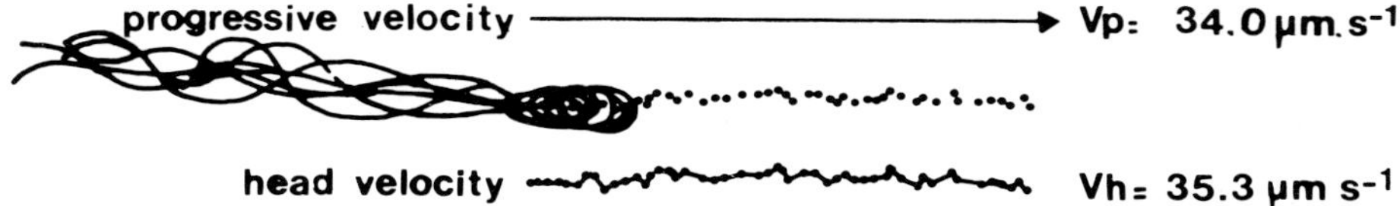


*Figure 3 : Sperm head velocities in cervical mucus,* 1s., *50 frames)*

2.2. Velocities measured by LDV

The frequency spectrum obtained by LDV analysis of sperm moving in ovulatory CM is a well defined peak shifted in frequency (Fig. 4).

It is in good agreement with the VP distribution measured by MC on the same sample (Fig. 4). The mean sperm velocity in CM measured in a serie of tests by MC and LDV at ambient temperature was respectively 19.1 $\pm$ 7.9 and 19.0 $\pm$ 2.7 $\mu ms^{-1}$ with a very good correlation between the two methods ($r = 0.87$, $n = 11$, $p < 0.001$) (SERRES and JOUANNET unpublished data)

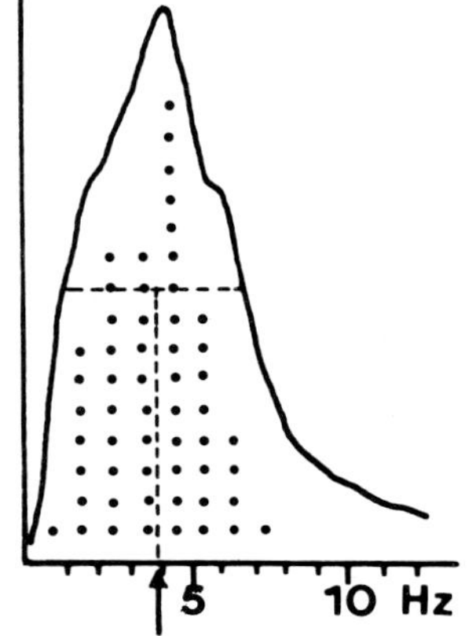


*Figure 4 : Sperm velocities in cervical mucus. Frequency spectrum measured by LDV (—). Arrow is the median point of spectral width. Points indicate VP of 50 spermatozoa measured by MC. 1 HZ = 4.5 $\mu ms^{-1}$*

## 3. DISCUSSION

Various velocities may be measured according to the methodology used. Microcinema analyses precisely quantitative and qualitative parameters of sperm movement, it is a very useful technique to realize studies at the cellular level. Nevertheless the fact that only a few cells by sample are analysed in most of the cases may be an handicap because of the heterogeineity of sperm velocities.LDV doesn't measures VP in seminal plasma as previously quotted (6) except when VP=VH. The velocity distribution measured by LDV and its mode are much closer to Vi and VH. The advantages of LDV are its swiftness (results are obtained in 3 minutes) and the great number of cells measured ($8.10^3$ when the sperm concentration is $10^8$/ml). LDV measures the velocities at the population level and is useful to evaluate any influence (chemical, physical, pharmacological) on sperm velocity but it doesn't give information on the cellular movement. In ovulatory CM, the local hydrodynamic interaction between flagellar beating and mucus microstructure (7) explains that VP and VH are almost equal. In this case LDV measures directly VP. Results of MC and LDV analysis confirm that heterogeneity of velocities is more important in SP than in MC (8,9) but also that the progressiveness ratio of sperm has various values in SP. The measurement of velocities is important to evaluate sperm function but the relation between various velocities and fertility has not been established yet. We had the opportunity to analyse the sperm motility of a man who was infertile for 5 years, had normal sperm values but a constant and unexplained inability to penetrate in ovulatory CM. The characteristic velocity measured in SP by LDV was 31.6 $s^{-1}$. MC results were : VP = 26.5 $\pm$ 10.2 $\mu m s^{-1}$, VH = 27.6 $\pm$ 10.5 $\mu m s^{-1}$ (n = 50). There was an absence of lateral displacement of the head in all spermatozoa and $\frac{VP}{VH}$ was 0.96 $\pm$ 0.06. The inability of spermatozoa swimming straight and fast to penetrate the cervical mucus suggests that other caracters have also to be looked for in order to define the fertile seminal spermatozoa.

## REFERENCES

1. KATZ D.F., OVERSTREET J.W., 1980 : Testicular development, structure, and function, pp 481-489, ed. A. Steinberger and E. Steinberger. Raven Press, New-York
2. DUBOIS M., JOUANNET P., BERGE P., VOLOCHINE B., SERRES C., DAVID G., 1975. Ann. Phys. Biol. Méd., 9, 19-41.
3. DAVID G., SERRES C., JOUANNET P., 1981, Gamete Research, 4, 83-95.
4. DUBOIS M., JOUANNET P., BERGE P., DAVID G., 1974, Nature 252,711-713.
5. SERRES C., FENEUX D., JOUANNET P., DAVID G., 1982 (This meeting)
6. FROST J., CUMMINS H.Z. 1981, Science, 212, 1520-1522.
7. KATZ D.F., MILLS R.N., PRITCHETT T.R., 1978, J.Reprod. Fert. 53, 259-265.
8. KATZ D.F., OVERSTREET J.W., HANSON F.W., 1981, J.Reprod. Fert.62, 221-228.
9. MAKLER A.,ITSKOVITZ J., BRANDES J.M., PALDI E., 1979, Fertil. Steril, 31,155-161

# LASER LIGHT SCATTERING STUDY OF SPERMATOZOA MOTILITY OF DOMESTIC ANIMALS

PAQUIGNON M., DACHEUX J.L.*, JEULIN C**, FAUQUENOT A.
I.N.R.A. Station de Physiologie de la Reproduction Nouzilly 37380 Monnaie
* Lab. Physio Comparée, Fac. Sciences, 37200 TOURS, FRANCE
** Lab. Histologie Embryologie Cytogenetique. C.H. 94270 Kremlin Bicetre France

## 1. INTRODUCTION

Determination of the percentage of motile sperm is the most widely used test of seminal quality but the quantitative measurement of sperm motility has been a long standing problem and visual assessement has become the only routine technique. An objective method for evaluating spermatozoal motility based on laser Doppler spectroscopy has been described (1) (2) and success fully used on human sperm (3-6).

The objective of this study was to correlate in sperm from domestic animals estimates of the percentage of motile sperm by Laser Doppler Velocymetry (LDV) and conventional visual laboratory tests, and to evaluate objectively changes in the percentage of motile sperm during physiological events such as epididymal maturation, aging and changes in season.

## 2. MATERIALS and METHODS

2.1. Sperm collection. Studies on ejaculated sperm were carried on with boar and horse spermatozoa. Spermatozoa were diluted, either in I.N.R.A.-ITP Medium for boar sperm or in I.N.R.A. medium for horse sperm, to a concentration of 30-50x$10^6$ cell/ml. Sperm from boars, horses, goats, rams, and bulls were collected from ten severed regions of the epididymis as shown

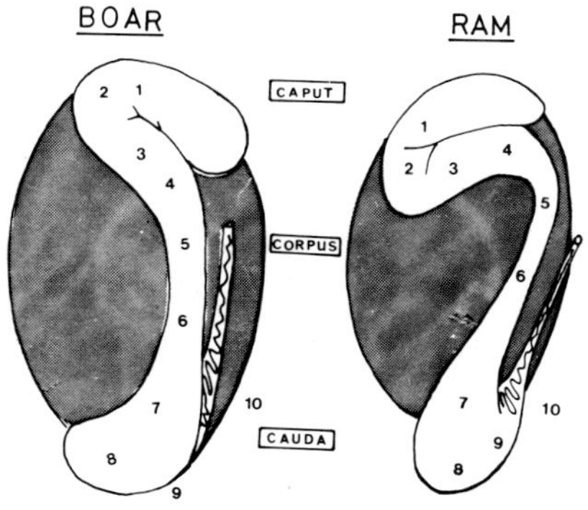


FIGURE 1. Localization of the epididymal origin for the boars and rams spermatozoa.

in Figure 1. The spermatozoa were diluted in Krebs Ringer bicarbonate (10-50 x $10^6$ cell/ml). The effects of seasonal variation on epididymal sperm motility was observed in samples collected from Ile-de-France rams (breeding season: July-December; non-breeding season: February-April) and the effects of animal age were observed in samples from Large White boars and Ile-de-France rams (adults: 1-3 years old; youngs: 6 months old).

2.2. Motility measurement. The motility of the spermatozoa was estimated as the percentage of mobile sperm after a 10 min incubation at 37°C. Objective measurement was assessed using a 0.1 µm depth cell in a Laser Doppler Velocimeter (Soro 200 Spermokinesimetre, Arcueil, F.). Subjective values were given after microscopic phase contrast observation at 37°C of same sample using the 0.1 µm depth cell or a glass slide and glass coverslip.

## 3. RESULTS

3.1. Correlation between the subjective and LDV measurements of the % of motile cells. Linear correlations presented in fig. 2 for boar and horse sperm were highly significant (boar: r = 0.83 ; horse: r = 0.90, P<0.01).

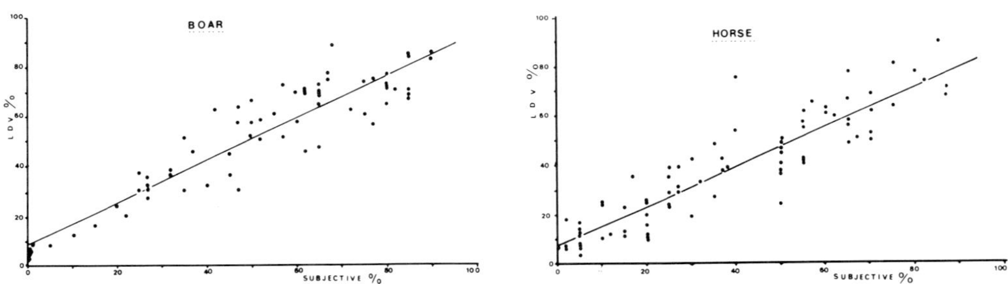


FIGURE 2: Correlation between objective (LDV) and subjective observations of the percentage of motile cells.

3.2. Percentage of motile sperm in epididymis. Results summarized in Table 1 show that there was no motile sperm in the anterior epididymal caput for all species studied,the percentage motile sperm increased in zones 4 to 7 (posterior caput and corpus) and the maximum percentages were observed in the cauda of the epididymis.

Table 1. Percentage of motile sperm in epididymis (m ± sem).

| | epididymal region | | | | | | | | | |
|---|---|---|---|---|---|---|---|---|---|---|
| Species | 1 | 2 | 3 | 4 | 5 | 6 | 7 | 8 | 9 | 10 |
| Ram (n = 18) | 2±1 | 4±1 | 13±3 | 46±3 | 54±6 | 63±5 | 72±4 | 77±3 | 84±2 | 82±3 |
| Boar (n = 13) | 3±1 | 5±1 | 10±4 | 33±5 | 60±6 | 74±3 | 75±4 | 81±5 | 81±5 | 77±5 |
| Bull (n = 4) | 5±1 | 6±2 | 19±11 | 23±13 | 36±18 | 56±16 | 47±9 | 61±6 | 84±6 | 83±1 |
| Goat (n = 2) | 6±1 | 8±3 | 22±1 | 48±6 | 40±20 | 54±19 | 73±13 | 77±1 | 83±5 | 83±1 |
| Horse (n = 3) | - | 9±5 | 17±2 | 27±8 | 55±4 | 68±7 | 77±6 | 84±4 | 83±4 | 90±5 |

3.3. Percentage of motile spermatozoa in the epididymis according to the age of the animals. There was no significant difference between the post-pubertal and adult animals, either boars and rams, in the epididymal gradients of percentage of motile spermatozoa.

Table 2. Percentage of motile sperm in epididymis according to the age (m ± sem).

| Species | Age | epididymal region | | | | | | |
|---|---|---|---|---|---|---|---|---|
| | | 1 | 3 | 4 | 6 | 8 | 9 | 10 |
| Ram | adult (n = 9) | 3±1 | 22±5 | 45±8 | 61±7 | 71±8 | 79±4 | 80±4 |
| | lamb (n = 7) | 4±1 | 18±9 | 24±8 | 51±6 | 74±6 | 71±3 | 73±8 |
| Boar | adult (n = 13) | 3±1 | 10±5 | 32±5 | 73±5 | 82±4 | 83±6 | 74±8 |
| | young (n = 5) | 4±1 | 15+6 | 27±15 | 62±11 | 79±7 | 85±1 | 67±5 |

3.4. Seasonal variation in the percentage of epididymal motile sperm. In Ile-de-France ram there was no seasonal variation in the percentage of motile cells in the epididymis (Table 3).

Table 3. Seasonal variation in percentage of epididymal motile sperm in the Ile-de-France ram (m ± sem).

| | epididymal region | | | | | | | | | |
|---|---|---|---|---|---|---|---|---|---|---|
| Season | 1 | 2 | 3 | 4 | 5 | 6 | 7 | 8 | 9 | 10 |
| breeding (n = 18) | 2±1 | 4±1 | 13±3 | 46±6 | 54±6 | 63±5 | 72±4 | 77±3 | 84±2 | 82±3 |
| non breeding (n = 9) | 3±1 | 4±1 | 22±5 | 45±8 | 51±8 | 61±7 | 62±6 | 71±8 | 79±4 | 80±4 |

3.5. Variation between rams in the percentage of motile cells in the posterior caput epididymis.

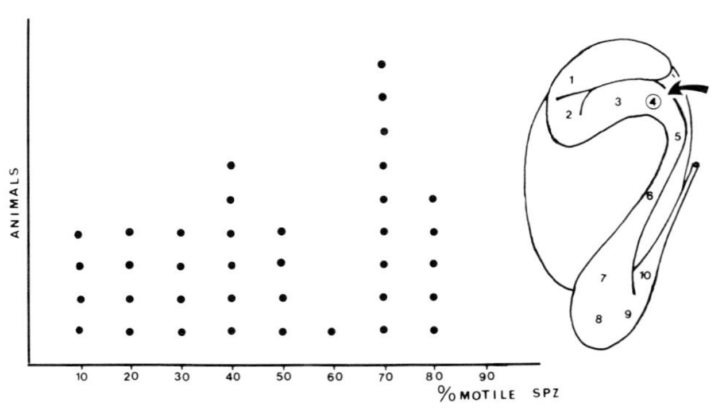


FIGURE 3. Partitioning of the percentage of epididymal motile sperm from zona 4 in the ram.

The partitioning of the different percentages of motile spermatozoa from the zona 4 of the epididymis in the ram indicate that the values are very heterogeneous between animals and it is evident that the development of motility is not uniform for all males.

## 4. DISCUSSION

The percentage of motile sperm assessed by the LDV method are highly correlated with the subjective visual assessment for boar and horse sperm, and also for samples from other species which have low or high percentages of motile cells. The objective measurement of percentage of motile sperm has given us the opportunity to compare epididymal motility gradients for several species, and in all five species observed the increase in motility begins in the posterior caput and in the corpus of the epididymis. With the objective technique, evaluation of the motility of sperm samples during an experiment covering long periods, such as seasonal variation or animal aging, may be done without the problems of subjectivity and variation between the observors.

## REFERENCES

1. Berge P, Dubois M. 1973. Rev. Physiol. Appl. 8: 89.
2. Dubois M, Jouannet P, Berge P, Volochine B, Serre C, David G. 1975. Ann. Physiol. Biol. & Med. 1975. 9: 19-41.
3. Jouannet P, Volochine B, Deguent P, Serres C, David G. 1977. 9: 36-49.
4. Jeulin C, Serres C, Jouannet P. 1982. Reprod. Nutr. Develop. 22: 99-109.
5. Lee WI, Blandau RJ. 1979. Fert. Ster. 32: 320-323.
6. Woolford MW, Woolhouse JK, Philipps DSM, Hawley SA, Harvey JD, Shannon P, Curson B. 1980. Proceed. N.Z. Soc. Anim. Prod. 40: 124-129.

# COMPUTER QUANTITATION OF IRRADIATED HAMSTER SPERM MOTILITY

DR. L. FRANKLIN
Department of Biology, University of Houston, Houston, Texas 77004, U.S.A.

## INTRODUCTION

The pattern of hamster sperm motility changes during capacitation (Yanagimachi, 1970), and is sensitive to the conditions and composition of suspending media in in vitro incubation.

The sperm assume a circular path in the direction of the curved acrosomal extension (Fig. a) when placed in droplets of RPMI or similar media on a glass slide and then covered with a coverslip. This pattern changes within a few minutes to a more linear path as the amplitude of the flagellar beat increases and its frequency decreases. A computer program was written to provide quantitative comparison of the motility of hamster sperm from different males, or following different treatments. Because we are using gamma irradiation to 'mark' sperm (eggs fertilized by the irradiated sperm undergo developmental arrest at the 1-4 cell stage, 7,000 r), preliminary tests of the program were made with hamster sperm exposed to gamma irradiation from a cobalt source (Gamma Cell 220, Atomic Energy of Canada, 29,150 r/min at the time of the trials).

## PROCEDURE

Sperm from the caudal epididymides were dispersed with gentle agitation into normal saline to a concentration of 20-40 X $10^7$/ml. Aliquots were placed in open culture tubes (12X75mm), covered with a few drops of paraffin oil and exposed to a cobalt source for various periods up to 1 1/2 min (174,900r).

One μl samples from each aliquot were inoculated into 20 μl RPMI (GIBCO) on a glass slide, dispersed by tapping the slide, then covered with a glass coverslip. The initial motility patterns for each irradiated sample and its corresponding unirradiated control were recorded on video tape.

The tape was played back in single frame mode, and eleven successive positions of sperm heads were marked on a plastic overlay on the TV monitor (0.1 sec intervals). The points, which represent the sperm head movement during 1.0 sec were transferred to graph paper, and x, y coordinates were assigned to each. The coordinates were then run through a program designed to yield the following information:

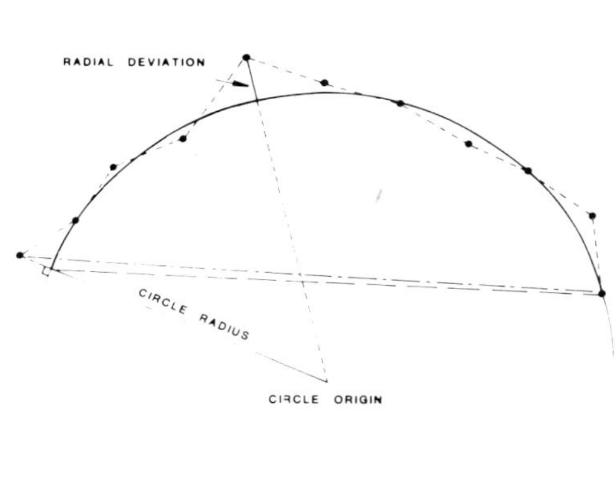


1. Circular path best fit for the 11 points from 165 possible combinations of groups of three points.
2. Radius, and coordinates of circle origin.
3. Sum of squares (variation), and standard deviation of radial deviation of points other than the three used to construct the best fit circular path.
4. Total distance, point-to-point.
5. Vector distance.
6. Distance along calculated circular path.
7. Ratio of the point-to-point distance/distance along calculated circular path.

The 20X microscope objective which we normally use to videotape hamster sperm produces a 496X monitor image. Consequently each mm of screen distance represents approximately 2.0 μ real distance. Motility data were obtained for sperm which received

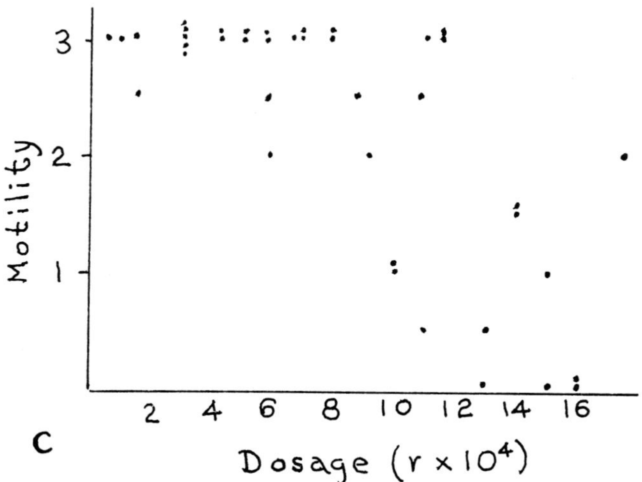


Table 1. Sperm Velocity and yaw

| Treatment | n | Velocity (μ/sec) | Ratio Point-to-Point Track/Calculated Curved Path (yaw) | % Live |
|---|---|---|---|---|
| Control | 10 | 73.28±15.04 | 1.33±0.15 | 47.0 |
| 58,300r | 10 | 66.20±15.90 | 1.38±0.23 | 50.0 |
| Q(z) | | 0.855 | 0.353 | |
| Control | 8 | 92.26±11.46 | 1.21±0.29 | - |
| 87,450r | 9 | 75.64±19.48 | 1.28±0.79 | - |
| Q(z) | | 0.983 | 0.217 | |
| Control | 10 | 73.82±9.52 | 1.28±0.20 | 29.0 |
| 174,900r | 10 | 16.36±8.68 | 3.89±1.77 | 34.8 |
| Q(z) | | 1.000 | 0.000 | |

REFERENCES

1. Yanagimachi, R. 1970. The movement of golden hamster Spermatozoa before and after capacitation. J. Reprod. Fert. 23: 193-196.

2. Supported in part by NIH grant HD 13236. I wish to acknowledge Mr. J. Patrick Finnigan for development of the computer program.

29,150r, 58,300r, 87,450r, 174,900r, and from their corresponding controls. The samples were video-taped as soon as possible after irradiation (< 1/2 hr). Eight to ten sperms were plotted for each sample, and means of the measurments for each experimental group were compared with those of the controls by the Wilcoxon-Mann-Whitney rank-sum test (95% level of confidence).

RESULTS AND DISCUSSION

None of the motility parameters calculated for samples irradiated 29,150r and 58,300r were significantly different from their controls (two-tailed test, 0.025>Q(z)>0.975 to reject null hypothesis). In contrast, all parameters were affected significantly at 174,900r. Although both the point-to-point distance and the calculated curve distance traveled per sec were greatly reduced in these sperm, the ratio of point-to-point distance/circular path distance was greater, i.e. the heads of the irradiated sperm moved a greater distance across the calculated circular path (yaw) than along the path. At 87,450r, the sperm velocity was significantly affected but the ratio of point-to-point distance/circular path distance and other parameters (not shown in Table 1) were not. Table 1 shows the mean values and the results of the Rank-Sum tests (Q(z)) for single 58,300r, 87,450r and 174,900r treatments.

The absence of adverse effects on motility at 58,300r (two trials with different males on separate days) is consistent with visual evaluation of overall motility taken independently by Ms. Cisneros (Fig. C). Although the control for the 174,900r sample (Table 1) had a lower than normal percentage of live sperm (50% live is not unusual for hamster epididymal sperm), the survival of the control sperm was unchanged at 1 hr whereas the heavily irradiated sperm died. The marked decline in motility in this instance was recorded within 10 min of irradiation.

The program and system described represent an inexpensive approach to quantitating sperm motility, and can include linear regression analyses suitable for human sperm.

# THE EVOLUTION OF MAMMALIAN SPERM MOTILITY IN THE MALE AND FEMALE REPRODUCTIVE TRACTS

DAVID F. KATZ
Dept. of Obstetrics and Gynecology, School of Medicine, University of California, Davis, Ca. 95616, U.S.A.

Spermatozoa recovered from the mammalian male and female reproductive tracts are seen to exhibit a variety of movements. These different types of movement appear related to a number of factors intrinsic and extrinsic to the spermatozoa. For example, sperm maturation in the male and capacitation in the female are associated with changes in the patterns of flagellar bending. Our current knowledge of sperm transport in the female suggests that the different types of sperm motility observed may be directly related to the mechanisms of transport. Indeed, the notion has developed that the female tract modulates sperm motility so as to insure the presence of viable fertile gametes at the site and time of fertilization.

In considering sperm motility it must be appreciated that the details of flagellar bending and the swimming trajectories that are their hydrodynamic consequence are influenced by the conditions of observation. Virtually all that we know is based upon observations _in vitro_, and the relationship between the movements of spermatozoa under glass and their true biological behavior _in vivo_ is not clear. Nonetheless, controlled and standardized observations _in vitro_ are suitable for comparative analysis, and it is primarily upon such work that this brief review focuses.

Spermatozoa recovered in epididymal fluid of mammals may or may not be motile, depending upon the species and conditions of sperm recovery. It is believed that such spermatozoa achieve the capacity for "progressive" (space-gaining) movement during transit from the caput to caudal portions of the epididymis. Caput sperm exhibit asymmetric, large-amplitude bending, which produces a convoluted tortuous swimming path; or else they display a very low amplitude bending producing little flagellar thrust and therefore a very low swimming speed. In contrast, caudal sperm can propagate relatively symmetrical waves of moderate-to-large amplitude, and these propel the sperm body considerably farther per beat. Two phenomena have already been identified that may contribute to this transition in sperm movement. A number of sperm structures become stabilized by disulfide bond formation during epididymal transit. These include flagellar structures that presumably contribute to bend formation and propagation, viz. the coarse outer fibers of the axoneme, the fibrous sheath of the principal piece, the connecting piece, and the outer mitochondrial membranes (1, 2). It has also been shown, in the bull, that a protein secreted by the epididymis is capable of changing the swimming patterns of spermatozoa recovered from its caput portion (3, 4). Initial studies here focused solely on swimming paths, and demonstrated that addition of this "forward motility protein" results in a straightening out of the trajectories, making them more "caudal-like". In a very recent study utilizing high-speed cinemicrography we have looked in greater detail at this phenomenon, concluding that caput spermatozoa in the presence of forward motility protein do not swim identically to caudal sperm (5). The latter exhibit an additional reduction in the maximum curvature of flagellar bending and straightness of the swimming trajectory, which could therefore be the consequence of the ultra-structural stabilization. Thus, changes in motility during epididymal maturation could be due to an increase in flagellar stiffness as well as a protein-membrane interaction. These are both phenomena that are known, in other contexts, to be capable of altering flagellar bending, and epididymal transit thus provides an excellent opportunity to study the contraction mechanism of mammalian sperm.

The anatomical and physiological details of insemination vary among mammalian species. Nonetheless, during the initial stages of interaction with the female tract post ejaculation, sperm motility appears to be vigorous and progressive. Regardless of the site of semen deposition (vagina, cervix, uterus) it is likely that spermatozoa immediately experience the physicochemical influence of female tract secretions. The semen may simply be diluted, e.g., by the fluids of the vagina or uterus; or the spermatozoa may enter a very different environment, i.e., the cervical mucus. In some species, sperm collected from the vagina remain motile for a number of hours post insemination. However, the biological relevance of this extended motility is limited since, in all probability, such spermatozoa do not participate in fertilization.

The interaction of primate and ruminant spermatozoa with cervical mucus is a perplexing phenomenon that has received much attention. Mucus is generally collected from the cervix by aspiration. In periovulatory mucus of "good quality" the majority of spermatozoa are seen to swim progressively. The details of flagellar bending are different from those prior to mucous entry (6, 7). The amplitude of the beat is lower, and the principal bending is confined to the distal half of the flagellum. Consequently the swimming trajectory is rather straight, with little yaw of the sperm head. These changes in flagellar movement are the consequence of biophysical interations between the flagellar contraction mechanism and the macromolecular structure of the mucus, which envelops the sperm body quite intimately (8). Indeed, if the mucus is stretched so as to align macromolecules in local regions of the microstructure, spermatozoa will swim in parallel to each other along the line of strain, like automobiles on a highway.

These, and many other _in vitro_ observations are quite intriguing. But we remain confronted by the fact that we have virtually no knowledge of the organization of the mucous microstructure _in vivo_. Aspiration of mucus from the cervix is undoubtedly a "homogenizing" process, and therefore limits our ability to identify even the general region from which a collected sample originates. The biological relevance of this problem relates to the popular notion that the cervix is a "reservoir" for spermatozoa. Sperm are thought to be "stored" in the cervix, and a time-release mechanism is said to provide them to the upper tract (9). If this is so (and direct evidence is lacking), could there be a localized suppression of sperm motility as part of the storage mechanism? Obviously, an examination of sperm behavior in localized regions of the cervix is long overdue.

Virtually all observations of uterine spermatozoa have indicated that these cells are vigorously motile at times prior to fertilization. However, there is striking evidence in the rabbit that once through the uterotubal junction, these sperm become virtually immobilized in the lowermost portion of the tubal isthmus (10, 11). Rabbit spermatozoa collected in lower isthmic fluid exhibit a very low amplitude beat and swim very slowly if at all. Only during the periovulatory period are viable sperm recovered from the upper isthmus or ampulla. It has accordingly been hypothesized that the isthmus is a sperm reservoir in this species, and that the quiescent motility contributes (perhaps with sperm-epithelium adhesion) to sperm retention (12).

The rabbit spermatozoa that subsequently arrive at the site of fertilization, when ovulation is in progress, exhibit strikingly different motile behavior. Not only are they at least twenty-fold more vigorous (as measured by power output), but they undergo episodes of very large amplitude, high-curvature flagellar bending (13). Such movement is often metaphorically referred to as "whiplash" motility, and the biological state of the sperm is called "activation" or "hyperactivation". Among mammals, hyperactivation has thus far been identified in the hamster, guinea pig, rabbit mouse, dog and sheep (14). The geometrical details of hyperactivated movement may show some species variation, but there remains the common basic characteristic of flagellar bending with high-curvature, especially in the midpiece.

The consequent swimming trajectories of hyperactivated sperm are highly dependent upon the conditions of observation. Under glass, the spermatozoa generally swim in circles marked by considerable yaw of the sperm head (13). Within the transilluminated hamster oviduct we have seen hyperactivated sperm glide progressively along epithelial surfaces (15). Hyperactivated sperm are seen within the cumulus, although its viscous matrix can transiently suppress the large amplitude flagellar bending. Indeed, the biological significance of hyperactivation may well be in the increased ability of the sperm to thrust against and through the solid and semisolid egg investments. For example, once the sperm head is bound to the zona pellucida, such high-amplitude bending persists, and may play an important role in zona penetration (16).

In summary, evidence is mounting that sperm motility "evolves" during transport in the male and female. The changes observed in sperm movement appear closely related to the necessity for spermatozoa to remain within local anatomical compartments or to depart from them. Motility is obviously necessary, albeit not sufficient, for successful transport and fertilization. Its further study should provide new insights into the mechanisms responsible for these events.

REFERENCES


1. Calvin HI, Bedford JM. 1971. Formation of disulfide bands in the nucleus and accessory structures of mammalian spermatozoa during maturation in the epididymis. J Reprod Fert Suppl 13:65-76.
2. Bedford, JM, Calvin HI. 1974. Changes in -s-s- linked structures of the sperm tail during epididymal maturation, with comparative observations in sub-mammalian species. J Exp Zool 187:181-204.
3. Acott TS, Hoskins DD. 1978. Bovine sperm forward motility protein: partial purification and characterization. J Biol Chem 253:6744-6750.
4. Hoskins DD, Johnson D, Brandt H, Acott TS. 1979. Evidence for a role for a forward motility protein in the development of sperm motility. In: The Spermatozoon. DW Fawcett and JM Bedford (Eds.). Urban & Schwarzenberg, Baltimore. pp. 43-53.
5. Acott TS, Katz DF, Hoskins DD. 1982. Submitted for publication.
6. Katz DF, Mills RN, Pitchett TR. 1978. The movement of human spermatozoa in cervical mucus. J Reprod Fert 53:259-265.
7. Katz DF, Bloom T, BonDurant R. 1981. The movement of bull spermatozoa in cervical mucus. Biol Reprod 25:931-937.
8. Katz DF, Berger SA. 1980. Flagellar propulsion of human sperm in cervical mucus. Biorheology 17:169-175.
9. Tredway DP, Settlage DSF, Nakamura RM, Motoshima M, Umezaki CV, Mishell DR. 1975. Significance of timing for the post coital evaluation of cervical mucus. Amer J Obstet Gynec 121:387-393.
10. Overstreet JW, Cooper GW. 1975. Reduced sperm motility in the isthmus of the rabbit oviduct. Nature (London) 258:718-719.
11. Cooper GW, Overstreet JW, Katz DF. 1979. Sperm transport in the reproductive tract of the female rabbit. Sperm motility and patterns of movement at different levels of the tract. Gamete Res 2:35-42.
12. Overstreet JW, Cooper GW, Katz DF. 1978. Sperm transport in the reproductive tract of the female rabbit. II. The sustained phase of transport. Biol Reprod 19:115-132.
13. Suarez SS, Katz DF. 1982. Movement characteristics and acrosomal status of spermatozoa recovered at the site and time of fertilization. Biol Reprod 26, Suppl 1, 146A.
14. Yanagimachi R. 1981. Mechanisms of fertilization in mammals. In: Fertilization and Development. L Mast, JD Biggers and WA Saddler (Eds.). Plenum Press, New York.
15. Katz DF, Yanagimachi R. 1980. Movement characteristics of hamster spermatozoa within the oviduct. Biol Reprod 22:759-764.
16. Katz DF, Yanagimachi R. 1981. Movement characteristics of hamster and guinea pig spermatozoa upon attachment to the zona pellucida. Biol Reprod 25:785-791.

# MOVEMENT CHARACTERISTICS OF HAMSTER SPERMATOZOA

H. MOHRI*, S. ISHIJIMA** AND Y. HIRAMOTO**
*Department of Biology, University of Tokyo and **Department of Biology, Tokyo Institute of Technology, Tokyo, Japan

## 1. INTRODUCTION

Mammalian spermatozoa exhibit sequential changes in their motility characteristics during their passage through the male reproductive tract and during capacitation process in the female reproductive tract. These changes in their motility characteristics are due to the change in the beating pattern of sperm flagella. In order to investigate the mechanism underlying the change in the beating pattern, it is important to describe the flagellar movement of the spermatozoa as exactly as possible. In the present study, we used hamster spermatozoa as material and recorded the flagellar movement of testicular, caput epididymal, cauda epididymal and capacitated spermatozoon, after holding it at its head with a micropipette.

## 2. MATERIAL AND METHODS

Golden hamster spermatozoa collected from the testis, the caput epididymis and the cauda epididymis were suspended in Tyrode solution. These spermatozoa were also demembranated with an extraction medium containing 0.1% Triton X-100 and reactivated with 1 mM ATP (1). Capacitation of the cauda epididymal spermatozoa was made according to Yanagimachi (2) with slight modification. For motility observation, a spermatozoon was attached at its head to the tip of a sucking micropipette held with a micromanipulator. The orientation of the spermatozoon was adjusted by tilting the micropipette and turning it about on its axis so as to bring the entire length of the flagellum into the focal plane of the microscope. Differential interference microscope was used for observation. The images of the flagellar movement taken by the video camera were recorded on a video tape by illuminating with stroboscopic flashes synchronized with the video camera. All the experiments were done at 37°C.

## 3. RESULTS AND DISCUSSION

The flagella of the intact testicular spermatozoa suspended in Tyrode solution were rather straight and repeated weak movement with waves of very small amplitude for several hours. The movement of the intact spermatozoa from the caput epididymis was similar to that of the testicular ones, except for slightly higher beat frequency. On the other hand, the spermatozoa from the cauda epididymis showed vigorous movement. When the cauda epididymal spermatozoon was fixed at its head with the micropipette, almost entire length of the flagellum could be observed. This fact indicated that the flagellar movement of the spermatozoa is almost planar. When the fixed head was rotated by 90°, the beating of the flagellum was seen as a train of spots corresponding to regions of the flagellum in focus. From these recordings of the train of spots, it is confirmed that the flagellar beating of the cauda epididymal spermatozoa was a planar movement. The bending wave of the cauda spermatozoa, however, was asymmetrical. Namely, the curvature of the "reverse" bend (Cf.(3)) was much larger than that of the "principal" bend. The mean values of beat frequency, amplitude and wavelength of the bending wave of the cauda epididymal spermatozoa were 12 Hz, 30 μm and 80 μm, respectively.

When the testicular spermatozoa were demembranated with the extraction medium and exposed to ATP, almost planar symmetrical bending wave could be observed. If such a spermatozoon was fixed with the micropipette, the planar and symmetrical bending wave continued only for about ten seconds, soon the wave becoming asymmetrical and of low beat frequency, and the flagellum was arrested within ten minutes. The mean values of the rather symmetrical bending wave obtained in this experiment were 4 Hz for beat frequency, 40 μm for amplitude and 95 μm for wavelength. The demembranated spermatozoa from the caput epididymis moved actively with waves of almost planar and large amplitude immediately after reactivation with ATP. After the fixation with the micropipette, the flagellar movement was gradually reduced. Especially, the movement of the proximal region of the flagellum stopped within a short time, whereas the distal region of the flagellum continued to beat. For symmetrical bending wave of these spermatozoa, the mean values of beat frequency, amplitude and wavelength were 7 Hz, 50 μm and 90 μm, respectively.

The demembranated cauda epididymal spermatozoa moved quite actively when ATP was added to the extraction medium, and such a movement continued for 20-30 minutes even after the sperm head was held with the micropipette. The obtained value of the beat frequency, 6 Hz, was somewhat lower than that of the intact cauda spermatozoa, whereas amplitude (55 μm) and wavelength (100 μm) were greater than those of the intact ones. The bending wave of such spermatozoa quite resembled the waveform of ciliary type, the curvature of the "reverse" bend being much greater than that of the "principal" bend. When the bending wave was observed from the direction parallel to the beating plane, the displacement from the plane at the distal region of the demembranated cauda sperm flagellum was larger than that of the intact cauda sperm flagellum. There was a tendency that the displacement from the plane of the distal region was smaller in the flagella showing symmetrical bending wave than in those exhibiting asymmetrical and irregular bending wave.

Next, the spermatozoa from the cauda epididymis were suspended in the capacitation medium, and their movement was observed after capacitation. When such a spermatozoon was captured with the micropipette, the active movement continued for about ten seconds and then rapidly ceased. The mean values of beat frequency, amplitude and wavelength were 10 Hz, 38 μm and 110 μm, respectively. These values lies between those of the intact and the demembranated cauda epididymal spermatozoa. Some of the capacitated spermatozoa, however, showed the wave pattern of a much greater amplitude, although such a pattern could be also observed in a few of the intact and demembranated cauda spermatozoa.

In the present study, it has been revealed that the bending

wave of hamster sperm flagella has some resemblance to that of cilia. As we pointed out previously (4,5), it appears that the "principal" bend in golden hamster spermatozoa corresponds to the effective stroke in gill cilia of mussels and the "reverse" bend to the recovery stroke, although the doublet No. 1 in the axoneme leads the "principal" bend in the former, whereas the doublets Nos. 5 and 6 lead the effective stroke in the latter. In sea urchin spermatozoa, the principal bend has a larger bend angle and beats toward outside if the spermatozoon swims in a circular path (6). If the same terminology is applied to hamster spermatozoa, the so-called "reverse" bend would corresponds to the principal bend in sea urchin sperm flagella.

This presentation was partly supported by the Yamada Science Foundation.

REFERENCES

1. Mohri H, Yanagimachi R. 1980. Characteristics of motor apparatus in testicular, epididymal and ejaculated spermatozoa. A study using demembranated sperm models. Exp Cell Res 127, 191-196.
2. Yanagimachi R. 1978. Current topics in developmental biology (ed Moscona AA, Monroy A). Vol 12, p83. Academic Press, New York.
3. Wooly DM. 1977. Evidence for "twisted plane" undulations in golden hamster sperm tails. J Cell Biol 75, 851-865.
4. Mohri H, Yano Y. 1980. Analysis of mechanism of flagellar movement with golden hamster spermatozoa. Biomed Res 1, 552-555.
5. Mohri H, Yano Y. 1982. Reactivation and microtubule sliding in rodent spermatozoa. Cell Motility Suppl 1, 143-147.
6. Gibbons IR, Gibbons B. 1974. Properties of flagellar "rigor waves" formed by abrupt removal of adenosine triphosphate from actively swimming sea urchin sperm. J Cell Biol 63, 970-985.

# DO CAPACITATED HUMAN SPERMATOZOA SHOW AN "ACTIVATED" PATTERN OF MOTILITY?

David Mortimer, Anne Marie Courtot, Yves Giovangrandi and Claudette Jeulin

## INTRODUCTION

A characteristic "activated" pattern of motility has been described for the spermatozoa of several mammalian species either recovered from the tubal ampulla around the time of ovulation or after in-vitro capacitation.[4] To date no similar phenomenon has been reported for the spermatozoa of our own species.

This study was designed as a preliminary attempt at quantifying subjectively observed changes in the motility pattern of human spermatozoa associated with their migration into, and incubation in, synthetic media known to support in-vitro capacitation.

## MATERIALS AND METHODS

Sperm motility was quantified by microcinematography[2] ('film') at 10 frames/s, 'photo'-micrography[7] and Laser Doppler Velocimetry.[3] One ejaculate was used from each of 5 men, and sperm motility assessed in 5 treatment samples:
*Initial:* semen diluted with cell-free seminal plasma prepared by centrifugation to give a suitable sperm concentration for evaluation (20-40 x$10^6$/ml).
*Post-migration (P-M)* by the "swim-up" method (1 h @ 37°C into either BWW medium[1] with 3 mg/ml HSA or B2 medium[6] with 10 mg/ml HSA.
*Post-incubation (P-I)* for either 4 h (B2) or 5 h (BWW) in humidified 5% $CO_2$ in air. Incubation periods were chosen as appropriate for in-vitro capacitation.[8,9]
Film and photo assessments were made on ∿20 µm deep preparations (15 µl under 22x32 mm coverslips). All slides etc were kept at 37°C prior to use and measurements made as quickly as possible; slides showing "flow" were discarded. LDV measurements were made at 37°C in a 200 µm deep cell. Whenever possible for film and photo methods 100 spermatozoa in each sample were counted and classed as either immotile, non-progressively motile (NPM: stationary with weakly beating tails or active but with no net linear progression), or progressively motile (PM). Trajectory parameters were measured for all PM spermatozoa: velocity of linear progression (Vp µm/s) and amplitude of lateral

head displacement (Ah um). LDV can only give the % motile (irrespective of progressivity) and the modal instantaneous velocity (Vm um/s). Vm is more or less equivalent to the velocity of the head along its trajectory ("Vh", see[2]).

RESULTS

Percent motile spermatozoa *P-M* was, as expected bb virtue of the migration process, greatly increased for all 5 subjects in both media by film and photo methods. However, the values given by LDV were more closely correlated to the film and photo % PM rather than the % motile (PM + NPM) as would have been expected. A number of the NPM spermatozoa *P-M* were seen by the film method to be moving with large lateral displacements (even reflexions) of the head but with no actual linear progression.

Trajectory parameters *P-M* were found to be significantly altered in both media but with appreciable inter-subject variation. Many individual changes reached statistical significance, and the overall mean values showed highly significant modifications (Table 1). With the exception of B2 'film' (essentially due to only one subject) progressive spermatozoa *P-M* generally moved with increased Vp and Ah. As would therefore be expected, the LDV Vm was also increased *P-M*.

*Table 1: Overall population mean values (± s.d. [n] for* A *velocity (μm/s) and* B *Ah (μm). *indicates significant changes initial→P-M or P-M→P-I within media.*

| | Sample | Film† | Photo† | LDV† |
|---|---|---|---|---|
| A: | Initial | 35.97 ±13.33 [242] | 24.43 ±11.28 [251] | 48.75 ±15.65 [4] |
| | *P-M* BWW | 40.86 ±16.79 [430]* | 28.32 ±12.52 [387]* | 84.38 ±15.65 [4]* |
| | B2 | 31.21 ±11.40 [296]* | 28.16 ±10.25 [363]* | 99.38 ± 8.96 [4]* |
| | *P-I* BWW | 47.47 ±16.91 [409]* | 35.11 ±16.10 [382]* | 76.50 ±16.42 [4] |
| | B2 | 46.48 ±18.71 [392]* | 39.83 ±16.74 [423]* | 99.63 ±17.38 [4] |
| B: | Initial | 5.14 ± 2.39 [242] | 4.30 ± 2.12 [251] | |
| | *P-M* BWW | 6.78 ± 3.18 [430]* | 5.68 ± 2.34 [387]* | |
| | B2 | 7.25 ± 3.32 [296]* | 5.29 ± 1.99 [363]* | |
| | *P-I* BWW | 7.22 ± 3.25 [409] | 6.38 ± 2.69 [382]* | † Vp for 'film' & 'photo', Vm for LD |
| | B2 | 6.98 ± 2.89 [392] | 5.90 ± 3.37 [423]* | |

Trajectory parameters *P-I* showed less variation both between subjects and between media than the *P-M* results. Vp was further increased, as was Ah by 'photo' (but not 'film'). The LDV Vm did not change during incubation.

Differences between 'film' and 'photo' were seen in the absolute measured values of both Vp and Ah, with 'photo' giving lower results. There were, however, close correlations between the qualitative findings over the samples of the two methods. Observed discrepancies may have been of technical origin due, perhaps, to differing sensitivities in detecting changes (as, for example, with Ah, see above).

Analysis of frequency distributions of the trajectory parameters (Figure 1) showed that while some of the observed differences between initial and *P-M* sperm populations may have been due to the selection of the more highly motile spermatozoa, *P-M* values of Ah were often greater than those seen in the initial samples. Furthermore, the changes in the distributions during incubation clearly indicated that changes did occur in the patterns of sperm movement with both Vp and Ah tending to increase.

*Figure 1: Mean % frequency distributions of Vp (left) and Ah (right) by 'film' and 'photo' methods over the five subjects.*

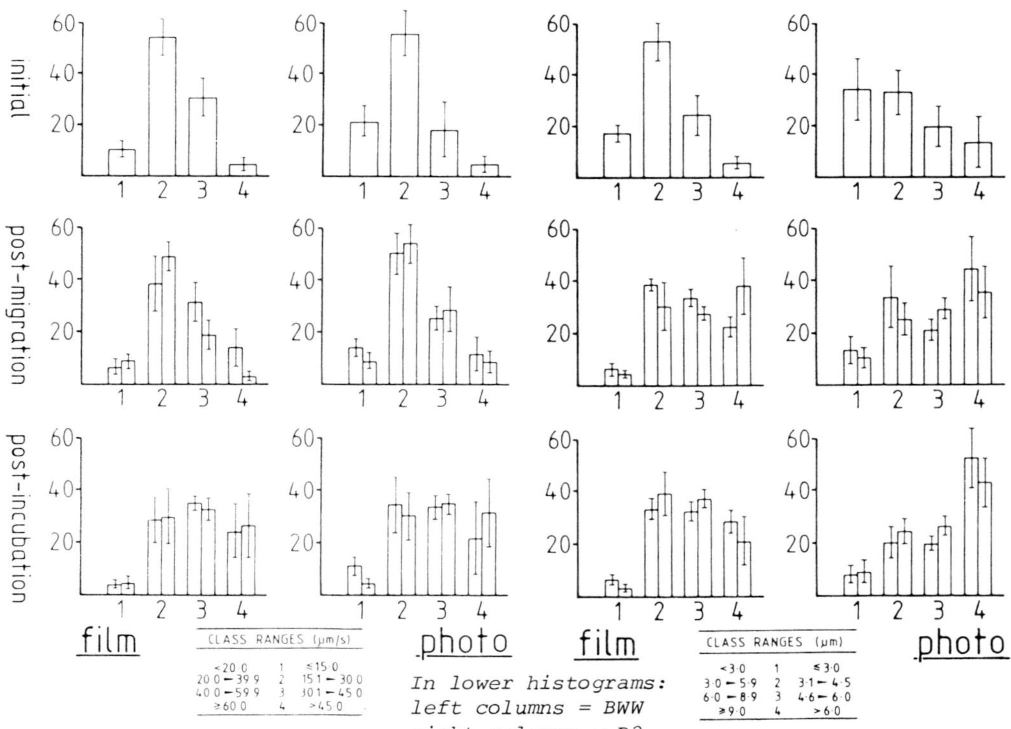


CONCLUSIONS

The existence of modifications in the patterns of movement of human spermatozoa associated with their migration from seminal plasma into synthetic media and their subsequent incubation under such conditions has been clearly demonstrated. Human sperm capacitation has generally been considered to require some 5-6 hours, although the whole process of capacitation, acrosome reaction and zona penetration may be achieved in <2½ h.[5] The heterogeneity of the sperm populations in human semen in terms of maturity will probably result in differences between spermatozoa in the evolution of their ability to embark upon capacitation. There may also be variations in the time needed to acquire the capacitated state. Furthermore, as the phenomenon of capacitation consists of several different processes, the achievement of an "activated" state of motility need not necessarily equate with attaining the fully capacitated state

Therefore while the limited numbers of active yet non-progressive gametes seen in the present study may have been exhibiting an equivalent form of motility to that reported in other mammalian species, the population phenomenon of "whiplash"-type motility may not be developed. Other changes in movement, seen even in the post-migration spermatozoa, may well constitute at least a part of the "activated" pattern for human spermatozoa.

REFERENCES

1. Biggers JD, Whitten WK & Whittingham DG. 1971. The culture of mouse embryos *in vitro*. In *Methods in Mammalian Embryology*, pp 86-116. Ed. JC Daniel Jr. Freeman & Co., San Francisco.
2. David G, Serres C & Jouannet P. 1981. Kinematics of human spermatozoa. Gamete Res., 2: 35-42.
3. Dubois M, Jouannet P, Berge P, Volochine B, Serres C & David G. 1975. Méthode et appareillage de mesure objective de la mobilité des spermatozoïdes humains. Ann. Phys. Biol. Med., 9: 19-41.
4. Katz DF & Overstreet JW. 1980. Mammalian sperm movement in the secretions of the male and female genital tracts. In *Testicular Development, Structure and Function*, pp 481-489. Eds A Steinberger & E Steinberger. Raven Press, New York
5. McMaster R, Yanagimachi R & Lopata A. 1978. Penetration of human eggs by human spermatozoa in vitro. Biol. Reprod., 19: 212-216.
6. Menezo Y. 1976. Milieu synthétique pour la survie et la maturation des gamètes et pour la culture de l'oeuf fécondé. C.R. Acad. Sci. Paris, 284: 1967-
7. Overstreet JW, Katz DF, Hanson FW & Fonseca JR. 1979. A simple inexpensive method for objective assessment of human sperm movement characteristics. Fert. Steril., 31: 162-172.
8. Overstreet JW, Yanagimachi R, Katz DF, Hayashi K & Hanson FW. 1980. Penetration of human spermatozoa into the human zona pellucida and the zona-free hamster egg: a study of fertile donors and infertile patients. Fert. Steril., 33: 534-542.
9. Testart J, Thébault A & Frydman R. 1980. La fécondation humaine ≤in vitro≥. Premiers résultats. J. Gyn. Obst. Biol. Reprod., 9: 319-324.

ACKNOWLEDGEMENTS: The authors would like to thank Professor G. David for his enthusiastic support and encouragement in these investigations.

ADDRESS: Laboratoire d'Histo-Embryologie, CHU de Bicêtre, 78 rue du Général Leclerc, 94270 le Kremlin-Bicêtre, France.

# DIRECTION OF ROLLING IN MAMMALIAN SPERMATOZOA

DAVID M. PHILLIPS, The Population Council, 1230 York Avenue, New York, New York, USA

Mature mammalian spermatozoa rotate around a central axis when they swim in an unconstrained environment. This rotation, termed rolling, may compensate for the asymmetrical beat to allow sperm to traverse straight paths (Phillips, 1973). Rolling is easy to document since the mammalian sperm head is flattened and, when viewed with dark field, phase or interference contrast optics, one observes flashes of light from the sperm head.

Since the compound microscope does not permit observation of the third dimension, it is difficult to determine in which direction spermatozoa roll. That is, figuratively speaking, if you were to view a sperm swimming towards you, would a point on the side of the sperm rotate in a clockwise or counterclockwise direction? Several investigators have addressed this problem using various types of approaches. Some have concluded that spermatozoa roll clockwise when viewed head-on (Linnet, 1981; Blokhuis, 1961), and others have concluded that the direction of the roll is counterclockwise (Woolley, 1977) or that sperm may roll either way (Drake, 1977).

In this investigator's opinion, it is possible to determine the direction of the roll by direct examination of films taken with a high speed motion picture camera. I, therefore, showed films to the participants of this conference and asked them to indicate by a show of hands their opinion as to the direction of the roll. A ten-minute film containing segments of swimming squirrel, chinchilla, mouse and rat sperm, as well as a short segment of the sperm of an insect (orthoptera) sperm was shown. Films were taken with a locam camera (Red Lake Labs, Santa Clara, CA), using Zeiss dark field or Nomarski optics. Frame rates were from 50 to 500 frames/sec and the film was shown at 24 frames/sec to the audience.

After they had seen the film, the spermatologists were requested to

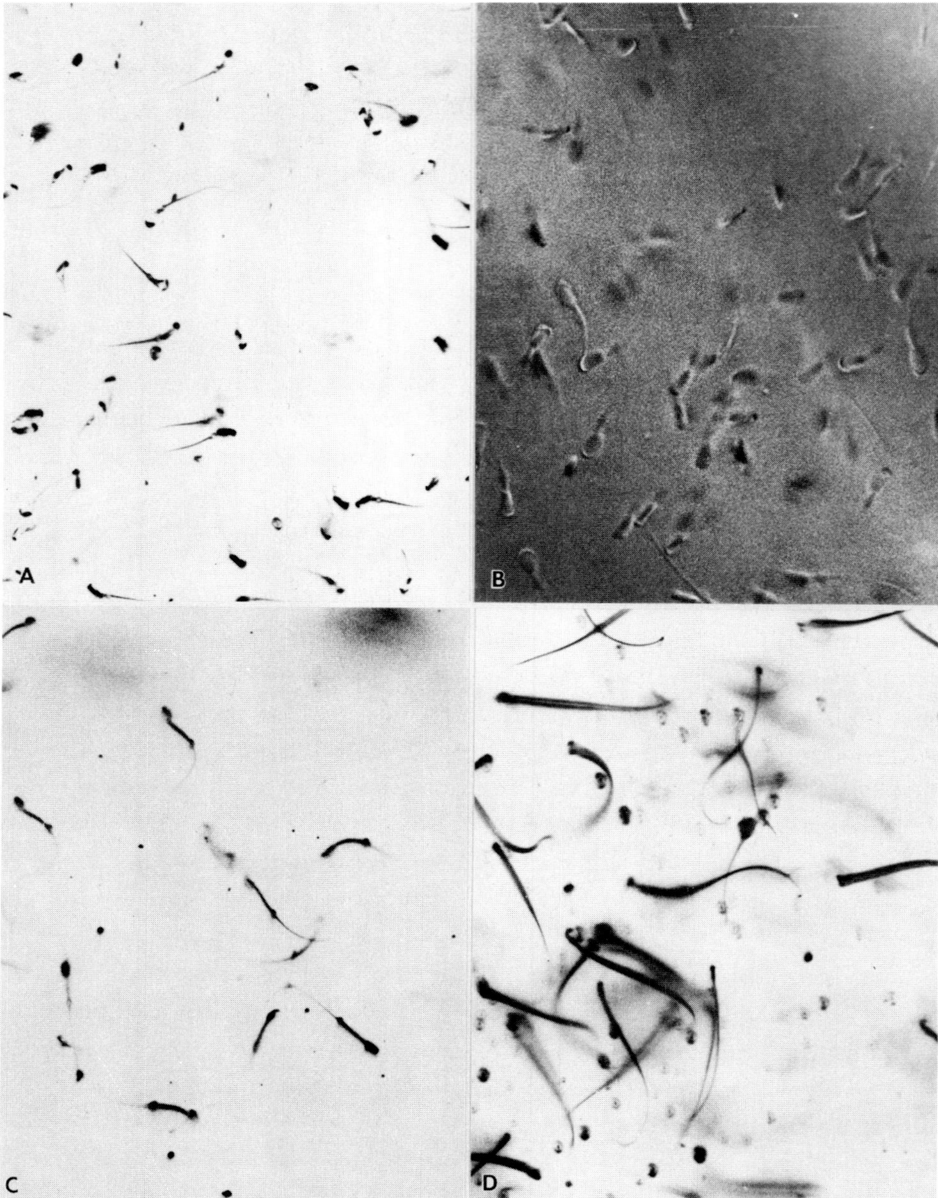


FIGURE 1. Prints of the film that was shown to eighty spermatologists at Siellac. Participants voted ten to one that the roll of the spermatozoa is equivalent to clockwise when viewed head-on. (A) Squirrel, dark field; (B) Chinchilla, Nomarski; (C) Mouse, dark field; (D) Rat, dark field.

indicate by a show of hands if they could discern whether the direction of rolling was equivalent to clockwise or counterclockwise when the sperm were viewed head-on. About 80 people were in the audience and about half indicated that they felt the spermatozoa rolled clockwise. Only three participants indicated that the direction of the roll was counterclockwise. The remainder of the audience either could not discern the direction of the roll, felt that spermatozoa rolled both ways, or abstained for some other reason*.

The fact that over 90% of those who felt they could discern the direction of rolling indicated that the sperm rolls clockwise strongly suggests that the predominant roll of mammalian spermatozoa (viewed head-on) is clockwise and not counterclockwise.

## REFERENCES

Blokhuis EWM. 1961. Optical investigations on the movement of bull spermatozoa. In: Proceedings of 4th International Congress on Animal Reproduction (The Hague), Vol. 2.

Drake AD. 1974. Observations on bull sperm rotation. Biol Reprod 10:78-84.

Linnet L. 1981. The three-dimensional nature of the sperm tail wave studied using heterophil head-to-head agglutinins. Am J Primatology 1:221-229.

Phillips DM. 1972. Comparative analysis of mammalian sperm motility. J Cell Biol 53:561-573.

Woolley DM. 1977. Evidence for "twisted plane" in undulations in golden hamster sperm tails. J Cell Biol 75:851-865.

*The film was shown in the evening after a delicious meal and generous quantities of excellent French wine and champagne. Participants had even carried bottles of champagne into the lecture room. Some of these spermatologists may not have been able to focus clearly on my presentation.

# OBSERVATION OF THE TAILWAVES OF MOUSE SPERMATOZOA

D.M. WOOLLEY, I.W. OSBORN, E.J.H. OLIVER & D.W. REA
Department of Physiology, The Medical School, University of Bristol, Bristol BS8 1TD, England.

## 1. INTRODUCTION

It is widely accepted that the waveform of the mammalian sperm flagellum is three-dimensional (1). The aim of this work is to discover the rules which govern the organization of bends into three dimensional configurations. By compiling a catalogue of the geometries recorded from free-swimming spermatozoa, we hope that a 3-D model will emerge to resolve the conflicting reports of the actual kinematics (2,3,4,5). We have used a new technique for 3-D reconstruction, involving two-colour darkfield micrography, coupled with computer-graphics (6,7). We present here some of the patterns seen in mouse spermatozoa at two stages of maturity: progressive cells from the cauda epididymidis, and cells 'activated' in vitro by the method of Fraser (8).

## 2. MATERIALS AND METHODS

C3H mice were stunned and killed by cervical dislocation. Spermatozoa were immediately flushed from the vas deferens and distal cauda epididymidis in a modified Tyrode solution (8) and the suspension was allowed to stand for 10 mins. Progressively swimming populations were withdrawn from this suspension, avoiding the sedimented aggregations. To induce activated motility, samples were incubated under liquid paraffin at 37°C for 2½ hours in an atmosphere of 5% $CO_2$ in air (8).

For photomicrography, aliquots were introduced into pre-warmed chambers approximately 80 um deep (6) and viewed with a Leitz Ortholux II microscope fitted with a heated stage (37°C), a Heine condenser adjusted to give true dark-field, and a x40 objective (type Pv, N.A. 0.70). Illumination was provided by a specially constructed xenon flash gun designed to run in a high frequency modelling mode and then be triggered to discharge 200 Wsec. of input energy in a single flash lasting 500 usec. (supplied by Noblelight Co. Ltd., Cambridge, England). This unit was mounted behind a Leitz Mecablitz condenser-housing and synchronized with a Wild MPS motor-driven 35 mm camera. Unambiguous 3-D information was recovered by placing an orange filter (Wratten 23a) over half the field and taking the pictures on Kodak E400 reversal film (6).

Thus far, the images have been studied in three ways: (i) by subjective inspection and physical modelling in wire; (ii) by tracing back-projected images, deriving sets of 3-D co-ordinates using a Graf-pen digitizing tablet (about 60 sets per flagellum), and then undertaking a graphics analysis based essentially on a 3-D to 2-D transformation and rotation algorithm, which allowed the flagellum to be inspected from any angle; (iii) by plotting from the stored data stereo-pair orthographic projections of the flagellum as seen from suitable viewpoints in advance of the sperm head. All three approaches were compared in interpreting the flagellar configurations.

## 3. OBSERVATIONS AND DISCUSSION

The progressive spermatozoa (38 clear examples) were swimming horizontally and rolling about the longitudinal axis. The incubated samples contained about 30% of activated cells characterized by exaggerated bend angles and non-progressive motility. Yet, although these activated cells were easy to photograph (Fig. 1f), only 8 of the images could be interpreted because of the frequent and confusing superimposition of one part of the flagellum on another.

In assessing the progressively swimming flagella, the greatest weight will be given to examples where the 3-D component is obvious and large. The reason for this caution is that, although the optical method is apparently very sensitive, the error associated with the measurement of a given image cannot be assessed. The most obvious conclusion then, is that the plane of bending of the flagellum rotates as the bends pass along the principal piece. This rotation has been seen in the proximal principal piece (Fig. 1a,f) but has a greater magnitude more distally (Figs. 1b,c). The most likely description is that the torsion occurs in the interbend regions, can be in either angular direction and commonly reaches about $90^{o}$. (From a single image, however, one cannot distinguish absolutely between, for example, an interbend $90^{o}$ CCW rotation and a midbend CW rotation of the same angle.) This description is in accord with that obtained for hamster spermatozoa by rapid freezing (5) except for the angular direction being variable. Our present observations would seem to deny any fundamental importance to the angular direction.

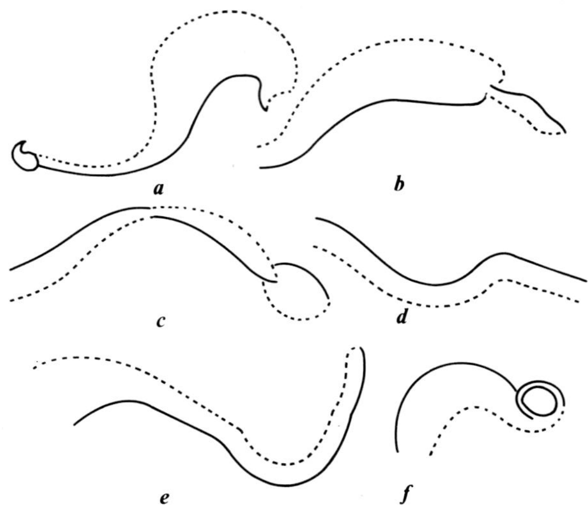


FIGURE 1·a-f   Tracings of representative flagellar images. The width of the image increases linearly with the magnitude of displacement in the optical axis. One edge appears orange in the micrographs and dotted in these drawings: by convention, where the dotted line is uppermost the flagellum is beneath the plane of the page, away from the viewer; and vice versa. The tracings are arranged so that the sperm head is always at the left extremity, but it is not drawn unless in focus. The distal part of Fig. 1f was lost from view. (See text for interpretations.)

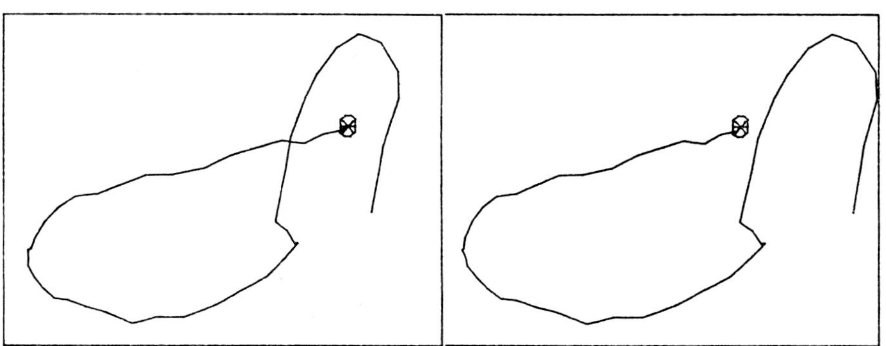

FIGURE 2. Data derived from Fig. 1c presented as stereo-pair orthographic (i.e. without perspective) projections. Using a stereo-viewer, the sperm should appear to be swimming towards the observer. This viewpoint is particularly helpful in visualising the overall geometry.

Apart from this characteristic change in the plane of bending in the principal piece, the bends themselves remain plane or nearly so. This is well known to be true of the most proximal bend but we have regularly seen that distal bend sequences, beyond the region of rotation, are also plane (Fig. 1, especially 1 e,f). Thus distal segments of the flagellum do not themselves, necessarily, have a 3-D action. Indeed there are a few examples of entirely plane momentary configurations in rotating spermatozoa (Fig. 1d); the existence of these is evidence against a mandatory 3-D output from the axonemal motor. Finally, the plane flagellar segments are often surprisingly non-smooth (Fig. 1e). This feature, which deserves further study, may be a manifestation of the current 'arc-line' interpretation of flagellar geometry (9); it was not seen in activated flagella.

In conclusion, the darkfield method has revealed the great complexity and variety of form of the active flagellum. While we cannot from our 'static' data reach a truly synthetic view, we have clear descriptive evidence that the plane of flagellar bending rotates progressively as the bends propagate along the principal piece. The cause of this rotation remains to be discovered.

4. ACKNOWLEDGEMENT This research has been supported by the A.R.C.

REFERENCES

1. Woolley DM. 1979. Interpretations of the pattern of sperm tail movements. In The Spermatozoon (edited by DW Fawcett & JM Bedford). Urban & Schwarzenburg, Baltimore-Munich. pp 69-79.
2. Gray J. 1958. The movement of the spermatozoa of the bull, J. Exp. Biol. 35, 96-108.
3. Rikmenspoel R. 1965. The tail movement of bull spermatozoa. Observations and model calculations. Biophys. J. 5, 365-92.
4. Denehy MA, Herbison-Evans D, and Denehy BV. 1975. Rotational and oscillatory components of the tailwave in ram spermatozoa. Biol. Reprod. 13, 289-97.
5. Woolley DM. 1977. Evidence for "twisted-plane" undulations in golden hamster sperm tails. J. Cell Biol. 75, 851-65.
6. Woolley DM. 1981. A method for determining the three-dimensional form of active flagella, using two-colour darkground illumination. J. Microscopy, 121, 241-44.
7. Osborn IW and Woolley DM. 1982. Flagellar torsions in motile spermatozoa as revealed by a two-colour darkfield method for three-dimensional reconstruction.. J. Physiol. in the press.
8. Fraser LR. 1977. Motility patterns in mouse spermatozoa before and after capacitation. J. Exp. Zool. 202, 439-44.
9. Brokaw CJ. 1965. Non-sinusoidal bending waves of sperm flagella. J. Exp. Biol. 43, 155-169.

# INFLUENCE OF FLAGELLAR WAVE DEVELOPMENT ON HUMAN SPERM MOVEMENT IN SEMINAL PLASMA

C. SERRES, D. FENEUX, P. JOUANNET and G. DAVID

*Laboratoire Histologie-Embryologie-Cytogénétique*
*CHU KREMLIN-BICETRE (France)*

## 1. INTRODUCTION

The heterogeneity of human sperm movement in seminal plasma(1) may be used as an approach to analysing flagellar beating. Various parameters characterizing the movement of spermatozoa have been described by microcinematography(2). The present study, which is a further analysis of the parameters of flagellar beating by the same technique, investigates the influence of these parameters on the pattern of the sperm trajectories.

## 2. MATERIAL AND METHODS

12 normal semen samples ($N=20-127 \times 10^6$/ml, motility $>$ 50 %) were used. Within 1 h of ejaculation a 25 $\mu$l aliquot of each sample was placed between slide and and 22x32 mm coverslip (preparation depth $\simeq$ 30 $\mu$m). Each sample was filmed at ambient temperature (Nomarski optics, x 25 objective, 50 frames/s). The frame-by-frame analysis was performed as previously described(2).

Head movement was characterized by the parameters Vp (progression velocity) and Ah (mean amplitude of the lateral head displacement). Flagellar beating was described by $\alpha$ (supplementary angle formed by the 2 tangents to the wave) and c (distance from the base of the sperm head to the wave center (o)).(Fig. 1).

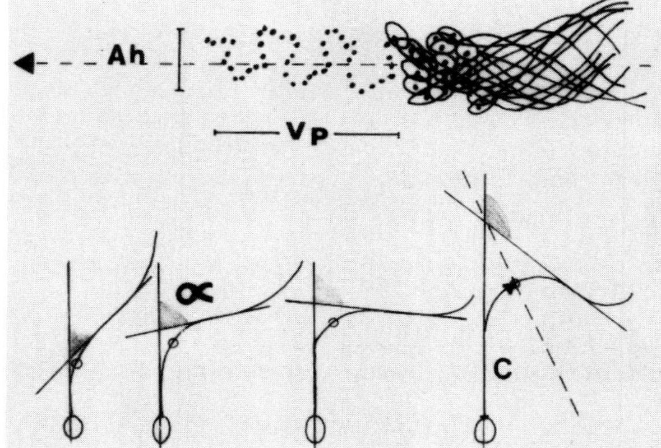


FIGURE 1. Sperm head trajectory during 1s, and flagellar wave analysis in 4 successive images.

## 3. RESULTS AND DISCUSSION

3.1. Head trajectory analysis. 57 spermatozoa, whose trajectories were typical of the heterogeneous seminal population, were considered. Four types of trajectories were defined according to the medians of the Vp (29.5 μm/s, range = 7-47 μm/s) and Ah (5.5 μm, range = 2-10 μm) distributions.

3.2. Flagellar beating analysis. Only that part of the flagellum beating in the plane of the head (the proximal 25-30 μm, i. e. ≃1/2 the flagellar length) could be observed and analysed.

The values of $\alpha$ and c increase linearly during beat development (Fig. 2a) as previously observed in the sea urchin by GOLDSTEIN(3). 3 spermatozoa showed delays in the wave propagation (Fig. 2b). This observation would suggest a possible dissociation between initiation and propagation of the wave, as hypothesized by LINDEMANN for bull spermatozoa(4).

The propagation velocity (Vc) and the angular velocity of wave development (Vd) were calculated from the slopes of $\alpha$ and c. The values of Vc and Vd were found to be different between the 57 spermatozoa (Vc : 91-499 μm/s, Vd : 0.6-3.3°/ms). The curvature of a wave varied between spermatozoa, e. g. at a point 15 μm along the flagellum (c), $\alpha$ varied between 60 and 150° (alternatively, values of $\alpha$ = 130-140° occurred at positions 17-27 μm along the flagellum).

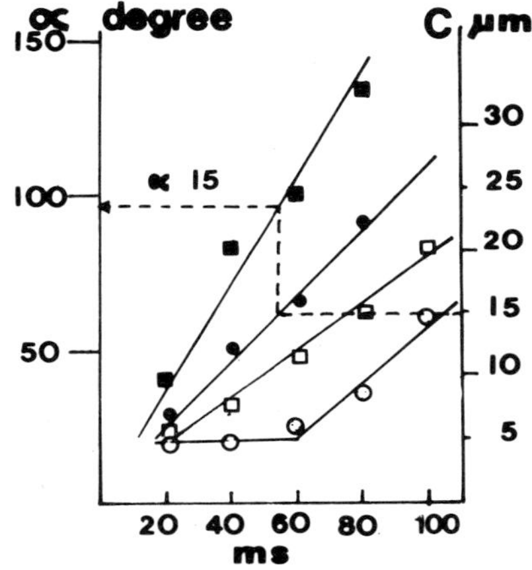


FIGURE 2. Variations of $\alpha$ (■ and □) and c (● and ○) during 1 beat. Example a : ■ ● ; example b : □ ○

The greatest values attained before the wave moved out of the plane of the head were $\alpha$ = 150° and c = 25 m. While the maximum values of $\alpha$ in man are identical to those measured in the sea urchin(3), they occur further along the flagellum (man : 17-25 μm vs sea urchin : 10 μm). This may be due to the presence of dense fibres around the axoneme in man, which would diminish the flexibility of the flagellum(5). The size of the dense fibres has previously been correlated with flagellar inflexibility(6).

3.3. Relationship between flagellar beating and head trajectory. Vp was correlated with VcxVd (r=0.67, $P<0.001$), Ah with Vd/Vc (r=0.42, $P<0.01$) and Ah with $\alpha_{15}$ (r=0.52, $P<0.01$). The 4 trajectory types correspond to different flagellar

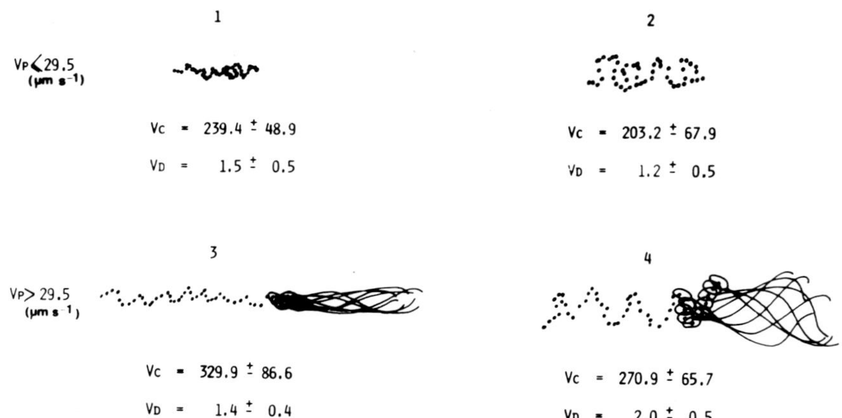


FIGURE 3. Flagellar parameters in the 4 groups of trajectories.

characteristics (Fig. 3). The mean values of Vc and Vd differed significantly between the 2 groups of fast-swimming spermatozoa (groups 3 & 4) : $P<0.01$). The group 3 with small Ah values had greater values of Vc, but smaller values of Vd, than had spermatozoa with large Ah (group 4). Vc and Vd were not significantly different between the 2 groups of slow-swimming spermatozoa (groups 1 & 2). Therefore, the various types of trajectories appear to depend mainly upon the 2 flagellar parameters Vc and Vd, which are both necessary for ensuring good progression. In the bull, RIKMENSPOEL(7) related progression velocity with rotation frequency and the square of the wave amplitude.

3.4. <u>Relationship between Vc and Vd</u>. The values of Vc and Vd were not correlated in the 57 spermatozoa (Fig. 4). There was a limiting value of Vd (2°/ms) above which it decreased as Vc further increased ($>$300 µm/s). The development of the wave and its propagation should not be independent, although experimental evidence has indicated that these two processes result from different mechanisms which may be modified independently. The reason for a limiting value of Vd is not clear. It may be due to biochemical mechanisms and structural differences in the processes of flagellar wave propagation and development(8,9).

The most progressive spermatozoa (Vp$>$40 µm/s) were found to be situated around the peak of the relationship between Vc and Vd. The majority of these spermatozoa (8/11) belonged to group 4. It would therefore appear that those spermatozoa with the most effective propulsion are dependent upon an optimal relationship between Vc and Vd. However, it is not yet possible to correlate a spermatozoon's pattern of movement to its fertilizing ability.

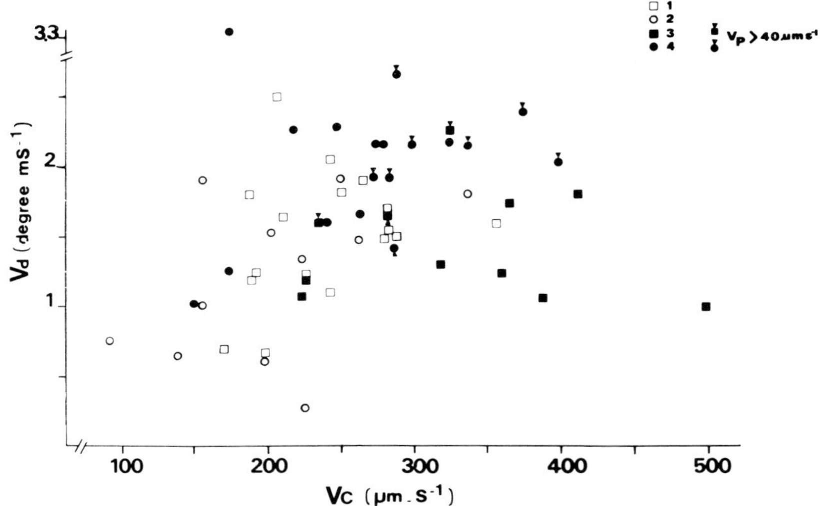


FIGURE 4. Relationship between Vc and Vd.

REFERENCES


1. OVERSTREET JW, KATZ DF, HANSON FW, FONSECA FR. *1979*. A simple inexpensive method for objective assessment of human sperm movement characteristics. *Fertil Steril 31*, 162-172.
2. DAVID G, SERRES C, JOUANNET P. *1981*. Kinematics of the human spermatozoa. *Gamete Res 4*, 83-95.
3. GOLDSTEIN S. *1977*. Assymmetric waveform in echinoderm sperm flagella. *J exp Biol 71*, 157-170.
4. LINDEMANN CB, FENTIE I, RIKMENSPOEL R. *1980*. A selective effect of $N_i^{2+}$ on wave initiation in bull sperm flagella. *J Cell Biol 87*, 420-426.
5. WOOLEY DM. *1979*. Interpretations of the pattern of sperm tail movements. *In : The Spermatozoon*, pp 69-79, *eds* DW Fawcett & JM Bedford. Baltimore, Munich, Urban & Schwarzenberg, Inc.
6. PHILIPPS D, OLSON G. *1975*. Mammalian sperm motility : Structure in relation to function. *In : Functional Anatomy of the Spermatozoon,* pp 117-126, *ed* BA Afzelius. Oxford, Pergamon Press.
7. RIKMENSPOEL R, VAN HERPEN G, EIJKHOUT P. *1960*. Cinematographic observations of the movements of bull sperm cells. *Phys Med Biol 5,* 167-181.
8. WARNER FD, SATIR P. *1974*. The structural basis of ciliary bend formation : radial spoke positional changes accompanying microtubule sliding. *J Cell Biol 63,* 35-63.
9. BROKAW CJ. *1980*. Elastase digestion of sea urchin sperm flagella. *Science NY 207,* 1365-1367.

# INITIATION OF SPERM FLAGELLAR MOVEMENT USING RAT DEMEMBRANATED SPERM MODEL: NUCLEOTIDE SPECIFICITIES

MONTRI CHULAVATNATOL and NAPA TREETIPSATIT

Department of Biochemistry, Faculty of Science, Mahidol University, Rama VI Road, Bangkok 10400, Thailand

## 1. INTRODUCTION

Sperm motility depends on ATP as the energy source and is controlled by cAMP (1). This has been supported not only by studies correlating cAMP content and sperm motility (2) but also by recent studies using demembranated sperm models (3,4,5). ATP is hydrolyzed by dynein ATPase of the sperm axoneme to generate the motility force (6), while cAMP controls the sperm motility by activating the cAMP-dependent protein kinase to phosphorylate certain sperm proteins (7-11). The increase in sperm cAMP is suggested to initiate the sperm motility (12). To further the understanding of the intracellular events involved in the initiation of sperm motility, we have employed rat demembranated sperm model (13,14) to investigate the specificities of the active site of dynein ATPase and of the binding site of the cAMP-binding protein controlling the sperm motility.

## 2. MATERIALS AND METHODS

Sperm from rat epididymis were freshly prepared as described previously (15). A small aliquot containing $10^6$ sperm was treated with 1 ml of the demembranating (DM) solution (4) for 30 sec and then reactivated with ATP and cAMP or their analogs. After a specified time, a small volume of the mixture was quickly smeared onto a glass slide. Differential counting of the beating sperm and non-motile ones in a random field to a total of 200 was immediately performed under a light microscope. The counting normally took 5-15 min. All steps were carried out at room temperature.

## 3. RESULTS

When the sperm were treated with the DM solution, all became non-motile within 30 sec. Upon reactivating with 1 mol $m^{-3}$ ATP and $10^{-1}$ mol $m^{-3}$ cAMP, instant beating of nearly all demembranated sperm was observed (Table 1). However, no forward motility was seen, suggesting the possible absence of

Table 1. Effect of the composition of demembranating (DM) solution on the % beating of demembranated rat sperm (mean ± SD; n=5). The beating was initiated by adding 1 mol $m^{-3}$ ATP and $10^{-1}$ mol $m^{-3}$ cAMP and determined at 1 and 30 min after the addition. The complete DM solution consisted of 0.1% Triton X-100, 2 x $10^{2}$ mol $m^{-3}$ sucrose, 2.5 x 10 mol $m^{-3}$ potassium glutamate, 1 mol $m^{-3}$ $MgSO_4$, 1 mol $m^{-3}$ dithiothreitol and 2 x 10 mol $m^{-3}$ Tris-HCl buffer, pH 7.9 (4).

| DM solution | Caput Epididymal Sperm | | Cauda Epididymal Sperm | |
|---|---|---|---|---|
| | 1 min | 30 min | 1 min | 30 min |
| Complete | 95 ± 3 | 99 ± 1 | 96 ± 3 | 99 ± 1 |
| -Triton X-100 | 67 ± 14 | 45 ± 18 | 73 ± 19 | 71 ± 21 |
| -Sucrose | 22 ± 12 | 71 ± 13 | 23 ± 19 | 52 ± 36 |
| -Glutamate | 81 ± 4 | 83 ± 5 | 66 ± 9 | 43 ± 32 |
| $-MgSO_4$ | 0 ± 1 | 1 ± 1 | 0 | 0 |
| -Dithiothreitol | 6 ± 6 | 29 ± 18 | 3 ± 3 | 0 |

the forward motility protein (12). When the DM solution without one of its components was used, the beating was reduced partially or completely (Table 1), suggesting that all components, particularly $Mg^{++}$ and dithiothreitol, were required for establishing the rat demembranated sperm model.

The beating cannot be initiated by cAMP alone or by cAMP and other natural ribonucleoside triphosphates (Table 2). Among ATP analogs tested, only that without 2'-OH (2'-deoxy-ATP) was found capable of initiating the beating (Table 2). The ineffective ATP analogs were those with non-hydrolyzable γ-phosphoryl group, with substitutions at C-8 or N-6 of the adenine ring, without 3'-OH or with 2'- β-OH (arabinosyl-ATP) in the ribose moiety. When ATP was replaced by ADP, the beating initiation was slow, due to the time needed for the conversion of 2ADP into ATP + AMP catalyzed by myokinase. Reactivation of the sperm beating by ATP alone was also slow due to the time needed to form cAMP from the added ATP catalyzed by adenylate cyclase. However, as little as $10^{-3}$ mol $m^{-3}$ of cAMP was sufficient to shorten the time for the complete reactivation to within 1 min (Table 3). The cAMP analogs with substitutions at C-8, N-6 of the adenine ring were partially effective in initiating the beating (Table 3). Two modifications on the cAMP structure leading to the complete loss of its reactivating capability were the removal of 2'-OH from the ribose moiety and change of the cyclic phosphate from 3',5'-position to 2',3'-position. Among the natural cyclic nucleotides, the order of effectiveness was: cAMP > cGMP > cCMP > cUMP, with cTMP being ineffective (Table 3).

Table 2. Effects of nucleotides on the % beating of demembranated rat sperm (mean ± SD; n=5). The beating was initiated by adding $10^{-1}$ mol $m^{-3}$ cAMP and one of the listed compounds and determined at 1 and 30 min after the addition.

| Addition (1 mol $m^{-3}$) | Caput Epididymal Sperm | | Cauda Epididymal Sperm | |
|---|---|---|---|---|
| | 1 min | 30 min | 1 min | 30 min |
| None | 0 | 0 | 0 | 0 |
| ATP | 95 ± 3 | 99 ± 1 | 96 ± 3 | 99 ± 1 |
| ADP | 13 ± 8 | 87 ± 4 | 6 ± 3 | 93 ± 3 |
| β,γ-Imido-ATP | 0 | 0 | 0 | 0 |
| β,γ-Methylene-ATP | 0 | 0 | 0 | 0 |
| 3'-Deoxy-ATP | 4 ± 5 | 0 | 0 | 0 |
| 2'-Deoxy-ATP | 90 ± 2 | 95 ± 1 | 88 ± 3 | 97 ± 1 |
| Arabinosyl-ATP | 0 | 0 | 0 | 0 |
| 8-Bromo-ATP | 0 | 0 | 0 | 0 |
| 8-6(Aminohexyl)-amino-ATP | 0 | 0 | 0 | 0 |
| $N^6$-[(6-Aminohexyl) carbamoyl-methyl]-ATP | 0 | 0 | 0 | 0 |
| GTP | 0 | 0 | 0 | 0 |
| UTP | 0 | 0 | 0 | 0 |
| CTP | 0 | 0 | 0 | 0 |
| TTP | 0 | 0 | 0 | 0 |

Table 3. Effects of cyclic nucleotides on the % beating of demembranated rat sperm (mean ± SD; n=5). The beating was initiated by adding 1 mol $m^{-3}$ ATP and one of the listed compounds and determined at 1 min after the addition.

| Addition | Caput Epididymal Sperm | | | | Cauda Epididymal Sperm | | | |
|---|---|---|---|---|---|---|---|---|
| Conc. (mol $m^{-3}$): | $10^{-1}$ | $10^{-2}$ | $10^{-3}$ | $10^{-4}$ | $10^{-1}$ | $10^{-2}$ | $10^{-3}$ | $10^{-4}$ |
| cAMP | 95±3 | 93±3 | 89±5 | 39±7 | 96±3 | 94±2 | 91±4 | 40±10 |
| 8-Azido-cAMP | 94±4 | 89±3 | 62±17 | 44±1 | 97±2 | 91±2 | 66±17 | 43±2 |
| 8-Bromo-cAMP | 97±1 | 79±8 | 63±10 | 31±2 | 97±1 | 80±5 | 62±10 | 29±2 |
| 8-Chlorophenyl-thio-cAMP | 97±1 | 79±6 | 58±7 | 32±5 | 97±1 | 76±6 | 60±6 | 31±2 |
| $N^6$-$O^{2'}$-Dibutyryl-cAMP | 92±4 | 42±6 | 10±3 | 3±2 | 93±3 | 42±8 | 8±4 | 3±2 |
| $N^6$-Butyryl-cAMP | 85±12 | 49±22 | 21±10 | 0 | 73±16 | 46±16 | 18±3 | 1±1 |
| $O^{2'}$-Butyryl-cAMP | 85±7 | 83±10 | 40±19 | 2±2 | 88±9 | 77±13 | 25±15 | 1±1 |
| 2'-Deoxy-cAMP | 3±2 | 0 | 0 | 0 | 1±1 | 0 | 0 | 0 |
| 2',3'-cAMP | 0 | 0 | 0 | 0 | 0 | 0 | 0 | 0 |
| cGMP | 61±13 | 16±2 | 6±2 | 1±1 | 55±18 | 17±6 | 6±5 | 2±3 |
| cCMP | 54±16 | 0±1 | 0 | - | 40±11 | 1±1 | 0 | - |
| cUMP | 40±7 | 3±3 | 0 | - | 23±8 | 2±2 | 0 | - |
| cTMP | 5±4 | 0 | 0 | - | 1±1 | 0 | 0 | - |

## 4. DISCUSSION

The rat demembranated sperm model appears suitable for studying the dynein ATPase and the cAMP control of flagellar movement because it is highly responsive to exogenous ATP and cAMP (Table 1). The rat sperm model has revealed the high substrate specificity for the dynein ATPase (Table 2). as contrast to a broad specificity for the cAMP-binding protein controlling the sperm motility (Table 3). Since models from the caput and the cauda epididymal sperm respond nearly identically to reactivation by nucleotides (Tables 2 and 3), the motile apparatus of the rat sperm is probably unchanged during epididymal transit. This conclusion disagrees with that reported previously (4). The contents of ATP (15) and cAMP (16) in rat epididymal sperm, calculated with the assumptions that the water content of the sperm is 50% (17) and that the packed cell volume of $10^9$ sperm is 0.5 ml, are 1-3 and (4-8) x $10^{-3}$ mol $m^{-3}$ respectively. So, the normal concentrations of both nucleotides seem to be high enough for full flagellar movement. Thus, the suggestion that increase in cAMP initiates the sperm motility (12) appears inaccurate.

Supported by IDRC (3-P-79-0132). We thank T. Vajrodaya for the typing.

## REFERENCES


1. Lindermann CB. 1980. Testicular Development, Structure and Function. Ed: A. Steinberger and E. Steinberger. New York, Raven Press, pp 473-479.
2. Garbers DL, Kopf GS. 1980. Adv. Cyclic Nucleotide Res. 13, 251-307.
3. Lindermann CB. 1978. Cell 13, 9-18.
4. Mohri H, Yanagimachi R. 1980. Exp. Cell Res. 127, 191-196.
5. Morisawa M, Okuno M. 1982. Nature 295, 703-704.
6. Warner FD, Mitchell DR. 1980. Intern. Rev. Cytol. 66, 1-43.
7. Huacuja L, Delgada NM, Merchant H, Pancardo RM, Rosado A. 1977. Biol. Reprod. 17, 89-96.
8. Brandt H, Hoskins DD. 1980. J. Biol. Chem. 255, 983-987.
9. Chulavatnatol M, Panyim S, Wititsuwannakul D. 1982. Biol. Reprod. 26, 197-207.
10. Tongkao D, Chulavatnatol M. 1979. The Spermatozoon. Ed: D.W. Fawcett and J.M. Bedford. Baltimore-Munich, Urban and Schwarzenberg. pp129-134.
11. Tash JS, Means AR. 1982. Biol. Reprod. 26, 745-763.
12. Hoskins DD, Brandt H, Acott TS. 1978 Fed. Proc. 37, 2534-2542.
13. Fentie IH, Lindermann CB. 1978. J. Cell Biol. 79, 290a.
14. Bouchard P, Penningroth SM, Cheung A, Gagnon C, Bardin CW. 1981. Proc. Natl. Acad. Sci. USA 78, 1033-1036.
15. Chulavatnatol M, Hasibuan I, Yindepit S, Eksittikul T. 1977. J. Reprod. Fertil. 50, 137-139.
16. Delrio AG, Raismans R. 1978. Experientia 34, 670-671.
17. Drevius LO. 1972. J. Reprod. Fertil. 28, 15-28.

# DIRECT EVIDENCE THAT CYCLIC AMP IS AN INDISPENSABLE FACTOR FOR INITIATION OF SPERM MOTILITY IN SALMONID FISHES

M. MORISAWA[1], M. OKUNO[2] and S. MORISAWA[3]

[1]Ocean Research Institute, University of Tokyo; [2]Department of Biology, University of Tokyo and [3]St. Marianna University, School of Medicine, Japan.

## 1. INTRODUCTION

Although much information has been accumulated concerning the mechanism of sperm motility, the initiation mechanism of sperm motility has been remained unclear. Recently we have demonstrated that sperm motility of salmonid fishes such as chum salmon, rainbow trout, masu salmon and char is suppressed in the sperm duct by potassium (1,2) which is contained at higher concentrations in the seminal plasma (3). In the rainbow trout, during natural spawning, release of suppression by potassium in fresh water induces intraflagellar cyclic AMP and the endogenous cAMP to convert the axoneme from immotile state to motile one, thereby allowing axonemal movement i.e. sperm motility to occur (4,5). Here we show that cAMP is also the internal factor indispensable for initiation of sperm motility in another salmonid fish, chum salmon. We further discuss the possibility that cAMP is the general factor initiating sperm motility in other animals.

## MATERIALS AND METHODS

Semen was collected from mature chum salmon *Oncorhynchus keta* and rainbow trout *Salmo gairdneri* by inserting a pippete into the sperm duct. The plasma membranes of spermatozoa were then removed on ice with extracting medium containing 0.15 M KCl, 0.5 mM $MgSO_4$, 5 mM $CaCl_2$, 0.5 mM EDTA, 4 mM DTT, 2 mM Tris buffer pH 8.2 and 0.04 % Triton X-100, 0.1 % saponin or 50 % v/v glycerol and the demembranated spermatozoa were reactivated at room temperature by previously described procedure (4,5). For electron microscopy, demembranated sperm were fixed with 2 % glutaraldehyde in

0.12 M phosphate buffer at pH 7.7 following 1 % osmium tetroxide in 0.12 M phosphate buffer and then embedded in epoxy resin and observed with JEOL-100B. In the case of chum salmon, we used 1 % glutaraldehyde.

## 3. RESULTS

3.1. Electron micrographs of demembranated spermatozoa. As shown in Fig. 1, when spermatozoa were treated with extracting medium containing 0.04 % Triton X-100, the plasma membrane, which surrounds a bullet shape head, midpiece including small mitochondria and flagellum, was completely extracted in rainbow trout(a,b)and chum salmon (data not shown). Mitochondria also removed completely (a) in both species (compare with intact sperm in ref. 6). In the spermatozoa treated with 0.1 % saponin, the plasma membrane did not solubilize but it remained as sheet structure which detached from axoneme in the suspension. Hexagonal structure are arranged on the surface of the flagment of the plasma membrane. In the chum salmon, the outer doublet of microtubules disappeared in both Triton X-100 and saponin-treated spermatozoa.

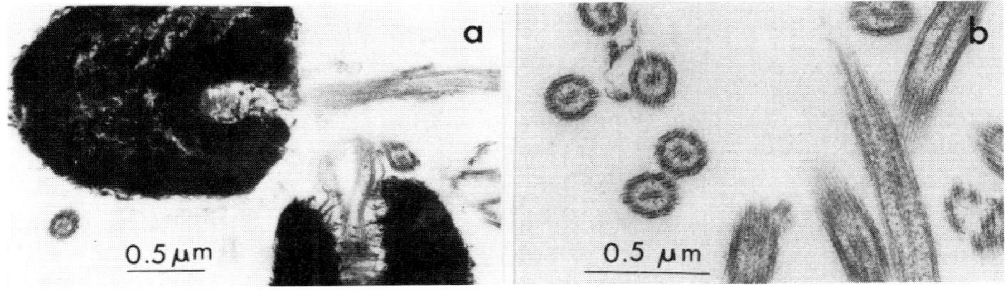


FIGURE 1. Electron micrographs of demembranated spermatozoa in rainbow trout.

3.2. Effect of cAMP on the demembranated spermatozoa.

When spermatozoa of chum salmon and of rainbow trout were demembranated with extracting medium containing 0.04 % Triton X-100, 0.1 % saponin or 50 % v/v glycerol and then suspended in reactivating medium containing 0.2 mM ATP in the absence of cAMP, axonemal movement did not occur; all spermatozoa were completely quiescent. However, when the demembranated sperm were suspended in reactivating medium containing $MgATP^{2-}$ and 20 μM cAMP, axonemal movement occurred ; 100 % of spermatozoa demembranated with Triton X-100 or saponin revealed active movement. In glycerol-treated spermatozoa, only head vibration was observed. These results suggest that an energy supply alone is not enough to induce motility, but that the axoneme of chum salmon as well as rainbow trout is functionally immotile and the immotile axoneme is converted to the motile state by exposure to cAMP resulting initiation of sperm motility (Table 1).

Table 1. Motility of spermatozoa demembranated with Triton X-100, saponin or glycerol in the presence or absence of cAMP.

| Extracting conditions | Species | Absence of cAMP | Presence of cAMP |
|---|---|---|---|
| Triton X-100 | Rainbow trout | 0 % | 100 % |
| | Chum salmon | 0 | 100 |
| Saponin | Rainbow trout | 0 | 100 |
| | Chum salmon | 0 | 100 |
| Glycerol | Rainbow trout | 0 | head vibration |
| | Chum salmon | 0 | head vibration |

Table 2. Effects of nucleotides on flagellar movement of demembranated spermatozoa in chum salmon and rainbow trout*

| | In the presence of 20 μM cAMP and 0.2 mM nucleotides | | In the presence of 0.2 mM MgATP and 0.2 mM nucleotides | |
|---|---|---|---|---|
| Nucleotides | rainbow trout | Chum salmon | Rainbow trout | Chum salmon |
| AMP | - | - | - | - |
| ADP | - | - | - | - |
| ATP | +++++ | +++++ | - | - |
| GMP | - | - | - | - |
| GDP | - | - | - | - |
| GTP | - | - | - | - |
| cAMP | - | - | +++++ | +++++ |
| cGMP | - | - | +++++ | ± |
| none | - | - | - | - |

* Quote from a paper by Okuno and Morisawa(5).

When the demembranated spermatozoa of chum salmon or rainbow trout were suspended in the solution containing 20 μM cAMP and 0.2 mM nucleotides, they were reactivated only in the presence of ATP (Table 2). When sperm motility was observed in the presence of 0.2 mM nucleotides and 0.2 mM $MgATP^{2-}$, only cAMP and cGMP were able to initiate motility. A much higher concentration of cGMP was required to initiate motility especially in chum salmon suggesting that cAMP rather than cGMP is the factor in initiating salmonid sperm motility.

## 4. DISCUSSION

Here, we show that cAMP is internal factor in initiating sperm motility in chum salmon as well as rainbow trout (see ref. 4,5). Because sperm motility of other salmonid species is also regulated by a change of external potassium (2), internal cAMP which may be induced by the release of suppression by potassium is the indispensable factor in initiating sperm motility in Salmonidae.

In gold fish, sperm motility is regulated by a change of external osmolality during natural spawning (1,7) and our preliminary results indicate that cAMP may be the factor in initiating sperm motility (8). Recently cAMP was suggested to be the initiation factor also in tunicate (9) and sea urchin(10) spermatozoa. Thus cyclic nucleotide seems to be the factor indispensable for initiation of sperm motility in invertebrates and vertebrates.

## REFERENCES


1. Morisawa M, Suzuki K. 1980. Science. 210, 1145-1147.
2. Morisawa M, Suzuki K, Morisawa S. in submitting, J.Exp. Biol.
3. Morisawa M, Hirano T, Suzuki K. 1979. Comp. Biochem. Physiol. 64A, 325-329.
4. Morisawa M, Okuno M. 1982. Nature. 295, 703-704.
5. Okuno M, Morisawa M. 1982. in press, Proc. Ooji Seminor in Tokyo. New York and London, Academic Press.
6. Morisawa S. 1981. Bull.St. Marianna Univ. School Med. 10, 151-169.
7. Morisawa M,Suzuki K, Shimizu H, Morisawa S, Yasuda K. in submitting. J. Exp. Biol.
8. Morisawa M, Okuno M, Suzuki K, Morisawa S, Ishida K. 1982. in press, J. Submicroscop. Cytol.
9. Brokaw CJ. Cell Motility. Suppl. 1, 185-189. 1982.
10. Murofushi. 1982. in press. Proc. Ooji Seminor in Tokyo. New York and London, Academic Press.

# PROTEIN-CARBOXYL METHYLATION IN SPERMATOZOA

C. Gagnon[1], P. Bouchard[2] and C.W. Bardin[3]

1- Unité de Biorégulation Cellulaire et Moléculaire, CHUL Québec, Canada and Département de Pharmacologie, Faculté de Médecine, Université Laval, Québec, Canada.

2- Service d'Endocrinologie, Hôpital Bicêtre, 94270 Le Kremlin-Bicêtre, France.

3- Population Council, 1230 York Ave., New York, USA.

## Introduction

The protein-carboxyl methylation system is composed of three components: a methylating enzyme, protein-carboxyl méthylase (PCM); a demethylating enzyme, protein methylesterase (PME); and methyl acceptor proteins (MAP). This enzymatic system reversibly modifies, by methylation, the carboxyl groups of protein substrates thus affecting their charge, structure and function.

## Localization of the protein-carboxyl methylation system in testes and mature spermatozoa

PCM has been shown to be present both in enkaryotes and prokaryotes (for reviews, see 1-3). PCM is very concentrated in testes and in brain (4). In testes, the cellular localization of PCM has been studied following collagenase digestion and cell separation on a 2-4% bovine serum albumin gradient. PCM was primarily concentrated in germ cells with the highest specific activity detected in spermatids (5). The MAP were also measured in different cell types and, as for PCM, the highest concentration of MAP was observed in spermatids (5).

Both PCM and MAP were also observed in mature spermatozoa and the ratio MAP/PCM increases by as much as 20 fold in rat and 60 fold in rabbit as the spermatids mature to spermatozoa and acquire full motility in the cauda of the epididymis (6). Recently, we have shown that all mammalian tissues contain a PME, an enzyme that hydrolyses protein-methyl esters formed by PCM to yield the protein substrate under its original form and methanol (7). The complete protein-carboxyl methylation system is depicted in Fig. 1. The enzyme PME has also been found in mature spermatozoa where, by opposition to PCM, more than 90% of the activity was associated with detergent insoluble flagellar components (6).

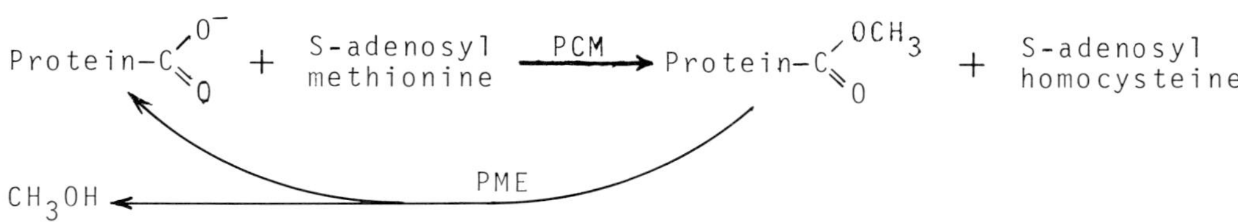


FIGURE 1. The protein-carboxyl methylation system. PCM: protein-carboxyl methylase PME: protein methylesterase.

Functions of the protein-carboxyl methylation system

On a functionnal aspect, the protein-carboxyl methylation system has been shown to be involved in exocytotic secretion in exocrine and endocrine glands (1,8), in leukocyte chemotaxis (1,9), and in bacterial chemotaxis (2). In all these biological processes, the interaction of ligands with their specific receptors caused a rapid increase in protein-carboxyl methylation.

Possible role of protein-carboxyl methylation in spermatozoa

The relationship between the protein methylating system and various forms of cellular motility and the increase in MAP/PCM ratio as spermatozoa transit from the testes to the cauda of the epididymis prompted the study of PCM levels in non-motile spermatozoa of infertile patients. Al patients were fully virilized, sexually active and had normal levels of LH, FSH, testosterone, estradiol and prolactin. The characteristics of these patients have been described in details elsewhere (10). The non-motile spermatozoa from these patients had a PCM activity of (17±3.4 pmol/mg protein) a level not significantly different from that of the cellular debris (10±1.4 pmol/mg protein) found in semen of vasectomized patients but much lower than that of normal volunteers (69±6 pmol/mg protein). Other enzymes such as lactate dehydrogenase and the enzymes involved in the mitochondrial protein synthesis were normal in the infertile patients. This indicates that their non-motile spermatozoa were alive (10).

TABLE 1 EFFECTS OF PROTEIN METHYLATION INHIBITORS ON SPERM MOTILITY

| | TREATMENT | MOTILITY IN μ/S 1 H | 2 H | 3 H |
|---|---|---|---|---|
| 1 | CONTROL | 46 ± 3 | 42 ± 5 | 41 ± 6 |
| 2 | 3-DA (50 μM) | 43 ± 4 | 41 ± 2 | 28 ± 6 |
| 3 | HOMOCYST (200 μM) + ADE (100 μM) | 49 ± 2 | 42 ± 4 | 36 ± 7 |
| 4 | 3-DA + HOMOCYST + ADE | 31 ± 5 | 8 ± 6 | 0 |

Washed rabbit spermatozoa were incubated in Gey's medium with various compounds. At 1, 2, and 3 hours, the velocity was determined by a multiple exposure photographic method. A minimum of 20 spermatozoa were measured at each time point. 3-DA: 3-deazaadenosine; HOMOCYST: homocysteine thiolactone; ADE: adenosine.

The involvement of protein methylation in sperm motility was also investigated by determining the effects of methylation blockers. Rabbit spermatozoa were incubated with an S-adenosylhomocysteine hydrolase inhibitor, 3-deazaadenosine (3-DA, 50 μM) (11), in the presence and absence of adenosine (ADE, 100 μM) and homocysteine thiolactone (HOMOCYST , 200 μM). This combination of products are known to increase the intracellular level of S-adenosylhomocysteine, the natural inhibitor of methyl transferases. After one hour of incubation, the combination 3-DA + ADE + HOMOCYST decreased the motility by about 33% whereas at 2 and 3 hours, the motility was reduced by 81% and 100% respectively (table 1). 3-DA alone or the combination ADE + HOMOCYST had only marginal effect on sperm motility.

## Conclusion

All components of the protein-carboxyl methylation system have been found in spermatozoa and protein-$^3$H methyl esters have been shown to be formed in intact spermatozoa incubated with ($^3$H methyl) methionine (10). These results, the lack of PCM activity in non-motile spermatozoa and the inhibition of sperm motility by methyl-transferase blockers suggest a role for protein-methylation in sperm motility.

Acknowledgement: This research was supported by the MRC of Canada and by the N.I.H. of U.S.A.

## References

1. Gagnon C. and Heisler S. 1979. Protein-carboxyl methylation: role in exocytosis and chemotaxis. Life Sciences 25: 993-1000.
2. Springer M.S., Goy M.F. and Adler J. 1979. Protein methylation in behavioural control mechanisms and in signal transduction. Nature: 280: 279-284.
3. Paik W.K. and Kim S.K. 1980. Protein methylation. John Wiley and Sons, New York.
4. Kim S. 1977. in The Biochemistry of S-adenosylmethionine, Salvatore, S., Borek, E., Zappia, V., William-Ashman, H.C. and Schlenk, S., F. eds. Columbia Univ. Press, NY; 415-434.
5. Gagnon C., Axelrod J., Musto N., Dym M. and Bardin C.W. 1979. Endocrinology 105: 1440-1445.
6. Bouchard P., Gagnon C., Phillips D.M. and Bardin C.W. 1980. The localization of protein-carboxyl methylase in sperm tails. J. Cell Biol. 86: 417-423.
7. Gagnon C. 1979. Presence of a protein methylesterase in mammalian tissues. Biochem. Biophys. Res. Commun. 88: 847-853.
8. Povilaitis V., Gagnon C. and Heisler S. 1981. Stimulus-secrection coupling in exocrine pancreas: role of protein-carboxyl methylation. Amer. J. Physiol. 240: G199-G205.
9. O'Dea R.F., Viveros O.H. and Diliberto E.J. Jr. 1981. Protein-carboxyl methylation: role in the regulation of cell functions. Biochem. Pharmacol. 30: 1163-1168.
10. Gagnon C., Sherins R.J., Phillips D.M. and Bardin C.W. 1982. Deficiency in protein-carboxyl methylase in immotile spermatozoa of infertile men. N.E.J. Med. 306: 821-825.
11. Cantoni G.L., Chiang P.K. and Richards H.H. 1979. Inhibitors of S-adenosylhomocysteine hydrolase and their role in the régulation of biological methylation. in Transmethylation, Usdin, E. Borchardt, R. and Creveling, C. eds, Elsevier/North-Holland, New York, pp. 155-164.

# LOSS OF ACTIVITY OF MOTILE COMPONENTS IN RAT SPERMATIDS DURING DIFFERENTIATION TO VIRTUALLY IMMOTILE TESTICULAR SPERMATOZOA

H. WALT and Chr. HEDINGER, Institute of Pathology of the University of Zurich, University Hospital, CH-8091 Zurich, Switzerland

## INTRODUCTION, MATERIALS AND METHODS

Two distinctly different kinds of motile components were recently found in stage III-VI (1) rat spermatids (2). One of them, the early flagellum, a naked 9+2 axoneme produces wave-like motions and torsions (Figs. 1a and 1b). Later stages where a spindle-shaped body is recognizable are restricted to bending (Fig. 2). Testicular spermatozoa are equipped with a hooked head and immotile (Fig. 3). Modification of flagellar motion seems to be due to thickening of the flagellum by formation of accessory structures such as a mitochondrial sleeve, dense fibers and fibrous sheath (Figs. 4 and 1b)(3). Earlier investigations demonstrated elastic skeletal function of the accessory elements in mammalian spermatozoa, where the axoneme is the only component of active motility (4). Dense fibers and the fibrous sheath differentiate in absence of the axoneme but their arrangement is not symmetric (5). Ultrastructural analysis of dynein-lacking human spermatozoa revealed asymmetric arrangement of the longitudinal columns of the fibrous sheath, possibly a result of disturbed flagellar motility already at the level of spermatids (6). In fact, early human spermatid flagella display motions similar to those found in the rat (7).

The second motile component, namely rhythmical cytoplasmic movements in the apical region of some spermatids seems to be related to the activity of contractile proteins such as myosin for the orientation of early spermatids (Fig. 4) (8). A further kind of transitory motion was found in even earlier stages than spermatids. These motions related to intercellular bridges, which connect germinal cells at different stages of development such as spermatogonia, spermatocytes and spermatids. All data from living germ cells were collected after fine mechanical fragmentation of seminiferous tubules in Ringer's or Hank's solution by use of a phase contrast microscope equipped with a TV-camera and recording system.

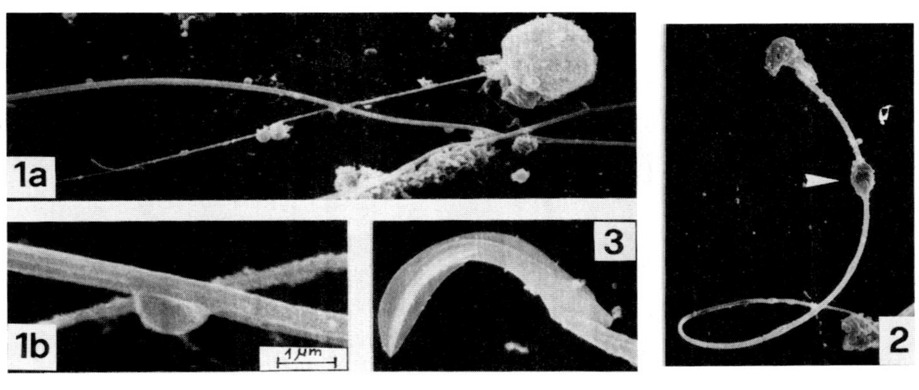


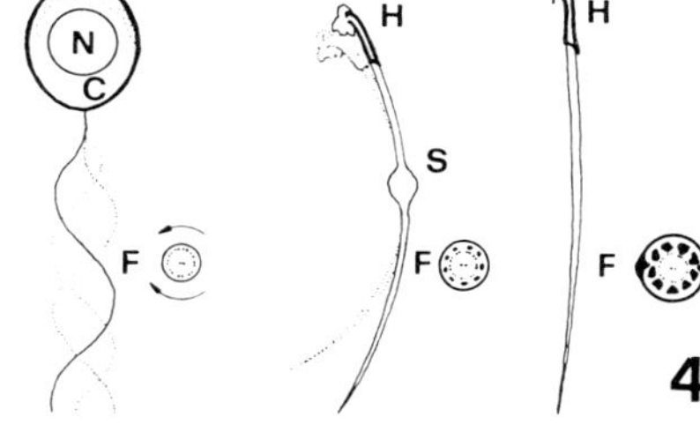


FIGURE 1. a) Scanning electron micrograph of an early rat spermatid. Its fine flagellum crosses one of a late stage. b) Area of crossed flagella indicating unequal thickness.
FIGURE 2. Late rat spermatid with a spindle-shaped body (arrow). Some cytoplasm is still present at the head region.
FIGURE 3. Hooked head of mature rat spermatozoon.
FIGURE 4. Schematic drawing of an early spermatid (left), late stage (middle) and a spermatozoon. Dotted lines, arrows and areas indicate locations of motility. C=cytoplasm; F=flagellum; H=head; N=nucleus; S=spindle-shaped body. Aspect of flagellar transverse sections is given to the right of each stage.

## RESULTS AND DISCUSSION

During preparation of living seminiferous tubules of adult albino rats, intercellular bridges between germ cells became stretched (Fig. 5) or disrupted (Fig. 6a). The latter are recognizable by stem-like protrusions. Some of them displayed to- and fro motion (Fig. 6a and 6b), neither flagellar in origin nor similar to the kind of motion found in the spermatid cytoplasm.

It could be the result of microtubular action especially since some intercellular bridges between germ cells show parallel microtubules in the electron microscope (Fig. 7). After interruption of bridges, remaining stem-like protrusions may act as primitive cilia-like structures, lacking a 9+2 arrangement and related structures. At terminal phases of mitosis, where intercellular bridges still exist, microtubules and motion were detected (9). What intercellelur bridges are concerned a situation comparable to the terminal mitotic phase is here induced artificially.

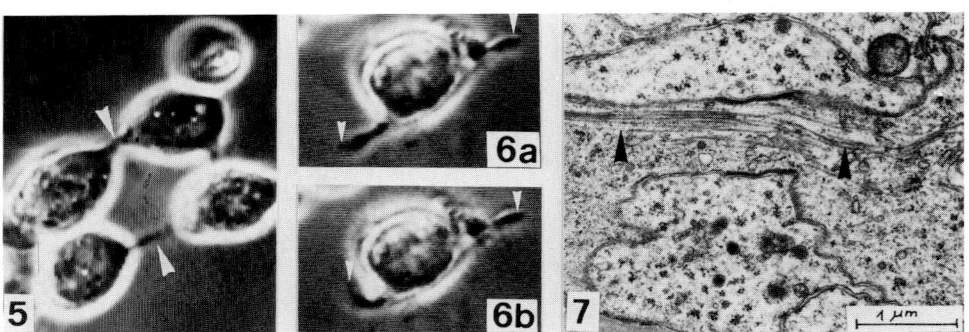


FIGURE 5. Phase contrast micrograph, taken from the TV-Screen. Slightly stretched intercellular bridges (arrows) are visible between germ cells (probably spermatogonia).
FIGURE 6. a) Germ cell (probably spermatogonium) with two interrupted intercellular bridges remain as stem-like protrusions (arrows). b) Both protrusions move but only the left one is shown in to and fro motion, once every two seconds.
FIGURE 7. Transmission electron micrograph of two adjacent human spermatogonia, connected by an intercellular bridge with microtubules (arrows).

Flagellar motility of living human and rat spermatids seems to be necessary for the normal arrangement of spermatozoa, the liberation of mature forms, and the transport of spermatozoa. Rhythmical cytoplasmic movements in early rat spermatids could play a role in orienting spermatids to point their flagella to the tubular lumen (2,8). Movements of stem-like cellular protrusions, originating from preexistent intercellular bridges, finally suggest that these are locations of high cytoplasmic transfer, aided by the action of microtubules and probably other related contractile proteins. During differentiation to spermatozoa, these components of motility lose their activity progressively. Loss of flagellar motility is thought to be the result of formation of accessory sperm structures (3), whereas cytoplasmic components will be resorbed by Sertoli cells. The kinds of motility demonstrate the complexity of the spermatid, where most probably an actin-myosin and a microtubule-dynein mechanism exist (10). A short time during differentiation, structural aspects resemble those of aggregated flagellates or cells which form a flagellated cell layer. One could speculate therefore that similar to ontogeny, spermiogenesis allows fragmentary insight into early phases of evolution, where protozoa such as flagellates represent an early phase in animal development. After a long period of time, once a Keimbahn exists, fragments of early behaviour are still conserved.

REFERENCES

1. Leblond CP, Clermont Y. 1952. Definition of the stages of the cycle of the seminiferous epithelium in the rat. Ann. N.Y. Acad. Sci. 55, 548-573.
2. Walt H. 1981. Motile components in early rat spermatids. Cell. Diff. 10, 157-161.
3. Walt H, Hedinger Chr. 1982. Motile components in spermatids as related to transport of spermatids and spermatozoa. Andrologia, manuscript in press.
4. Baccetti B, Burrini AG, Pallini V, Renieri T. 1981. Human dynein and sperm pathology. J. Cell. Biol. 88, 102-107, for review.
5. Baccetti B, Burrini AG, Pallini V. 1980. Spermatozoa and cilia lacking axoneme in an unfertile man. Andrologia 12, 526-532.
6. Walt H. Campana A, Balerna M, Domenighetti G, Hedinger Chr, Jakob M. Pescia G, Sulmoni A. Mosaicism of dynein in spermatozoa and cilia and fibrous sheath aberrations in an infertile man. Manuscript in preparation.
7. Walt H. Motile spermatid flagella during differentiation. Manuscript in preparation.
8. Walt H, Saremaslani P, Heizmann CW, Hedinger Chr. 1982. Myosin in early rat spermatids. Virchows Arch. (Cell Pathol.) 38, 307-310
9. Dustin P. 1978. Microtubules. Springer-Verlag New York. Inc. New York, 366.
10. Afzelius BA. 1981. Pathology of cell motility. In: Daems WTH et al. (eds). Cell biologic aspects of disease. Martinus Nijhoff publishers bv, The Hague, Boston, London, 145-150.

# THE MOVEMENT OF DISTAL ENDS OF SPERM FLAGELLA WITH AND WITHOUT A TERMINAL FILAMENT

C. K. OMOTO and C. J. BROKAW

## 1. INTRODUCTION

A thin terminal filament, about 5 μm in length, was noted by Gray (1955) at the end of the sea urchin sperm flagellum. When examined by electron microscopy, the terminal filament has been found to contain only a subset of the 9+2 complement of microtubules that is characteristic of the sperm axoneme. We describe here new observations on the ultrastructure of the terminal filament, based on a more extensive examination than has been published previously. We have used spermatozoa from the tunicate, *Ciona intestinalis* and the sea urchin, *Lytechinus pictus*. These spermatozoa have long flagella and make very planar bending waves, and we have exploited these features to obtain high-resolution photographs showing the bending behavior of the terminal filament and the bending behavior of the 9+2 region of the flagellum in the presence and absence of the terminal filament.

Materials and methods are detailed in a separate publication (Omoto and Brokaw, 1982).

## 2. PROCEDURE

### 2.1. Results

2.1.1. *The ultrastructure of the terminal filament*. Electron micrographs of terminal filaments of *Ciona* are shown in Fig. 1. Fig. 1a, a platinum replica, shows an intact terminal filament. A negatively stained image (Fig. 1b) indicates that it contains a subset of microtubules found in the 9+2 region. This subset includes the central pair of microtubules, identifiable by their connection together near the 9+2 region and by the cap structure which joins them at the distal tip, as shown in a platinum replica of demembranated terminal filament (Fig. 1c). These indicate that the major portion of the *Ciona* terminal filament consists of 2-4 microtubules. However, *Lytechinus* flagella have more and longer A tubules extending into the terminal filament. This difference between *Ciona* and *Lytechinus* can be demonstrated by a count of microtubules at a fixed distance distal to the termination of the 9+2 region. At 1 μm distance,

Lytechinus flagella have a broad distribution of microtubule number, ranging from 3 to 9 with over 70% having 5 or more microtubules. At the same distance, Ciona flagella have 2 to 6 microtubules with 80% having 4 or fewer microtubules (Omoto and Brokaw, 1982).

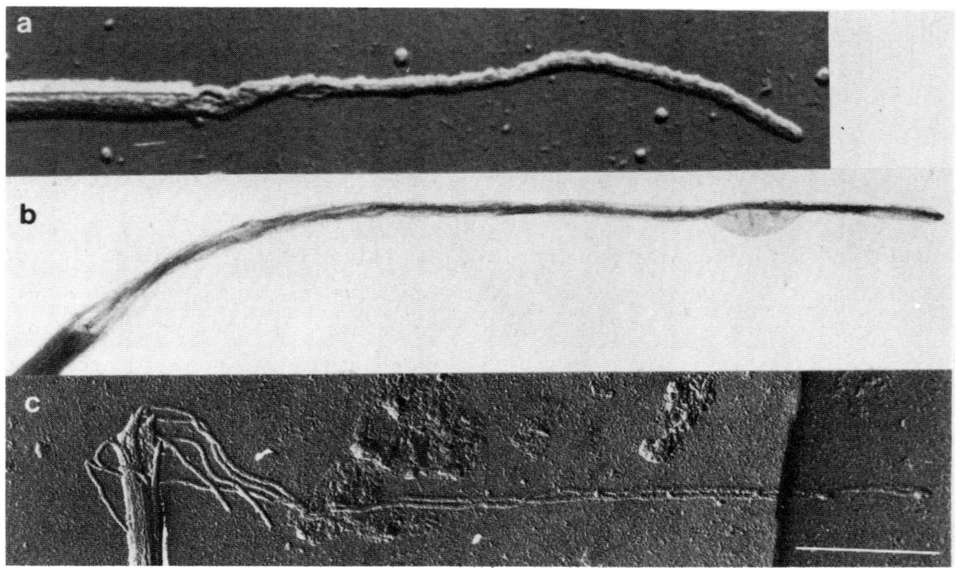


FIGURE 1. Electron micrographs of Ciona terminal filaments. a) intact, platinum replica. b) intact, negatively stained. c) demembranated, platinum replica. Scale bar in c represents 1 μm.

2.1.2. Bending behavior of the distal end. Examination of the photographs of sperm flagella with and without a terminal filament (Fig. 2) leads to the following conclusions: 1) When the terminal filament is present, bends propagate smoothly off the end of the flagellum, with no obvious discontinuity in bending behavior at the transition between the 9+2 region and the terminal filament. 2) When the terminal filament is absent, bends at the end of the flagellum decrease rapidly in curvature, causing an obvious "end effect". The presence of the terminal filament thus modifies the behavior of the distal end of the 9+2 region of the flagellum. 3) The bending behavior of the terminal filament does not show an end effect similar to that seen when the flagellum is terminated in the 9+2 region. Instead, bends extending into the terminal filament retain a nearly normal curvature until they propagate off the end of the filament.

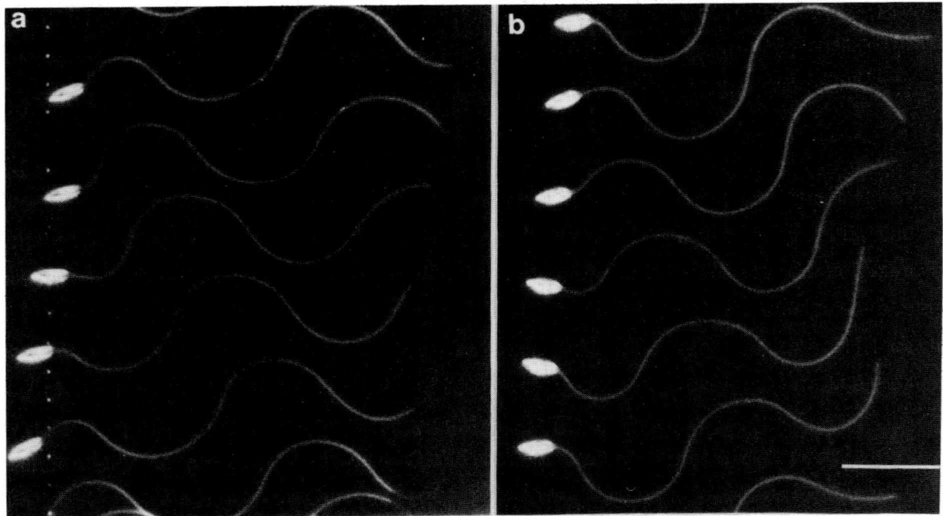


FIGURE 2. Multiple-flash photographs on moving film showing Ciona spermatozoa swimming in sea water. The sequence of images is from top to bottom. In A, a terminal filament is present at the end of the flagellum. In B, the terminal filament has been removed from the flagellum. The flash rate is 200 Hz, and the scale bar represents 10 µm.

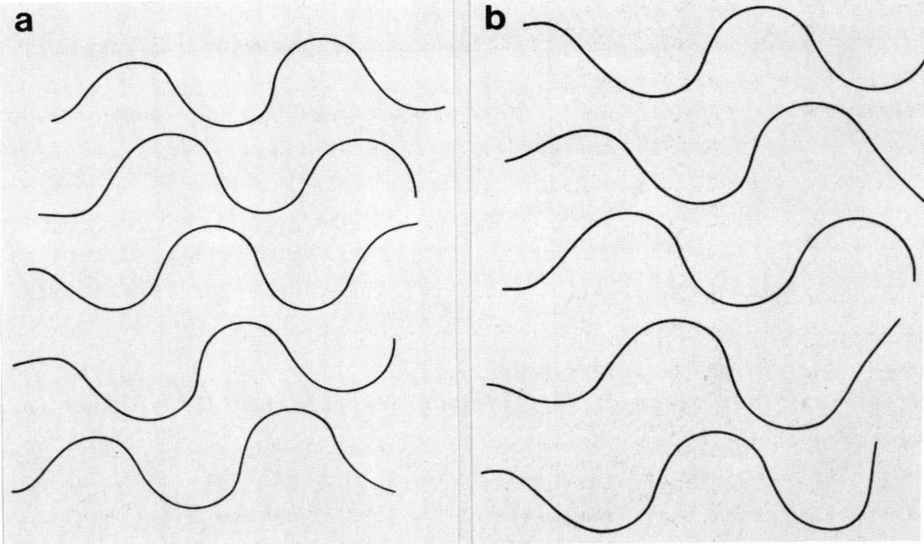


FIGURE 3. Output from computer simulations of model flagella. a) with a terminal filament with 5 joints between six 1 µm segments. The elastic bending resistances at these joints, proceeding distally, were 0.6, 0.4, 0.25, 0.16 and 0.08 times the value of elastic bending resistance of the 9+2 region ($0.2 \times 10^9$ $pNnm^2$). b) without a terminal filament.

2.1.3. Computer simulation. Output from computer simulations of flagellar models are shown in Fig. 3. The model without a terminal filament shows a distinct end effect (Fig. 3b), in which bends which have reached the end of the flagellum straighten out instead of propagating off the end of the flagellum with constant curvature. Fig. 3a shows a model in which the terminal filament has the following properties: no active moment, no shear resistance and a tapered elastic bending resistance. In this case, the end effect is eliminated and the curvature is constant throughout the terminal bend region.

2.2. Discussion

Our observations on the terminal filaments of *Ciona* and *Lytechinus* (Omoto and Brokaw, 1982) sperm flagella and those of others suggest that the terminal filament contains a variable number of microtubules which lack any mechanism for active sliding or bending. The major portion of the terminal filament usually contains the central pair of microtubules and may also contain additional single microtubules, which are extensions of the A tubule of outer doublets. From this structural information, we might conclude that a precise length or number of microtubules in the terminal filament is not important for its function in producing viscous drag.

2.2.1. The end effect and the effect of the terminal filament. The end effect seen in the absence of the terminal filament, both in real flagella, Fig. 2b and in computer simulations (Fig. 3a), involves a rapid unbending of a bend that has reached the distal end of the flagellum. When a terminal filament is present, the additional viscous drag which it introduces at the end of the flagellum will retard the unbending of a distal bent region when it reaches the point where no active shear moment is available to maintain its bent configuration. This provides a simple explanation of the effect of the terminal filament in eliminating the end effect.

REFERENCES


1. Gray, J. (1955) The movement of sea-urchin spermatozoa. J. exp. Biol. 32, 775-801.
2. Omoto, C. K. and Brokaw, C. J. (1982) Structure and behavior of the sperm terminal filament. (Submitted).


Division of Biology, California Institute of Technology, Pasadena, California 91125, U.S.A.

# CHAPTER 6 SPERM STRUCTURE IN RELATION TO FUNCTION AND PHYLOGENY

## SPERM STRUCTURE IN RELATION TO FUNCTION AND PHYLOGENY

B. A. Afzelius
Wenner-Gren Institute, University of Stockholm, S- 113 45 Stockholm

Research on male reproduction can be said to have begun with the thesis by A. Koelliker in 1840. He was first to show that the semen of all animals contains motile spermatozoa and that these have to get in contact with the egg for a fertile union to occur. Koelliker also stated that the shapes of spermatozoa vary within rather narrow limits and that the movements displayed by the various types of spermatozoa also are characterized by uniformity. Koelliker´s work has recently been reviewed by Roosen-Runge (35).

The situation today differs from that of last century only insofar that the store of facts of spermatology has grown enormously and that some facts seem puzzling and even conflicting. The purpose of this article is to call attention to some unsolved problems in spermatology.

### Why are there so many spermatozoa?

According to the laws of evolutionary biology the only traits to survive are those that have a positive value with regard to the survival of the individual, or at least have no negative value. Natural selection hereby has gradually led to a considerable efficiency of animals and plants in their proper environment and to a great efficiency of every part of the animal or plant. It is hence meaningful to inquire in which way this or that trait in an animal gives it improved chances to survive and reproduce. It is proper to ask: Why are so many spermatozoa formed?

Some examples will be given of the figures for sperm production. In our own species at least 50,000 - 100,000 spermatozoa per minute are formed in a healthy and fertile man or $10^7$ - $10^8$ per day. Figures for boars or bulls are much higher (10,25). The two testes of a spermaceti whale may together weigh 1 ton (3) and they produce an uncounted number of spermatozoa, each weighing about $10^{-12}$ g. There is hence, in this species, a difference in weight between the sperm-producing organ and the cell produced amounting to no less than eighteen orders of magnitude. Considering that only a few spermatozoa per year, at the most, will ever reach the egg, it is obvius that the inefficiency of the fertilizing mechanisms in the spermaceti whale is monumental.

In a few cases the efficiency in sperm production and sperm transfer is nearly optimal. According to Weismann (41) males of the small water flea *Daphella brachyura* produce one spermatozoon in the authumn, transfer it to the female and fertilize her one egg. Something similar has been observed a hundred years later regarding the minute featherwing beetle *Bambara invisibilis*. The mated female can be seen carrying anything between one and twentyeight spermatozoa (15). Examples of such low sperm numbers are rare however.

Cohen has defined the term 'sperm redundancy' as the ratio between number of spermatozoa offered and number of eggs fertilized. The highest values recorded for the sperm redundancy ( $10^4$ - $10^{10}$) were found in the large mammals in which the number of crossovers during spermiogenesis is high. The lowest values ( two to fifty) were recorded from bees, fruitfly, and some mite species in which there is no crossover during spermiogenesis. Cohen noted that in general there was a direct relation between the number of crossovers and values for sperm redundancy. He concluded that there were errors in the crossover process ( chiasma formations) and that these errors have necessitated a high sperm redundancy. Most spermatozoa are believed by him to be genetically defective ( see further below).

Another explanation to the apparent wastefulness in sperm handling will be offered here, but a related problem will first be treated:

Why are there so many males?

It can be argued that it is wasteful for the species to produce equally many males and females, as one male would suffice to impregnate a large number of females. In modern animal breeding one bull is used to sire several thousands of calves. In nature, the fallow buck will impregnate about 25 does per day for some weeks (9).

Although a small number of males are sufficient in a large population, there is a tendency to produce both sexes in equal number. Charles Darwin (12) was unable to provide an explanation hereof and wrote that ' the whole problem is so intricate that it is safer to leave its solution for the future.' The solution came, sixty years later, (17) and was presented in non-mathematical terms by Bodmer and Edwards (7). They claim that the reproductive value of males in a generation of offspring has to be equal to the reproductive value of the females because each sex must supply half the ancestry of future generations. The sex ratio will adjust itself under the influence of natural selection. If the total parental expenditure of effort incurred in respect of children of each sex were not equal, then parents genetically inclined to producing males in excess would, for the same expediture, produce a greater amount of reproductive value, with the result that in future generations more males would be produced. The sex ratio will hence be stabilized near the value one-half by natural selection.

Let us now return to the problem of sperm redundancy. If the spermatozoon is regarded as a sperm cell - a cell among many other cell types within the animal body - then the problems remains. If, however, the spermatozoon is regarded as a little animal in its own right, or perhaps as a separate link in the chain of events called evolution, then the problem dissipates. ( It should be reminded here that the term spermatozoon means semen animal ).The regular alternation between the haploid sex cells and the diploid animal can be regarded as something similar to the alternation between generations, and both links in the chain have to be equally strong. Could it be that the expenditure on the gamete 'generation' has to approach that of the diploid 'generation'?

The unique position of the spermatozoa can also be noted in the literal sense of the word. Sperm production in all animals takes place within compartments, which are sealed off from the remainder of the body (34). Spermatozoa are thus formed in splendid isolation within what has been called the 'blood-testis barrier' and which has equivalents even in animals lacking a blood system (26). This type of respectable treatment is otherwise bestowed only the mammalian brain by the blood-brain barrier.

Samuel Butler formulated last century his famous epigram ' A hen is only an egg's way of making another hen'. I believe that Butler was correct in directing our attention to the gametes rather than to the diploid generation. The idea that the gametes - or rather their load of genetic material - is the most valuable part of the chain of life, and that the somas are nothing but the machineries for its replication and propagation, has been more fully developed by Dawkins in his book on the 'selfish gene' (14). Orgel and Crick (32) have extended this idea even further in their claim that the eukaryotic nucleus carries much junk DNA that has no effect on the phenotype and no function in the animal carrying it, but which exists only because it has developed a possibility to exist. Junk DNA gets a frèe ride in the genome; it is the ultimate parasite.

The problem of sperm excess is so intricate, that it may be wise to leave its solution for the future.

## How does natural selection influence sperm morphology?

If the gamete 'generation' carries equal weight to the diploid animal generation, then it may be inquired why spermatozoa have remained at a relatively primitive stage. During a period when diploid animals have evolved from medusa to man, their spermatozoa ( not to mention their ova) have stopped at one of the lowest rungs of the ladder of life. Most spermatozoa would be classified as flagellates, if found free. The explanation to the problem seems to be that there is no simple and direct relation between the properties of the individual spermatozoon and its genome.

An investigator watching spermatozoa in their race to the egg gets a vivid impression of the competition, where the ablest will win. The features of the winning spermatozoa are not determined by the haploid genome in its nucleus, however, they are determined by the diploid genome of the spermatocyte preceding it and this set of genes is shared by all spermatozoa from the same male. This is, I believe, the probable reason why evolution of spermatozoa has lagged behind that of the animal soma. Natural selection has got no handle to steer the evolution of the spermatozoa.

What has been said above on the relation between sperm genotype and sperm phenotype may not be 100 % correct. It is possible that some proteins of the spermatozoon are manufactured on instructions given at the spermatid stage and based on the haploid genome - the spermatozoon will then be said to show a haploid gene expression. RNA-synthesis stops very soon after the second meiotic division in the mouse (28) and the amount of proteins which can show a haploid gene expression must be very low. The best candidate for a protein with haploid gene expression ( in healthy individuals; the T-locus system in mice is another trait with a haploid gene expression) seems to be the HL-A antigen in man ( and corresponding antigens in other mammals) (19). This protein is also called the transplantation antigen and it is a surface marker of great specificity. If a man has the HL-A phenotype 1,2 and if haploid gene expression occurs, his spermatozoa will display on their surface whether they carry the HL-A 1 gene or the HL-A 2 gene within their genome. This would give the egg or the female ducts an opportunity to select compatible spermatozoa through a recognition of cell surface markers, and thus to choose a spermatozoon with a known genome. If this does occur - and we have no evidence that it does - then the mendelian laws will not be obeyed for this particular gene. It will not be easy to find out whether such cases of non-random fertilization indeed occur. It has been estimated that 26,546 observations are needed, if we wish to be 90 % certain of a 1 % deviation from an expected frequency of 50 % (30). It has been suggested that 'non-random fertilization is rarely discovered, not because they are actually rare, but because they are surprisingly difficult to detect' (30).

In the hypothesis by Cohen, mentioned above, all spermatozoa are supposed to carry markers on their surface indicating whether they are genetically normal or carry defects from errors in the crossovers. Only a tiny fraction are supposed to be normal. The long sex ducts in the mammals are assumed to provide a long steeple-chase for this selection. However recent data from test tube fertilization indicate that these unselected spermatozoa give rise to normal children.

It has been remarked above that some DNA is transcribed into RNA in the spermatid stage and that some proteins are synthesized. These proteins will show a haploid gene expression if,and only if, the transcription is from the nuclear genome. In reality most transcription and protein synthesis is based on mitochondrial DNA rather than on the nuclear DNA (5, 20). This brings us to the next question, which is:

## Sperm mitochondria – why are they there?

Twenty years ago Jean André described the aggregations and transformations exhibited by spermatid mitochondria from many species (2). In some species the mitochondria also fuse. It is by no means clear why they do so. One hypothesis is that the transformations and fusions serve to bring mitochondrial DNA-strands from separate mitochondria together for recombination. This would make sense if the paternal mitochondrial DNA would be incorporated into the pool of cytoplasmic DNA of the zygote and would then be beneficial to the new individual. However, this seems not to be the case. Although sperm mitochondria enter the egg at fertilization in nearly all animal species ( ascidians being the exception) it seems that the paternal mitochondria degenerate in the egg and the paternal DNA gets lost. As a result the mitochondrial DNA strictly follows a maternal inheritance. Mules will hence have horse mitochondrial DNA and hinnies have donkey mitochondrial DNA (23). Maternal inheritance of mitochondrial DNA has also been demonstrated for *Xenopus*-crosses (13) and a uniparental inheritance for crosses in yeasts (6).

Recently the mitochondrial DNA from 14 laboratory strains of the mouse were analysed and compared to similar data from 20 mice caught in nature. Whereas there was a great variability between the wild mice, all 14 laboratory mice had the same type mitochondrial DNA (16). This was interpreted as a result of a common foremother of all laboratory mice. They are all descendants of the same single female although both she and her offspring have been cross-bred with many different males. The foremother has provided mitochondrial genes to all her descendants but only a few of her nuclear genes remain in them.

The degeneration of the paternal mitochondrial DNA in the zygote seems to be effected by some restriction enzymes, which have the task to digest any exogenous DNA, which may enter the cell (36). Such enzymes could selectively destroy the paternal mitochondrial DNA and may also eliminate chloroplast DNA from the pollen in plants and also the foreign chromosomes within somatic cell hybrids, e.g. between man and mouse (8).

Rarely, a gene seems to be able to escape the action of such restriction enzymes and to be able to jump from one species to another. Such unique events seems to exist among the histones in some sea urchin species (31).

The problem remains: Why do spermatid mitochondria transform and aggregate? It is interesting in this connection that the small haploid plantlets, which can be grown from uninucleate grass pollen, have mitochondria that aggregate much the same way as sperm mitochondria do (40). Maybe the transformations are just the by-products of other events or they are signs of a degenerative process? Yet, it is not only something simple as a fusion of the mitochondria; their DNA-strands also undergo changes (5,20). A portion of the mitochondrial DNA amounting to about 5 % of the total becomes eliminated from each strand at spermiogenesis. More information here has meant more confusion.

What factors have influenced sperm structure?

Koelliker stated that the shapes of spermatozoa vary within rather narrow limits. His claim would be hard to defend today (4). There are minute and inconspicuous spermatozoa in some animals, for instance in some gastrotrichs (42). There are giant spermatozoa, as long as or longer than the animal producing them (1,39) ( And here is an unsolved problem of sperm motility – why do these long spermatozoa wriggle and twist and move, when the distance they have to move is much shorter than their own length?)

It seems that spermatozoa of the type called 'primitive spermatozoa' (18) have a structure which is well suited for one of their tasks, namely to swim a few centimeters in water. But the sundry sperm types found in crustaceans, insects, nematods, worms, and fish could they all be said to be optimally adapted to the technical problems involved in getting to the egg and into it? This seems unlikely. It is hence legimate to inquire what forces have been at play to model them to the shapes they have. A playful creator would seem as good a guess as any other one.

Differences in fertilization biology certainly is the explanation for the transformation from the primitive sperm type in many marine and fresh-water animals to the modified and aberrant sperm types found in animals with internal fertilization (18). In most other cases it seems that random drift in morphology has led to the diversity in shapes and sizes. This makes it possible, in some cases, to use sperm morphology as an indicator of phylogenesis.

In a different context Mossman has written (29): 'When compared to organ systems ordinarily employed as criteria for taxonomic and phylogenetic studies of mammals, the characters of the male and female reproductive systems .. are more conservative.' Sperm characters could well be included in the components which are useful indicators for phylogenetic studies.

If all animal spermatozoa were described, classified, and arranged in a phylogenetic tree without previous knowledge about the systematics of the animals producing them, this tree would only partially resemble the phylogenetic tree we are familiar with. The phylogenetic tree of spermatozoa would have a broad stem of similar forms, primitive spermatozoa according to Franzén (18). Branches of this tree would probably be the same as those constituting the familiar animal phyla and classes except that primitive representatives of the classes would be found to have spermatozoa that would belong to the stem rather than to the branches ( this is one way to express the notion that primitive representatives from most animal phyla and classes have spermatozoa of the primitive type.)

Some branches would consist of a fairly uniform lot of sperm forms, where the corresponding animals would consist of widely different forms. This would be the case of the mammals. Other branches would make up a large portion of the tree with ramifying twigs, where the corresponding animals form a uniform group. This would be the case of the small gall-midges examined by Dallai (11). It is not evident why this little group has so widely variable and diverse sperm forms. Maybe the aberrant meiotic divisions are responsible.

Although the use of the electron microscope for studies on sperm phylogeny still is in its beginning there are already some studies which demonstrate that sperm morphology can solve problems of classification and taxonomy. The following studies are examples hereof. The animals studied are marsupials (21), amphibians (33), elopomorph fish (27), lower crustacea (44), gallmidges (11), chelicerates (24,43), turbellarians (22), gnathostomulids (38), and anthozoa (37). It is a safe prediction to say that we will see more studies of such kind in the future.

REFERENCES

1. Afzelius BA, Baccetti B, Dallai R. 1976. The giant spermatozoon of Notonecta. J Submicr Cytol 8, 149-161.
2. André J. 1962. Contribution à la connaissance du chondriome. J Ultrastr Res Suppl 3, 1-185.
3. Arvy L. 1977. Asymmetry in cetaceans. In: Investigations on Cetacea G. Pilleri (ed.) Vol VIII, 162-212.
4. Baccetti B, Afzelius BA. 1976. The biology of the sperm cell. S Karger Publ Basel.
5. Bartoov B, Fisher J. 1980. Uniqueness of sperm mtDNA as compared to somatic mtDNA in ram. Intern J Androl 3, 594-601.
6. Birky CW Jr. 1975. Zygote heterogeneity and uniparental inheritance of mitochondrial genes in yeast. Molec Gener Genet 141, 41-58.
7. Bodmer WF, Edwards AWF. 1960. Natural selection and the sex ratio. Ann Human Genet (London) 24, 239-244.

8. Clayton DA, Teplitz RL, Nabholz M, Dovey H, Bodmer W. 1971. Mitochondrial DNA of human-mouse cell hybrids. Nature 234, 560-562.
9. Cohen J. 1971. The comparative physiology of gamete population. Adv Comp Physiol Biochem 4, 267-380.
10. Cohen J. 1977. Reproduction. Butterworths, London.
11. Dallai R. 1979. An overview of atypical spermatozoa in insects. In: The spermatozoon. Fawcett DW, Bedford JM (eds.) Urban & Schwarzenberg, Baltimore. pp 253-265.
12. Darwin C. 1871. The descent of man, and selection in relation to sex. John Murray, London.
13. Dawid IB. 1972. Cytoplasmic DNA. In: Oogenesis. Biggers JD and Schuetz (eds.) University Park Press, Baltimore. pp 215-226.
14. Dawkins R. 1976. The selfish gene. Oxford University Press Oxford.
15. Dybas LK, Dybas HS. 1981. Coadaptation and taxonomic differentiation of sperm and spermathecae in featherwing beetles. Evolution 35, 168-174.
16. Ferris SD, Sage RD, Wilson AC.1982. Evidence from mtDNA sequences that common laboratory strains of inbred mice are descended from a single female. Nature 295, 163-165.
17. Fisher RA. 1930. The genetic theory of natural selection. Clarendon Press, Oxford.
18. Franzén Å. 1956. On spermiogenesis, morphology of the spermatozoon, and biology of fertilization among invertebrates. Zool Bidrag Uppsala 31, 355-482.
19. Halim K, Wong DM, Mittal KK. 1982. The HL-A typing of human spermatozoa by color fluorescence. Tissue Antigens 19, 90-91.
20. Handel MA, Papaconstantinou J, Allison DP, Julku EM, Chin ET. 1973. Synthesis of mitochondrial DNA in spermatocytes of Rhynchosciara hollaenderi. Dev Biol 35, 240-249.
21. Harding HR, Carrick FN, Shorey CD. 1979. Special features of sperm structure and function in marsupials. In: The spermatozoon. Fawcett DW, Bedford JM (eds.) Urban & Schwarzenberg, Baltimore. pp 289-303.
22. Hendelberg J. 1977. Comparative morphology of turbellarian spermatozoa studied by electron microscopy. Acta Zool Fennica 154, 149-162.
23. Hutchinson CA, Newbold JE, Potter SS, Edgell MH. 1974. Maternal inheritance of mammalian mitochondrial DNA. Nature 251, 536-538.
24. Juberthie C, Manier JF. 1978. Etude ultrastructurale comparée de la spermiogenèse des opilions et son intérêt phylétique. Symp Zool Soc London 42, 407-416.
25. Mann T, Lutwak-Mann C. 1981. Male reproductive function and semen. Springer Verlag Berlin-Heidelberg-New York.
26. Marcaillou C, Szöllösi A. 1980. The 'blood-testis' barrier in a nematode and a fish: a generalizable concept. J Ultrastr Res 70, 128-136.
27. Mattei C, Mattei X. 1975. Spermiogenesis and spermatozoa of the elopomorpha ( teleost fish). In: The functional anatomy of the spermatozoon. Afzelius BA (ed.) Pergamon Press, Oxford. pp 211-221.
28. Monesi V. 1965. Synthesis activities during spermatogenesis in the mouse. Exp Cell Res 39, 197-224.
29. Mossman HW. 1953. The genital system and the fetal membranes as criteria for mammalian phylogeny and taxonomy. J Mammol 34, 289-298.
30. Mulcany DL, Kaplan SM. Mendelian ratios despite nonrandom fertilization. Amer Natural 113, 419-425.
31. Newmark P. 1982. Cancer genes - processed genes - jumping genes. Nature 296, 393-394.

32. Orgel LE, Crick FHC. 1980. Selfish DNA:the ultimate parasite. Nature, 284, 604-607.
33. Pugin-Rios E. 1980. Etude comparative sur la structure du spermatozoïde des amphibiens anoures. Thesis Université de Rennes Ser 6, 35:15.
34. Roosen-Runge EC. 1977. The process of spermatogenesis in animals. Cambridge University Press, Cambridge.
35. Roosen-Runge EC. 1981. Koelliker´s role in the history of research on male reproduction. Fortschr Androl 7, 1-9.
36. Sager R, Kitchin R. 1975. Selective silencing of eucaryotic DNA. Science 189, 426-433.
37. Schmidt H, Zissler D. 1979. Die Spermien der Anthozoen und ihre phylogenetische Bedeutung. Zoologica (Stuttgart) 44, 1-97.
38. Sterrer W. 1972. Systematics and evolution within the Gnathostomulida. System Zool 21, 151-173.
39. Taylor VA, Luke BM, Lomas MB. 1982. The giant sperm of a minute beetle. Tissue & Cell 14, 113-124.
40. Vaughn KC, DeBonte LR, Wilson KG, Schaffer GW. 1980. Organelle alteration as a mechanism for maternal inheritance. Science 208, 196-198.
41. Weismann A. 1880. Samen and Begattung der Daphnoiden. Z Wiss Zool 33, 55-270.
42. Weiss MJ, Levy DP. 1979. Sperm in 'parthogenic' freshwater gastrotrichs. Science 205, 302-304.
43. Weygoldt P, Paulus HF. 1979. Untersuchunen zur Morphologie, Taxonomie und Phylogenie der Chelicerata. Z Zool System Evolut.forsch 17, 85-116.
44. Wingstrand KG. 1978. Comparative spermatology of the Crustacea Entomostraca. I Subclass Branchiopoda. Kongl Dansk Vid Selskab Biol Skr 22, 1-67.

# SPERM SIZE, BODY MASS AND REPRODUCTION IN MAMMALS

J.M. CUMMINS
Reproductive Biology Group,
Department of Veterinary Anatomy,
University of Queensland,
St Lucia, Australia 4067

## INTRODUCTION AND METHODOLOGY

That sperm size seems to be inversely related to body mass in mammals has been regarded as a scientific curiosity for many years, but does not appear to have been rigorously investigated. This is surprising, given the intense interest in the evolution of anisogamy and sexuality in general (1).

Data on sperm dimensions and body masses were obtained from the literature, from personal observations, and from communications with other scientists. For more than one species within a genus, the data were pooled, and calculations were based on the means. A full analysis is forthcoming (2).

## RESULTS AND DISCUSSION

1 *General Observations*. The relative proportions of head, midpiece and principal piece are remarkably consistent within Eutherian and Marsupial spermatozoa, while the sauropsid-like spermatozoa of Monotremata renders this group apart (3). Proportionality in sperm structures will be discussed elsewhere (2): its relevance here is to justify the use of total sperm length— rather than, say, midpiece volume— in analysing the evolution of sperm size. Total sperm length is also the most frequently recorded parameter in the literature.

2 *Sperm Lengths and Body Masses*. The data are summarised in Table 1, and points for individual genera are plotted in Figure 1. In all, there are data for 195 species of mammals, representing 13% of all genera and 53% of all families. Information is not available for 5 of the 19 orders (Dermoptera, Hyracoidea, Pholidota, Sirenia and Tubulidentata).

Total sperm lengths vary from a minimum of 33.49 μm for Hippopotamus amphibius to a maximum of 356.4 μm for the marsupial Tarsipes rostratus. I estimate that the sperm of Tarsipes has a volume of c. 1600 $\mu m^3$, contrasting strikingly with figures of 10 to 25 $\mu m^3$ for man and domestic animals (4). The volume ratio between the largest and smallest spermatozoa is thus c.160:1, which is somewhat larger than that for ova (excluding the Monotremata)(5).

TABLE 1. SPERM LENGTHS AND BODY MASSES WITHIN MAMMALIAN ORDERS

| *EUTHERIA* | Total no. genera | Genera with data | Species with data | Sperm length µm. 'L' Range | Mean | Body mass, 'M' $Log_{10}$ gm. Range | Mean | r, M × L |
|---|---|---|---|---|---|---|---|---|
| ARTIODACTYLA | 82 | 22 | 23 | 33-80 | 49 | 4.36-6.57 | 5.23 | -.0737NS |
| CARNIVORA | 101 | 8 | 8 | 43-69 | 56 | 3.20-5.40 | 4.07 | -.1022NS |
| CETACEA | 38 | 5 | 5 | 45-74 | 62 | 4.80-7.63 | 6.20 | -.9258* |
| CHIROPTERA | 178 | 12 | 17 | 50-106 | 69 | 0.78-2.51 | 1.31 | +.6367* |
| EDENTATA | 14 | 1 | 1 | — | 68 | 3.60-3.85 | 3.78 | — |
| INSECTIVORA | 63 | 5 | 5 | 64-122 | 87 | 1.65-3.00 | 2.36 | -.7186NS |
| LAGOMORPHA | 10 | 1 | 1 | 55-58 | 57 | 3.57-3.61 | 3.59 | — |
| PERISSODACTYLA | 6 | 1 | 2 | 60-64 | 62 | 4.30-6.00 | 5.78 | — |
| PROBOSCIDEA | 2 | 2 | 2 | 50-62 | 58 | 6.70-6.88 | 6.75 | — |
| PINNIPEDIA | 20 | 2 | 2 | 53-74 | 63 | 5.31-6.36 | 5.84 | — |
| PRIMATES | 60 | 14 | 20 | 52-94 | 64 | 1.80-5.32 | 3.82 | -.1447NS |
| RODENTIA | 352 | 37 | 87 | 35-250 | 86 | 0.78-3.90 | 2.28 | -.3256* |
| *ALL EUTHERIA* | 935 | 110 | 174 | 33-250 | 64 | 0.78-7.63 | 3.20 | -.4815*** |
| MARSUPIALIA | 80 | 13 | 19 | 83-356 | 139 | 1.02-4.57 | 3.28 | -.6808*** |
| MONOTREMATA | 3 | 2 | 2 | 112-123 | 116 | 3.12-3.60 | 3.36 | — |
| *ALL MAMMALIA* | 1018 | 125 | 195 | 33-356 | 72 | 0.78-7.63 | 3.46 | -.4397*** |

NS = Not Significant * P < .05 *** P < .001

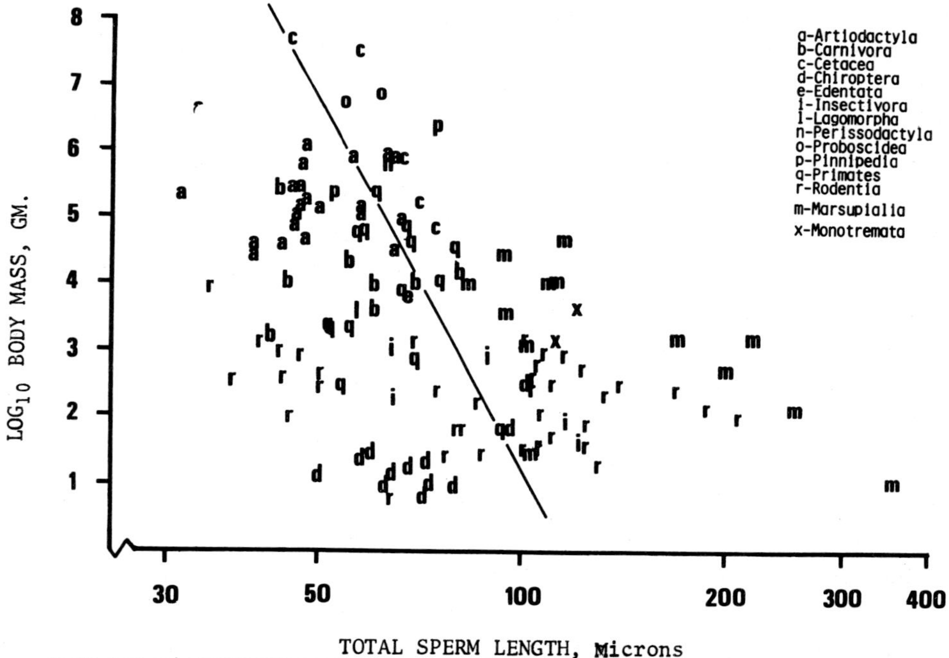


*Supported by a grant from the Australian Wool Corporation. The interest of all my colleagues, but particularly Dr Peter Woodall, is appreciated.*

When all data are considered, there is a highly significant negative correlation between sperm length (L) and body mass (M). The regression line in Figure 1 is for all data, although the regressions differ significantly between Eutheria and Marsupialia (2). Negative correlations are also found within orders— except Chiroptera— but are only significant for 4 out of 9. The Chiroptera are positive for L×M, and also tend to have small spermatozoa in relation to M. However, this group is generally reproductively unusual, with factors such as sperm storage in the female tract, and delays in fertilization to be taken into account (6). If one excludes the Chiroptera as a special case, then the inverse relationship between L and M in the other mammals is even more striking. Negative allometry of this sort is highly unusual; the only other good examples being metabolic rate, and surface area/volume ratios (7).

*3 How can we explain the inverse relationship between L and M?* Models of the evolution of anisogamy usually assume that organisms tend to invest in a given 'mass' of gametic tissue: either many, small gametes or few, large (1). This concept is useful in analysing sperm size. It is apparent that in most mammalian orders sperm sizes are relatively low and uniform, with figures exceeding 100 μm only in Chiroptera, Insectivora, Rodentia, Marsupialia and Monotremata. The central position of the Insectivora in Figure 1 could be significant, as they are commonly assumed to be close to the ancestral type of mammal. It may be that the L-M relationship in this order is also ancestral, in which case mammals may have evolved either 'microspermy' or 'macrospermy' in radiating from this position. While this may be fallacious, it nevertheless gives a starting point for analysis.

*4 Why microspermy?* The production of large numbers of very small male gametes is widespread. Among plants, a downgrading of sperm size is associated with increased dispersion distance (8) reaching an extreme in the pollen grains of flowering plants (9). Internal fertilization would appear to bypass the problems of dispersion, but in fact most organisms using this mode produce and deliver far more spermatozoa than are needed for fertilization: so-called 'Gamete Redundancy' (GR)(10). Parker (11) argued convincingly that this results from intersexual competition, with competitive mating favouring the male delivering the most gametes, at relatively trivial cost. This may be complicated by increased genetic variation— possibly deleterious— among spermatozoa, as Cohen (10) has shown a striking correlation between GR and chiasma frequency.

5 *Why Macrospermy?* There is no immediately obvious reason for the production of very large spermatozoa in some small mammals, particularly as the relationship is not obligate, eg Chiroptera and some small rodents. Large spermatozoa may be archaic, as seen in the Dasyurid marsupials, but this is unlikely to be the case in the recently evolved and highly successful rodents. One clue may lie in recent findings on fertilization in rodents, in which it is clear that fertilization is effected by the first spermatozoa to enter the ampulla of the oviduct (12). Here we can see the immediate selective benefits of a long, vigorous flagellum. Maximum attainable velocity is a function of total length, and in fact the fastest known sperm is that of the hamster (500μm/sec:13) If intersperm competition of this order is indeed significant, then it could help to explain, not just sperm size, but other 'bizarre' features such as large acrosomes. It is noteworthy that sperm morphology is highly heritable, and is a trait closely linked with the germ cell line (14). The genes for sperm morphology may indeed be sex-linked, in which case, like testis size, one would expect the potential for rapid selection (15).

## CONCLUSIONS

The optimum sperm size for any species probably results from a balance of opposing forces. To produce many, small spermatozoa most probably results from inter-male competition in competitive mating situations, particularly when losses of spermatozoa through temporal or spatial dispersion in a large female tract are significant. To produce few, large spermatozoa is an acceptable investment only when such dispersion is minimal, and when competition between spermatozoa at the site of fertilization favours large, vigorous gametes.


1. Bell G. 1982. The Masterpiece of Nature. London & Canberra, Croom Helm.
2. Cummins JM, Woodall P. In Preparation.
3. Carrick FN, Hughes RL. 1982. Cell Tissue Res. 222:127-141.
4. van Duijn C. 1975. Biblphy Reprod. 25:121-248.
5. Hartman CG. 1929. Q Rev Biol. 4:373-388.
6. Racey PA. 1975. Biol J Linn Soc 7 Suppl 1:385-413.
7. Gould SJ. 1966. Bio Rev 41:587-640.
8. Paulillo DJ. 1981. BioScience 31:367-373.
9. Linskens HF. 1980. In: The Cell Surface eds S. Subtelny, NK Wessens pp. 113-126. London, Academic Press.
10. Cohen J. 1969. Sci Prog Oxford, 57:23-41.
11. Parker GA. 1970. Biol Rev. 45:525-567.
12. Cummins JM, Yanagimachi R. 1982. Gamete Res 5:239-256.
13. Katz DF, Yanagimachi R. 1980. Biol Reprod 22:759-764.
14. Beatty RA. 1970. Biol Rev 45:73-119.
15. Short RV. 1980. In: Reprod in Mammals, eds CR Austin & RV Short, v 8: 1-33, Cambridge University Press.

# THE NECK REGION OF MOUSE SPERMATIDS: A MORPHOLOGICAL AND FUNCTIONAL ANALYSIS.

P. ESPONDA
Instituto de Biología Celular (CSIC) Madrid-6. Spain.

## 1. INTRODUCTION

Spermiogenesis in mammals has been amply studied using conventional methods for electron microscopy (1). The ultrastructure of the neck region of spermatids has been analysed in some species (2) and appears formed by three main structures: the proximal centriole(PC), the distal centriole (DC) and the centriolar adjunct (CA) which is an extension of the PC composed by atypical triplets. All these microtubular (MT) structures have been specially analysed by means of conventional electron microscopy and only few analyses have been carried out using other methods, as cytochemical procedures, cell dissociation techniques,etc. For this reason in this communication the observations made on the neck structures after the application of cytochemical and dissociation procedures are reported. Moreover the observation of dissociated neck structures in spread and whole mount preparations has permited to apply the "in vitro" assembly of MT using purified MT proteins, thus testing the capacity of such structures to act as microtubule organizing centers (MTOC).

## 2. MATERIAL AND METHODS.

2.1. Conventional methods: testes from adult males were dissected and fixed in 2% glutaraldehyde in Phosphate buffer (pH 7.2). After post-fixation in $OsO_4$ (2%, pH 7.2) samples were embedded in Epon and ultrathin sections stained with Uranyl acetate and lead citrate. 2.2. Cytochemical methods: the ethanolic phosphotungstic acid method (E-PTA) was used for the localization of basic proteins (3) and the EDTA procedure (4) as a preferential staining for ribonucleoproteins. 2.3 Cell dissociation procedures: the technique was recently described (5) and sometimes buffer MES was used instead TBS. Dissociated material was picked out in carbon coated grids, and stained with Uranyl acetate or shadowed with platinum. 2.4. Preparation of the MT protein: the method described by Shelanski et. al (6) was used. 2.5. Polymerization of tubuline onto the dissociated structures: A solution of dissasembled porcine brain MT protein (1mg/ml, 1mM GTP, pH 6.5) in MES buffer cointaining 1mM $MgCl_2$ was mixed with an equal volumen of the dissociated cells. Then incubated for 30 minutes at 32°C and incubation stopped with glutaraldehyde (3% in MES). After, the material was picked out in the grids and stained with Uranyl acetate.

3. RESULTS AND DISCUSSION

Early spermatids show neck structures similar to those described previously for other mammals (2). PC and DC appear as dense structures and CA as a prolongation from the PC. The CA shows a dense and diffuse material related to his end. This material is composed by fine fibers (Fig.1a). A rounded and sometimes vacuolated body appears associated to the DC. After using the E-PTA method PC and DC appear deeply stained as well as the rounded body, nevertheless the CA is poorly stained (Fig. 1b). Striated columns and capitulum of the connecting-piece as well as the annulus are deeply stained by the E-PTA, as it has happened in other cases (7). After using the EDTA procedure chromatin appear as if bleached and the neck structures are preferentially stained. In the CA the surrounding diffuse material is stained (Fig. 1c) as well as the round body associated to the DC and structures of the connecting-piece. These cytochemical results suggests that proteins are the main components of the structures of the neck region. Basic proteins (7) seem to be present in all neck structures with the exception of the CA, which apparently show a different concentration of such proteins. The stainability of the PC and DC can be correlated to the material which is associated to them which participates in the formation of the connecting-piece (1,2). The positivity after EDTA staining of the material surrounding the CA suggests a ribonucleoprotein composition. Interphasic and mitotic centrioles show surrounding material which seems to be a MTOC and which is positive after EDTA staining (8). Nevertheless the possibility that EDTA stains other types of proteins (9) must be taken into account.

Dissociated material observed in spread and whole mount preparations shows the different spermatid structures. The neck structures are clearly observed (Figs. 1d,e) maintaining similar sizes to that showed in sectioned material while the round body remains attached to the DC. This observation suggests that the neck structures are strongly attached between them appearing as a whole. The dissociation procedure appears as an advantageous method for the more accurate analysis (chemical and functional) of spermatid structures. When the dissociated material is incubated with MT proteins a great amount of newly assembled MT can be observed attached to the centriolar structures, in particular to the CA and the PC (Figs. 1f,g). This fact suggests that such structures are like a true MTOC. Similar observations were realized on centrioles of starfish spermatozoon (10) centrioles of other cell types and chromosome kinetochores (11). In the present case a possible functional implicance of these structures during fertilization and post-fertilization events can be infered.

## 4. REFERENCES


1.- Fawcett DW. 1975. The mammalian spermatozoon. Develop. Biol 44,394-436
2.- Fawcett DW, Phillips DM. 1969. The fine structure and development of the neck région of the mammalian spermatozoon. Anat. Rec.165,153-184
3.- Sheridan WF, Barrnett RJ. 1969. Cytochemical studies on chromosome ultrastructure. J. Ultrastruct. Res.27, 216-229
4.- Bernhard W. 1969. A new staining procedure for electron microscopical cytology. J. Ultrastruct. Res. 27, 250-265
5.- Krimer DB, Moreno ML, Esponda P.1980. Manchette microtubules as visualized by whole mount electron microscopy. Cell Biol Int Rep 4,941-946
6.- Shelanski ML, Gaskin F, Cantor CR.1973. Microtubule assembly in the absence of added nucleotides. Proc. Natl Acad Sci US 70,765-768
7.- Courtens JL, Loir M. 1981. A cytochemical study of nuclear changes in boar,bull,goat,mouse,rat and stallion spermatids. J.Ultrastruct. Res. 74,327-340
8.- Rieder CL.1980. The fine structure of the mammalian kietochore and of the centrohelidian centroplast as revealed by Bernhard's staining. In "Microtubule and microtubule inhibitors" (M.DeBrabander and J. DeMey,eds) p. 311-324. North Holland. New York.
9.- Wassef M.1979. A cytochemical study of interchromatin granules. J. Ultrastruct. Res. 69,121-133
10.- Kuriyama R, Kanatani H. 1981. The centriolar complex isolated from starfish spermatozoa. J.Cell Sci 49,33-49
11.- Dustin P. 1978. Microtubules. Springer Verlag. New York.

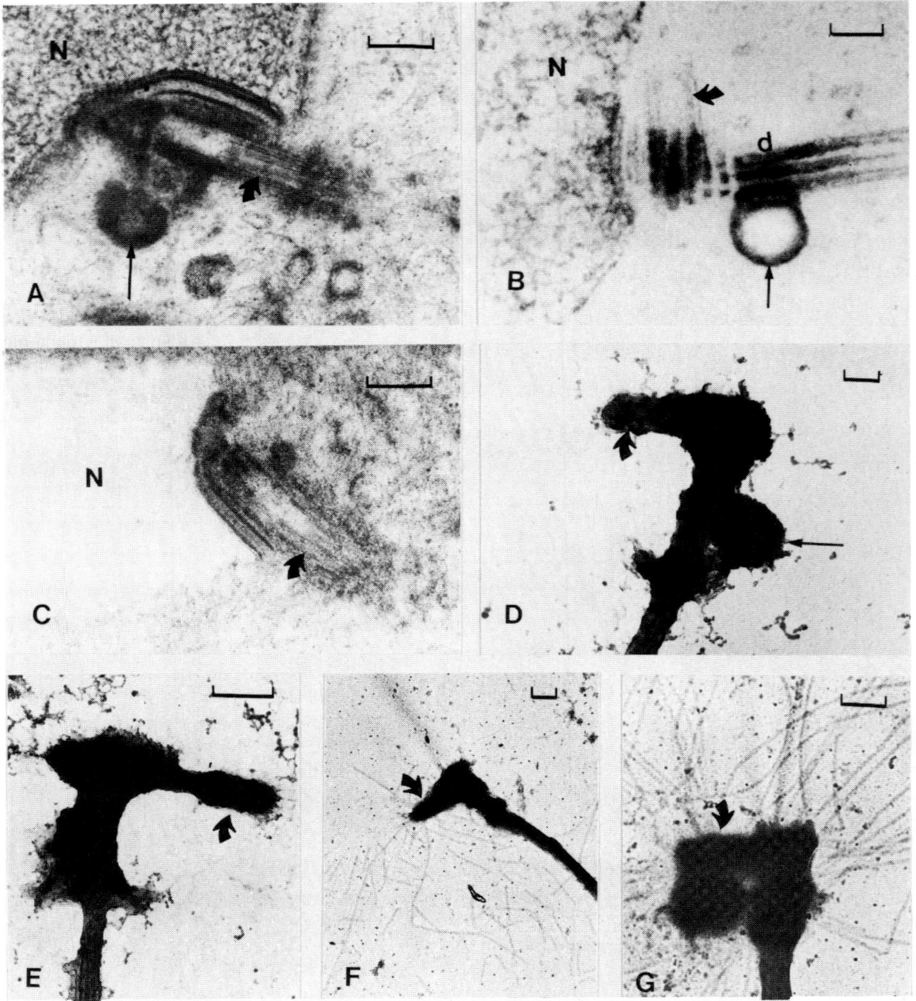


FIGURE 1. a) Neck region after uranyl lead staining.Note the diffuse material at the end of the CA. b) E-PTA staining. Centrioles and round body appear deeply stained, but the CA is poorly contrasted. c)EDTA stain. The chromatin appears bleached. Elements of the connecting-piece and the material that surrounds the CA are deeply stained. d)A tail observed after dissociation. The round body is clearly seen attached to the DC. e)Neck structures are indicated in a dissociated tail. f-g) Newly assembled MT appears preferentially connected to the CA and PC.
(Bars : a-b-c-d and g = 0,25um ; e and f = 0,5um)

(N:Nucleus; d: distal centriole ; Arrow: centriolar adjuct ; fine arrow: round body).

# STUDY OF SPERMATOZOA IN LEMUR HYBRIDS. (PROSIMIANS - PRIMATES).

Y. RUMPLER, M.C. ANDRIANJAKA, A. CLAVERT, C. CRANZ AND S. WARTER*
**Institut d'Embryologie, Faculté de Médecine, Strasbourg, France.**

Among the Malagasy Lemurs, numerous groups are still engaged in the process of speciation, so that the reproductive barriers are undefined. Within the genus Lemur, this problem concerns mainly the species L.fulvus and L.macaco. This group has evolved by centric fusions so that the hybrids obtained in captivity between L.fulvus (2N=60) and L.macaco (2N=44), show 8 trivalents during meiosis (Dutrillaux and Rumpler,1977). In spite of such a high number of centric fusions, all hybrids obtained are not sterile (Hamilton and Moses, 1979 , Albignac et al., 1971 , Rumpler and Dutrillaux, 1980). A spermatogenetic study of hybrids shows that many germinal cells degenerate after the pachytene stage, so that fewer spermatozoa are produced than in the pure species, but these spermatozoa show enormous individual diversity (Ratomponirina et al., in preparation). We thought it would be interesting to study this qualitative modification apart from the existing quantitative modifications specially as the spermatozoa morphology differs in the two pure species (Martin et al., 1975).

## MATERIAL AND METHODS

A morphometrical study was carried out on sperm from 5 Lemur fulvus (2N=60), 2 Lemur macaco (2N=44) and 3 hybrids : (1 hybrid L.fulvus fulvus (2N=60) X L. macaco (2N=44), 1 hybrid L.fulvus collaris (2N=52) L. macaco (2N=44), and 1 compound hybrid resulting from a cross between 2 different hybrids : L. fulvus fulvus X L.macaco and L. fulvus fulvus X L. fulvus collaris).

*skilled technical assistance of J.L. Maetz.

Spermatozoa were obtained by electro-ejaculation under general anaesthesia according to the technique developed by Martin et al. (1975). The ejaculate was spread on slides, fixed with alcohol-ether, stained with light-green haemalun-eosine. The photographs were taken with the aid of a Zeiss phase-contrast microscope, at a magnification of X 100. The spermatozoa heads (nucleus + acrosome) were measured from enlarged prints with the aid of a graphics tablet.

Electron microscopic studies were carried out on the ejaculates from 3 L.fulvus and 1 hybrid of L. macaco X L. fulvus, and on testicular biopsies from 2 L.fulvus and the 3 hybrids. The samples were obtained under general anaesthesia, fixed in glutaraldehyde at 3 % for the biopsies and 1.5 % for the ejaculates, post-fixed in osmic acid, the grids were stained with lead citrate and finally examined with the aid of a Siemens Electron microscope.

RESULTS

Spermatozoa from L.fulvus are longer in form than those from L.macaco due to the position of the acrosome, however, the surfaces measured are nearly the same for the two species. On the other hand, the surface of the spermatozoa for the 3 hybrids is clearly reduced. This reduction seems to be partly due to the diminished surface (see fig. 1) of acrosome in many spermatozoa which thus take on a more rounded appearance.

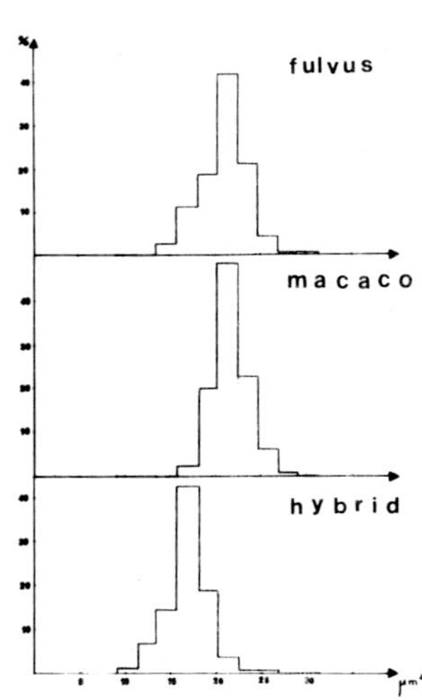

FIGURE 1.

Spermatozoa of L.fulvus show clearly a pronouncedly tapered acrosome with the electron microscope, which explains the elongated form when viewed by the light microscope (Martin et al., 1975). Less than 10 % teratospermia is observed on the 3 ejaculates and 2 testicular biopsies examined. This is mainly caused by a destruction of the acrosomial membrane (pseudo-capacitation) in the three hybrids studied, where teratospermia is important and polymorphic. Between 50 % and 80 % of spermatozoa show abnormal forms. Teratospermia affects both nucleus and acrosome. The nucleus does not show the typical reference form when observed in sagittal section, the thickness is clearly increased. There are many vacuoles present in the chromatin and condensation is irregular. The acrosome is shortened with localised thickening, showing evaginations at internal and/or external membranes.

## DISCUSSION

In all hybrids studied we noted a significant teratospermia affecting the nucleus and acrosome. The observed modifications often cause an irregular appearance of the nucleus and a shortening of the acrosome. The abnormal forms seen frequently in the 3 hybrids under study cause the significant reduction in the mean value of the spermatozoa surface.

The pattern of teratospermia encountered is aspecific, it may inconstantly associate the anomalies of the nucleus and the acrosome. It does not vary according to the type of meiotic abnormality present, as two of the hybrids studied during meiosis show trivalents, while the third one shows a complex crossing-over of the long chain (Dutrillaux and Rumpler, 1977). Furthermore, it was not even possible to correlate the number of centric fusions with the importance of teratospermia, considering the scarcity of hybrids available for this study.

However, it has been established that the 3 hybrids studied show this teratospermia, offering a good model for the study of the importance of chromosomal rearrangements on teratospermia in Primates.

## REFERENCES

1. Albignac R, Rumpler Y, Petter J. J. 1971. L'hybridation des Lémuriens de Madagascar. Mammalia, T.35, n°3, 358-368.

2. Dutrillaux B, Rumpler Y. 1977. Chromosomal evolution in Malagasy Lemurs. II. Meiosis in intra- and interspecific hybrids in the genus Lemur. Cytogenet. cell Genet. 18, 197-211.
3. Hamilton A.E, Moses M.J. 1979. Fertility in Lemur macaco X Lemur fulvus hybrids. Am. J. phys. Anthrop. 50, 413-496.
4. Martin D.E, Gould K.G, Warner H. 1975. Comparative morphology of primate spermatozoa using scanning electron microscopy. I. Families Hominidae, Pongidae, Cercopithecidae and Cebidae. J. Human Evol., 4, 287-292,
5. Ratomponirina Ch, Andrianīvo J, Rumpler Y. In preparation. A study on spermatogenesis in several intra- and interspecific hybrids of the Lemur (Malagasy prosimians). Submitted to J. Reprod. Fert.
6. Rumpler Y, Dutrillaux B. 1980. Chromosomal evolution in Malagasy Lemurs. V. Chromosomal banding studies of L.fulvus albifrons, L.rubriventer and its hybrids with L.FULVUS fulvus. Folia Primatol., 33, 253-261.

# COMPARATIVE HISTO-ENZYMATIC STUDY OF SPERMATOZOA IN MONKEY AND MOUSE

A. ABOU-HAÏLA and M.A. FAIN-MAUREL.
Laboratoire de Biologie Cellulaire, Université René Descartes, 75270 Paris, France.

## 1. INTRODUCTION

Enzyme activities have been demonstrated in mammalian spermatozoa (2,3,5,6) but no comparative study has yet been made on changes in these activities during spermatozoon epididymal transit. The aim of this study is to evaluate some enzymatic activities of spermatozoa during maturation in a rodent and a primate.

## 2. MATERIALS AND METHODS

Testis and samples at different levels along the length of the epididymis of *Macaca fascicularis* and of the mouse (*Swiss OF 1*) were juxtaposed on the same block and frozen in liquid nitrogen for 10 seconds. Eight micron thick sections of these blocks and longitudinal sections of the entire mouse epididymis were treated uniformly for a systematic comparison of 24 enzyme activities involved in various metabolic pathways and for 2 macromolecular complexes: PAS positive substances and sudanophilic lipids (Cf. GABE 1969, PEARSE 1968, 1972).

## 3. RESULTS AND DISCUSSION

1°) The results demonstrate that common enzymatic activities occur in spermatozoa of the monkey and the mouse.

| Reactions | MONKEY | | MOUSE | |
|---|---|---|---|---|
| | Testis | Epididymis | Testis | Epididymis |
| Succinate-D | 3,5 | 4 | 3,5 | 4 |
| Glycerol-P-D | 3 | 4 | 4 | 4 |
| β-OH-butyrate-D | 1,5 | 1,5 | 1,5 | 1,5 |
| Uridine-di-P-G-D | 0,5 | 1 | 0,5 | 1 |
| Glucose-6-Pase | 1 | 1 | 1 | 1 |
| β-glucuronidase | 1 | 2 | 1 | 2 |
| Aryl-sulfatase | 1,5 | 2,5 | 1,5 | 2,5 |

*Semi-quantitative evaluation of the reaction intensity in spermatozoa by grades (-) to 4: negative to strong.*

These activities are acquired in the testis. They persist, or increase, in the epididymis of the two species in the same way.

With the techniques used, spermatozoa of the monkey and the mouse appeared to lack glucose-6-phosphate-dehydrogenase, glycogen phosphorylase, 5-nucleotidase, ATPase-$Mg^{2+}$ (whose activities are positive in the epithelium), cholinesterase, as well as glycogen and lipids. Acid phosphatase activity is seen in the acrosome of monkey spermatozoa. In the two species, a weak hydroxysteroid dehydrogenase appears in the lumen of the epididymis.

2°) Some differences are noted between the monkey and the mouse concerning the distribution of the following reactions:

| Reactions | MONKEY | | | | MOUSE | | | |
|---|---|---|---|---|---|---|---|---|
| | Testis | Epididymis | | | Testis | Epididymis | | |
| | | Ep | Em | Ed | | Ep | Em | Ed |
| Isocitrate-D | - | | 0,5 | | - | -to 1,5 | | 1,5 |
| Malate-D | 0,5 | | 0,5 | | 0,5 | 1 | 1,5 | 1,5 |
| $NADH_2$-D | 1 | | 1,5 | | 1,5 | 2 | 2,5 | 2,5 |
| $NADPH_2$-D | 1 | | 1,5 | | 1 | | 1,5 | 1,5 |
| Glutamate-D | 0,5 | | 1 | | 1 | 1 | 1,5 | 1,5 |
| Lactate-D | 2 | | 3,5 | | 2 | 2,5 | 3,5 | 3,5 |
| ATPase-$Ca^{2+}$ | - | 1 | 1,5 | 1,5 | - | | 1 | |
| PAS | 0,5 | 1 | 2 | 2,5 | 0,5 | 1 | 2 | 2,5 |

*Ep, Em, Ed: proximal, medial and distal parts of the epididymis.*

Generally, these modifications appear more obvious in the mouse epididymis than in the monkey. The increase of PAS positivity has also been noted in *Macaca mulatta* (1,2).

Thus this study shows numerous common enzymatic activities during mammalian spermatozoon maturation which essentially concern the respiratory enzyme chain localized in the midpiece.

3°) However, there are specific differences concerning the nature and the distribution of the following reactions:

| Reactions | MONKEY | | | | MOUSE | | | |
|---|---|---|---|---|---|---|---|---|
| | Testis | Epididymis | | | Testis | Epididymis | | |
| | | Ep | Em | Ed | | Ep | Em | Ed |
| Alkaline Pase | 2,5 | | 2,5 | | - | - | 1 | 2,5 |
| Indoxyl-esterase | - | | - | | - | - | 1 | 2 |
| Naphtyl-esterase | - | | - | | - | - | 1 | 1,5 |

In spermatozoa of the monkey, no new enzymatic activity occurs in the epididymis. Alkaline phosphatase is acquired in the testis and non-specific esterase remains negative.

On the contrary, in spermatozoa of the mouse, as in the camel (6), new enzymes appear in the epididymis. Alkaline phosphatase and non-specific esterase first appear at the beginning of the body. Reaction intensities increase progressively up to the end of the duct.

The question arisesas to whether these activities, which exclusively appear in the mouse epididymis, are modulated either by factors intrinsic to the spermatozoon or by the functional activity of the epididymal epithelium.

REFERENCES

1. Alsum D.J. and Hunter A.G. 1978. Regional histology and histochemistry of the ductus epididymis in the rhesus monkey (*Macaca mulatta*). Biol. Reprod., 19, 1063-1069.
2. Arora-Dinakar R., Dinakar N., Kumar P., Talesara C.L. and Prasad M.R.N. 1977. Histochemical distribution of phosphatases in different regions of the epididymis of the adult rhesus monkey *Macaca mulatta*. Ind. J. Exp. Biol., 15, 859-864.
3. Arora-Dinakar R., Dinakar N., Kumar P., Talesara C.L. and Prasad M.R.N. 1977. Histochemical distribution of metabolites and lipolytic enzymes in different regions of the epididymis of the adult rhesus monkey *Macaca mulatta*. Ind. J. Exp. Biol., 15, 870-873.
4. Dinakar N., Arora-Dinakar R., Kumar P., Talesara C.L. and Prasad M.R.N. 1977. Histochemical localization of dehydrogenases in different regions of the epididymis of the adult rhesus monkey *Macaca mulatta*. Ind. J. Exp. Biol., 15, 865-869.
5. Holt W.V., Jones R.C. and Skinner J.D. 1980. Studies of the deferent ducts from the testis of the African elephant, *Loxodonta africana*. II. Histochemistry of the epididymis. J. Anat., 130, 367-379.
6. Tingari M.D. and Moniem K.A. 1979. On the regional histology and histochemistry of the epididymis of the camel (*Camelus dromedarius*). J. Reprod. Fert., 57, 11-20.

FIGURE LEGENDS

Fig. 1: $NADH_2$-D in Ed; Figs. 2, 3: Naphtyl-esterase in Ep (2) and Ed (3); Fig. 4: Indoxyl-esterase in Ed; Fig. 5: PAS in Ed (x 125).

Figs. 6,11: Alkaline phosphatase in the testis; Figs. 7, 12: Alkaline phosphatase in Ed; Figs. 8, 9: Indoxyl-esterase in Ep (8) and Ed (9) (x 312); Figs. 13, 14: Indoxyl-esterase in Ep (13) and Ed (14) (x 625); Figs. 10, 15: PAS in Ep (x 312).

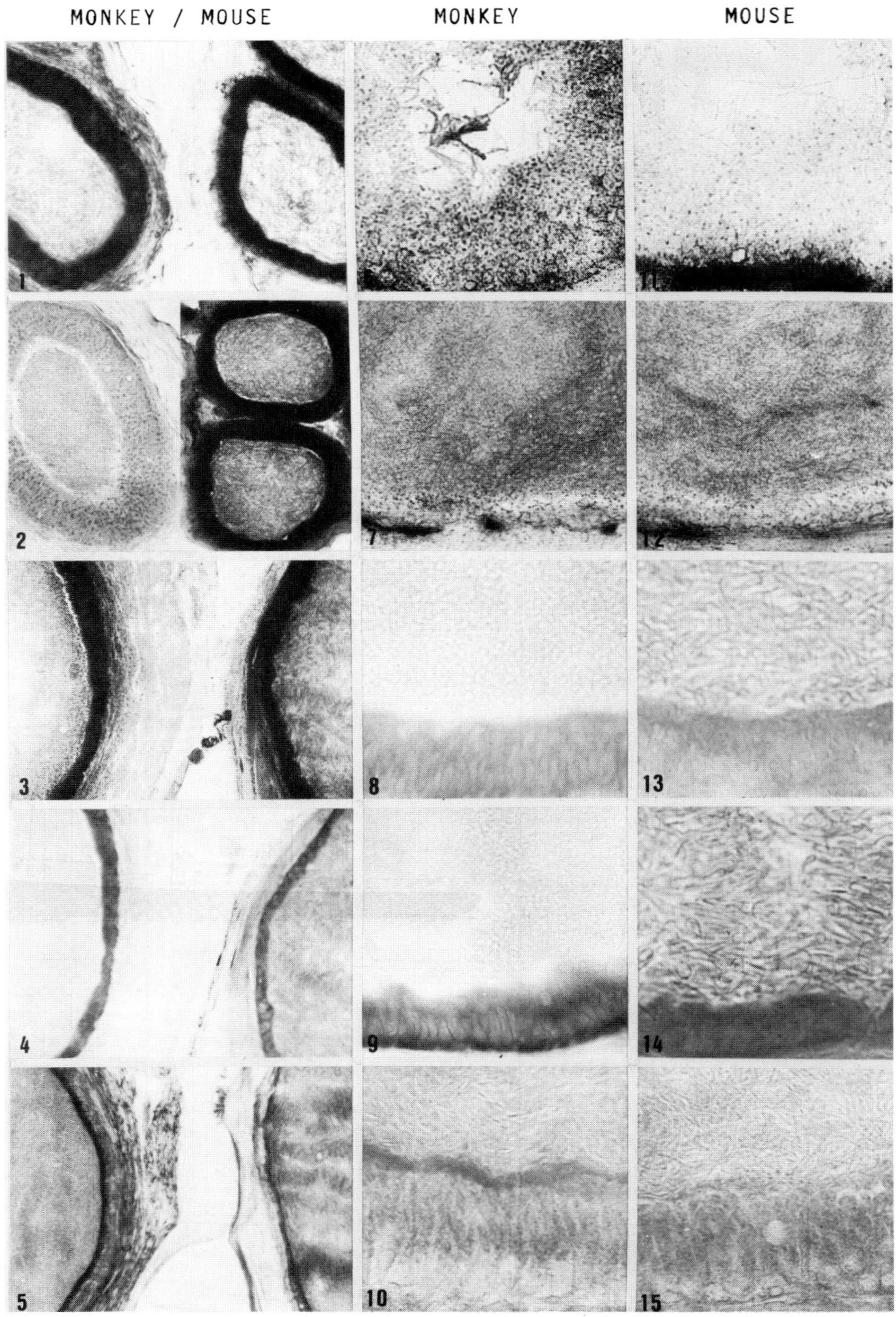
MONKEY / MOUSE
MONKEY
MOUSE
1
2
3
4
5
7
8
9
10
11
12
13
14
15

# ACROSOME DEVELOPMENT DURING SPERMIOGENESIS AND EPIDIDYMAL SPERM MATURATION IN AUSTRALIAN MARSUPIALS

H.R. HARDING, F.N. CARRICK & C.D. SHOREY
Department of General Studies, University of New South Wales, Kensington, N.S.W. Australia; Department of Veterinary Anatomy, University of Queensland; and Department of Histology and Embryology, University of Sydney, Australia.

In previous papers (1,3) we have shown that a distinct acrosomal maturation process occurs as sperm move down the epididymis in some but not all Australian marsupial species. We first described this process in sperm of the Australian Brushtail Possum, *Trichosurus vulpecula* (1), and in attempting to explain its occurrence we examined acrosomal development during spermiogenesis in this species(2).

Figure 1 illustrates the previously published acrosomal changes observed in *T. vulpecula*. In mid stages of spermiogenesis the acrosome forms a thin granular layer covering the entire dorsal nuclear surface and bounded by the nuclear ring (Fig. 1a). Later, the distal extremity of the nuclear ring moves to occupy a more rostral position and simultaneously, the acrosome still bounded by the nuclear ring, is thrown into folds (Fig. 1b). On the basis of these observations we suggested that acrosomal maturation could be considered as a continuation of a process initiated during spermiogenesis: the relocation of the acrosomal material over a more restricted area of the dorsal nuclear surface than that occupied in mid-stage spermatids (1,2). Consequently we hypothesised that since the acrosome covers the entire dorsal nuclear surface in mid-stage spermatids of all marsupials we had examined, then other factors being equal, acrosomal maturation should be a pronounced feature of epididymal sperm maturation in those species in which the 'mature' acrosome covers only a relatively small proportion of the dorsal nuclear surface, but should be less marked or lacking in those in which it covers a relatively large proportion.

To test this hypothesis we have examined 24 Australian marsupial species from 6 families with the following results. For species in which the acrosome covers approximately one-third of the dorsal nuclear surface in mature sperm, an acrosomal maturation process is pronounced in *T. vulpecula* (Phalangeridae), *Macropus parma, M. robustus, M. rufus, Potorous tridactylus* (Macropodidae), but is not apparent in *Perameles nasuta* or *Isoodon macrourus* (Peramelidae).

From the hypothesis, acrosomal maturation should be obvious in the two peramelid species. Sapsford, Rae and Cleland have examined spermiogenesis in detail for *P. nasuta* and they describe (4) a nuclear ring contraction and acrosomal compression similar to that in *Trichosurus*, except that the acrosomal folds do not persist and the acrosome displays its mature appearance at spermiation. In mid-stage spermatids of *Perameles* the granular component of the acrosome occupies only part of the acrosomal vacuole, the remainder appearing electron lucent (4). In similar stage spermatids of *Trichosurus* a thin evenly condensed layer of granular material fully occupies the acrosomal vacuole. We suggest that this difference allows for rapid accommodation of the granular material within the reduced acrosomal area in *Perameles*, whilst in *Trichosurus* and the macropodids no such vacuolar accommodation is available for rearrangement of the granular material, and thus the acrosomal folds persist at spermiation necessitating acrosomal epididymal maturation. In support of this idea we note that in *Perameles* the rapid disappearance of the acrosomal folds and formation of the mature acrosome in late spermatids is accompanied by loss of the electron lucent component of the acrosomal vacuole and its total occupation by the granular component (4).

Among species with a relatively large acrosomal coverage of the dorsal nuclear surface the following observations were made. In 13 species of the family Dasyuridae in which the acrosome thinly covers approximately four-fifths of the dorsal nuclear surface, acrosomal maturation is not apparent; this is likewise true of *Petaurus breviceps* and *P. norfolcensis* (Petauridae) in which the coverage is two-thirds. However, in *Pseudocheirus peregrinus* (Petauridae) with a similar proportional but much thicker acrosomal coverage to *Petaurus*, a maturation process is obvious, as it is in *Tarsipes spenserae* (Tarsipedidae). From the hypothesis, the acrosomal maturation in *Pseudocheirus* and *Tarsipes* (with two-thirds nuclear coverage) should be less pronounced than in those species with one-third nuclear coverage, all other factors being equal. Whilst this is true for *Tarsipes*, acrosomal maturation is more pronounced in *Pseudocheirus* than might be expected. Other factors may account for this difference. Acrosomal maturation in *Tarsipes* involves relocating a relatively thin layer of acrosome material in two-thirds the length of a nuclear surface measuring 10.6µm x 3µm whilst that in *Pseudocheirus* requires relocation of a relatively thick layer of acrosomal material into two-thirds the length of a nucleus measuring 5µm x 3.2µm.

It remains to explain the absence of acrosomal maturation in *Petaurus* despite its occurrence in *Pseudocheirus* and *Tarsipes*. Late spermatids in *Petaurus* are notable for the relatively uncondensed and sparse nature of the granular acrosomal component (Fig.2) at a stage when it appears evenly condensed and fully occupies the acrosomal vacuole in *Pseudocheirus*. As in the peramelids, this may allow the granular component of the acrosome in *Petaurus* spermatids to form into the mature acrosome within its reduced location over the nuclear surface without the production of persistent acrosomal folds.

Thus whilst the observations for *Trichosurus*, the macropodids, *Tarsipes* and the dasyurids support our original hypothesis, the evidence from the peramelids and petaurids appears non-supportive. We believe our explanations for these seeming inconsistencies among the latter groups still enable the hypothesis to stand. However, the potential importance of other factors such as the nature of the acrosomal vacuole and its granular component in mid-stage spermatids, the thickness of the acrosomal layer, and the ratio of total nuclear length to proportion of acrosomal coverage of the nuclear surface in mature sperm, should not be overlooked.

The interesting question arising from this study concerns the reason for the relocation of the acrosomal material into a reduced area of the nuclear surface in some Australian marsupial species, since the consequences of this process involve an extremely complex epididymal maturation of the acrosome which imposes a seemingly fragile condition on it en route. In view of this, it seems likely that this process must have a strong functional role in terms of the end product - the definitive acrosome. Thus the question becomes: is there a functional (e.g. fertilization) reason for the location of the acrosome over a relatively small proportion of the dorsal nuclear surface in some marsupial species, whereas it can perform its function when spread over a much larger area in others?

REFERENCES

1. Harding HR, Carrick FN, Shorey CD. 1976. Cell Tiss. Res. 171, 61-73.
2. Harding HR, Carrick FN, Shorey CD. 1976. Cell Tiss. Res. 171, 75-90.
3. Harding HR, Carrick FN, Shorey CD. 1979. p.289-303. In, The Spermatozoon Ed. DW Fawcett, JM Bedford. Urban and Schwarzenberg Inc., Baltimore-Munich.
4. Sapsford CS, Rae CA, Cleland KW. 1969. Aust. J. Zool. 17, 195-292.

FIGURE 1. Diagrammatic representations of acrosomal changes during spermiogenesis and epididymal transit in *Trichosurus vulpecula*. Sertoli cell components not shown. a) Mid to late stage spermatid. Acrosome covers length of dorsal nuclear surface. b) Later stage spermatid. Nuclear ring has contracted; acrosome thrown into folds. c) Spermatid at spermiation. Sub-acrosomal space eliminated; acrosomal projections persist. d) Sperm from middle segments of epididymis showing acrosomal projections in process of consolidation onto dorsal nuclear surface. e) 'Mature' compacted acrosome located over rostral one-third of dorsal nuclear surface. A acrosome; D cytoplasmic droplet; M manchette; N nucleus; NR nuclear ring; SA sub-acrosomal space.

FIGURE 2. Late spermatid of *Petaurus norfolcensis* showing sparse and relatively uncondensed nature of granular acrosomal component (AG). AV acrosomal vacuole; M manchette; N nucleus.

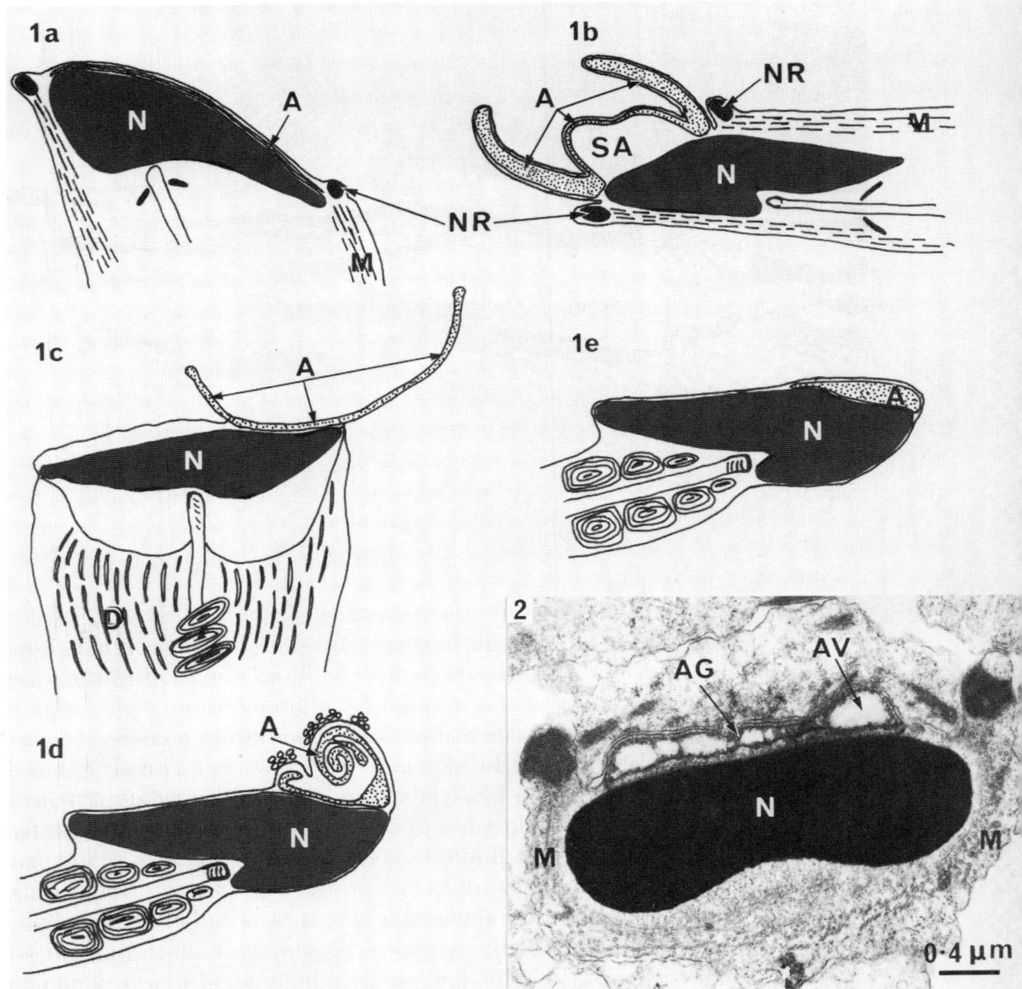

# EVOLUTION OF THE MAMMALIAN EPIDIDYMIS

RUSSELL C. JONES and DANIEL DJAKIEW
Department of Biological Sciences,
University of Newcastle, N.S.W., Australia.

## 1. INTRODUCTION

In order to assess the evolutionary development of the mammalian epididymis, its structural differentiation and the functions of each structurally distinct region are being studied in selected animals from the major groups: the Prototheria (the echidna, *Tachyglossus*) which may be archetypic for the Mammalia, the Metatheria (tammar wallaby, *Macropus eugenii*), and the testicond (African elephant and elephant shrew, *Macroscelides*) and scrotal (rat) eutherians.

## 2. RESULTS AND DISCUSSION

2.1. *Ductuli efferentes*. These are characterized by a low, ciliated epithelium which has a uniform ultrastructure within the Mammalia. The principal cells are structurally similar to the proximal convoluted tubules of the kidney, and like them, the ductuli are differentiated into two structurally distinct zones (13).

Although the absorptive role of the ductuli has been recognized for some time, due to technical difficulties in satisfactorily collecting luminal fluids, the constituents of the absorbate and the relative role of the ductuli and the initial segment of the ductus epididymidis in reabsorbing fluid have not been resolved (10,15). Our recent micropuncture studies (19) on the anaesthetized echidna (6) show that the ductuli efferentes reabsorb 85% of the fluid leaving the testis, and the initial segment of the ductus epididymidis reabsorbs 22%. We also confirmed (14) that the ductuli reabsorb specific protein.

2.2. Initial segments of the ductus epididymidis. The initial segment proper, is characterized by a narrow lumen containing few sperm, and a very tall secretory epithelium (2), however, the concentration of sperm gradually increases along the segment (see above, 3, 5 and zone 1 & 2, Fig. 2 in 17). Narrow cells are unique to the epithelium (18), and the principal cells have a chatacteristic ultrastructure which is dependent on luminal testosterone (7). Other initial segments (7), which differ in structure to the initial segment proper, but which are dependent on luminal testosterone, are also present in the rat epididymis.

We have identified the initial segment proper in all species studied and consider it significant that it may have a different embryological origin to the post-initial segments (16). It is also considered noteworthy that it is much longer in the echidna than rat (1104 vs 134 mm, 5) or tammar wallaby.

In the echidna, the initial segment is the source of all the luminal epididymal specific soluble proteins (Table 2), and it is also the main site of sperm maturation in the epididymis (Table 1). The increase in mean motility scores as sperm enter the terminal segment (Table 1) is correlated with the formation of sperm into bundles (3).

Table 1. Incidence of mature sperm as assessed by the location of the cytoplasmic droplet, and mean scores of progressive motility (scale 0-4) after dilution in Ringer

| Location | Droplet (%) | | Motility Score | | |
|---|---|---|---|---|---|
| | Echidna (n=3) 6 | Wallaby (n=5) | Echidna (n=3) 6 | Wallaby (n=5) | Rat (n=4) |
| Testis | 0 | - | 0.0 | - | - |
| Ductuli efferentes | 0 | 0 | 0.7 | 0.1 | 0.0 |
| Ductus epididymidis: | 0 | | | | |
| distal initial segments | 100[b] | 0 | 2.0 | 0.6 | 0.4 |
| distal middle segments[a] | - | 84 | - | 1.4 | 2.5 |
| terminal segment | 100[c] | 89 | 4.0 | 3.4 | 3.1 |
| *s.e.m.* | *2.5* | *1.8* | *0.22* | *0.15* | *0.26* |

*a, distal corpus epididymis  Droplet absent in 90% (b) & 100% (c)*

2.3. Middle segments of the ductus epididymis. This region occurs between the initial and terminal segments in the therian mammals and is absent in the echidna. It is usually subdivided into a number of segments in scrotal eutherians (e.g. zones 3-5 in the rat, 17), but is undifferentiated in the elephant, elephant shrew and tammar wallaby.

Glover and Nicander (8) proposed that the region is involved in sperm maturation. Table 1 confirms this for the therian mammals and Table 2 shows that in the rat the segments are also involved in the secretion of considerable specific protein. However, it is not clear why this region has evolved to take over functions carried out by the initial segment in the echidna.

Table 2. Apparent molecular weights ($x10^{-3}$ daltons) of proteins which are secreted by the epididymis of the echidna and rat and which are not present in blood or rete testis fluid.

| Echidna (6) | Rat (9) |
|---|---|
| 82.5 ± 3.5* | 80.0* |
| 48.5 ± 2.5*(G) | 47.0 |
| 39.0 ± 1.5* | - |
| 32.0 ± 1.0*(G) | 32.0(G) |
| 20.5 ± 0.5*(G) | 23.0*(G) |
| 19.0 ± 0.5*(G) | 19.0(G) |
| - | 18.5(G) |

** secreted by the initial segment (G) = stains with PAS*

2.4. Terminal segment of the ductus epididymidis. Glover and Nicander (8) described this as a "sperm store" where the duct is wide, the epithelium low and sperm are densely packed. Since the flow of sperm must be constant through the epididymis presumably sperm "storage" is a consequence of the relationship which describes flow through a tube i.e. velocity is inversely proportional to the radius squared. However, as the lumen of much of the ductus epididymidis may be wide it is suggested that a more objective definition than "sperm store" would be: the region where mature sperm are stored accessible for ejaculation. This is characterized (1) by thick as well as thin periductal muscle fibres i.e. the proximal end is identified by a thickening of the periductal muscle layer.

Jones and Djakiew (12) noted that the evolution of the scrotal epididymis has involved the storage in the terminal segment of an increased proportion of the extragonadal sperm.

## 3. CONCLUSION

It is suggested that the initial segment(s) is present in the epididymis of all mammals and that it is an important site of sperm maturation. There may be homologous proteins secreted by the mammalian epididymis (Table 2).

## ACKNOWLEDGEMENTS

Australian Research Grants Scheme and Internal Research Assessment Committee, University of Newcastle.

## REFERENCES

1. Baumgarten HG, Holstein AF, Rosengren E. 1971. Z. Zellforsch. mik. Anat. 120, 37-79.
2. Benoit J. 1926. Arch. d'anat., d'histol. d'embryol. 5, 173-412.
3. Djakiew D, Jones RC. 1981. J. Anat. 132, 187-202.
4. Djakiew D, Jones RC. 1982. J. Anat. (in press).
5. Djakiew D, Jones RC. 1982. Aust. J. Zool. 30(6).
6. Djakiew D, Jones RC. unpublished data.
7. Fawcett DW, Hoffer AP. 1979. Biol. Reprod. 20, 162-181.
8. Glover TD, Nicander L. 1971. J. Reprod. Fertil. Suppl. 13, 39-50.
9. Jones R, Brown CR, Von Glós KI, Parker MG. 1980. Biochem. J. 188, 667-676.
10. Jones RC. 1980. J. Reprod. Fertil. 60, 87-92.
11. Jones RC, Brosnan MF. 1981. J. Anat. 132, 371-386.
12. Jones RC, Djakiew D. 1978. Aust. Zool. 20, 201-210.
13. Jurd KM, Jones RC. 1982. Structure and Function of the ductuli efferentes of the rat. J. Reprod. Fert. (submitted).
14. Koskimes AI, Kormano M. 1975. J. Reprod. Fertil. 43, 345-348.
15. Levine N, Marsh DJ. 1971. J. Physiol.(Lond.) 213, 557-570.
16. Marshall FF, Reiner WG, Goldberg BS. 1979. Invest. Urol. 17, 78-82.
17. Reid BL, Cleland KW. 1957. Aust. J. Zool. 5, 223-246.
18. Sun EL, Flickinger CJ. 1979. Amer. J. Anat. 154, 27-56.
19. Ullrich KJ, Fromter E, Baumann K. 1969. Laboratory Techniques in Membrane Biophysics. N. Passon and R. Stämpfli, eds. Springer, Berlin, N.Y.

# SYNTHESIS OF MITOCHONDRIAL PROTEINS FROM MOUSE EPIDIDYMAL SPERMATOZOA DURING SPERMATOGENESIS

NORMAN B. HECHT AND ELIZABETH KENNINGTON

## INTRODUCTION

In the mammalian testis, marked changes of morphology of mitochondria are seen during spermatogenesis (1-3). Initially, the mitochondria of pre-meiotic cells appear structurally similar to those found in somatic tissues but during meiosis "germ cell" mitochondria with diffuse and vacuolated matrices appear. A third morphologically distinct form of mitochondria with condensed matrix replaces the "germ cell" mitochondria in spermatozoa. Concomitant with morphological changes, changes in the density and protein composition of the differentiating mitochondria occur, with several new polypeptides being found in high abundance in ejaculated spermatozoa (4). In this report we define the temporal synthetic sequence of mitochondrial polypeptides during spermatogenesis in mouse. We demonstrate that mitochondria from epididymal and ejaculated spermatozoa contain similar "sperm specific" polypeptides and that most of the polypeptides present in spermatozoal mitochondria are synthesized during both meiosis and spermiogenesis.

## METHODS

### Labeling and collection of spermatozoa

Groups of sexually mature male CD-1 mice were anaesthetized with ether and injected intratesticularly with ($^{35}$S)methionine (50 uCi/testis). At 3-day intervals, groups of 3-5 mice were sacrificed and cauda epididymal spermatozoa (C.E.S.) collected by gentle agitation of slit open cauda epididymidis in phosphate buffered saline (5). Following filtration and several washes in 0.45% saline to lyse somatic cells, 2-5x$10^{6}$ spermatozoa were recovered from each animal.

### Isolation of mitochondria from spermatozoa

Mitochondria were isolated by our modification of the method of Pallini (4, 6).

Polyacrylamide gel electrophoresis

Mitochondrial pellets, solubilized in 30 ul of 0.05 M Tris HCl pH 6.8, 2% SDS, and 0.035 M 2-mercaptoethanol, were analyzed on 15% polyacrylamide slab gels containing 0.1% SDS and 5.5 M urea with a 5% polyacrylamide stacking gel containing 2.7 M urea. Electrophoresis was carried out at 200 volts for 24 hours.

RESULTS AND DISCUSSION

Gel electrophoresis was used to determine whether the novel mitochondrial proteins seen in ejaculated spermatozoa (ES) are present in cauda epididymal spermatozoa (C.E.S.). After staining with Coomassie brilliant blue R, similar polypeptide profiles with species ranging from 13,000 to 70,000 daltons were present in mitochondria from both sources of spermatozoa while testicular mitochondria differed (Fig. 1). The mitochondrial polypeptides of spermatozoa appear to be highly conserved and can be readily compared between mouse, rat, and bull. Of particular interest is a polypeptide of molecular weight 17,000, a molecule that comigrates with the cysteine-rich structural stabilizing component of the external layer of sperm mitochondria (7-9).

When the amount of ($^{35}$S)methionine incorporation into mitochondria of C.E.S. was quantitated over a 33 day period following testicular injection, maximal incorporation was observed during mid spermiogenesis with lesser but substantial synthesis during meiosis (Fig. 2). In contrast, the majority of head proteins were synthesized during late spermiogenesis with lesser amounts during meiosis.

Aliquots of mitochondrial proteins from radiolabeled C.E.S. were analyzed by gel electrophoresis and fluorography to define their temporal synthesis pattern. In general, most mitochondrial polypeptides had long intervals of synthesis commencing during meiosis and continuing during spermiogenesis (Fig. 3). The cysteine-rich structural protein of the mitochondrial capsule (indicated by arrow) appears to start synthesis during meiosis but terminates or decreases synthesis to undetectable levels during mid-spermiogenesis. The extended, but not temporally identical, synthesis for many mitochondrial polypeptides argues for synthesis and insertion into the differentiating mitochondria at different stages of spermatogenesis. These results suggest that spermatozoal mitochondria differentiate by a sequential substitution of spermatozoal proteins instead of an abrupt synthesis of a new class of mitochondria at one specific interval during spermatogenesis, data in agreement with mitochondrial turnover studies in rat liver where different components have been shown to have different half-lives (10).

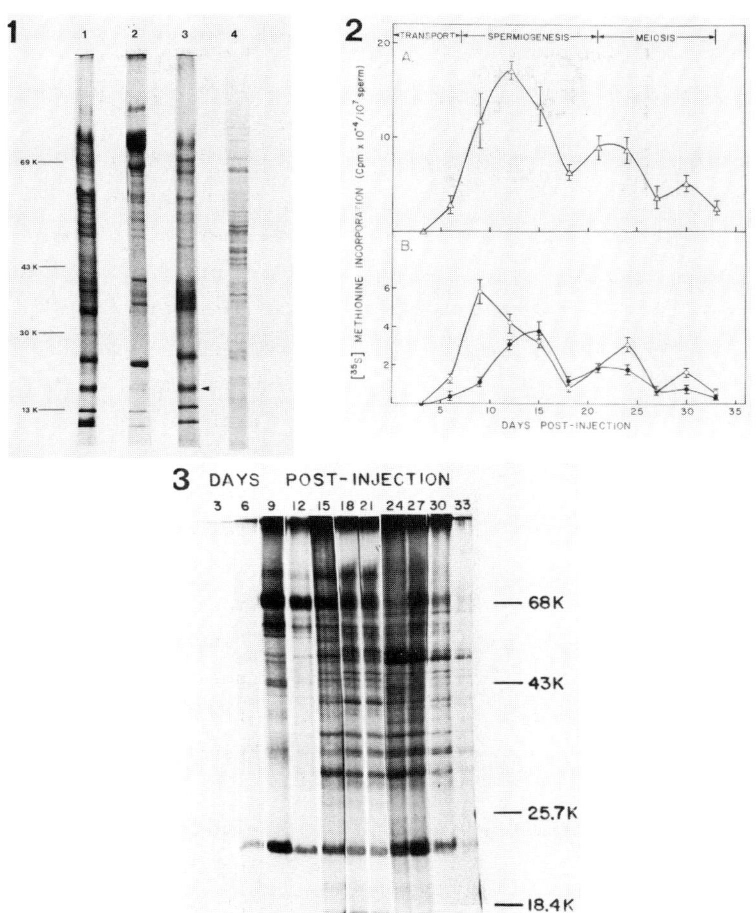


Figure 1. Gel electrophoresis of mitochondrial polypeptides. 1) 60 ug protein from bovine E.S. 2) 75 ug protein from mouse C.E.S. 3) 60 ug protein from rat C.E.S. 4) 60 ug protein from mouse testicular mitochondria. Arrow indicates electrophoretic mobility of mitochondrial capsule protein from rat (kindly provided by H. Calvin) (8).

Figure 2. ($^{35}$S)methionine incorporation into murine C.E.S. Aliquots in duplicate (50 ul) were precipitated in 10% TCA, filtered and counted for radioactivity. Each data point represents mean ±S.E. of at least 6 determinations. Time intervals are based on the kinetic data of Oakberg (11). A. (△) total spermatozoa. B. (○) head fraction. C. (●) mitochondrial fraction.

Figure 3. Fluorograph of ($^{35}$S)methionine labeled mitochondrial polypeptides from C.E.S.

REFERENCES

1. André, J. 1962. J. Ultra. Res. 3: 1-185S.

2. Machado de Domenech, E., Domenech, C.E., Aoki, A., Blanco, A. 1972. Biol. Reprod. 6: 136-147.
3. DeMartino, C. Floridi, A., Marcante, M.L., Malorni, W., Scorza Barcellona, P., Bellocci, M., Silverstrini, B. 1979. Cell Tiss. Res. 196: 1-22.
4. Hecht, N.B., Bradley, F.M. 1981. Gam. Res. 4: 433-449.
5. O'Brien, D., Bellvé, A.R. 1980. Dev. Biol. 75: 405-418.
6. Pallini, V. 1979. The Spermatozoan: Maturation, Motility, Surface Properties and Comparative Aspects. Baltimore-Munich. Urban and Schwarzenberg. pp. 401-410.
7. Calvin, H. 1978. J. Exp. Zool. 204: 445-452.
8. Calvin, H., Cooper, G.W. 1979. The Spermatozoan: Maturation, Motility, Surface Properties and Comparative Aspects. Baltimore-Munich. Urban and Schwarzenberg. pp. 135-140.
9. Pallini, V., Baccetti, B., Burrini, A.G. 1979. The Spermatozoan: Maturation, Motility, Surface Properties and Comparative Aspects. Baltimore-Munich. Urban and Schwarzenberg. pp. 141-151.
10. Lipsky, N.G., Pedersen, P.L. 1981. J. Biol. Chem. 256: 8652-8657.
11. Oakberg, E. 1956. Am. J. Anat. 99: 507-516.

SUMMARY

Mitochondria from ejaculated bovine spermatozoa contain a group of polypeptides not found in other bovine or murine testicular mitochondria (4). To establish when during epididymal transport, spermiogenesis, and/or meiosis these proteins were synthesized, the synthesis intervals for the mitochondrial proteins from cauda epididymal spermatozoa were established following intratesticular injection of ($^{35}$S)methionine. Mice were labeled every third day over a 33 day period and cauda epididymal spermatozoa were fractionated into mitochondrial and head components. Radioactivity in each fraction was monitored by liquid scintillation counting. Maximal incorporation was observed during spermiogenesis although substantial amounts of protein were synthesized during meiosis. Analysis of the polypeptides by gel electrophoresis revealed that many of the polypeptides were synthesized over prolonged intervals of spermiogenesis and meiosis rather than in a brief specific time period. These results suggest that spermatozoal mitochondria are produced by a sequential substitution of new proteins into the differentiating mitochondria rather than the abrupt appearance of a new class of mitochondria during spermatogenesis.

# SOME ASPECTS OF SPERMATOGENESIS AND SPERMATOLOGY IN TELEOST FISH.

R. BILLARD

I.N.R.A., Laboratoire de Physiologie des Poissons,

Campus de Beaulieu, 35042 Rennes, France.

Spermatogenesis in teleost fish has several original features not found in mammals. The wide diversity in testicular structure and spermatogenetic function in teleosts may be related to the large number of species and the wide differences in colonized biotopes. However, all the lower vertebrates and many groups of invertebrates have one peculiarity in common, i.e. the presence of cysts composed of aggregates of isogenic cells that develop synchronously. Moreover, all the germinal cells are included in Sertoli cells so that they are never in direct contact with the basal lamina of the seminiferous tubule.

There are two main types of testicular structure in the teleost fish, the tubular type described by Billard (1) in guppy Poecilia reticulata and extended to atheriniforms by Grier et al. (2), and the lobular type described for instance par Billard et al. (3), many variants of which are found in most of the other groups of fish. Basically different spermatogenetic processes characterize these two types of testicular structure Fig 1. In the tubular type, the seminiferous tubule, having no central lumen or regular Sertoli epithelium, radiates from the periphery to the center or the opposite side of the testis. At the blind peripheral apex of the tubule are found isolated spermatogonia which correspond to type A spermatogonia ($G_A$); after multiplication, they organize into cysts composed of cells corresponding to type B spermatogonia ($G_B$) having a Sertoli cell layer. While spermatogenesis continues, these cysts migrate to the opposite end of the tubule which opens into an efferent and testicular duct. In the lobular type, spermatogenesis takes place in a network of connective tissue originating from the testicular capsule (hence the term "lobular") with a permanent layer of Sertoli cells and a central lumen into which the spermatozoa flow on their way to be evacuated. This situation is similar to that in the higher vertebrates where the movement of the germinal cells is only limited and centripetal in the seminiferous tubule during spermatogenesis.

Other original characteristics are more or less closely related to these two main types of structure. First, there are fewer $G_A$ per tubule in the tubular than in the lobular-type structure but, at least in Poecilids, the cysts are larger, giving a higher number of spermatogonial divisions (14 in guppy) (1). However, spermatogenetic production in guppy (tubular-type testis) and trout (lobular-type testis) is comparable when expressed by the number of spermatozoa per g of testis per year (about 50 to $70.10^9$ spermatozoa). In other words, the higher number of spermatogonial divisions compensates for the reduced number of $G_A$. Moreover, spermatogenesis is a continuous process in guppy, while in trout, as in most species with lobular-type testes, it is seasonal.

Spermiogenesis is often more complex in the tubular than in the lobular-type structure. This process terminates in the formation of a more complex spermatozoon with an elongated head and highly condensed chromatin, a more marked transformation of the centriolar complex and a large mid-piece including glycogen stores. The anterior end of the sperm head and the Sertoli cells are closely related morphologically. In guppy, the spermatids are all oriented towards the periphery of the cyst so that its center is wholly occupied by flagella only. During spermiation, the sperm heads detach from the Sertoli cells and an original spermatozoal aggregates remains (spermatozeugma); there is almost no seminal fluid and the vas efferens is the only secretor. Spermatozoal aggregates occur in species with internal fertilization and seem to represent a privileged form of spermatozoon transport in the female genital tract. This transport is effected by means of the gonopod which is none other than a modified anal fin. The physiology of the spermatozoa is original in Poecilids since it includes *in vitro* motility times of several hours and survival of several months in the female genital tract.

Spermiogenesis is simple in the lobular-type testis; the spermatids are never in contact with the Sertoli cells and remain scattered throughout the cyst. The spermatozoa are always free in the genital tract, including the lobular lumen after the cysts open; they are accompanied by large amounts of seminal fluid (spermatocrit: 20-50%). The seminal fluid seems to originate in both the Sertoli cells, where secretion is high during spermiation, and in the epithelium of the vas deferens. After spermiation, the sperm ages in the genital tract. Fertilization is always external and spermatozoon motility time is short (30 sec to a few min) in fresh water, so that the spermatozoon penetrates the egg very soon after "ejaculation".

A structure for storing the spermatozoa, the testicular duct, is found in the tubular-type testes of Poecilids; the transit times is of several days in guppy. In lobular-type testes, there is no morphologically differentiated structure for storage but the spermatozoa are stocked throughout the lobular network and even in the vas deferens which can stretch considerably as in salmonids.

The length of spermatogenesis has been estimated in some atheriniform species; at 25°C the leptotene-spermatozoon interval is 14 days in Poecilia reticulata (4) and 12 days in Oryzias latipes (5). The part of spermatozogenesis (not including $G_A$) which occurs inside the cysts requires 36 days in guppy (1). Very little information is available on species with lobular testes; in the roach, the leptotene-spermatozoon interval is about 30 days at 10°C (Gillet and Billard, unpublished).

It is difficult to compare the endocrine control of spermatogenesis in the two groups because there are very few data on tubular-type species. Hypophysectomy has shown that spermatogenesis is always under pituitary control. The pituitary gonadotropins have high zoological specificity; carp pituitary extracts have no effect on the maintenance or restoration of gametogenesis in hypophysectomized guppy, while pituitary extracts of Gambusia are effective (6). Mammalian LH and FSH have little or no effect on teleost spermatogenesis but HCG stimulates spermiation.

Environmental temperature appears to be the determining factor in species with continuous spermatogenesis and belonging essentially to the atheriniform group. In other groups with seasonal spermatogenesis, the temperature plays a role concommittantly with the photoperiod and other factors such as the rainy season. Some recent reviews give more detailled informations on fish spermatogenesis (7, 8, 9) and spermatology (10, 11).

REFERENCES

1. Billard R., 1969. Ann.Biol.anim.Bioch.Biophys., 9, 251-271.
2. Grier H.J., Linton J.R., Leatherland J.F., De Vlaming V.L., 1980. Am.J.Anat., 159, 331-345.
3. Billard R., Jalabert B., Breton B., 1972. Ann.Biol.anim.Bioch.Biophys., 12, 19-32.
4. Billard R., 1968. C.R.Acad.Sci.Paris, série D, 266, 2287-2290.
5. Egami N., Hyodo-Taguchi Y., 1967. Exp.Cell.Res., 47, 665-667.
6. Breton B., Billard R., Jalabert B., 1973. Ann.Biol.anim.Bioch.Biophys., 13, 347-362.
7. Roosen-Runge E.C., 1977.Cambridge University Press, Cambridge, New York, Melbourne.
8. Grier H.J., 1981. Am.Zool., 21, 345-357.
9. Billard R., Fostier A., Weil C., Breton B., 1982. Can.J.Fish.Aquat.Sci., 39, 65-79.
10. Billard R.,1978. Actes de Colloque du CNEXO, 8, 59-73.
11. Billard R., 1980. 9th Inter.Congress Anim.Reprod.Artif.Insem., Vol II, 327-337.

FIGURE 1. Some morphological and physiological differences between the two main types of testicular structure in teleost (tubule type ex. guppy and lobule type ex. trout or carp)
ed : efferent duct
ic : interstitial cells
Spz : spermatozeugma

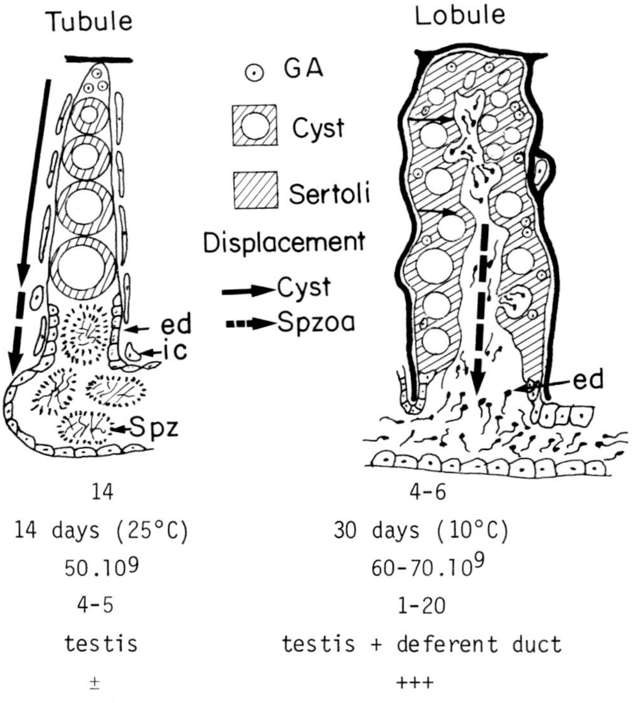

| | Tubule | Lobule |
|---|---|---|
| Spermatogonial divisions | 14 | 4-6 |
| Duration Leptotene-Spzoa | 14 days (25°C) | 30 days (10°C) |
| Spzoa/g testis/year | $50.10^9$ | $60\text{-}70.10^9$ |
| Gonadosomatic index % | 4-5 | 1-20 |
| Sperm storage | testis | testis + deferent duct |
| Seminal fluid | ± | +++ |
| Fertilization | internal | external |
| Sperm motility (*in vitro*) | few hrs | few min |
| Minimum ratio Spzoa/egg at insemination ($10^3$) | 2-5 | 30-300 |

# NUCLEUS AND ACROSOME OF TICK SPERM: POSITION AND STRUCTURE (WITH COMPARISONS BETWEEN IXODID AND ARGASID TICKS)

B. FELDMAN-MUHSAM and B.K. FILSHIE
Hebrew University, Jerusalem, Israel and CSIRO, Canberra, Australia

The sperm cell of ticks is not only relatively large, but of a very unusual structure, and undergoes a strange and peculiar series of transformations during its development, maturation and capacitation. These processes as well as the fine structure of the sperm have been studied for some species of ixodid and argasid ticks. Each of these studies sheds light on one or several morphological and physiological aspects; but they do not enable us to discern whether the findings are relevant only to the species under observation or are a general feature of the Ixodoidea.

Under the light microscope the structure of the sperm cell before spermateleosis (prospermium) and after spermateleosis (spermiophore) as well as the process of spermateleosis itself appear to be similar in the two groups, but the size of the sperm of argasids is almost twice that of ixodids (prospermium: 225-250 μm long in _Ornithodoros gurneyi_ and less than 100 μm in _Aponomma hydrosauri_; spermiophore: ca. 500 μm as against about 200 μm respectively). In spite of this, the nucleus is roughly the same size, viz. ca. 16 μm in both species.

In the following an attempt is made to compare certain details in the ultrastructure of the nucleus and the acrosome of two species of ticks, one (_A. hydrosauri_) belonging to Ixodidae and the other (_O. gurneyi_) to Argasidae.

In the two families the nucleus is attached to an acrosomal plate which is embedded in a cisternum of endoplasmic reticulum (e.r.). In the prospermium this e.r. constitutes the outermost sheath and in the spermiophore, the acrosomal canal. In the spermiophore of _Aponomma_ there are a number of electron dense platelets between the outer membrane of the cisternum in which the acrosomal plate is embedded and the membrane facing the lumen of the acrosomal canal (Fig. 1). These platelets were described in the prospermium of O. gurneyi (3). Slightly oblique sections show that these platelets are actually elongated straps which run parallel to the axis of the sperm cell (Fig. 1)

As in Ornithodoros, the acrosomal plate of Aponomma consists of two relatively thick granular layers. In Ornithodoros the inner layer, i.e. the layer adjacent to the nucleus, is electron lucent, and the outer layer is electron dense, and both are of about the same thickness. In Aponomma, the inner layer is electron dense, and the outer lucent, and the latter is thicker than the former (Fig. 1).

In both families the locus of attachment of the nucleus to the acrosomal plate appears in SEM as a nipple (Fig. 4). In Ornithodoros the nucleus is attached to the acrosomal plate by means of a complex structure of which a central stalk is the main element (1 and 3). This stalk is a bundle of electron dense parallel fibres surrounded by several layers of e.r. In Aponomma a stalk of a similar structure connects the nucleus to the acrosomal plate: but the stalk is centrally located in the nucleus only for part of its length and otherwise is either off the center or even outside the nucleus (Fig. 1).

In the prospermium and spermiophore of both species, several layers of e.r. envelop not only the stalk but also the nucleus (Fig. 1). In addition, large amounts of e.r. were found in the spermiophore of Aponomma, sometimes filling it almost completely. At some places the e.r. folds in upon itself several times, back and forth, as shown by a serpentine pattern in some cross sections (Fig. 3). However, in most sections the image of a meandrine body appears (Fig. 2).

In the spermiophore of Aponomma amorphous accumulations of electron dense material appear at many places outside the cell membrane under the cellular processes (Figs. 1 and 2). Similar accumulations were described in Amblyomma hebraeum as "peripheral granules" (2). In Aponomma they are not granules and they tend to be spread over the whole surface of the cell.

REFERENCES

1. Breucker H., Horstmann E. 1968. Die Spermatozoen der Zecke Ornithodorus moubata (Murr.) Z. Zellforsch Mikrosk Anat 88: 1-22
2. El Said A., Swiderski Z. 1980. Regional specialization of the sperm membrane in the tick Amblyomma hebraeum K. Cell Tissue Res. 208: 34-45
3. Feldman-Muhsam B., Filshie B.K., 1979. The ultrastructure of the prospermium of Ornithodoros ticks and its relation to sperm maturation capacitation in The Spermatozoon ed.: Fawcett, Bedford. Urban & Schwarzenberg, Baltimore-Munich.

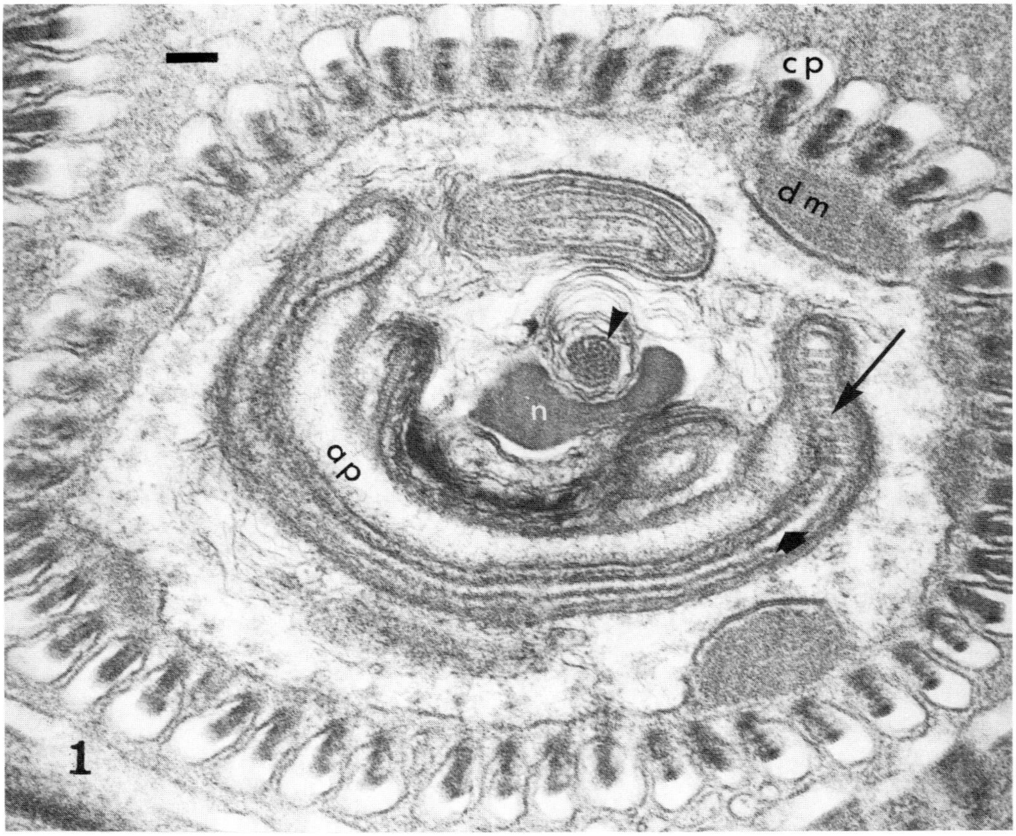

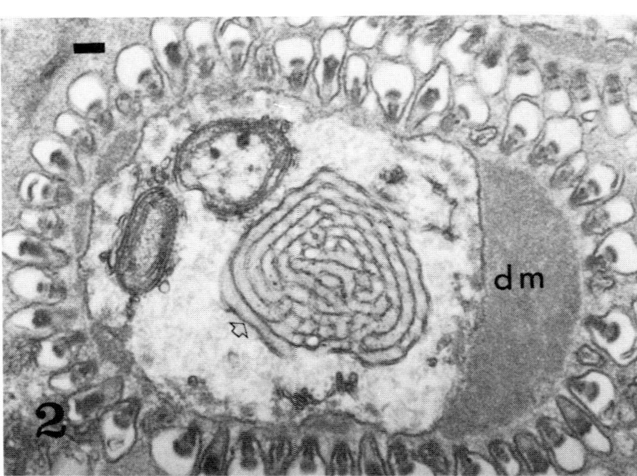


Figs. 1 and 2.
Spermiophores of <u>Aponomma</u>.
Nucleus (n); stalk (➤);
acrosomal canal (⬟);
acrosomal plate (ap);
diagonally cut
platelets (⟶)
accumulation of electron
dense material (dm);
meandrine body (⇧);
cellular process (cp).

Bar = 100nm

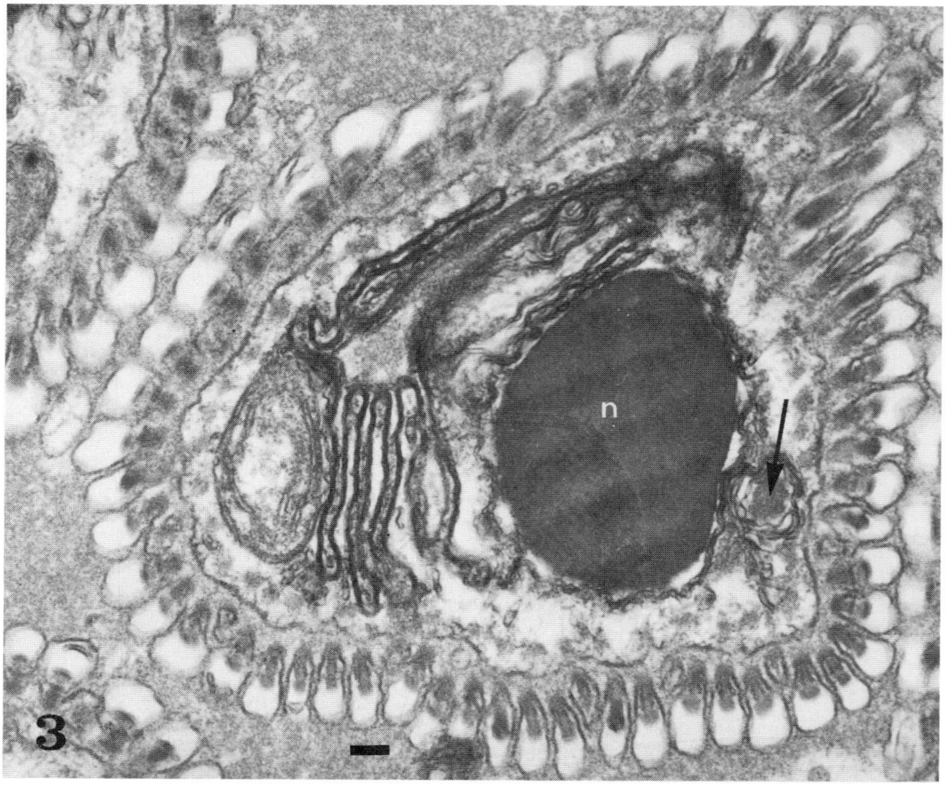

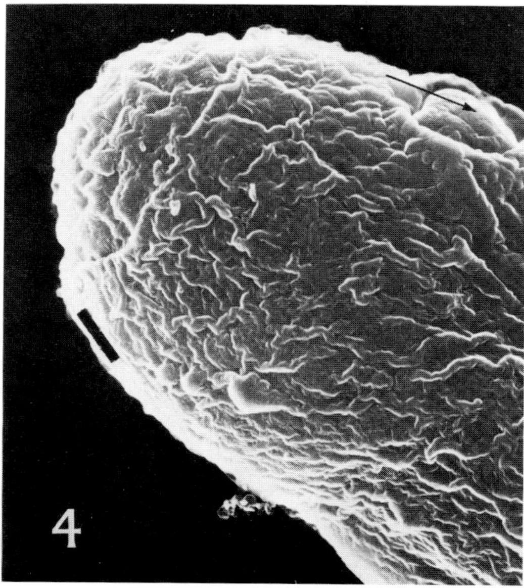

Fig. 3.
Spermiophore of Aponomma.

Serpentine pattern of endoplasmic reticulum: nucleus (n); stalk (↖).

Bar = 100nm

Fig. 4.
Prospermium of O. gurneyi.

Nipple (↖).

Bar = 1μm.

# THE GIANT ACROSOME OF *Geogarypus* (ARACHNIDA, PSEUDOSCORPIONES) AND PHYLOGENETIC CONSIDERATIONS

R. DALLAI and G. CALLAINI
Institute of Zoology, via P. A. Mattioli 4, 53100 Siena, Italy

In Pseudoscorpions reproduction is known to be carried out through the indirect transfer of spermatozoa (1,2) by means of spermatophores. At the end of stereotipate and after more or less evolved courtships (3) which differ in the several families, males deposit a drop of sperm on the apex of a stem (4). Spermatozoa in these spermatophores, as well as at the end of the male deferentes are always enclosed within a cyst, which offers protection against dehydration during the interval between deposition of the structure and its reception by the female.

Spermiogenesis in Pseudoscorpions has been studied in recent years (4, 5,6,7), but some doubts still remain as to mitochondrial transformations, axonemal structure and acrosomal evolution.

In this note we make a comparison between the formation of the acrosome in a common species and that seen in *Geogarypus*, which is characterized by an unusually large acrosomal structure.

## OBSERVATIONS

In every species of Pseudoscorpions studied so far, the acrosome arises as a result of an active secretion by a large Golgi complex. After the formation of the primitive acrosomal vesicle, which is located at the anterior end of the nucleus and which contains filamentous material, a kind of ribbon starting from the acrosomal vesicle begins to surround the nucleus in helicoidal fashion, moving towards its basal end. This acrosomal helix, triangular in section, increases in length by the continuous activity of the Golgi complex. The definitive organization of the acrosomal helix seems to occur concomitantly with condensation of the nuclear content; it becomes flat and the material within appears well condenced at maturity. During the acrosomal helix formation, the apical portion of the acrosome that is located just over the nucleus also undergoes conspicuous trans= formations. First the acrosomal material progressively increases its density in the axial part and pushes off the acrosomal membrane in such a way that a kind of hook is formed; secondly the acrosomal membrane in close contact with the nuclear membrane forms an invagination in its central region; at the same time the nuclear membrane also begins to form an endonuclear cavity which in the final stage of maturation is very long. This sub= acrosomal formation, according to Baccetti (8), corresponds to the "per= foratorium". In this endonuclear cavity amorphous dense material accumulates progressively. At the end of spermiogenesis the acrosomal complex of *Chthonius* is constituted of three parts: a) the helix surrounding the nucleus; b) the perforatorium and c) the apical region. We cannot identify the last as an extra-acrosomal structure because it still derives from transformation of the primitive acrosomal vesicle; however in some images, where granular material seems to be associated with this structure, it is difficult to follow the acrosomal membrane and therefore the presence of extra-acrosomal material may be possible.

Let us now consider spermiogenesis in *Geogarypus*; the first events of acrosome formation are almost the same as in the preceeding species. Minor differences concern the nuclear condensation, the number of the acrosomal helices around the nucleus and the length of the endonuclear perforatorium and of the nucleus itself. As to the nuclear condensation, it usually involves only the axial region of the nucleus, whilst the marginal area contains only electron transparent material; in *Geogarypus*, however, the nuclear membrane of the basal region forms large swellings which sometimes reach the periphery of the spermatid.

The greatest difference in the acrosome formation of *Geogarypus*, compared with that in other Pseudoscorpion spermatozoa, begins to appear when the chromatin is already condenced in an axial cylinder internally crossed by the perforatorium. At this stage the apical region of the acrosome, filled

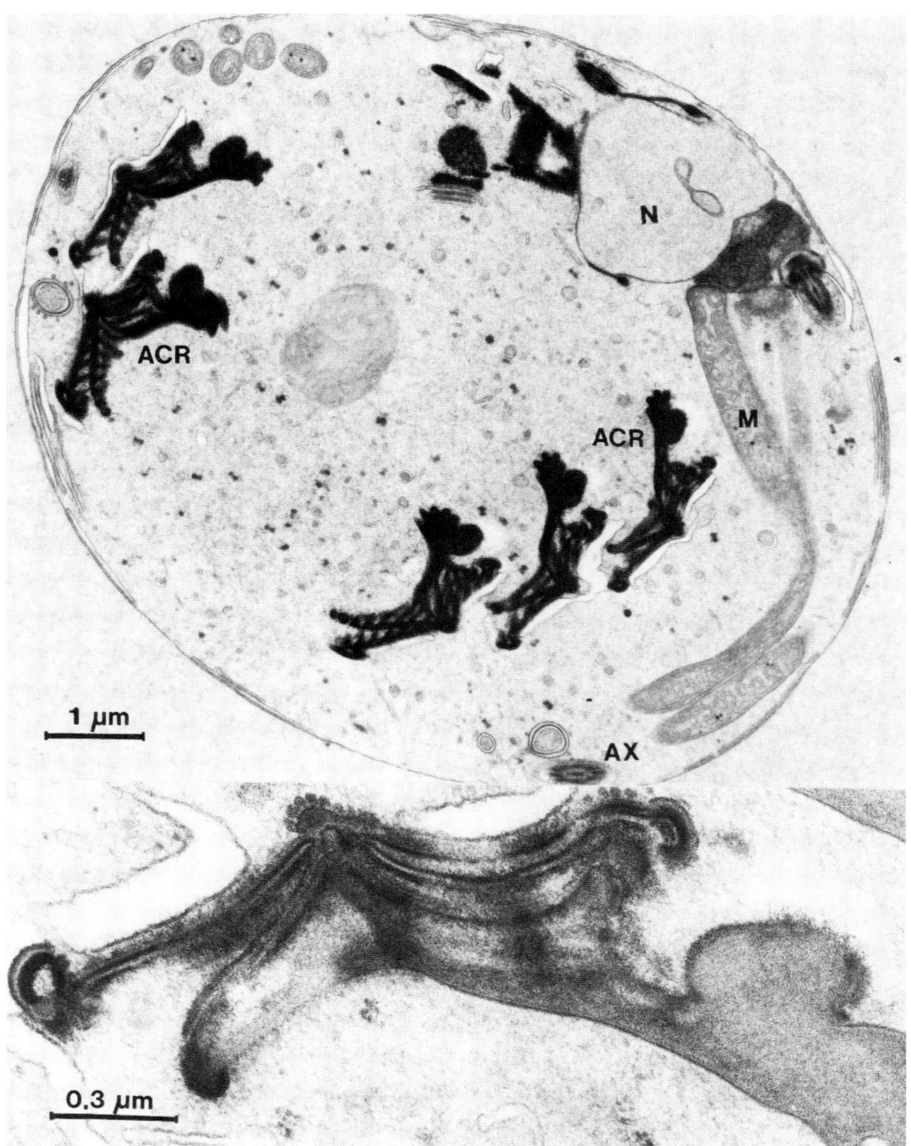


FIGURE 1. Cross section of a *Geogarypus* encysted spermatozoon. ACR, acrosome; AX, axoneme; M, mitochondrial derivatives; N, nucleus.
FIGURE 2. Cross section of an acrosomal coil.

with dense material secreted by an expanded Golgi complex that occupies the greater part of the cell, increases its dimensions and becomes elongated and spindle-shaped. The upper part of this acrosomal structure produces many laminar projections representing the basal support for a series of arches formed in the meanwhile. These extremely elaborate formations also derive from the activity of the Golgi cisternae, engaged either in the secretion of bundles of filamentous material to be associated with the apical acrosomal structure, or in the apposition of long parallel cisternae on the outside of the acrosome. The concomitant condensation of these two Golgi products will give origin to the peculiar structure that is observed during the final stages of spermio= genesis. This structure, up to 74.5 μm long, forms at least three coils in the encysted cell, and is the most evident structure in the maturing spermatid cytoplasm. All the other components, such as the nucleus, the mitochondria and the flagellum, are in fact relatively less developed (fig.1). In cross section, this last-formed acrosomal structure, 2 μm wide, is reminiscent of a crown (fig.2). At the periphery a series of layers can be observed; these layers differ in density and probably derive from the condensation of tightly juxtaposed Golgi cisternae and of the material between them.

## DISCUSSION AND CONCLUSIONS

The presence of a giant acrosomal complex is not a common event. Examples of species with a large acrosome have been found in a variety of groups, apparently with no clear relationship to the systematic position of the taxa. In none of these examples, however, does the acrosome complex attain the remarkably elaborate structure of *Geogarypus*. Within the Pseudoscorpions order, once established that the group has a true acrosome as in other Arachnida and that certainly is not represented by the multi-layered structure reported by Legg (4), we can observe that a giant acrosome is present only in *Geogarypus*. Hence, from a phylogenetic point of view, this fact is of little interest. Not even a relationship between sperm morphology and pattern of reproduction or of spermatophore deposition (3) leads to a positive conclusion because Gary= pidae, to which *Geogarypus* belongs, and Neobisiidae share common reproductive habits and in this family aberrant spermatozoa are not described.

Interesting considerations instead arise when we compare the development of the sperm organelles. This analysis evidences that the giant acrosome of *Geogarypus* occurs together with extreme reduction of the nuclear and axonemal length. This might be only a mere coincidence, but we cannot exclude that it may depend on a reciprocal influence between fundamental spermatozoan structures, which realize a sort of compensation in order to make the cell efficient for the function in which it is involved. Certainly the reduction of the axonemal length is related to the inferior motility of the *Geogarypus* spermatozoon, and it may be considered as a sign of an evolutive tendency towards the acquisition of an aflagellate spermatozoon. This is of some importance for phylogenetic considerations. Weygoldt and Paulus (9), on the basis of a supposed presence of a flagellate spermatozoon within Solifugae, suggested a relationship of this group with Pseudoscorpions. The recent findings of Alberti (10) on the aflagellate spermatozoa in *Eusimoniae mirabilis* (Solifugae) and on the Acarina Actinotrichida (11), however, seemed to indicate, on the basis of commonly derived characteristics in the two orders, a certain relationship between these groups. Perhaps the sperm of *Geogarypus*, although flagellate but with a short axoneme showing little or no motility, anticipate the aflagellate condition realized by Solifugae. In light of this view it might well be valid to consider the Solifugae as evolutionarily derived from Pseudoscorpions (12).

REFERENCES


1. Schaller, F. (1965) Proc. 12th Int. Congr. Entomol. Lond., 297-298.
2. Schaller, F. (1979) In: Arthropod Phylogeny, pp. 587-608, ed. by A. P. Gupta. Van Nostrand Reinhold Co., New York.
3. Weygoldt, P. (1966) Z. Morphol. Okol. Tiere, 56, 39-92.
4. Legg, G. (1973) J. Zool., Lond., 170, 429-440.
5. Boissin, L. and Manier, J. F. (1966) Bull. Soc. Zool. Fr., 91, 697-706.
6. Boissin, L. and Manier, J. F. (1967) Bull. Soc. Zool. Fr., 92, 705-712.
7. Boissin, L. (1970) Thèse, 210 pp.
8. Baccetti, B. (1979) In: The Spermatozoon, pp. 305-329, ed. by D. W. Fawcett and J. M. Bedford, Urban & Schwarzenberg, Inc. Baltimora-Munich.
9. Weygoldt, P. and Paulus, H. F. (1979) Z. zool. Syst. Evolut. -forsch., 17, 177-200.
10. Alberti, G. (1980a) Zool. Anz., Jena, 204, 345-352.
11. Alberti, G. (1980b) Zool. Jb. Anat., 104, 144-203.
12. Savory, T. (1977) Arachnida. Academic Press, London - New York - San Francisco, 340 pp.

# A COMPARATIVE STUDY ON THE SPERMATOGENESIS OF TRICHOPTERA AND LEPIDOPTERA

MICHAEL FRIEDLANDER Department of Biology, Ben Gurion University
Beer Sheva, Israel

## 1. INTRODUCTION

The insect orders of Trichoptera (caddisflies) and Lepidoptera (moths and butterflies have a strong phylogenetic affinity as indicated by aboundant taxo-morphological data (1, 2). There is, however,disagreement concerning the origin of the divergence of the two orders from the common ancestors, mainly due to lack of clarity on the relative phylogenetic position of Zeugloptera. Zeugloptera is considered by many taxonomists as the most primitive systematic group of Lepidoptera. However, since Zeugloptera has so many "trichopteroid" characteristics others have recognized it as a separate order which might be more archaic than either Trichoptera or Lepidoptera (3). The determination of the phylogenetic position of the lower Lepidoptera is further impeded by paucity of fossils and by disagreement concerning the comparative karyological analysis (4, 5). Comparative spermatology has not been yet applied to the Lepidoptera-Trichoptera evolution.

Spermatogenesis of the higher Lepidoptera is dichotomous and produces concomitant eupyrene (nucleate) and apyrene (anucleate) spermatozoa (6, 7). We found no published data concerning the lower Lepidoptera. In Trichoptera it is unclear whether the spermatogenesis is dichotomous. Early light microscope studies indicated the presence of apyrene spermatozoa in Trichoptera (5, 8) but ultrastructural analysis of one trichopteran family revealed only nucleated spermatozoa (9). In the present study, the presence of the eupyrene-apyrene dichotomy was studied in several families of the two suborders of Trichoptera, in the lower lepidopteran family Hepialoidea and in the suborder Zeugloptera.

## 2. MATERIAL AND METHODS

The following species were studied. Trichoptera: *Polycentropus sp.* (polycentropodidae), *Chimarra florida* (Philopotamidae), *Glossosoma sp.* (Glossosomatidae), *Rhyacophila minora* (Rhyacophilidae), *Pychnosyche sonso* (Limnephilidae), *Goera calcarata* (Goeridae), *Agrypnia vestita*(Phryganeidae), *Psilotreta frontalis* (Odontoceridae). Lepidoptera: *Hepialus sequoiolus* (Hepialidae), *Epimartyria pardella* (Micropterigidae, Zeugloptera). Testes and female genital ducts were fixed in 3% of glutaraldehyde in 0.1M cacodylate buffer pH 7.2 for 4-5 hours and subsequently post fixed in 1% $OsO_4$, dehydrated and embedded in Epon. Ultrathin sections were contrasted with lead citrate and uranyl acetate.

## 3. RESULTS

3.1. Higher Lepidoptera. The eupyrene spermatozoon has an elongate nucleus and two apposed mitochondrial derivatives showing poorly defined cristae in transverse sections. From the surface of the cell protrude the lacinate appendages which appear in transverse sections as rays of alternating electron-opaque, electron-lucid bars (Fig. 1) and the reticular appendage which is a rod of small spheres embedded in electron-opaque material. The apyrene sperm has an electron-opaque cone in place of the nucleus and two mitochondrial derivatives showing in transverse sections a V configuration and clearly discernible electron-lucid cristae. The mitochondrial derivatives are more electron-opaque and smaller than those of the eupyrene cells (7, 10).

3.2. Lower Lepidoptera. In Zeugloptera and Hepialidae both eupyrene and apyrene spermatozoa are present (Fig.2). They resemble the corresponding spermatozoa of the Higher Lepidoptera. However, no appendages protrude from the surface of the eupyrene spermatozoa of the lower Lepidoptera (Fig. 3).

3.3 Trichoptera. The 8 species studied which belong to the two sub-orders of Trichoptera contain only nucleate spermatozoa. In each species only one kind of spermatozoa was found (Fig. 4).

## 4. DISCUSSION

Our results indicate that the dichotomous eupyrene-apyrene spermatogenesis have evolved only in Lepidoptera and that, from this point of view, Zeugloptera is a typical lepidopteran. The development of the sperm dichotomy, therefore, began after Trichoptera and Lepidoptera branched from the common ancestors. This indicate that Zeugloptera can not be more

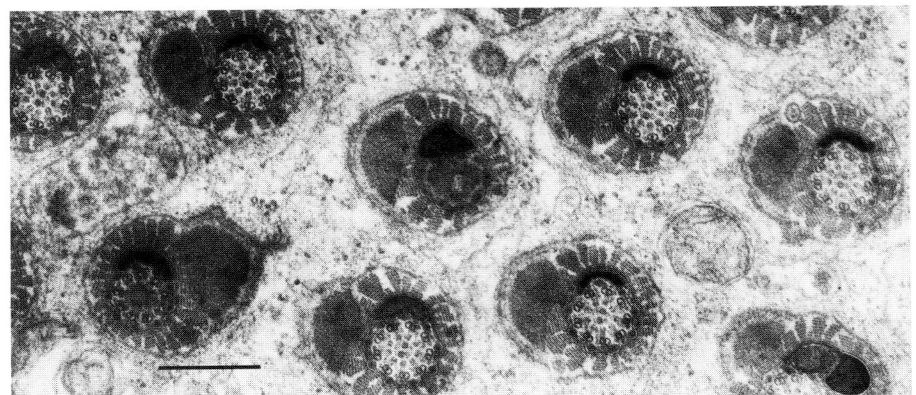

FIGURE 1. Transverse sections of eupyrene sperm of the moth Ephestia cautella. The rays of the lacinate appendages protrude from the surface of the cells. Scale line= 0.5 μm.

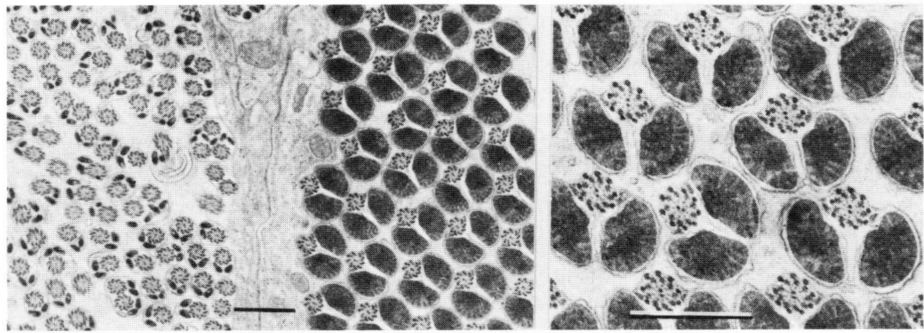

FIGURE 2. Transverse sections of apyrene (left) and eupyrene (right) sperm of the lower lepidopteran Hepialus sequoiolus. Scale line= 1 μm.
FIGURE 3. Enlargement of Fig. 2. The eupyrene cells lack the appendages found in higher Lepidoptera. Scale line= 1 μm.

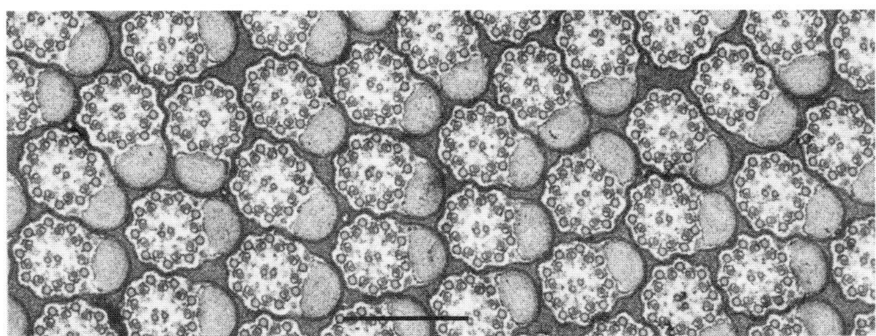

FIGURE 4. Transverse sections of sperm of the caddisfly Goera calcarata. No other kind of spermatozoa could be found. Scale line= 0.5 μm.

archaic than Trichoptera, in spite of the many trichopteroid characteristics retained in this lepidopteran suborder. The trichoptera studied here belonged to species which are considered to be very distant phylogenetically. The early reports on the presence of apyrene sperm in species of Trichoptera could stem from the relatively low resolution of the light microscope or from the study of diapausing specimens in which spermatocytes and spermatids degenerate.

The lack of appendages in eupyrene sperm of Zeugloptera and Hepialidae has to be a primitive characteristics since they are present throughout the higher Lepidoptera. Therefore, our data confirm that these two systematic groups belong to the Lepidoptera, close to the branching point of Lepidoptera and Trichoptera from their common ancestors.

REFERENCES


1. Ross HH. 1967. Ann. Rev. Entomol. 12: 169-206.
2. Common IFB. 1975. Ann. Rev. Entomol. 20: 183-203.
3. Hinton HE. 1946. Trans. Roy. Entomol. Soc. London 97:1-37.
4. Kiuta B. 1971. Ergeb. Wiss. Unters. Schweitz Natn. Park 9: 174-185.
5. Suomalainen E. 1969. Chromosomes Today 2: 132-138.
6. Meves F. 1903. Arch. mikrosk. Anat. 61: 1-84.
7. Friedlander M and Gitay H. 1972. J. Morphol. 138: 121-130.
8. Klingstedt H. 1931. Acta Zool. Fenn. 10: 1-69.
9. Friedlander M and Morse JC. 1982. J. Ultrastruct. Res. 78: 84-94.
10. Phillips DM.1971. J. Ultrastruct. Res. 34: 567-585.

# SPERMATOZOA AND DIPTERA PHYLOGENY

R. DALLAI and M. MAZZINI

Institute of Zoology, via P. A. Mattioli 4, 53100 Siena, Italy

It is generally believed that the Diptera originated at the latest during Permian, probably from a Mecopteroid ancestor or from a group close to Neuroptera.

No doubt seems to exist concerning the phylogenetic position of Nematocera, which clearly represent the more primitive section of Diptera, from which the Brachycera must have arisen at a later date. Systematics within the suborder, however, have been very much discussed, mainly concerning which taxonomic rank is to be assigned to the different groups and their relationships.

The numerous data collected in these years on the sperm morphology of Diptera anable us to make some phylogenetic considerations.

## OBSERVATIONS

On morphological ground, the crane-flies (Tipulomorpha) are considered the most archaic type of living Diptera so it is their sperm structure that we have to take into consideration for evaluating the modifications in the sperm model of the other Nematoceran groups.

Within Tipulomorpha we have studied species belonging to Trichoceridae and Tipulidae. They show a quite conventional Insect sperm with a monolayer=ed acrosome, a condensed nucleus and a long tail with a 9+9+2 axonemal complex flanked by one or two mitochondrial derivatives partially filled with crystallized material (1).

Among Mycetophilidae sperm we found as common characteristic the presence of only one mitochondrial derivative, not yet completely crystallized. With regard to the accessory microtubules the situation is more complex; in two species of *Keroplatus*, in fact, a single 9+2 axonemal pattern is still present; in *Tarnania* externally to the 9+2 axoneme there are, for a long section, only two microtubules, while in the related species *Exechia* the number of accessory tubules is seven. In *Mycetophila* as well as in *Lepto=morphus* the situation is the typical 9+9+2.

The two families Sciaridae and Cecidomyiidae are the most aberrant in the group. Within Sciaridae, spermatozoa have an acrosome, a large mitochon= drial derivative and a giant axoneme; this last is made up of a spiral of many doublets, each showing two dynein arms and connected to an accessory tubule (2,3,4). In Cecidomyiidae the spermatological situation is even more complex. Two phylogenetic trends have been seen in the subfamily Cecidomyiidae; to the first, named "*Sciara*-like", belong species whose motile spermatozoa are reminiscent of the *Sciara* sperm morphology. Clear differences are however present: lack of acrosome and accessory tubules, presence of only the external dynein arm on each doublet and one mitochondrial derivative still with internal cristae and without crystalline material. The species having "9+0" immotile spermatozoa belong to the second trend; usually the spermatozoa are short and in addition the axoneme is surrounded by a more or less conspicuous layer of microtubules (5,6,7,8). Axonemal doublets lack dynein arms as well as any intra-axonemal connections and are oriented in a disordered manner along the cell. The species belonging to the other two subfamilies of Cecidomyiidae, Lestremiinae and Porricondylinae, exhibit a different image. *Lestremia* spermatozoa superficially resamble those of the above mentioned *Sciara*-like trend; nevertheless *Lestremia* spermatozoa have two dynein arms and the mitochondrial derivatives show a beginning of crystallization. *Dicerura* and *Claspettomyia*, both belonging to the subfamily Porricondylinae, have a 9+3 axonemal model, longitudinally crossing the nucleus, and mitochondrial derivatives partially crystallized.

The species of the family Simulidae are characterized by sperm having a 9+9+3 axoneme which reaches the anterior end of the spermatozoon where a short acrosome is located (9).

Two species of Bibionidae have been studied so far: *Plecia nearctica* has a 9+0 axonemal model (10), while *Bibio* (7) has a 9+9+0. According to

Justine and Mattei (11), these models have to be considered atypical 9+9+"1", because of the unusual size of the central axonemal element.

A similar axonemal model is present also in the representatives of the family Culicidae (*Aedes, Culex, Anopheles, Culiseta*: 12,13), even though the axial element in the axoneme has a different aspect, since it appears to be full of dense material.

The spermatozoa of the species belonging to the Ceratopogonidae and Chironomidae have a conventional 9+9+2 axoneme pattern.

The representatives of the Psychodidae family have aflagellate, bifurcate spermatozoa (14).

All the species belonging to the suborder Brachycera so far examined show a common model of spermatozoon. In every species, in fact, in addition to an apical, monolayered acrosome, a compact nucleus and fully crystallized mitochondrial derivatives, a 9+9+2 axoneme has been observed. This axoneme shows well developed coarse fibers, in particular in the portion closer to the accessory tubules (15,16,17 and unpublished).

## CONCLUSIONS

Analysis of the sperm morphology within Diptera, and Nematocera in particular, confirms the variability found with different cytological studies (18). Sperm structure, however, reveals a more high degree of variation and enables us to suggest some new relationships.

The presence of a simple acrosome in almost the whole order is of little help for evolutive considerations; only in Cecidomyiidae has the acrosome been lost, indicating that a new adaptive strategy of fertilization must have been devised. The nucleus has a constant aspect: it is long, cylindrical and in every group it shows condensed chromatin. Mitochondria have more or less transformed their content; generally Brachycera have more evolved mitochondrial derivatives, with crystalline material which occupies all the internal space. In Nematocera, on the contrary, the intramitochondrial content shows an incomplete crystallization. The axoneme is the feature showing the greatest variability and we may base phylogenetic conclusion

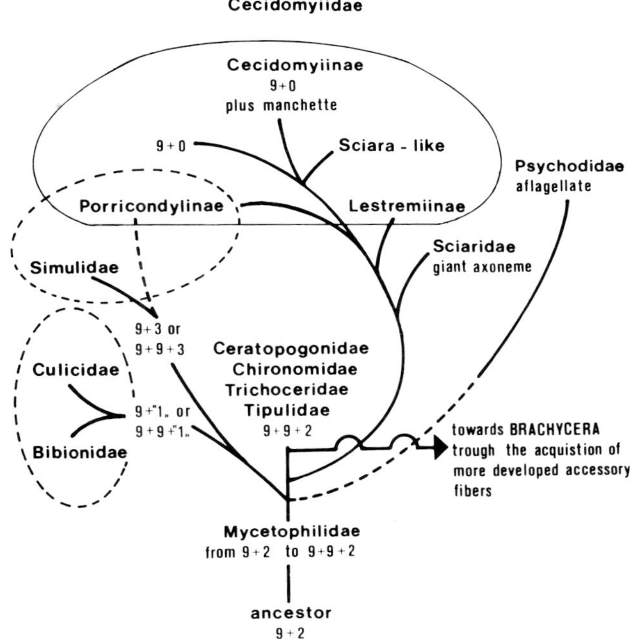


FIGURE 1. Graphic representation of Diptera spermatological trends.

on its structure. Among Nematocera a few families still retain a conventional axoneme 9+9+2 (Trichoceridae, Tipulidae, Ceratopogonidae and Chironomidae). This model has evolved from a simpler 9+2 axonemal pattern still retained by some primitive Mycetophilidae. In light of this view the family may be considered as the most primitive of the suborder. Besides, within Myceto= philidae we have the evidence of the gradual acquisition of the accessory microtubular set. Except for these groups, all the others show an aberrant form. Bibionidae and Culicidae have in common a derived 9+9+"1" axonemal pattern indicating a relationship not previously supposed. A similar situation can be repeated for Simulidae and the subfamily Porricondylinae among Cecidomyiidae, both having a 9+3 (or 9+9+3) sperm axoneme. This last finding is very important because it once more underlines the great heterogeneity of the family Cecidomyiidae. This family in fact comprises at present three subfamilies: Porricondylinae, Lestremiinae and Cecidomyiinae. The first, as above mentioned, has a 9+3 axoneme, the second has a giant axoneme reminiscent of that of *Sciara*, while species of the third subfamily can be arranged in two main spermatological lines, one motile and the other immotile. The many species investigated so far can well be listed in these two lines, sometimes also independently from their systematic position in the family. External characteristics, in fact, often are strongly differentiated in order to make the species adapted to the various environments they leave. Indeed sperm morphology seems capable of discovering relationships within a group not previously suggested by the external morphological characteristics. Also Sciaridae, which share with Cecidomyiidae a similar aberrant meiosis (18), have unusual spermatozoa with a large number of axonemal doublets (2). As to sperm morphology, Psychodidae seem to be the most specialized group among Nematocera with their aflagellate immotile spermatozoa (14) and this is in good agreement with the systematics classification. Nevertheless it is not yet possible to decide from which Nematoceran group Psychodidae have arisen because their affinities are still being debated. This comparative analysis of sperm morphology in Nematocera leads to the reasonable conclusion that the group, starting from a conventional 9+2 pattern, has evolved many times producing several models (fig.1). It appears that often the evolved models are less efficient, leading to a decrease or loss of sperm motility. Nevertheless we cannot exclude that a parallel evolution has taken place also in the female gametes or/and in the whole genital apparatuses in order to permit fertilization to occur. The variability found in Nematocera is in contrast to the situation present in Brachycera, where there is uniform sperm model showing mainly variations involving only the mitochondrial derivatives.

Two general questions still remain to be considered. Can sperm morphology give some information on the origin of Diptera? Can sperm morphology contribute to explain the origin of Brachycera? To the first question, on the basis of the simple 9+2 axoneme flanked by a mitochondrial derivative observed in Mycetophilidae which is also shared by the Scorpion-flies (20) as well as by the fleas (21),we can suggest the origin of Diptera from a Mecopteran ancestor in agreement with the recent classification (22). Moreover it is interesting to put in evidence the impressive resemblance between the spermatozoa of *Exechia* and *Mycetophila* with those of caddis-flies (13,19). Both these Mycetophilidae and Trichoptera, in fact, have stout accessory tubules which push off the plasma membrane. As to the second question it is possible that from the conventional sperm model in Tipulomorpha may have arisen that of Brachycera, through the acquisition of more developed accessory fibers.

## REFERENCES

1. Baccetti B. 1979. In: A.P. Gupta (ed.) Arthropod Phylogeny, pp. 609-644, Van Nostrand Reinhold Co., New York.
2. Phillips DM. 1966. J. Cell Biol., 30, 499-517.
3. Shay JW. 1972. J. Cell Biol., 54, 598-608.
4. Dallai R, Bernini F, Giusti F. 1973. J. Submicr. Cytol., 5, 137-145.
5. Baccetti B, Dallai R. 1976. J. Ultrastruct. Res., 55, 50-69.
6. Dallai R. 1979. In: D.W. Fawcett and J.M. Bedford (eds.). The sperma= tozoon, pp. 253-265, Urban & Schwarzenberg, Inc. Baltimore, Munich.
7. Dallai R, Mazzini M. 1980a. J. Ultrastruct. Res., 70, 363-368.
8. Dallai R, Mazzini M. 1980b. Int. J. Insect Morphol. Embryol., 9, 383-393.
9. Baccetti B, Dallai R, Giusti F, Bernini F. 1974. J. Ultrastruct. Res., 46, 427-440.
10. Trimble JJ, Thompson SA. 1974. Int. J. Insect Morphol. Embryol.,3,425-432.
11. Justine JL, Mattei X. 1981. J. Ultrastruct. Res., 76, 89-95.
12. Breland OP, Gassner G, Riess RW, Biesele JJ. 1966. Can. J. Genet. Cytol., 8, 759-773.
13. Phillips DM. 1969. J. Cell Biol., 40, 28-43.
14. Baccetti B, Dallai R, Burrini AG. 1973. J. Cell Sci., 12, 287-311.
15. Baccetti B, Bairati A. 1964. Redia, 49, 1-29.
16. Perotti ME. 1969. J. Submicr. Cytol., 1, 171-196.
17. Warner FD. 1971. J. Ultrastruct. Res., 35, 210-232.
18. White MJD. 1949. Evolution, 3, 252-261.
19. Baccetti B, Dallai R, Rosati F. 1970. J. Ultrastruct. Res., 31, 212-228.
20. Baccetti B, Dallai R, Rosati F. 1969. J. Microscopie, 8, 233-248.
21. Baccetti B. 1968. Redia, 51, 153-158.
22. Kristensen NP. 1975. Z. zool. Syst. Evolut. forsch., 13, 1-44.

# TWO DIFFERENT LINES IN Tubifex tubifex spermiogenesis

M. FERRAGUTI, P. BRAIDOTTI, A. TRIGARI. Department of Biology, University of Milano, Italy.

## 1. INTRODUCTION

The presence of two sperm lines in tubificids has been discovered in the last years in Limnodrilus hoffmeisteri by Block and Goodnight (1980) who described the two sperm types and summarized the main spermatogenetic events, and in Tubifex tubifex by Braidotti et al. (1980). In a subsequent paper, Braidotti and Ferraguti (1982) found that within the sperm bundles (spermatozeugmata) the conventional, fertilizing spermatozoa were parallely arranged and made up the core of the structure, whereas the modified sperms were helically arranged and tightly packed around the core by means of a series of septate junctions linking the sperm tails. The main differences between the two sperm types are in acrosome structure, nuclear length and shape, dimentions of mitochondria, and structure of the flagellum (Fig. a). Such different models have raised the problem of the cytological mechanisms leading to the differentiation of the two lines.

## 2. MATERIALS AND METHODS

2.1. Optical microscopy. The seminal vesicles of mature Tubifex tubifex have been pierced, and the coelomatic fluid containing the different stages of germ cells maturation has been observed under the phase contrast microscope.

2.2. Electron microscopy. Male segments of Tubifex tubifex were cutted, fixed in Karnovsky's solution, washed in cacodylate buffer postfixed on 1% osmium tetroxide, dehydrated and embedded in Spurr's resin.

## 3. OBSERVATIONS

3.1. Spermiogenesis. In Tubifex, the spermatogonia produced by the testes, are released in the seminal vesicles where undergo a series of mitotic divisions without cytodieresis leading to "morulae" of 2,4,8,16,32... cells. Under the phase contrast microscope it is possible to observe a double population of morulae, mainly distinguished by the diameter both of the morula and of the cells

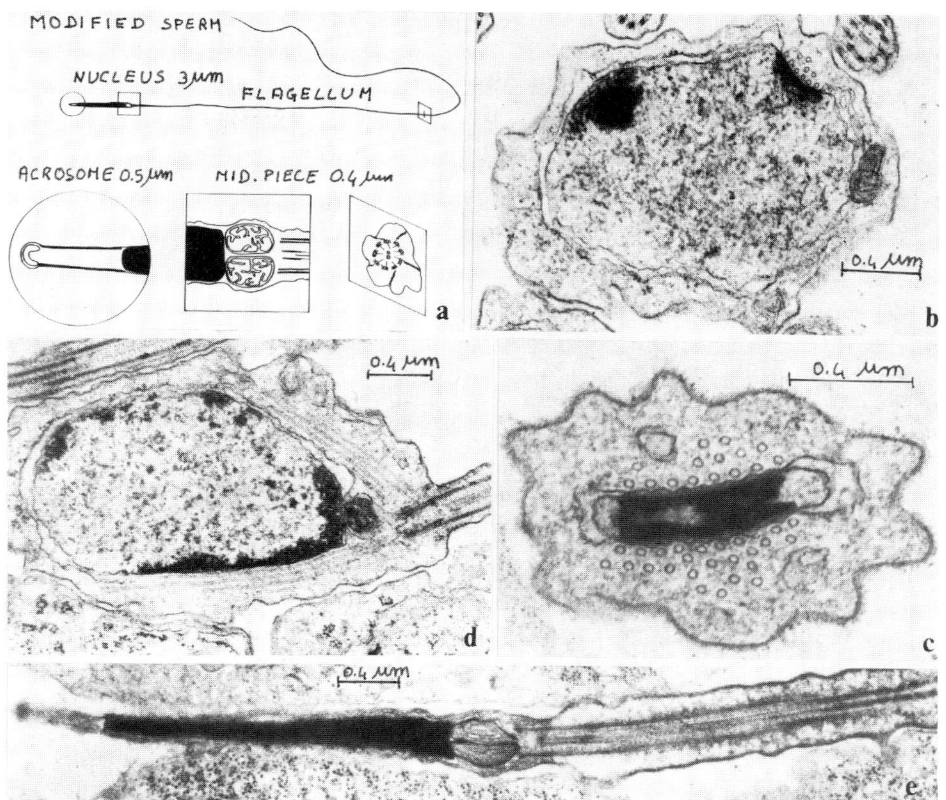


The morulae of modified germ cells consistently showed a higher diameter. Furthermore, the modified morulae produced 1024 spermatids, whereas the conventional morulae stopped at 128. The diameter of the modified spermatids was consistently lower (2 µm or less)then the one of conventional spermatids (3 µm or more).

3.2. Spermiohystogenesis. The first event of the spermiohystogenesisis the polarization of the early spermatid, marked by the emission of the flagellum at the opposite pole with respect to the collar linking the cell to the cytophore. The flagellum arise from the distal centriole, the only one left at this stage (Fig. d). The centriole is curiosly asymmetric, having the shape of a cylinder obliquely cut at one extremity. The centriole is connected to the cell membrane by an anchoring apparatus reminding, at this stage, the one of marine spermatozoa. This structure will be deeply modified during spermiogenesis, and finally disappears. Around the centriole are many microtubules (MTs), disorderly arranged at this stage. The mitochondria are reduced to tthe final number (two) and placed at the centriolar pole (Figs. d,e). An extended Golgi apparatus lie in the same area. In the next stage, an electron-dense pro-acrosomal vesicle is formed by the Golgi and takes contact with the plasma membrane of the spermatid. Under this vesicle the other structures of the mature acrosome are progressively formed. Subsequently, the MTs risen from the ccentriolar zone surround the nucleus of the spermatid forming a structure reminding the manchette of the conventinal spermatids. The MTs, however, do not form a continuous shell around the nucleus, but are irregularely grouped (Fig.b). Chromatin condensation occours where mitochondria are in close proximity of the perinuclear cisterna, and, casually, in other places of the nucleus (Fig.d). In the majority of cases, however, where the manchette is irregular, no chromatin condensation occours. Chromatin condenses at first in irregular masses, then forms anastomizing filaments wich in a second time, adhaere. Contemporarely with manchette formation begins nuclear elongation. The manchette remains incomplete, and the nucleus assumes a pyramidal shape with a rectangular base. MTs are present along the major sides; at the minor sides, devoid of manchette, the nuclear cisterna forms blebs.(c) The mature nucleus is 3 um long, and is conical (Fig.e). In the maintime the other organelles are redistributed. Golgi apparatus disappears after the formation of the proacrosomal vesicle. Mitochondria are packed together and placed between nucleus and centriole when the manchette grows. Proacrosome mooves to its final position at the top of the nucleus at an advanced stage of nuclear maturation. Later the manchette disappears completely.

The flagellum shows early in the spermiogenesis the plasma membraneseparated from the axoneme, wich is an evident characteristic of the modified sperms, but never forms cell junctions with the other flagella, as observed in the spermatozeugmata (Fig.e).

## 4. DISCUSSION

The presence of a double population of morulae pertaining to the conventional or modified line is in accord with the data on Limnodrilus. The early differentiation of the two lines must hence be studied at the gonial stage.

The mechanisms involved in tubificid spermiogenesis have been discussed by Ferraguti and Lanzavecchia (1971), Jamieson and Daddow (1979), and Block and Goodnight (1980). With slight variations all point to a central function of the manchette in chromatin condensation and/or in nuclear shaping. In Tubifex modified spermatids, the connection between manchette and chromatin condensation is less evident. The presence od condensed chromatin in connection with mitochondria and in scattered zones of the nucleus, where MTs are absent prevents the application of the model established for Tubifex conventional spermatids to the modified ones. On the contrary, the presence of MTs only along the major faces of spermatid nucleus, suggests a connection between MTs and nuclear morphogenesis.

The absence of cell junctions between the tails of seminal vesicles modified sperms rise the problem of the origin of spermatozeugmata cell junctions. Since spermatozeugmata are formed in the spermathecal duct, it can be suggested that a sort of "modification" takes place during the passage through the male ducts, wich modifies the sperm membranes and enables them to build junctions.

REFERENCES


1. BLOCK EM., GOODNIGHT CJ. 1980. Spermiogenesis in Limnodrilus hoffmeisteri (Annelida, Tubificidae). A morphological study of the development of two sperm types. Trans. Amer. Micros. Soc. 99:368-384.
2. BRAIDOTTI P., FERRAGUTI M, FLEMING TP. 1980. Cell junctions between spermatozoa flagella within the spermatozeugmata of Tubifex tubifex (Annelida:Oligochaeta). J. Ultrastruct. Rés., 73:299-309.
3. BRAIDOTTI P., FERRAGUTI M. 1982. Two sperm types in the spermatozeugmata of Tubifex tubifex (Annelida, Oligochaeta). J. Morphol. 171:123-136.
4. FERRAGUTI M., LANZAVECCHIA G. 1971. Morphogenetic effects of microtubules. I. Spermiogenesis in Annelida Tubificidae. J. Sumicr. Cytol. 3:121-137.
5. JAMIESON B G M., DADDOW L. 1979. An ultrastructural study of microtubules and the acrosome in spermiogenesis of Tubificidae (Oligochaeta). J. Ultrastruct. Res. 67:209-224.

# TRENDS IN THE EVOLUTION OF FLATWORM SPERMATOZOA

JAN HENDELBERG
Department of Zoology, University of Göteborg, Göteborg, Sweden

## 1. INTRODUCTION

Preliminary results are reported from studies of the ultrastructure of the spermatozoa of two groups of turbellarian flatworms: the Proseriata and the Kalyptorhynchia. Some main characteristics of the spermatozoa of different groups of turbellarians are summarized, and the occurrence of parallel evolution (or convergence) is discussed.

## 2. MATERIAL AND METHODS

Two species were studied of the Proseriata, *Monocelis* cf. *lineata* (Müller) and *Coelogynopora biarmata* Steinböck, and three of the Kalyptorhynchia, *Gyratrix hermaphroditus* Ehrenberg, *Phonorhynchus helgolandicus* (Meczn.) and *Prognathorhynchus typhlus* L'Hardy. The material was collected at marine biological stations in Norway, Sweden and Finland, and studied with transmission electron microscopy according to methods described earlier (1).

## 3. RESULTS

Some of the features of the types of spermatozoa found in the Proseriata and Kalyptorhynchia are demonstrated in Fig. 1.
In *M. lineata* (Fig. 1 a) the mitochondria form a row along the elongated nucleus. The axoneme pattern of the two free sperm flagella found in this species and in *C. biarmata* was found to be of the aberrant so-called '9+1' pattern found earlier in other groups of 'higher' flatworms. In the kalyptorhynchs the mitochondria were found to form a rodlike derivative, the flagella are incorporated into the sperm body, and their axoneme pattern is '9+1' (cf. Fig. 1 b).

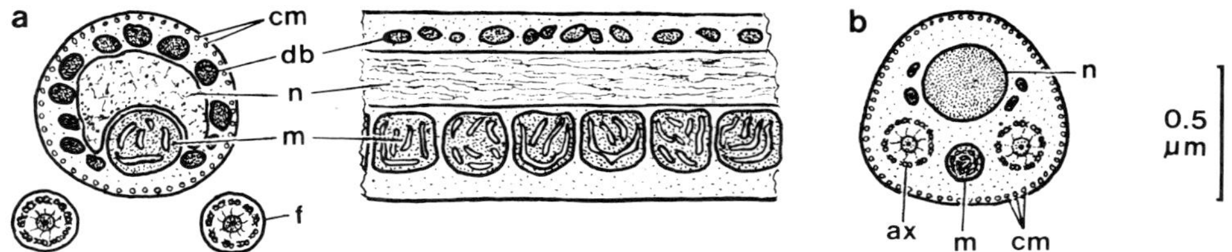


FIGURE 1. a) Cross section of spermatozoon and part of longitudinal section of sperm body of *Monocelis* cf. *lineata* (Proseriata). b) Cross section of spermatozoon of *Gyratrix hermaphroditus* (Kalyptorhynchia). ax, axoneme of incorporated flagellum; cm, cortical microtubules; db, dense body; f, flagellum; m, mitochondrion; n, nucleus.

The results reported here are indicated in the diagram, Fig. 2, in which the occurrence among turbellarian taxa of some main sperm characteristics are summarized. The diagram is based on results of other authors and of my own (1, 3, 4, 5, 6 and references in these papers).

## 4. DISCUSSION AND CONCLUSIONS

The diagram, Fig. 2, is tentative, since the information on the organization at the ultrastructural level of turbellarian spermatozoa is still poor for many of the subgroups.

The differences in position and ultrastructure of the sperm flagella of turbellarians have been dealt with earlier (1). The results preliminarily reported here from some Proseriata and Kalyptorhynchia agree with the assumption that the '9+1' axonemal pattern is a synapomorphy of the 'higher' flatworms.

It is noteworthy that Rieger (4) on the basis of a number of different results of ultrastructural studies asserts that the Nemertodermatida and the Acoela have evolved along a common line separate from that leading to 'higher' turbellarians, that is, according to the alternative (A) in Fig. 2. If this is true, the doubling of the flagellum and the aberrant '9+1' axonemal pattern are consequences of a parallel evolution in the two lines. The finding (1) in the spermatozoa of some Acoela of a '9+1' pattern which seems to be different from that in other flatworms may support the assumption that '9+1' axonemal patterns have evolved independently at different times. However, the possibility exists

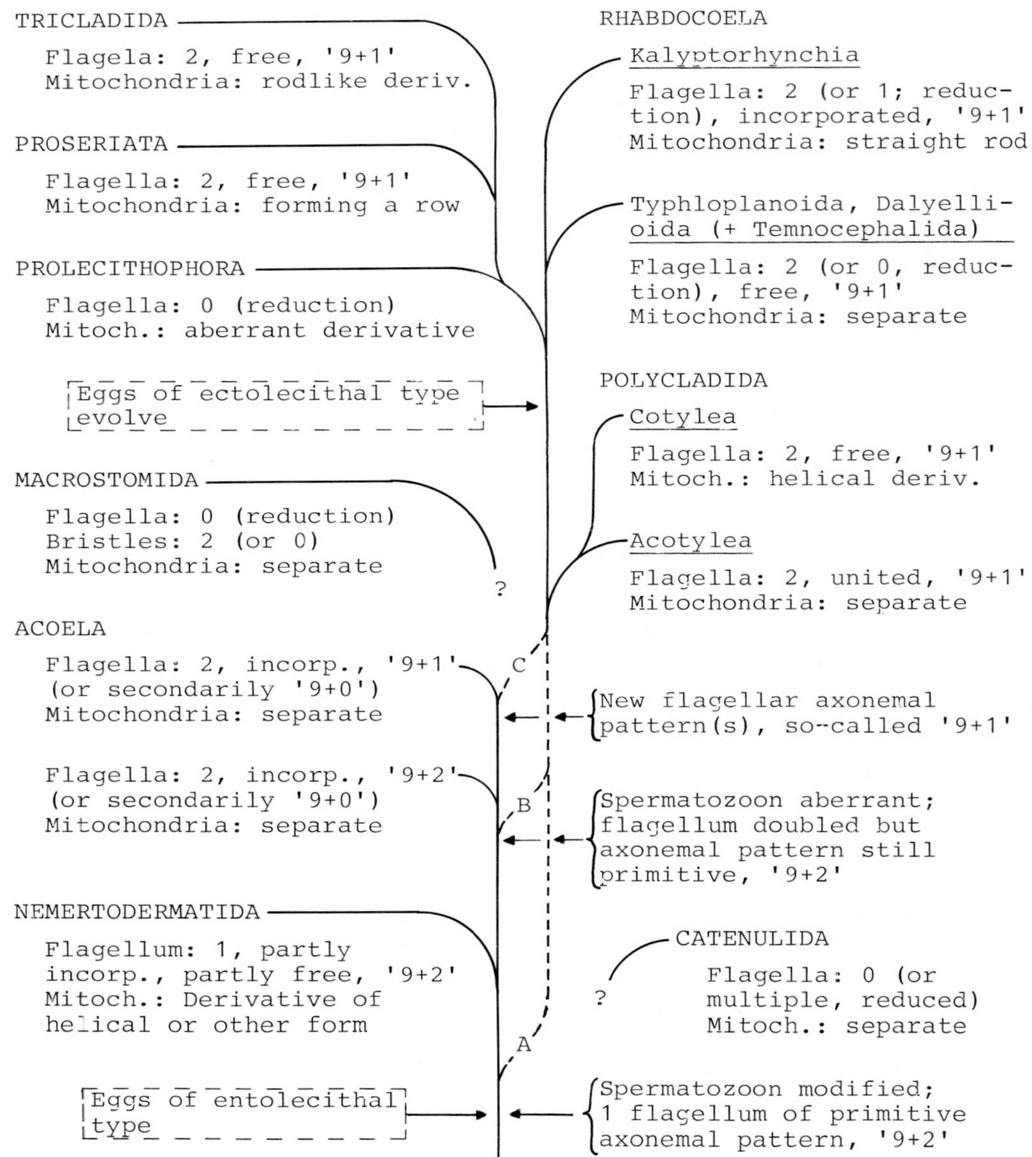


FIGURE 2. Sperm characteristics (and egg types) of the main groups of turbellarian flatworms. Number of flagella, their position (free, incorporated into the sperm body, or united with the sperm body running along it) and axonemal pattern ('9+2', '9+1' or '9+0') are indicated, as well as arrangement of mitochondria. The phylogenetic tree is (mainly) in agreement with Karling (2) and Rieger (3).

that the '9+1' patterns found in different lines of evolution are the results of a common change of other structures before the lines diverged. Thus, e.g., a change to an abberant basal body is found in some '9+2' and '9+1' flatworm spermatozoa studied in this respect (1, 4, 5).

New in the diagram (Fig. 2) if compared with one I published earlier (1) is the indication of mitochondrial arrangement.

A number of single, separate mitochondria may be a primitive feature but, if derivated, it can easily be explained by the mitochondria retaining their separate occurrence in the early spermatids. The rodlike or ribbonlike (6) mitochondrial derivatives, straight or helical, are most probably formed by fusion of single mitochondria as found in a few species studied in this respect (5). Thus the difference is small between the row of mitochondria found in, e.g., the Proseriata, and the mitochondrial rod reported from the closely related Tricladida. A mitochondrial rod is found now in the Kalyptorhynchia, indicating a convergent evolution.

A rodlike mitochondrial derivative and the '9+1' axonemal pattern are found in many spermatozoa of the parasitic groups of flatworms. These groups are most probably closely related to one or other of the turbellarian groups characterized by ectolecithal eggs.

REFERENCES


1. Hendelberg J. 1977. Comparative morphology of turbellarian spermatozoa studied by electron microscopy. Acta Zool. Fenn. 154: 149-162.
2. Karling TG. 1974. On the anatomy and affinities of the turbellarian orders, in Biology of the Turbellaria, Eds. Rieser NW, Morse MP, pp. 1-16. McGraw-Hill, New York.
3. Rieger RM. 1981. Morphology of the Turbellaria at the ultrastructural level. Hydrobiologia 84: 213-229.
4. Hendelberg J. 1977. Ultrastructure of the cytoplasmic region in the spermatozoa of Cryptocelides (Polycladida, Turbellaria). Zoon 5:107-114.
5. Hendelberg J. 1982, in press. Chapter 'Turbellaria', in Reproductive Biology of Invertebrates, Vol. II, Eds. Adiyodi KG, Adiyodi RG. John Wiley & Sons, London.
6. Ehlers, U. 1981. Fine structure of the giant aflagellate spermatozoon in Pseudostomum quadrioculatum (Leuckart) (Platyhelminthes, Prolecithophora). Hydrobiologia 84: 287-300.


Author's mailing address: Department of Zoology, University of Göteborg, Box 25059, 400 31 Göteborg, Sweden.

# A NEW LOOK AT MONOGENEA AND DIGENEA SPERMATOZOA

Jean-Lou JUSTINE

## 1. INTRODUCTION

The spermatozoa of parasitic platyhelminths (Cestoda, Monogenea, Digenea) are generally rather long (up to 400 µm), threadlike and very thin (ca. 1 µm). For this reason, they are usually studied in transverse sections; in any group of these, the overwhelming majority are sections of the so-called "middle region", showing dorsal and ventral microtubules, mitochondrion, two 9+"1" axonemes and nucleus. These are the sections most ofter described. Scattered among such sections are a very small proportion which do not conform to the general pattern (their relative infrequency reflects the shortness of the region involved). We have demonstrated (2,4) that some of these aberrant sections show the anterior region of the spermatozoon, on which we have concentrated our efforts, using ultrathin serial sections.

Spermiogenesis has been described by BURTON (1). Neither this nor other later studies have resolved two key problems: the existence and location of the centrioles and the orientation of the spermatozoon.

## 2. STUDIES OF DIGENEA

### 2. 1. Gonapodasmius (Didymozoidae) (2).

Diagram 1 shows the formation of the anterior region of the spermatozoon.

1) The zones of differentiation (hereinafter the ZD) are cylindrical; at their extremity they have one middle cytoplasmic process (MP) and two free flagella (FF) with centrioles (C) at their base. The plasma membrane possesses peripheral microtubules (white arrows). These microtubules coincide with external ornamentations on the membrane. The nucleus (N) inserts itself into the zone of differentiation.

2) The three processes fuse and lengthen. The nucleus penetrates

*Author's Address:* Laboratoire de Zoologie, Département de Biologie Animale, Faculté des Sciences, Dakar, Sénégal.

further into the future spermatozoon. The mitochondrion (M) likewise penetrates into it. The description thus far agrees more or less with that given by BURTON (1).

3) We have proven (2) that the spermatozoon is pinched off from the cytoplasmic mass in front of the centrioles (curved arrows). The microtubules with associated ornamentations are found in the mature spermatozoon; hence it is clear that the ZD has been incorporated into it.

4) The perimeter of the truncated anterior region (Fig. 1) is lined with microtubules coinciding with ornamentations (glycocalyx). The two centrioles or centriolar derivatives are made up of singlets; they look like distal extremities of flagella, but since the spermatozoon tail is forked, confusion is unlikely.

Behind the anterior region is a short "intermediary region" without microtubules, then the middle region (not shown) bearing a second set of microtubules, with posterior nucleus.

2. 2. Other digeneans

2. 2. 1. *Haematoloechus* (Haematoloechidae). In this much-examined case (1), the anterior region, originally the ZD, has external ornamentations coinciding with microtubules (4), as in *Gonapodasmius*. A second type of ornamentation appears further back. The centrioles are hard to see because of the cytoplasmic density. Only the front tip of the spermatozoon is motile.

2. 2. 2. In other species, we have found the same incorporation of the ZD; this would seem to be the general pattern among Digenea.

## 3. STUDIES OF MONOGENEA

3. 1. Polyopisthocotylean Monogenea

Their spermiogenesis seems to follow the digenean pattern.

3. 2. Monopisthocotylean Monogenea

Some differences had been foreseen here on the basis of earlier reports of spermatozoa without microtubules (6).

3. 2. 1. *Megalocotyle* (Capsalidae) (5). The evolution of the anterior region of the spermatozoon is shown in diagram 2.

1) The ZD is short, without the usual three processes. It bears two centrioles. The unornamented membrane possesses peripheral microtubules (white arrows). The nucleus (N) penetrates into the ZD.

2) The mitochondria fuse together to form a bead around the nucleus.

3) The beadlike mitochondrion moves along the nucleus, generating a

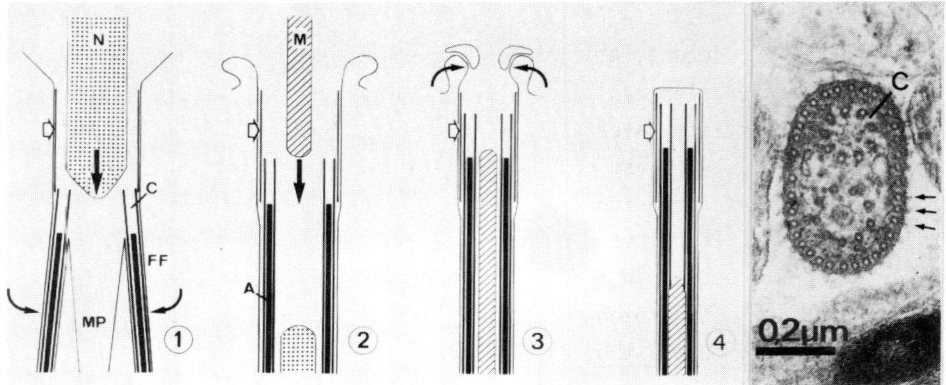


DIAGRAM 1. Evolution of the zone of differentiation and anterior region of the spermatozoon in a DIGENEAN (Gonapodasmius). Longitudinal sections. 32 spermatozoa are formed from each common cytoplasmic mass.

FIGURE 1. Transverse section of the anterior region of the spermatozoon of Gonapodasmius. C, centriole; arrows, external ornamentations.

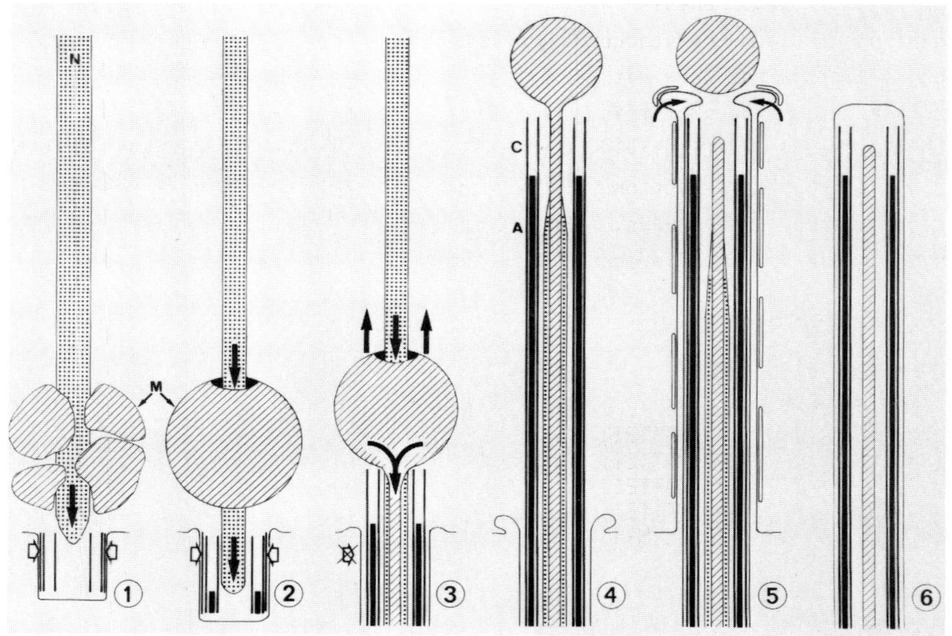


DIAGRAM 2. Evolution of the anterior region of the spermatozoon in a MONOPISTHOCOTYLEAN MONOGENEA (Megalocotyle). Quasi-longitudinal sections. 64 spermatozoa are formed from each common cytoplasmic mass.

streamer to its rear. As a result its bulk decreases. The ZD lengthens out about 100 µm (not shown). At this stage the peripheral microtubules disappear.

4) The mitochondrion has moved about 10 µm. The two parallel centrioles (C) formed of singlets have followed it and are continued as two axonemes (A). At this stage the cytoplasm contains 64 parallel four-part elements made up of nucleus, mitochondrion and two flagella.

5) Membranes now enclose each four-part element in a cytoplasmic canal. The pinching off occurs in front of the centrioles (curved arrows).

6) The mature spermatozoon is devoid of microtubules.

3. 2. 2. Other Monopisthocotylean Monogenea. Surprisingly, we have found some uniflagellate spermatozoa with microtubules and ornamentations (3).

## 4. CONCLUSION

In terms of their anterior region, Monogenea and Digenea display several different pattern of spermiogenesis. In Digenea and polyopisthocotylean Monogenea, spermiogenesis involves a ZD with three fusing processes. Megalocotyle, a monopisthocotylean monogenean, shows contrasting spermiogenesis; we do not know yet whether this pattern is the case for all other biflagellate members of this group. Finally, there is evidence of still other patterns, particularly among uniflagellate Monopisthocotylea.

While transverse sections of the "middle region" of filiform spermatozoa are relatively homogeneous, in-depth study of the anterior region may reveal significant differences of biological or phylogenetic interest.

## REFERENCES

1. BURTON, P. R. (1972) J. Parasitol. 58, 68-83.
2. JUSTINE, J-L. and MATTEI, X. (1982a) J. Ultrastruct. Res. (in press).
3. JUSTINE, J-L. and MATTEI, X. (1982b) Ann. Parasitol. Hum. Comp. (in press).
4. JUSTINE, J-L. and MATTEI, X. (1982c) J. Ultrastruct. Res. (in press).
5. JUSTINE, J-L. and MATTEI, X. *A study of* Megalocotyle *will be issued shortly*.
6. TUZET, O. and KTARI, M-H. (1971) Bull. Soc. Zool. France 96, 535-540.

We should like to thank William D. WHITE of the Peace Corps for his assistance in translating the text.

# THE REVERSE MOVEMENT OF A SNAIL SPERMATOZOON

M. G. SELMI and F. GIUSTI

Institute of Zoology, via P. A. Mattioli 4, 53100 Siena, Italy

The mature typical spermatozoon of *Theodoxus fluviatilis* (L.) (Gastropoda: Prosobranchia) is 66 μm in length and has a unique mode of locomotion, with the tail first and trailing back over the head. This previously undescribed type of movement seems to derive from its particular morphology (Giusti and Selmi, 1982) (Plate 1).

The acrosome (1 μm long) has a pointed conical shape and contains a fibrous acrosomal rod in its interior. The nucleus, a conical cylinder 15 μm long, is composed of a homogeneously condensed chromatin. In its interior runs a canal of 0.2 μm. This intranuclear canal is entirely filled with glycogen granules in its upper portion. 4 μm from its origin, it is occupied by the compact and cylindrical proximal centriole and by the distal centriole which gives rise to the flagellum. The last two-thirds of the intranuclear canal contain the initial portion of the axoneme.

The tail can be distinguished in two distinct portions, parallel to one another but running in opposite directions. The first portion of the tail, which originates at the base of the nucleus, is constituted by two distinct elements: the flagellum and a cytoplasmic sheath which extends backwards to encircle it. One half of the cytoplasmic sheath is occupied by two mitochondria kidney-shaped in cross section while the other half is filled with a double layer of glycogen granules.

At the end of this portion the flagellum has a peculiar structure which is gradually acquired during spermatogenesis and determined by the breaking of the axoneme into two parts (Giusti and Selmi, 1982). This point of breaking will be called "tail-joint". At the level of the "tail-joint" there is present a proteinaceous cap-like structure. In its concavity it contains the end of the first portion and the beginning of the second portion of the flagellum. Here the microtubules of the first portion of the axoneme are fused in a rootlet-like structure, which is embedded in a very electron-dense material.

The second portion of the flagellum, free of the cytoplasmic sheath, is folded back, lying parallel and in contact with the first portion. Its axoneme begins with doublets separated the one from the other. The normal 9+2 structure of the second portion of the axoneme is lost only at the end of the tail. Here the doublets and the two central tubules give origin to a ring of 20 single microtubules.

The locomotion of this spermatozoon is as peculiar as its morphology. In contrast with all other known spermatozoa, it moves backwards, with the "tail-joint" leading and the head trailing behind. Small waves originate at the "tail-joint" and are propagate backwards. The entire spermatozoon rotates on itself in helicoidal movements (fig.4).

In two spermatozoa that remained functional although attached to the glass slide, the first and the second portion of the tail appeared to be separate. Waves were seen to propagate only in the second part of the tail (fig.5). For further investigations of this movement, reactivation of the spermatozoa was attempted. They were first demembranated with 0.1% Triton X 100 and then treated with 0.1mM to 1mM ATP. Unfortunately reactivation has not been clearly observed.

Negative staining was performed on sperm untreated (fig.1) and treated (fig.2) with Triton X 100. This demonstrated that the microtubules of both the first and the second portions of the tail run parallel to one another. Negative staining of ATP-treated (fig.3) spermatozoa however, show the twisted microtubules in the second portion. This perhaps is a consequence of a momentary or partial reactivation with the formation of waves and therefore twisting of the microtubules.

## CONCLUSIONS

Although our results at present do not allow us to exclude the possibility that both parts of the tail move, they do confirm the presence of mobility in the second part of it. The waves originate at the level of the "Tail joint" and propagate backwards to the end of the tail. From this it seems to follow that the reverse movement of the spermatozoon is not due to the inversion of the flagellar beat but rather to an inverted dislocation of the second part of the tail.

## REFERENCE

1. Giusti F , Selmi MG. 1982. J. Ultrastruct. Res., 78, 166-177.

## EXPLANATION OF PLATES AND FIGURES

Plate 1. FP, first portion of the tail; SP second portion of the tail. AC, acrosomal cone; AR, acrosomal rod; GG, glycogen granules; N, nucleus; PL, proximal centriole; AX, axoneme; M, mitochondria; SR, spiral ribbon; CS, end of cytoplasmic sheath; CL, cap-like structure; OB, oblongue body. A, apical section of the nucleus; B, section through the middle portion of the nucleus; C, sections of the first portion of the tail; D, section near the tail-joint; E, the flattened tip of the tail; F, atypical spermatozoa; G, sections trough the tail-joint.

Figures 1,2 and 3. See explanations in the text.

Figure 4. Eight successive frames of a moving mature spermatozoon. The tail--joint (T) and the head (H) are clearly visible (24 frames/sec).

Figure 5. Eight successive frames of a motile although attached to the glass slide spermatozoon. Only the second portion of the tail is moving (SP) (24 frames/sec).

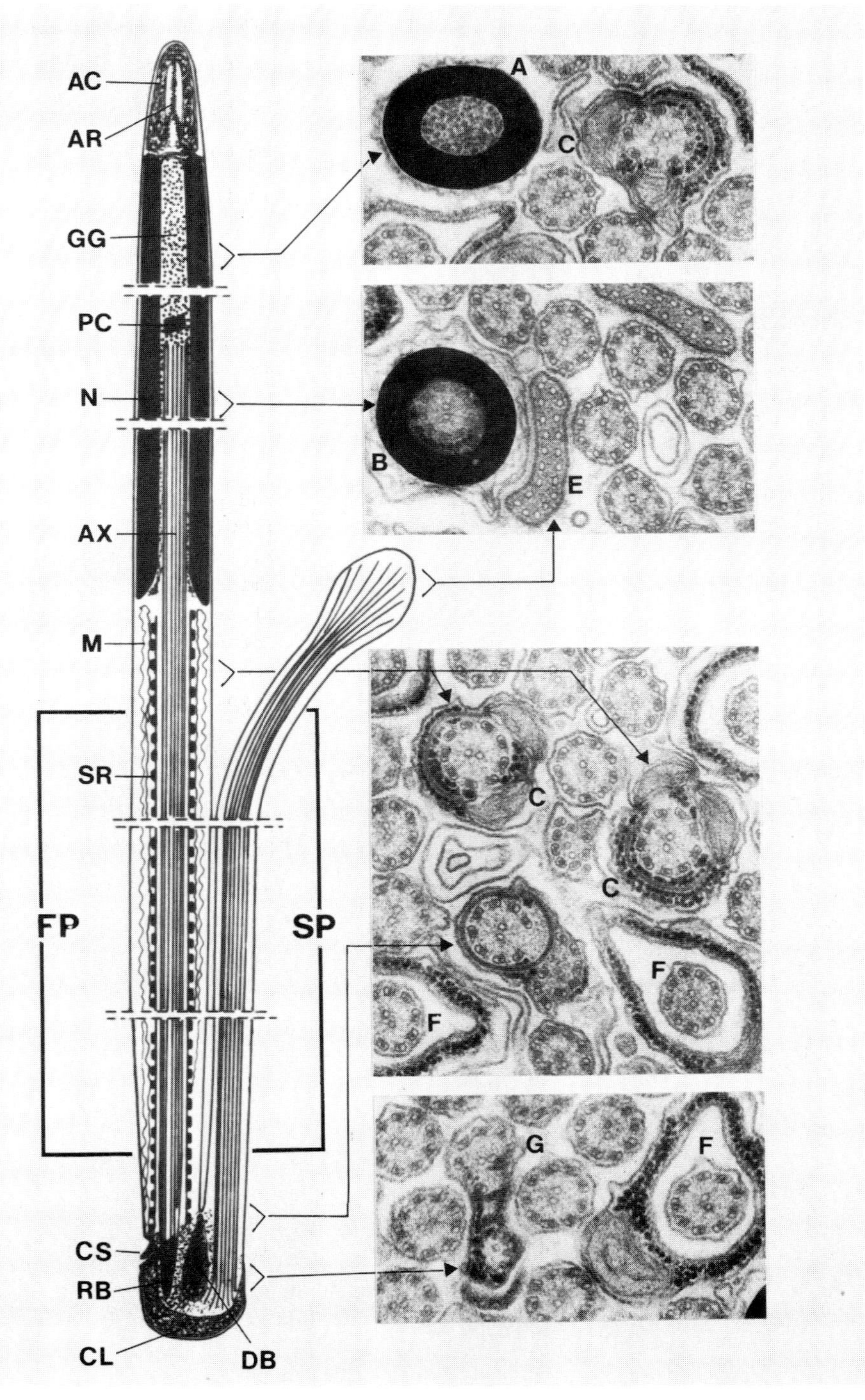
AC
AR
GG
PC
N
AX
M
SR
FP
SP
CS
RB
CL
DB
A
C
B
E
C
C
F
F
G
F

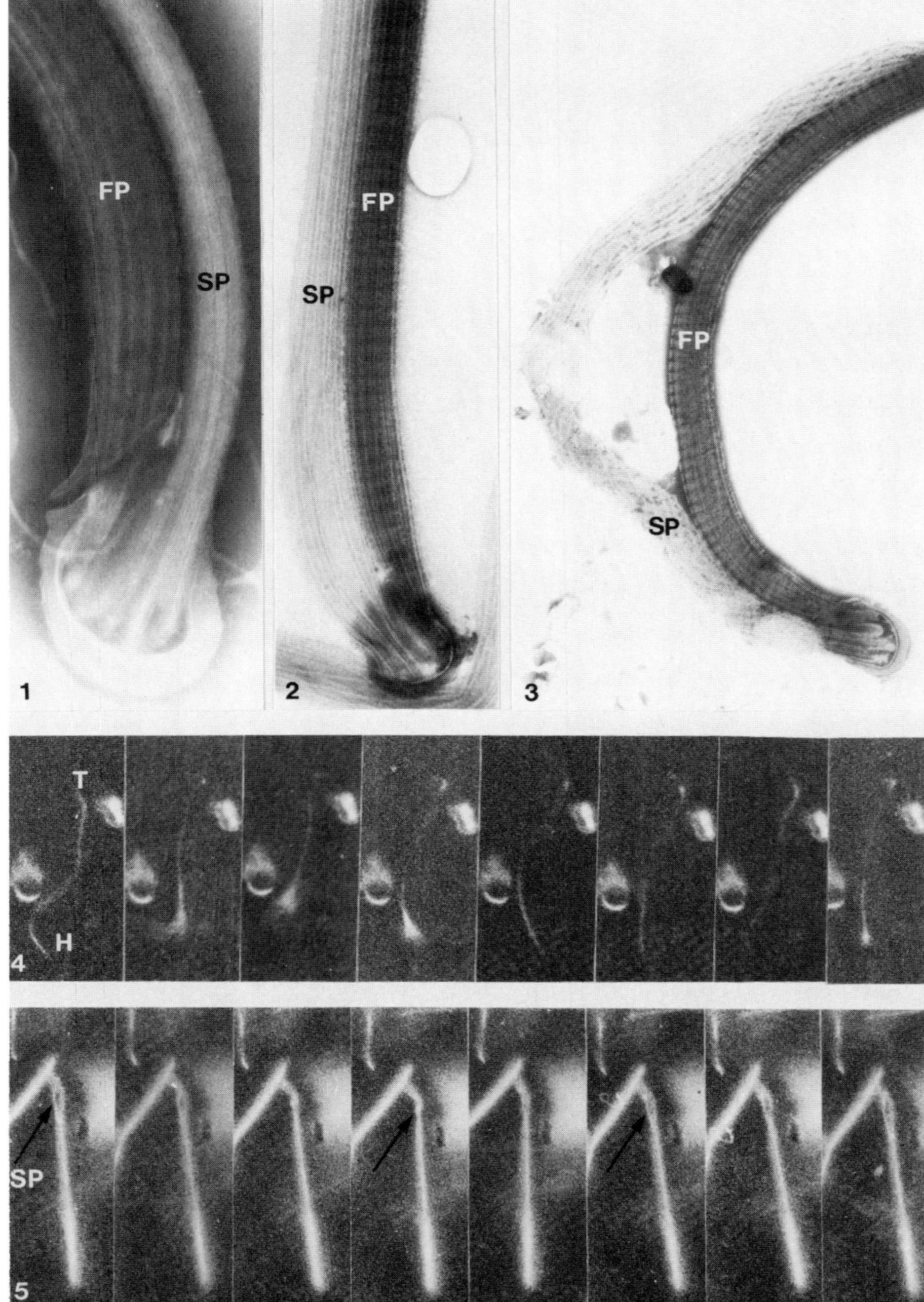
FP
SP
1
FP
SP
2
FP
SP
3
T
H
4
SP
5

# AUTHORS'INDEX